NON-VIRAL GENE THERAPY

GENE DESIGN AND DELIVERY

Kruger Brentt Publishers

NON-VIRAL GENE THERAPY

GENE DESIGN AND DELIVERY

VOLUME-1

Merita Rumano
Diana Avetyan

Kruger Brentt
Publishers
2025

Kruger Brentt Publishers UK. LTD.

Company Number 9728962

Regd. Office: 68 St Margarets Road, Edgware, Middlesex HA8 9UU

ISBN: 978-1-78715-395-0 (Vol. 1)

For information on all our publications visit our website at http://krugerbrentt.com/

PREFACE

"Non-Viral Gene Therapy: Gene Design and Delivery" embarks on an illuminating journey into the innovative realm of non-viral gene therapy, offering readers a comprehensive understanding of the principles, methodologies, and applications of gene design and delivery strategies. This book serves as an essential resource for researchers, clinicians, students, and industry professionals engaged in the development and implementation of non-viral gene therapies for a wide range of diseases and disorders.

Gene therapy holds tremendous promise for the treatment of genetic disorders, cancer, infectious diseases, and other conditions by harnessing the power of nucleic acids to modulate gene expression, repair genetic defects, and enhance therapeutic outcomes. While viral vectors have traditionally been used as delivery vehicles for gene therapy, non-viral vectors offer several advantages, including safety, scalability, and ease of manufacturing, making them attractive alternatives for gene delivery.

The primary objective of this book is to provide readers with a comprehensive overview of non-viral gene therapy, covering key topics such as gene design principles, vector engineering strategies, delivery systems, cellular uptake mechanisms, and therapeutic applications. Each chapter offers insights, theoretical frameworks, experimental techniques, and clinical perspectives to help readers navigate the complexities of non-viral gene therapy research and development.

Contributions from leading experts in the field explore various aspects of non-viral gene therapy, including the design and optimization of nucleic acid sequences, the development of novel delivery vehicles such as nanoparticles, liposomes, and polymers, and the translation of preclinical findings into clinical trials and therapeutic interventions. Special emphasis is placed on emerging technologies and innovations that hold promise for advancing the field and overcoming existing challenges in gene delivery and expression.

As non-viral gene therapy continues to evolve and expand its therapeutic potential, interdisciplinary collaboration and knowledge exchange are essential for driving progress and realizing the full clinical impact of gene-based therapies. By fostering dialogue and collaboration among researchers, clinicians, industry stakeholders, and regulatory agencies, we can accelerate the development and adoption of safe, effective, and accessible gene therapies for patients in need.

We extend our sincere gratitude to the contributors whose expertise and dedication have enriched this book. It is our hope that "Non-Viral Gene Therapy: Gene Design and Delivery" will serve as a valuable resource and catalyst for innovation in the field of gene therapy, paving the way for new treatments and cures that improve human health and quality of life.Laura Sanchez

Merita Rumano
Diana Avetyan

CONTENTS

VOLUME-2

Chapter 1 GENE THERAPY VECTORS THAT DO NOT INVOLVE VIRUSES AND CARRY GENOMIC CONSTRUCTS

1. INTRODUCTION: AN OVERVIEW

Gene therapy is the utilization of genes or DNA to treat diseases. In the therapy of hereditary illnesses, functional DNA containing a gene is inserted into the cells of a patient to correct the flaw caused by the defective natural gene. Prior genetic characterisation of the disease and cloning of the causative gene are essential. Typically, the cDNA of the therapeutic gene is inserted into a bacterial plasmid using a powerful foreign promoter, generally derived from a virus. Nevertheless, these creations, known as mini-genes, do not possess introns, promoters, enhancers, or long-range regulatory elements that intricately regulate the timing and location of expression of the natural gene.

In order to effectively treat some diseases using gene therapy, it is crucial to attain precise levels of expression for the therapeutic gene. Inadequate expression at suboptimal levels may not be effective in rectifying the deficiency, whereas excessive expression at elevated levels could lead to unwanted consequences. Tissue-specific expression can be quite significant in certain instances. The regulatory elements that govern the precise and tissue-specific expression of a gene are often located within the introns and the sequences upstream and downstream of the gene. Thus, the utilization of genomic constructs that include the introns and surrounding DNA of the therapeutic gene is anticipated to be more efficient than mini-gene constructs in gene therapy for specific genetic disorders that necessitate precise levels of the gene product (as discussed by Pérez-Luz & Díaz-Nido, 2010). Bacterial Artificial Chromosomes (BACs), derived from the human genome project, consist of genomic regions that are roughly 180 kb in size on average and span the entirety of the human genome (Osoegawa et al., 2001). These ordered bacterial artificial chromosomes (BACs) have the capacity to accommodate a wide range of genes, including their regulatory components. They can be utilized as effective instruments in gene therapy by employing genomic constructs.

Gene therapy, as a contemporary therapeutic technique, aims to achieve a lasting cure for patients by ensuring the sustained presence and activation of the introduced gene. This can be accomplished through either the integration of the transgene into the native chromosomes or through other mechanisms that facilitate its replication and nuclear retention.

The selection of the vector for gene therapy is crucial as it determines how the corrected gene will be delivered and expressed in the specific cells. The present vectors can be classified into two distinct categories: viral and non-viral. In addition to dictating the mode of distribution, the type of vector also dictates the outcome of the therapeutic gene within the cells. For example, the vector may possess the capacity to exist outside of the chromosome. Non-viral artificial chromosome vectors, as well as adeno-associated viral, adenoviral, Herpes viral, and EBV vectors, are all instances of this category. On the other hand, retroviral and lentiviral vectors have the ability to incorporate themselves into the genetic material of the host (as discussed by Verma & Weitzman, 2005).

Most gene therapy methods now rely on viral vectors because of their exceptional ability to transport genetic material into cells. Several effective viral gene therapy clinical trials have demonstrated remarkable therapeutic benefits for patients (Cavazzana-Calvo et al., 2000). Nevertheless, viral gene therapy has faced criticism mostly due to two regrettable incidents. A patient with ornithine transcarbamylase deficiency, who was undergoing treatment with an adenoviral vector, died as a result of an immunological reaction being triggered (Raper et al., 2003). The occurrence of this death has brought to light a challenging safety concern, as it is difficult to anticipate human immunological responses in pre-clinical settings. Another instance involved patients with SCID-X1 who received treatment with a retroviral vector. These patients acquired a disease similar to leukemia because the integration of the vector disrupted a natural oncogene (Hacein-Bey-Abina et al., 2003a, 2003b). Insertional mutagenesis is a pervasive issue that affects all integrating viral vectors due to the random and unpredictable nature of vector integration.

Optimally, vectors appropriate for gene therapy should satisfy four characteristics. First and foremost, their safety should be ensured. In this particular situation, any vectors originating from non-human sequences have the potential to elicit negative immunological reactions and are therefore unsuitable. Furthermore, vectors that integrate into the host genome at arbitrary locations are likewise considered to be dangerous. Furthermore, they have the potential to be effectively transported into cells. Viral vectors possess a distinct edge in this regard, however, newly established physical techniques for delivering DNA without the need of viruses, as reviewed by Al-Dosari and Gao (2009), may demonstrate comparable effectiveness. Furthermore, it is crucial that they persist in the cells indefinitely and exhibit sustained expression of the transgene they contain. Since random integration is not allowed according to the first requirement, using extra-chromosomally maintained or site-specific integrating vectors could be considered as an alternative method. Furthermore, it is essential that their cloning capability is sufficiently elevated to enable them to bear fully operational genes accompanied by suitable regulatory components. Large intact genes or gene clusters may exceed the capabilities of typical molecular biology tools for manipulation. Hence, there is a requirement for novel technology.

This chapter will specifically concentrate on non-viral vectors that encompass complete genomic loci instead than mini-genes. The need to utilize these constructs will be evident through many examples of preclinical research involving the integration of vectors that provide position-independent expression from big transgenes. This presentation will focus on the transfer of big naked DNA molecules into patients utilizing physical means. The focus will be on transferring genomic constructions to grown mammalian cells using ex vivo methods. Two potential techniques, nucleofection and bactofection, will be examined for their ability to carry big complexes. This text will provide a comprehensive examination of various vectors that enable the maintenance of foreign DNA outside of the chromosome. The focus will be on the structure and potential uses of EBV-based episomes, Human Artificial Chromosomes, and Scaffod/Matrix Attachment Region (S/MAR) vectors, which are examples of non-integrating vectors that can be retained outside of the chromosome. Furthermore, this article will discuss two methods of precise integration at particular locations that are not linked to the development of cancer. This course will also address the utilization of advanced recombination-based techniques for manipulating huge vectors, known as recombineering. Lastly, an illustration will be provided on the creation of genomic DNA-containing vectors for the purpose of gene therapy in treating Cystic Fibrosis, utilizing recombineering techniques.

2. BENEFITS OF UTILIZING GENOMIC CONSTRUCTS IN GENE THERAPY

The rationale for use genomic constructs instead of cDNA in gene therapy is that they encompass all the introns and surrounding DNA, which can bestow complete regulation over gene expression.

The presence of introns is essential for the proper functioning of genes that encode for many products through differential splicing, such as the immunoglobulin genes. Promoters located in the adjacent DNA regions can be replicated and utilized in a complementary DNA (cDNA) structure. Nevertheless, the introns and adjacent DNA regions also encompass other components that can actively engage in regulating chromatin condensation, consequently exerting an impact on gene expression. Simple enhancers and locus control regions (LCRs) are the most crucial features among them (Lipps et al., 2003).

Enhancers can enhance transcription regardless of their orientation and distance. They have the ability to inhibit the silencing of transgenes, which often occurs when integration takes place in a condensed and hence repressive heterochromatin region. This phenomenon is known as position effect (Martin & Whitelaw, 1996). In addition, they have the ability to direct transgenes to transcriptionally active nuclear compartments and hinder their positioning in close proximity to heterochromatin (Francastel, et al., 1999). LCRs are characterized by their increased complexity and inclusion of enhancers (Li et al., 1999). One common feature they possess is the existence of DNAse I hypersensitive sites (DHS), which are regions where chromatin is not tightly packed and transcription factors can attach to their specific sequences, enabling transgene expression.

Genomic constructions have demonstrated their significance over cDNA or mini-genes in multiple genes, both in cultured cells and transgenic mice. In general, the inclusion of promoters, enhancers, LCRs, and other elements, situated either 5' or 3' of the gene or within introns, led to the expression of transgenes in a manner that was not dependent on their position. This resulted in the transgenes being expressed in the appropriate tissue, at the correct levels, and at the right times. This is in contrast to the use of small transgenes that carry foreign promoters. As an illustration, transgenic mice created using lengthy structures that encompassed all the identified DHS (DNase I hypersensitive sites) located upstream of the class II MHC Ea gene exhibited production of Ea that was not influenced by its position but was reliant on the number of copies present. Constructs that were shorter and did not have some elements of the DHS (Deoxyribonuclease Hypersensitive Site) were susceptible to position effects, as shown by Carson and Wiles in 1993. Comparable outcomes were achieved when a 250-kb YAC containing the genomic locus of the mouse tyrosinase gene was utilized to produce transgenic mice. The transgenic was expressed at levels that were similar to the indigenous gene, in the appropriate tissues, and in proportion to the number of copies, but not influenced by its location (Schedl et al., 1993). The human Huntington Disease gene has been successfully expressed physiologically in transgenic mice by a YAC, which has a targeted disruption of the endogenous gene. Huntigtin is crucial for development as its absence is fatal in mice. However, the introduction of the human transgene in the correct tissue, at appropriate amounts, and early in development successfully saved the mice from dying in the embryonic stage (Hodgson et al., 1996). In addition, it has been demonstrated that the human Friedrich ataxia (FRDA) gene, when produced from a 188-kb BAC, is capable of saving FRDA knockout mice from dying before birth (Sarsero et al., 2004). Transgenic mice were created using constructs that encompass the human β-globin locus, serving as a model system to investigate the developmental regulation of gene clusters. The locus comprises five genes, namely ε, Gγ, Aγ, δ, and β, whose expression is rigorously controlled in terms of both timing and location. The expression of ε-globin occurs throughout embryonic stages, whereas Gγ- and Aγ-globin are expressed in the fetal liver. In adults, δ-globin and β-globin are expressed in the bone marrow. The regulation of gene expression appears to be controlled by an upstream LCR (locus control region) and a downstream DHS (DNase I hypersensitive site). The researchers created transgenic mice by introducing a 160-kb BAC that contained the complete human β-globin locus. These mice demonstrated appropriate developmental control and tissue-specific expression of the globin genes, as reported by Huang et al. in

2000. Curiously, the expression of a human BCL2L10 (Boo) BAC transgene, which was specifically directed at the HPRT locus in transgenic mice, mimicked the pattern observed in humans. The study found that Human Boo mRNA was present in organs where there was no murine Boo mRNA. However, these organs were known to have Boo expression in humans. This suggests that there are human regulatory elements in the BAC (bacterial artificial chromosome) that are not present in the mouse endogenous gene. These human regulatory elements are capable of driving tissue-specific expression in the mouse background (Heaney et al., 2004).

Transgenic research has also been conducted in relation to cystic fibrosis (CF). The condition is a result of genetic abnormalities in the cystic fibrosis transmembrane conductance regulator (CFTR) gene. CFTR is typically expressed in certain tissues and its expression is tightly regulated by specific DNA regions known as DHS, which are located up to 80 kb upstream of the first exon (Smith, et al., 1995). Prior studies using limited mini-gene constructs demonstrated partial expression of CFTR in transgenic mice (Alton et al., 1993; Hyde et al., 1993) and temporary improvement in CF patients at a low level (Caplen et al., 1995). However, these structures were not adequately expressed in the relevant tissues to provide therapeutic benefit in patients. A 320-kilobase yeast artificial chromosome (YAC) containing the human CFTR gene successfully restored the function of CFTR in mice that lacked the CFTR gene (Manson et al., 1997). The expression of genes in the mouse followed the normal pattern, except in some organs such as the pyloric glands, Brunner's glands, epididymis, and sublingual glands. This deviation is likely caused by the absence of a specific DNA sequence in the YAC or the inability of mice transcription factors to recognize the regulatory elements from humans. In recent studies, researchers have successfully proven the accurate timing and location of the human COL6A1 gene (Xing et al., 2007), the human Brain-Derived neurotropic factor (BDNF) gene (Koppel et al., 2009), and the pig Growth Hormone gene (Tong et al., 2010) in BAC transgenic mice.

BAC transgenics have the unique advantage of being able to control the expression of genes in specific tissues and at certain times. This sets them apart from cDNA and mini-genes. As a result, BAC transgenics are a valuable tool for identifying regulatory regions within a gene's locus, in addition to knockout transgenics. For example, researchers have identified an enhancer in the Tyrp1 gene locus and a region responsible for tissue-specific expression in the Neurogenin1 gene locus in transgenic mice created using BACs (Murisier et al., 2006; Quiñones et al., 2010).

3. DELIVERY OF LARGE DNA MOLECULES

The use of gene therapy involving large genomic structures cloned on BAC vectors inevitably presents the challenge of delivering these constructs into specific target cells. Most recombinant viruses, which are routinely used to transport plasmid DNA, can elicit undesirable immunological responses and have a limited capacity for packing, making them unsuitable for delivering big BACs. One exception to this norm is the Herpes simplex virus 1 (HSV-1) -derived amplicon vector. It has been demonstrated that this vector can accommodate and transmit massive bacterial artificial chromosomes (BACs) of around 150 kb in size (Wade-Martins et al., 2003). Nevertheless, this auspicious vector relies on viral sequences and is susceptible to scrutiny over its safety. Thus, it is preferable for the delivery of genomic loci of therapeutic genes to be non-viral.

Regardless of their size, bare DNA molecules pose challenges for transfection into cells, both in vivo and in vitro, due to various obstacles associated with nearly all aspects of cellular biology. Barriers that impede the delivery of substances include the breakdown of molecules by enzymes called nucleases found in the blood and the extracellular matrix, the protective layer of the cell called the plasma membrane, the conversion of endosomes into digestive lysozomes after being taken up by the cell, and the barrier around the nucleus known as the nuclear envelope (Al-Dosari & Gao, 2009). Various physical techniques have been used to enhance the transfer of DNA without any protective covering into cells, achieving similar levels of efficiency as viral methods in certain circumstances. Additionally, chemical techniques utilizing predominantly cationic lipids and polymers have been created and employed for the purpose of transferring genes in both laboratory settings and living organisms. Nevertheless, much of the data that has been published focuses on the delivery of plasmid DNA. However, there is a need for more research to evaluate both physical and chemical approaches for delivering big BACs.

3.1 Methods of physically delivering treatments directly to patients

Physical techniques enhance the introduction of DNA into cells by temporarily disrupting the cell membrane using physical forces, including hydrostatic pressure, electric pulse, ultrasound, laser irradiation, magnetic fields, and particle bombardment. As of March 2011, naked plasmid DNA has been utilized in 18.7% (n=319) of clinical gene therapy trials, as reported in a study published on the Wiley website (http://www.wiley.com/legacy/wileychi/genmed/clinical/).

3.1.1 Injection

During first trials, undressed plasmid DNA (or RNA) was directly administered into various organs of diverse animals, resulting in the detection of reporter gene expression at the injection locations. This technique was employed in mice to transport and manifest reporter genes in several tissues, including as skeletal muscle (Wolff et al., 1990) and lung (Meyer et al., 1995). Following this, the method has successfully enabled the expression of therapeutic genes, leading to the initiation of human clinical trials for limb ischemia (Morishita et al., 2010), erectile dysfunction (Melman et al., 2007), and Duchenne/Becker muscular dystrophy (Romero et al., 2004). These trials have utilized the direct injection of naked DNA into tissues. While these trials primarily focused on evaluating safety, they did demonstrate modest transgenic expression and clinical improvement. However, the degree of expression observed was minimal. Jet injection is a non-invasive alternative to conventional needle injection. This technology utilizes a high-speed and focused stream of liquid that contains DNA, allowing it to effectively permeate the skin and underlying tissues. The device operates using compressed air and allows for precise control of tissue penetration by altering the air pressure. Currently, the primary use of jet injection technology has been in the advancement of DNA vaccines (Raviprakash & Porter, 2006). The efficiency of distribution using jet injection has been enhanced due to technological advancements, bringing it on par with other non-viral approaches. Recently, a clinical trial was conducted to test its effectiveness on patients with melanoma and breast cancer (Walther et al., 2008).

Advancements in the field have resulted in the creation of hydrodynamic injection, a highly effective non-viral gene delivery technique in mice (Al-Dosari & Gao, 2009). The hydrodynamic approach involves injecting a large volume of saline solution containing plasmid DNA into the tail vein at a fast speed. Preliminary research has indicated that this leads to elevated levels of gene expression in the liver (Liu, et al., 1999; Zhang et al., 1999). The hydrodynamic injection into the tail vein has demonstrated effective results when used with big BAC DNA (Hibbitt et al., 2007; Magin-Lachmann et al., 2004). In addition, researchers have successfully achieved local hydrodynamic administration into the liver of rabbits using catheter-assisted perfusion (Eastman et al., 2002), as well as pressure-mediated delivery to the kidneys of rats (Maruyama et al., 2002) and to the muscle of limbs in mammals (Hagstrom et al., 2004). It is still to be determined if comparable hydrodynamic techniques could be utilized in human patients.

3.1.2 Electroporation

In vivo electroporation involves the administration of electrical pulses after injecting DNA directly into the desired tissue. This transiently enhances the permeability of the cell membrane and promotes the uptake of DNA by a mechanism that is currently not well understood. Under ideal circumstances, the efficacy of plasmid DNA distribution using in vivo electroporation can reach levels comparable to viral techniques. However, the efficacy diminishes when larger DNA molecules are employed. Electroporation has demonstrated efficacy in delivering tiny DNA molecules to various tissues, such as muscle and lung, employing a range of electrodes, including needle and surface electrodes (Brown et al., 2008; Dean et al., 2003). However, there is limited data regarding the in vivo electroporation of big DNA molecules. Magin-Lachmann et al. (2004) conducted a study where they successfully delivered an 80-kb BAC into electroporated muscle. However, it was observed that the production of the reporter gene from the BAC was 5 times less efficient compared to a plasmid, as was expected. The advancement of knowledge and technology in electroporation has led to its use in clinical settings for treating melanoma in humans (Daud et al., 2008). Additionally, there are current clinical trials utilizing electroporation for the treatment of various types of cancer and for DNA vaccine.

3.1.3 Sonoporation

Sonoporation is a method that use high-intensity, low-frequency ultrasonic waves to induce transitory permeabilization of the plasma membrane, similar to electroporation. This technology aims to enhance the delivery of DNA into cells. However, the technique differs from electroporation. Ultrasound is thought to cause acoustic cavitation, which can temporarily break the cell membrane. When microbubbles, which are gas-filled vesicles coated with albumin, polymers, or phospholipids, are utilized together with it, cavitation and the subsequent uptake of DNA in a specific area are increased (Wells, 2004). Commercially accessible microbubbles have been demonstrated to directly impact the efficiency of in vivo sonoporation, as evidenced by Alter et al. (2009).

Multiple studies have demonstrated the in vivo administration of plasmids containing either reporter or therapeutic genes to various tissues, such as the lung, heart, and muscle (Xenariou et al., 2010; Alter et al., 2009; Sheyn et al., 2008). However, comparable data, when available, verified that the efficacy of sonoporation was markedly inferior to that of electroporation. Nevertheless, the potential clinical use of sonoporation in humans is still under consideration due to its non-invasive nature and reduced tissue damage compared to electroporation. Remarkably, the utilization of both electric pulses and ultrasound waves, known as electrosonoporation, resulted in a significantly higher gene expression in the skeletal muscle of mice compared to using electroporation or sonoporation alone. Specifically, the gene expression was 8 times higher with electrosonoporation compared to electroporation alone, and 1.6 times higher compared to sonoporation alone (Yamashita et al., 2002).

3.1.4 Other physical methods

Particle bombardment, performed using a gene gun initially intended for DNA delivery in plants, is a non-viral technique for transferring genes. This method involves using gold particles that are coated with DNA. The particles undergo acceleration through the use of compressed gas and are subsequently released into tissues. This method, known as ballistic DNA delivery, has been employed to transport transgenes to the skin, liver, and muscle tissues of rats and mice (Yang et al., 1990). It has also been utilized to administer DNA vaccines to the skin, muscles, and tumors in animal models and in human clinical trials. A recent study by Chang et al. (2008) demonstrated the efficacy of delivering a tiny reporter gene to the liver of mice in vivo. There is a scarcity of data regarding the efficacy of gene gun in delivering substantial DNA constructs in living organisms. A single study demonstrated the delivery of a DNA vaccine comprising a 183-kb BAC using a gene gun, which resulted in immunological protection in chickens (Tischer et al., 2002). A novel and auspicious technique for gene transfer, utilizing an infrared laser beam, has been devised and employed to transport a tiny transgene into the muscle of live mice (Zeira et al., 2003). This study found that the efficacy of delivery, as measured by the intensity and duration of transgene expression, was equivalent to that achieved by electroporation. However, the delivery method utilized in this study caused less damage to the tissue.

In conclusion, magnetic fields have been utilized to improve the efficiency of delivering genes to specific locations within living organisms. In the magnetofection approach, DNA is temporarily attached to superparamagnetic nanoparticles. These nanoparticles are then guided to the desired location using a strong magnetic field after being injected locally. The effectiveness of in vivo magnetofection has been demonstrated for delivering tiny plasmid DNA to the gastrointestinal system of rats, the blood arteries of the ear of pigs (Scherer et al., 2002), and the respiratory epithelium of mice (Xenariou et al., 2006).

3.2 Chemical methods for direct delivery to patients

Gene delivery utilizes a wide range of chemical vectors, with several reagents available (Al-Dosari & Gao, 2009). However, these vectors can generally be classified into two primary categories: cationic lipids and cationic polymers. These substances work by creating complexes with DNA, which is negatively charged. These complexes, called lipoplexes and polyplexes, serve to protect the DNA from enzymes called nucleases. Additionally, they enable the DNA to enter cells through processes such as endocytosis, pinocytosis, or phagocytosis. Once inside the cells, the complexes help the DNA to reach the nucleus by escaping from the endosomes. The processes involved in lipoplexes and polyplexes differ in their mechanisms, and the overall effectiveness is influenced by various factors. These factors include the chemical composition of the cationic lipids/polymers, the ratio of charge between the cationic lipids/polymers and the DNA, the size and structure of the lipoplexes/polyplexes, and the presence of helper lipids like DOPE and cholesterol in the complexes (Tros de Ilarduya et al., 2010).

Throughout the years, numerous cationic lipids with exceptional transfection efficiency have been created, making lipofection the prevailing technique for in vitro gene transfer. Regrettably, lipoplexes do not exhibit equivalent efficacy for in vivo administration due to their inactivation upon interaction with blood-borne substances. Nevertheless, there have been documented instances of effective in vivo DNA delivery using both systemic and local injection methods. Administering lipoplexes with a single intravenous injection in mice has been demonstrated to induce the expression of reporter transgenes in the lung, heart, liver, spleen, and kidney (Song et al., 1997). Remarkably, when cationic lipid/CFTR-plasmid-DNA complexes were administered locally in the form of an aerosol to the lungs of animals with cystic fibrosis, it successfully corrected the ion transport deficit (Hyde et al., 1993). Studies conducted in human patients showed little transgenic expression, which was not significant enough to yield a therapeutic advantage (Griesenbach and Alton, 2009).

The selection of the cationic lipid to be employed is dependent on the specific application. It is necessary to carefully optimize the transfection process, as lipoplexes have the potential to elicit undesired immunological responses. Lipofection has been employed in 6.4% (n=109) of gene therapy clinical studies, as reported in a study published on the Wiley website (http://www.wiley.com/legacy/wileychi/genmed/clinical/).

Cationic polymers, particularly polyethylenimine (PEI), have been widely utilized in experiments involving the transport of DNA. Polyplexes exhibit greater stability compared to lipoplexes, with their toxicity and transfection efficiency being influenced by their molecular weight (mw). According to Fischer et al. (1999), polymers with low molecular weight (mw) are more effective and less harmful than those with high molecular weight. Remarkably, the effectiveness of intravenous injection of PEI/BAC-DNA complexes in mice has been seen to be lower compared to alternative non-viral gene delivery techniques, such as electroporation and hydrodynamic injection (Magin-Lachmann et al., 2004).

3.3 *Ex vivo* delivery to cells

The ex vivo technique is an alternative to in vivo administration of DNA for gene therapy. This process involves the surgical extraction of certain cells from a patient, followed by the introduction of a suitable therapeutic gene into these cells in a controlled environment. Finally, the modified cells are reintroduced into the donor's body. Due to the utilization of autologous cells, immunosuppression is unnecessary when the cells are transplanted back into the patient. In the majority of ex vivo gene therapy cases, it is necessary for donated cells to successfully integrate into the specific tissue. The effectiveness of this process, known as "homing," varies depending on the kind of tissue.

Advancements in stem cell research have significantly transformed ex vivo therapy. Stem cells has the capacity to transform into many cell types under suitable conditions, both inside and outside the body. This unique feature makes them suitable for direct application in cell therapy (Abdulrazzak et al., 2010). A cell treatment technique utilizing Embryonic Stem Cells or foetal Mesenchymal Stem Cells outside of the body would not require the introduction of a therapeutic gene and would successfully overcome the challenge of low gene delivery efficiency. Nevertheless, it would be necessary to utilize it in an allogeneic manner, hence requiring the administration of immunosuppressive medications. An alternative and appealing approach would involve introducing a therapeutic gene into stem cells that are derived from the patient's own body, such as Mesenchymal Stem Cells (MSCs). These MSCs can be easily obtained from the bone marrow or adipose tissue of adults and have demonstrated a remarkable ability to develop into different types of cells (Abdallah & Kassem, 2008). Alternatively, Induced Pluripotent Stem Cells (iPSCs) can be created by reprogramming adult somatic cells (Yu et al., 2007; Takahashi et al., 2007).

Various non-viral techniques have been employed to introduce DNA into MSCs for the purpose of ex vivo gene therapy. The objective of the subsequent sections is to evaluate two methods, namely nucleofection and bactofection. The selection criteria for this decision are based on the capacity to successfully transport and preserve big DNA molecules without damage, as well as the ability to retain the differentiation potential of the MSCs after transfection.

3.3.1 Nucleofection

The Nucleofector technology, developed by Amaxa Biosystems, is a non-viral gene transfer approach that relies on electroporation. This technique utilizes a precise mix of electrical settings and cell-specific solutions to facilitate gene transfer. Unlike alternative non-viral transfection techniques, this method enables the direct transfer of DNA into the nucleus of the cell. The technique is highly advantageous for introducing genes into several primary cell types from different species that are typically challenging to transfect (Gresch et al., 2004). Aluigi et al. (2006) demonstrated that nucleofection is capable of effectively delivering plasmid DNA containing a reporter gene to MSCs without negatively affecting their ability to differentiate. Initial findings from experiments involving the introduction of large BACs into stable cell lines using nucleofection indicate that nucleofection can also be employed to transport genomic constructs. However, it should be noted that not all clones produced using this method possess undamaged and unaltered DNA, as reported by Cheung et. al. in unpublished data.

3.3.2 Bactofection

Bactofection, the process of directly delivering DNA into mammalian cells using invasive bacteria, is a promising technology for gene transfer. It has the potential to be used for both in vitro and in vivo delivery applications (Larsen et al., 2008). In this technique, the DNA is initially introduced into bacteria, either in the form of a plasmid or a BAC. These bacteria possess the capability to enter eukaryotic cells, and they are later employed to invade and transfer their DNA content into the desired target cells. Multiple bacterial mechanisms facilitating invasion of eukaryotic cells have been documented. The most optimal option is derived from the *E. coli* strain BM4573 (Laner et al., 2005). The BM4573 bacteria have been genetically modified to possess three specific traits. Firstly, they are engineered to consistently produce invasin, a protein derived from the inv gene of Yersinia pseudotuberculosis. This protein binds to β1-integrins found on mammalian cells, leading to the internalization of the bacteria. Secondly, these modified bacteria have a permanent impairment in their ability to synthesize the cell wall due to diaminopimelic acid (DAP) auxotrophy. This condition causes the bacteria to undergo lysis once they are internalized. Lastly, the modified bacteria stably express listeriolysin O, a pore-forming cytolysin derived from the hly locus of Listeria monocytogenes. This cytolysin enables the bacteria to escape from the vacuole after entering the host cell and release their DNA into the cytosol. As a result, there is a significant increase in the expression of the transgene. (Laner et al., 2005).

An advantage of this approach is that it eliminates the requirement for DNA purification before transfection, a procedure that remains technically difficult for extremely big DNA molecules. Another advantage is that DNA delivered by bacterial invasion is rarely rearranged. Reconfigurations are consistently a matter of apprehension while employing alternative DNA transportation techniques to transmit sizable BACs. For example, lipofection and electroporation have been effectively utilized to transport big BAC DNA into cultivated cells. However, it has been shown that some of the resulting clones experience rearrangements, as reported by Magin-Lachmann et al. in 2004 and by Cheung et al. in unpublished data. In contrast, several studies have demonstrated that large BACs are delivered intact by bacterial invasion albeit with low efficiency. Two stable cell lines have been created by bacterial invasion. One has integrations of a ~250-kb BAC carrying the human clotting factor VIII (FVIII) gene (Pérez-Luz et al., 2007), while the other contains integrations of a ~258-kb BAC carrying the human CFTR gene (Kotzamanis et al., 2009). Neither of the two studies have discovered any clones with altered DNA, and both studies have demonstrated the expression of the corresponding transgenes. Thus, bactofection is a desirable technique for transporting sizeable genomic DNA that contains complete structures. The method's capacity to transport high concentrations of BACs into MSCs without compromising their ability to differentiate has not been demonstrated yet.

4. EXTRA-CHROMOSOMAL VECTORS

In gene therapy protocols, it is necessary to ensure efficient retention and long-term expression of the therapeutic gene, regardless of its type (small cDNA or large genomic DNA) and the delivery method (viral or non-viral). This is important in order to avoid the need for repeated administrations. Integration into the host genome has been extensively utilized in gene therapy to meet this criterion. Nevertheless, the risks associated with integration caused by insertional mutagenesis have gained significant attention following a clinical trial that employed a retroviral vector to address X-linked severe combined immune deficiency (SCIDX1). During this experiment, certain individuals experienced the development of leukemia as a result of the disruption of the growth-stimulating LIM domain only 2 (LMO2) proto-oncogene produced by the integration of the vector (Hacein-Bey-Abina et al., 2003a, 2003b). The safety concerns regarding the unregulated integration of the therapeutic gene into the host genome have been reinforced by studies that the vectors have a tendency to integrate into the regulatory regions of genes that are actively transcribing (Bushman et al., 2005). Considering the requirement for prolonged gene expression and the challenges related to vector integration, the utilization of vectors that remain in the nucleus as episomes without integrating could offer significant benefits. Three distinct mechanisms have been utilized to accomplish the maintenance of vectors carrying the therapeutic gene outside of the chromosomes. These mechanisms include systems derived from the Epstein-Barr virus (EBV), artificial chromosomes, and systems based on scaffold/matrix attachment region (S/MAR). These systems possess a significant ability to clone and can be utilized in conjunction with extensive genomic constructions.

4.1 OriP/EBNA-1 episomal vectors

The most extensively studied strategy for maintaining episomes is based on sequences obtained from the Epstein-Barr virus (EBV) genome. EBV, short for Epstein-Barr virus, belongs to the herpesvirus family. It has a genomic size of 172 kilobases and is kept in a latent state as an independently replicating episome in a small portion of infected cells (Masucci & Ernberg, 1994). During the dormant phase of its cycle, DNA replication takes place at the origin of replication oriP. Only around 10 proteins are synthesized during this phase, and the Epstein Barr Nuclear Antigen-1 (EBNA-1) is the sole protein necessary for replication at oriP (Yates et al., 1985). The interaction between oriP and EBNA-1 facilitates the separation of the viral genome amongst the daughter cells by binding EBNA-1 to host metaphase chromosomes (Harris et al., 1985).

The characteristics of the EBV have been utilized to create a method for the sustained presence of foreign DNA introduced into cells. Research has demonstrated that plasmids containing oriP and expressing EBNA-1 have the ability to replicate independently once throughout each cell cycle when introduced into human cells. Additionally, these plasmids can separate by attaching themselves to the chromosomes of the host cells (Haase & Calos, 1991). The oriP/EBNA-1 system has demonstrated the ability to sustain long-term episomal maintenance of large human genes, such as CFTR, HPRT, and the β-globin gene, without the need for selection and expression. This has been supported by studies conducted by Huertas et al. (2000), Wade-Martins et al. (2000), and Black & Vos (2002). After obtaining encouraging results, a practical method for incorporating the oriP/EBNA-1 sequences into any BAC that already contains a therapeutic gene has been created (Magin-Lachmann et al., 2003) and will be examined at a later time.

The oriP/EBNA-1 retention system is user-friendly and capable of providing extrachromosomal maintenance to foreign DNA of several hundred kilobases that is introduced into cells. However, it does have significant drawbacks. The episomal vector is randomly segregated to daughter cells, leading to an unequal distribution. This causes a loss of the episomes at a rate of 2-8% every cell division (Sclimenti & Calos, 1998). The involvement of viral sequences, particularly from the EBV, which has been linked to various forms of human cancers (Cohen, 2000), restricts the safe use of EBV vectors in gene therapy.

4.2 Human artificial chromosomes

Human Artificial Chromosomes (HACs) are vectors that possess the ability to duplicate and separate alongside the natural chromosomes within human cells. In order to accomplish this, the chromosomes must possess the essential components necessary for their function, which include an origin of replication, telomeres, and centromeres (Pérez-Luz & Díaz-Nido, 2010).

An strategy to constructing HACs, known as the "top-down" approach, involves breaking down existing chromosomes and creating smaller mini-chromosomes that include only the three essential chromosomal elements. Multiple studies have demonstrated that mini-chromosomes can serve as a platform for hosting and enabling the expression of sizable therapeutic genes. These mini-chromosomes can be transferred across different mouse and human cell lines and can also be passed on through the germ line of mice (Kakeda et al., 2005; Shen et al., 2001; Voet et al., 2001). While mini-chromosomes provide advantageous characteristics for implementation in transgenics, their utilization in gene therapy is limited exclusively to an ex vivo methodology.

HACs with a significant capacity for utilization in in vivo gene therapy are produced by an alternative method known as the bottom-up technique. This process is analogous to the one used for YAC development in yeast. It entails the assembly of the functional chromosomal components and the creation of a HAC from scratch in human cells. Various approaches have been employed to create new HACs from scratch, with the most practical method being the introduction of a BAC containing a substantial α-satellite DNA array and a few flag genes into HT1080 cells by transfection. There is no evidence to suggest that telomeric sequences or an origin of replication are necessary. This is likely because circular HACs are generated and replication is initiated at origins located within the marker genes (Ebersole et al., 2000). HACs produced in this manner are present as individual (or low quantity) chromosomes within the nucleus and exhibit a high level of stability throughout cell division (near to 100%) in the absence of selective pressure.

The potential application of these vectors in gene therapy has been shown through the production of substantial therapeutic genes from them (Grimes et al., 2001; Mejía & Larin, 2000). Significant progress was observed with the development and utilization of effective techniques for altering extensive repetitive sequences, like alphoid DNA, to create human artificial chromosomes (HACs), which will be explored in detail later (Kotzamanis et al., 2005). However, there are some unresolved difficulties that must be addressed prior to any use in a clinical setting. Currently, HACs have demonstrated effective formation exclusively in HT1080 cells. The reason for their inability to form in other cell lines has not been determined yet, but it is currently restricting their application in in vivo gene therapy. Additionally, all HACs that have been created through de novo synthesis in HT1080 cells have ranged in size from 1 to 10 Mb. While they are smaller than natural chromosomes, they are larger than the initial DNA used, indicating that unexpected amplifications and rearrangements have taken place during their formation. This is undesirable for the development of safe gene therapy vectors. Additionally, alternative outcomes of the input DNA, aside from the creation of HACs, have been documented, and the possibility of integration has not been ruled out (Harrington et al., 1997). In ex vivo therapy, isolating specific clones that express the desired gene from a HAC would prevent this issue. However, in *in vivo* gene therapy, any interaction between the input DNA and the natural chromosomes could result in the same outcomes as viral vectors. Additional research is required to enhance the efficacy of HAC creation in order to prevent any integration events from occurring.

4.3 S/MAR vectors

S/MARs, or Scaffold/Matrix Attachment Regions, are a variety of sequences that exist in all eukaryotic genomes. They play a crucial role in several functions of chromatin, such as organizing chromatin into loops. This process appears to be facilitated by the connection between S/MARs and the nuclear matrix (Heng et al., 2004). The use of a plasmid vector bearing a specific S/MAR element from the human interferon β-gene in experiments has uncovered an additional characteristic of S/MARs: their capacity to sustain the presence of foreign DNA injected into cells in an episomal manner. The vector demonstrated the ability to reproduce and persist as an episome in CHO cells, with a minimal number of copies, for over 100 generations without the need for selection. Additionally, it exhibited a mitotic stability of 98% (Piechaczek et al., 1999). It was later confirmed that the mitotic stability of the vector was provided through the interaction of the S/MAR with the nuclear matrix (Baiker et al., 2000). Remarkably, the utilization of S/MAR appeared to hinder the integration of vectors into the host genome, as integration occurrences were detected in less than 1% of clones that underwent stable transfection (Jackson et al., 2006). The interferon β S/MAR has been shown to possess the capability of facilitating replication and episomal retention in several cell lines and primary cells (Papapetrou et al., 2006), as well as in genetically engineered pigs in vivo (Manzini et al., 2006). In addition, the researchers introduced the same S/MAR element into a BAC that contained 135 kb of the human low density lipoprotein receptor (LDLR) genomic locus using site-specific homologous recombination. They found that this element allowed for the low copy episomal maintenance of the BAC in CHO ldlr-/- cells for over 100 generations without the need for selection. Furthermore, the transgene expressed at sufficiently high levels to fully restore LDLR function in these cells over the long term (Lufino et al., 2007).

To summarize, it appears that S/MAR vectors accomplish their replication and segregation by enlisting and interacting with proteins from the host cell, in contrast to viral episomal vectors that require encoding viral components for their functioning. In addition, S/MAR vectors do not require amplifications and rearrangements in order to work, unlike HAC vectors. This means that S/MAR vectors have a specific and predetermined structure and composition. The distinctive safety characteristics of S/MARs make them highly appealing for utilization in gene therapy involving extensive genomic complexes.

5. VECTORS INTEGRATING AT SPECIFIC SITES

An alternative to the utilization of episomal vectors mentioned earlier, which fulfills the criteria for long-lasting expression of the transgene and prevention of harmful genetic effects, is the deliberate integration of the therapeutic construct at a precise location in the host genome where no active genes exist. Various vector systems have been devised to accomplish this, each with its unique set of constraints (Voigt et al., 2008). Out of the many vectors available, only two types will be discussed. These types are chosen based on their affinity for specific sequences already present in the human genome, their suitability for usage in living organisms, and their capacity to facilitate the integration of complex genomic structures.

5.1 Phage integrase based vectors

The Streptomyces phage ΦC31 integrase is an enzyme capable of facilitating site-specific recombination between a phage attachment site (attP) and a bacterial attachment site (attB). This process leads to the integration of ΦC31 into the bacterial genome and the initiation of the lysogenic phase of its life cycle (Groth & Calos, 2004). This integrase has been demonstrated to permanently incorporate a solitary instance of foreign DNA, which includes the attP site, into the human genome. This integration occurs at natural pseudo att sites located in the intergenic regions of human chromosomes, far away from known oncogenes (Chalberg et al., 2005).

Several studies have confirmed the effectiveness of ΦC31 integrase-based vector systems in ex vivo gene therapy. These studies have shown that these systems can successfully express various therapeutic genes in cultured cells, such as human embryonic stem cells (Thyagarajan et al., 2008). Additionally, in vivo gene therapy experiments have demonstrated the expression of different therapeutic genes in animal models, such as the dystrophin gene in dystrophic mouse muscle (Bertoni et al., 2006). The ability of the technology to be utilized in gene therapy using genomic constructs has been demonstrated through transgenic experiments conducted in Drosophila. In these experiments, the ΦC31 integrase enzyme was employed to integrate sizable DNA segments, measuring up to 133 kb, into the organism's genome (Venken et al., 2006). Nevertheless, additional research has raised concerns about the safety of these vectors. It has been demonstrated that the stable expression of ΦC31 integrase can lead to various chromosomal abnormalities in human cells (Liu et al., 2006). Furthermore, in certain instances,

ΦC31-mediated integration has been linked to chromosome rearrangements, likely resulting from recombination between hidden att sites (Ehrhardt et al., 2006). The authors of the study (Keravala et al., 2009) have created new versions of ΦC31 integrase that have improved efficiency and specificity in integrating genetic material. These advancements may address the safety issues associated with using ΦC31 integrase in gene therapy.

Site-specific integration can be achieved by utilizing transposase enzymes, such as Sleeping beauty and piggyBac, which have been extensively investigated. These enzymes enable the integration of foreign genes into specific sections of the genome that contain transposable elements. (Ivics & Izsvak, 2010). Nevertheless, these methods are not expected to be beneficial for the integration of extensive genomic constructs due to a substantial loss in their ability to transpose when the insert length is increased. This phenomenon is referred to as "length-dependence" (Atkinson & Chalmers, 2010).

5.2 Adeno-associated virus based vectors

The Adeno-associated virus (AAV) possesses some characteristics that can be utilized for targeted incorporation of external DNA. AAV is a harmless virus that has a genome made up of a single-stranded DNA molecule that is 4.7 kilobases long. This genome has two genes, rep and cap, which are surrounded by 145 base pair palindromic sequences called inverted terminal repeats (ITRs) (Srivastava et al., 1983). When a helper virus like adenovirus or herpes simplex virus is present, AAV is capable of replicating and entering its lytic cycle. However, in the absence of these helper viruses, AAV integrates into the human genome and remains in a dormant state. The ITRs encompass the necessary sequences for the replication, packaging, and integration of the virus. The rep gene encodes four regulatory proteins that are essential for facilitating the integration of the virus into the human genome during periods of latency. The process of integration takes place at a specific location on chromosome 19 known as AAVS1 (Kotin et al., 1992).

Research has demonstrated that human transgenes, which are surrounded by inverted terminal repeats (ITRs), can be inserted into the AAVS1 gene. This integration occurs when viral Rep is present in human cells grown in culture (Cortés et al., 2008), as well as in transgenic mice that carry the human AAVS1 gene (Liu et al., 2010; Recchia et al., 2004). Studies have demonstrated that the AAV system is capable of incorporating large genes, up to 100kb in length, into the AAVS1 region (Oehmig et al., 2007). These research have shown the efficacy of AAV-based vectors, however safety concerns have arisen. 10-30% of infected human cells in culture have been found to contain small transgenes integrated into their DNA. Approximately half of these integrations occur specifically at the AAVS1 site. This indicates that there is a 50% chance of insertional mutagenesis occurring in a gene therapy application. Furthermore, the continuous presence of the viral Rep protein is harmful and can lead to chromosome instability and movement of the transgene (McCarty et al., 2004).

6. METHODS FOR MODIFICATION OF LARGE DNA MOLECULES

Gene therapy involving genomic constructs involves the manipulation of extensive DNA regions, which are frequently repetitive in character. To illustrate, it is necessary to incorporate marker genes and other beneficial sequences into vectors that contain extensive genomic regions, while avoiding any alterations. In certain instances, the complete genomic region of a therapeutic gene may not be present in a single BAC vector, requiring the connection of the inserts from two or more BACs. The complexity involved in carrying out such modifications has impeded advancements in this subject for an extended period. Recently, engineers have devised efficient methods that enable the consideration of BACs containing genomic regions as gene therapy vectors.

6.1 Addition of marker and small genes

Typically, the process of cloning mammalian selectable markers and small reporter genes onto the vector region of a BAC that contains a genomic insert is hindered by the absence of easily accessible cloning sites on the vector, the potential existence of numerous restriction sites within the insert, and the challenge of manipulating large DNA molecules without compromising their integrity. Therefore, it is necessary to seek alternatives to traditional molecular biology procedures that rely on the use of restriction enzymes. An alternate method for incorporating reporter genes and short sequences onto BACs is through site-specific homologous recombination, which is mediated by the bacteriophage P1-derived Cre/loxP system.

The Cre protein exhibits high affinity and catalytic activity exclusively towards particular loxP sites, which are universally present on all BAC vectors. This characteristic makes the modification strategy based on these sites widely applicable. Figure 1 illustrates that the initial stage of this process involves creating a retrofitting plasmid (pRetro) that contains the desired gene, a loxP site, and a selectable marker, which will be introduced to the BAC. The replication of pRetro relies on the high-copy gamma origin (γ-ori), which is functional only in an E. coli host that expresses the π protein (produced by the pir gene).

After inserting the plasmid into the BAC in the DH10B E. coli host, which lacks the expression of π, the γ-ori loses its functionality and the BAC remains as a low-copy plasmid. The depicted system in Figure 1 employs an independent plasmid to facilitate the expression of Cre. This plasmid is co-transfected with the retrofitting plasmid and subsequently eliminated after the retrofitting procedure, as it lacks a loxP site and hence does not participate in the recombination process (Mejía & Larin, 2000). This plasmid harbors a fusion gene of LacZ and cre, enabling the expression of Cre upon IPTG induction. Additionally, it contains a temperature-sensitive origin of replication that is active exclusively at a temperature of 30°C. Figure 1 demonstrates that the BAC is retrofitted with pRetro in vivo at a temperature of 300C. Subsequently, the bacteria are cultivated at 420C to eliminate the Cre expressing plasmid and prevent any additional undesired recombination occurrences. Magin-Lachmann et al. (2003) have developed different retro plasmids that can be used to easily incorporate a G418 mammalian selectable marker, a GFP reporter gene, a luciferase reporter gene, and/or an OriP/EBNA-1 element into any BAC.

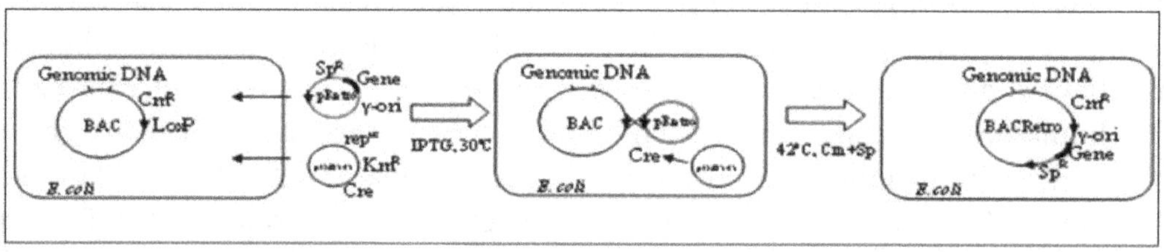

Fig. 1. Addition of small marked/reporter genes onto BACs by Cre/*loxP* recombination

1.2 Addition of large sequences

The process of creating a gene therapy vector that contains genomic DNA and can exist outside of the chromosomes may require incorporating long segments of DNA that are challenging to clone into a pRetro plasmid. These segments are then added to a BAC using Cre/loxP recombination, as explained in the preceding section. For instance, a 70-kilobase alphoid array has been demonstrated to be necessary for a PAC vector to generate human artificial chromosomes (HACs) in HT1080 cells (Ebersole et al., 2000). A technique called recombineering, which is based on homologous recombination in *E. coli*, has been developed for altering vast portions of DNA in specific applications (Copeland et al., 2001). In recombineering, the sequence to be inserted is bordered by two regions of similarity to the BAC. The length of these regions depends on the recombination system that facilitates the recombination reaction.

Most applications incorporate a selectable marker to enable the identification of properly modified clones in *E. coli*. The phage recombination systems RecET and Red include genes that encode proteins responsible for homologous recombination of the cryptic Rac prophage and bacteriophage λ, respectively. These systems exhibit high efficiency, do not necessitate extensive homology regions, and infrequently induce undesired recombination events during recombineering (Court et al., 2002; Muyrers et al., 2001).

The Red recombination technique was specifically utilized to include a 70-kb alphoid array into a BAC that already contained a 156-kb genomic insert containing the HPRT gene. This was achieved through recombineering, as shown in Figure 2. The successful expression of the HPRT gene from the resulting HACs was demonstrated in a study by Kotzamanis et al. in 2005.

Figure 2 demonstrates that recombination was specifically directed towards the chloramphenicol resistance gene, which is found in all BACs. This enabled the insertion of the 70-kb alphoid array into any BAC, leading to the creation of HACs in the suitable cells (Kotzamanis et al., 2005).

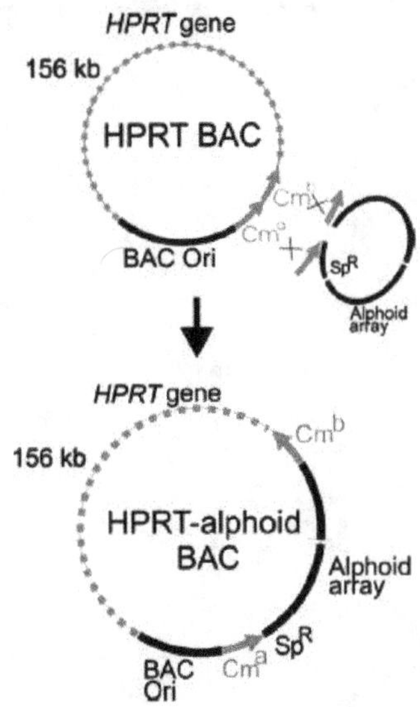

Fig. 2. Addition of large arrays of alphoid DNA onto BACs by recombineering

6.3 Linking of two overlapping BACs

There are numerous human genes that are as large as, or even larger than, the average size of the BAC libraries. It is often challenging to find a single BAC that covers the complete gene together with its related regulatory components. To perform gene therapy using the genomic loci of specific genes, it is necessary to combine various sequences into a single BAC clone. This is achieved by connecting many overlapping BAC clones that cover the appropriate region. Recombineering, facilitated by the Red system derived from the λ-prophage, has been employed to connect two partially overlapping Bacterial Artificial Chromosomes (BACs) without inducing any rearrangements or altering the reading frame of the therapeutic gene. The procedure involves utilizing two rounds of homologous recombination to connect the inserts of two overlapping BACs, as illustrated in Figure 3. During the initial stage, the BAC inserts are inserted into modified BAC vectors (pBACLink vectors that have been linearized by NotI) using homologous recombination. This recombination occurs at specific areas known as HomA, HomB, and HomC, which are first amplified through PCR and then cloned into the pBACLink vectors before they are linearized. In the second round, a modified BAC is cut into linear form using the rare cutting enzyme I-PpoI. This linearized BAC is then inserted into bacteria that are efficient at recombination and already contain another modified BAC. As a result, recombination occurs at specific sites called HomB and Cma, which are part of the chloramphenicol resistance gene found in all BACs. This process links the two inserts together, creating a single BAC. Additional overlapping BAC inserts can be incorporated by alternating the utilization of the two pBACLink vectors as outlined in the work conducted by Kotzamanis and Huxley in 2004.

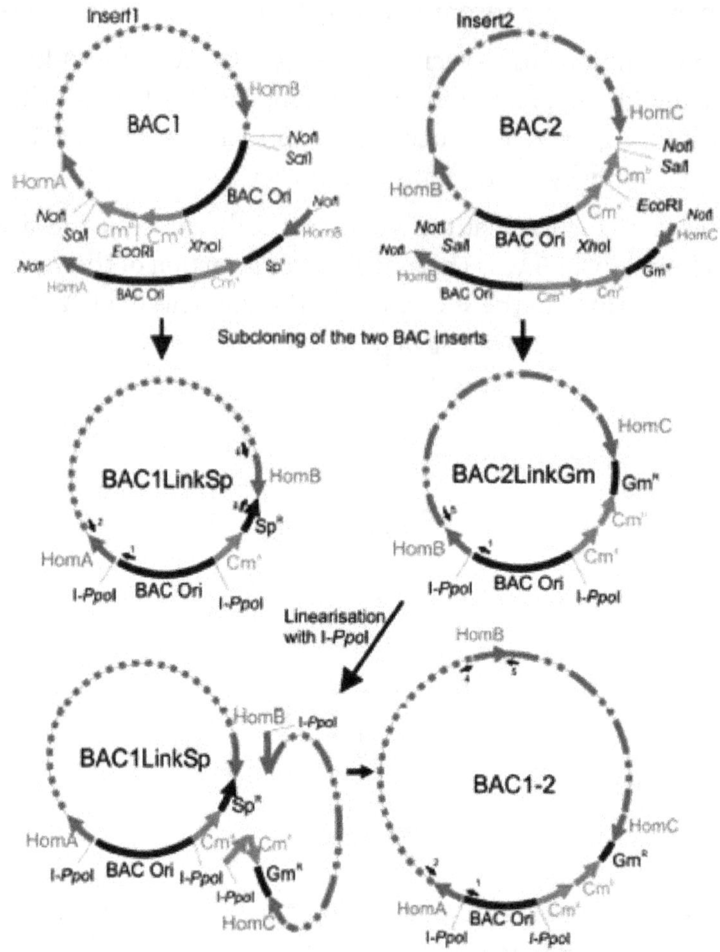

Fig. 3. Linking of two overlapping BACs into a single larger BAC

7. TOWARDS GENE THERAPY OF CYSTIC FIBROSIS USING A GENOMIC CONSTRUCT

Cystic Fibrosis is the most common genetic disease in the Caucasians caused by mutations in the CFTR gene which is 250 kb long and encodes a cAMP regulated transmembrane Cl^- ion channel in epithelial cells of several organs. The lungs are the site where the most severe consequences, ultimately resulting in mortality, occur (Boucher, 2002).

Cystic Fibrosis has been identified as an ideal candidate for gene therapy due to several factors. These include the convenient access to the respiratory tract without the need for invasive procedures, the successful cloning and characterization of the CFTR gene by Riordan et al. (1989) and Rommens et al. (1989), and the belief that even small amounts of gene expression could lead to therapeutic benefits, as suggested by Dorin et al. (1996).

Previous studies have used CFTR cDNA-heterologous promoter systems and various viral vectors to deliver and express the transgene. These studies have shown some expression in transgenic mice and temporary correction of Cl^- ion transport deficiency in patients, but no significant clinical improvement has been achieved (Griesenbach & Alton, 2009).

The precise regulation of the expression of the CFTR gene during specific developmental stages and in specific tissues, which is controlled by regulatory elements located before, after, and within the gene (McCarthy & Harris, 2005), suggests that using constructs that cover the entire genomic locus of the gene may have a greater therapeutic potential for treating Cystic Fibrosis. So far, the only transgene that has completely revived transgenic mice, which did not naturally produce CFTR and would typically perish, is the entire gene found on a YAC of around 300 kb in size (Manson et al., 1997). Nevertheless, YAC vectors possess the drawback of being arduous to transfer across cells and are intrinsically unstable, thus leading to their replacement by BACs. The CFTR gene, although big, has not been discovered in its complete form inside any of the sequenced BACs obtained from the Human Genome Project. The method described in section 6.3 was created and utilized to create a BAC vector that contains the entire CFTR gene and its associated regulatory elements (Kotzamanis et al., 2009). The successful transcription of the gene into a properly spliced mRNA has been proven in a mouse cell line. The BAC (Bacterial Artificial Chromosome) is the sole CFTR (Cystic Fibrosis Transmembrane Conductance Regulator) genomic system that is accessible on a user-friendly vector. It has the potential to serve as the foundation for non-viral gene therapy for Cystic Fibrosis.

Gene therapy for Cystic Fibrosis via viral methods is hindered by obstacles in delivering the genes, such as the lack of viral receptors in the respiratory epithelium. Additionally, there are safety issues related to the potential triggering of an inflammatory reaction. Both in vivo and ex vivo non viral gene therapy are appealing areas of research. Nevertheless, there are some unresolved difficulties that must be addressed prior to any use in a clinical setting. For example, the introduction of non-viral vectors into living organisms is hindered by their low effectiveness, which decreases further when larger structures are employed. The evidence presented by Wang et al. (2005) demonstrated that mesenchymal stem cells (MSCs) derived from bone marrow have the ability to transform into various cell types, including airway epithelial cells. This suggests that MSCs could be used in an ex vivo approach. However, the lack of sufficient data on the administration and successful integration of ex vivo corrected MSCs into the lungs of Cystic Fibrosis patients limits their potential application.

Genetic alteration using the CFTR BAC, whether through in vivo or ex vivo gene therapy, should ideally result in persistent transgenic expression to prevent the need for recurring gene or cell delivery. Therefore, any of the systems that enable the retention of genetic material outside of chromosomes or the specific insertion of genetic material into regions not linked with the development of cancer (as explained in sections 4 and 5) would need to be incorporated into the CFTR BAC. The procedure necessary to incorporate short sequences like OriP/EBNA-1 or S/MAR elements, as well as longer sequences like the 70-kb alphoid array, into the CFTR BAC is accessible (see to sections 6.1 and 6.2). In addition, various techniques for delivering the final construct to the respiratory epithelium, either in vivo or ex vivo, have been devised and are accessible for use with the genomic CFTR-locus carrying BAC (see to section 3).

8. CONCISE SUMMARY

Non-viral gene therapy utilizing the complete genomic locus of the therapeutic gene has two advantages compared to viral cDNA gene therapy. Firstly, it does not provoke undesired immune responses. Secondly, it enables the regulation of transgene expression at precise amounts and in specific tissues and timeframes. When used in conjunction with a method for maintaining extra-chromosomal elements or targeted integration, it can enable long-lasting expression of transgenes without producing harmful changes in the DNA sequence, making it highly regarded for its safety. Out of all the systems currently available, the S/MAR element appears to be the most effective, secure, and convenient. It can even be used in living organisms. The challenge of effectively manipulating large segments of DNA and delivering them into cells, both in laboratory settings and in living organisms, has long hindered the application of genomic constructs in gene therapy. However, recent advancements in recombineering and various physical and chemical techniques for gene transfer have begun to address this issue. Evidently, a universal delivery system will not be available, and the selection of the most suitable delivery technique will rely on the particular application and gene therapy approach.

REFERENCES

Abdallah, BM. & Kassem, M. (2008). Human mesenchymal stem cells: from basic biology to clinical applications. *Gene therapy*, 15(2), pp. 109-16.

Abdulrazzak, H., Moschidou, D., Jones, G. & Guillot PV. (2010). Biological characteristics of stem cells from foetal, cord blood and extraembryonic tissues.*Journal of Royal Society Interface*, 6(7), pp. 689-706

Al-Dosari, MS. & Gao, X. (2009). Nonviral gene delivery: principle, limitations, and recent progress. *AAPS journal*, 11(4), pp. 671-81.

Alter, J., Sennoga, CA., Lopes, DM., Eckersley, RJ. & Wells, DJ. (2009). Microbubble stability is a major determinant of the efficiency of ultrasound and microbubble mediated in vivo gene transfer. *Ultrasound in medicine & biology*, 35(6), pp. 976-84.

Alton, EW, Middleton, PG., Caplen, NJ., Smith, SN., Steel, DM., Munkonge, FM., Jeffery, PK., Geddes, DM., Hart, SL. & Williamson, R. (1993). Non-invasive liposome- mediated gene delivery can correct the ion transport defect in cystic fibrosis mutant mice. *Nature genetics*,Vol. 5, pp. 135-42.

Aluigi, M., Fogli, M., Curti, A., Isidori, A., Gruppioni, E., Chiodoni, C., Colombo, MP., Versura, P., D'Errico-Grigioni, A., Ferri, E., Baccarani, M. & Lemoli, RM. (2006). Nucleofection is an efficient nonviral transfection technique for human bone marrow-derived mesenchymal stem cells. *Stem cells*, 24(2), pp. 454-61.

Atkinson, H. & Chalmers, R. (2010). Delivering the goods: viral and non-viral gene therapy systems and the inherent limits on cargo DNA and internal sequences. *Genetica*, 138, pp. 485–498

Baiker, A., Maercker, C., Piechaczek, C., Schmidt, S. B., Bode, J, Benham, C. & Lipps, H. (2000). Mitotic stability of an episomal vector containing a human scaffold/matrix- attached region is provided by association with nuclear matrix. *Nature cell biology*, 2(3), pp. 182-4.

Bertoni, C., Jarrahian, S., Wheeler, TM., Li, Y., Olivares, EC., Calos, MP. & Rando, TA. (2006). Enhancement of plasmid-mediated gene therapy for muscular dystrophy by

directed plasmid integration. *Proceedings of the National Academy of Sciences USA*, 103(2), pp. 419-24.

Black, J., & Vos, JM. (2002). Establishment of an oriP/EBNA1-based episomal vector transcribing human genomic beta-globin in cultured murine fibroblasts. *Gene therapy*, 9(21), pp. 1447-54.

Boucher, RC. (2002). An overview of the pathogenesis of cystic fibrosis lung disease.*Advanced drug delivery reviews*, 54(11), pp. 1359-71.

Brown, PA., Khan, AS., & Draghia-Akli, R. (2008). Delivery of DNA into skeletal muscle in large animals. *Methods in molecular biology*, 423, pp. 215-24.

Bushman, F., Lewinski, M., Ciuffi, A., Barr, S., Leipzig, J., Hannenhalli, S. & Hoffmann, C. (2005). Genome-wide analysis of retroviral DNA integration. *Nature reviews. Microbiology*, 3(11), pp. 848-58.

Caplen, NJ., Alton, EW, Middleton, PG., Dorin, JR., Stevenson, BJ., Gao, X., Durham, SR., Jeffery, PK., Hodson, ME. & Coutelle, C. (1995). Liposome-mediated CFTR gene transfer to the nasal epithelium of patients with cystic fibrosis. *Nature medicine*, 1(1), pp. 39-46.

Carson, S., & Wiles, MV. (1993). Far upstream regions of class II MHC Ea are necessary for position-independent, copy-dependent expression of Ea transgene. *Nucleic acids research*, 21(9), pp. 2065-72.

Cavazzana-Calvo, M, Hacein-Bey, S., Saint Basile, G. de Gross, F., Yvon, E., Nusbaum, P., Seiz, F., Hue, C., Certain, S., Casanova, JL., Bousso, P., Deist, FL. & Fischer, A. (2000). Gene therapy of human severe combined immunodeficiency (SCID)-X1 disease. *Science*, 288(5466), pp. 669-72.

Chang, M.L., Chen, JC., Yeh, CT., Chang, MY., Liang, CK., Chiu, CT., Lin, DY. & Liaw, YF. (2008). Gene gun bombardment with DNA-coated gold particles is a potential alternative to hydrodynamics-based transfection for delivering genes into superficial hepatocytes. *Human gene therapy*, 19(4), pp. 391-5.

Cohen, JI. (2000). Epstein-Barr virus infection. *New England journal of medicine*, 343(7), pp.481-92.

Copeland, NG, Jenkins, NA, & Court, DL. (2001). Recombineering: a powerful new tool for mouse functional genomics. *Nature reviews. Genetics*, 2(10), pp. 769-79.

Cortés, ML, Oehmig, A., Saydam, O., Sanford, JD., Perry, KF, Fraefel, C. & Breakefield, X. (2008). Targeted integration of functional human ATM cDNA into genome mediated by HSV/AAV hybrid amplicon vector. *Molecular therapy*, 16(1), pp. 81-8.

Court, DL., Sawitzke, JA., & Thomason, LC. (2002). Genetic engineering using homologous recombination. *Annual review of genetics*, 36, pp. 361-88.

Daud, AI., DeConti, RC., Andrews, S., Urbas, P., Riker, A. I., Sondak, V. K., Munster, P., Sullivan, D., Urgen, K., Messina, J. & Heller, R. (2008). Phase I trial of interleukin-12 plasmid electroporation in patients with metastatic melanoma. *Journal of clinical oncology*, 26(36), pp. 5896-903.

Dean, DA., Machado-Aranda, D., Blair-Parks, K., Yeldandi, AV., & Young, JL. (2003). Electroporation as a method for high-level nonviral gene transfer to the lung. *Gene therapy*, 10(18), pp. 1608-15.

Dorin, JR., Farley, R, Webb, S., Smith, SN., Farini, E., Delaney, SJ., Wainwright, B., Alton, EW. & Porteous, DJ. (1996). A demonstration using mouse models that successful gene therapy for cystic fibrosis requires only partial gene correction. *Gene therapy*, 3(9), pp. 797-801.

Eastman, SJ., Baskin, KM., Hodges, BL., Chu, Q., Gates, A., Dreusicke, R., Anderson, S. & Scheule, R. (2002). Development of catheter-based procedures for transducing the isolated rabbit liver with plasmid DNA. *Human gene therapy*, 13(17), pp. 2065-77.

Ebersole, T., Ross, A., Clark, E., McGill, N., Schindelhauer, D., Cooke, H & Grimes, B. (2000). Mammalian artificial chromosome formation from circular alphoid input DNA does not require telomere repeats. *Human molecular genetics*, 9(11), pp. 1623-31.

Ehrhardt, A., Engler, JA., Xu, H., Cherry, A., & Kay, MA. (2006). Molecular analysis of chromosomal rearrangements in mammalian cells after phiC31-mediated integration. *Human gene therapy*, 17(11), pp. 1077-94.

Fischer, D., Bieber, T., Li, Y, Elsässer, H. P., & Kissel, T. (1999). A novel non-viral vector for DNA delivery based on low molecular weight, branched polyethylenimine: effect of molecular weight on transfection efficiency and cytotoxicity. *Pharmaceutical research*, 16(8), pp. 1273-9.

Francastel, C., Walters, MC., Groudine, M., & Martin, DI. (1999). A functional enhancer suppresses silencing of a transgene and prevents its localization close to centrometric heterochromatin. *Cell*, 99(3), pp. 259-69.

Gresch, O., Engel, F. B., Nesic, D., Tran, T., England, H., Hickman, E., Korner, I., Gan, L., Chen, S., Castro-Obregon, S., Hammermann, R., Wolf, J., Muler-Hartmann, H., Nix, M., Siebenkotten, G., Kraus, G. & Lun, K. (2004). New non-viral method for gene transfer into primary cells. *Methods*, 33(2), pp. 151-63.

Griesenbach, U., & Alton, EW. (2009). Cystic fibrosis gene therapy: successes, failures and hopes for the future. *Expert review of respiratory medicine*, 3(4), pp. 363-71.

Grimes, BR., Schindelhauer, D., McGill, NI., Ross, A., Ebersole, TA., & Cooke, HJ. (2001). Stable gene expression from a mammalian artificial chromosome. *EMBO reports*, 2(10), pp. 910-4.

Groth, AC., & Calos MP. (2004). Phage integrases: biology and applications. *Journal of molecular biology*, 335(3), pp. 667-78.

Haase, SB., & Calos, MP. (1991). Replication control of autonomously replicating human sequences. *Nucleic acids research*, 19(18), pp. 5053-8.

Hacein-Bey-Abina, S, Von Kalle, C., Schmidt, M, McCormack, MP., Wulffraat, N, Leboulch, P., (2003). LMO2-associated clonal T cell proliferation in two patients after gene therapy for SCID-X1. *Science*, 302(5644), pp. 415-9.

Hacein-Bey-Abina, S., Von Kalle, C., Schmidt, M., Le Deist, F., Wulffraat, N., McIntyre, E., Radford, I., Villeval, JL., Fraser, C., Cabazzana-Calvo, M. & Fischer, A. (2003). A serious adverse event after successful gene therapy for X-linked severe combined immunodeficiency. *New England journal of medicine*, 348(3), pp. 255-6.

Hagstrom, JE., Hegge, J., Zhang, G., Noble, M., Budker, V., Lewis, DL., Herweijer, H. & Wolf, J. (2004). A facile nonviral method for delivering genes and siRNAs to skeletal muscle of mammalian limbs. *Molecular therapy*, 10(2), pp. 386-98.

Harrington, JJ., Van Bokkelen, G., Mays, RW., Gustashaw, K., & Willard, HF. (1997). Formation of de novo centromeres and construction of first-generation human artificial microchromosomes. *Nature genetics*, 15(4), pp. 345-55.

Harris, A, Young, BD., & Griffin, BE. (1985). Random association of Epstein-Barr virus genomes with host cell metaphase chromosomes in Burkitt's lymphoma-derived cell lines. *Journal of virology*, 56(1), pp. 328-32.

Heaney, JD., Rettew, AN., & Bronson, SK. (2004). Tissue-specific expression of a BAC transgene targeted to the Hprt locus in mouse embryonic stem cells. *Genomics*, 83(6), pp. 1072-82.

Heng, HQ., Goetze, S., Ye, CJ., Liu, G., Stevens, JB., Bremer, SW., Wykes, S., Bode, J. & Krawetz, S. (2004). Chromatin loops are selectively anchored using scaffold/matrix-attachment regions. *Journal of cell science*, 117(Pt 7), pp. 999-1008.

Hibbitt, OC., Harbottle, RP., Waddington, SN., Bursill, CA., Coutelle, C., Channon, KM.

&Wade-Martins, R. (2007). Delivery and long-term expression of a 135 kb LDLR genomic DNA locus in vivo by hydrodynamic tail vein injection. *Journal of gene medicine*, 9(6), pp. 488-97.

Hodgson, JG., Smith, DJ., McCutcheon, K., Koide, HB., Nishiyama, K., Dinulos, MB., et al. (1996). Human huntingtin derived from YAC transgenes compensates for loss of murine huntingtin by rescue of the embryonic lethal phenotype. *Human molecular genetics*, 5(12), pp. 1875-85.

Huang, Y., Liu, DP., Wu, L., Li, TC., Wu, M., Feng, DX., et al. (2000). Proper developmental control of human globin genes reproduced by transgenic mice containing a 160-kb BAC carrying the human beta-globin locus. *Blood cells, molecules & diseases*, 26(6), pp. 598-610.

Huertas, D., Howe, S, McGuigan, A., & Huxley, C. (2000). Expression of the human CFTR gene from episomal oriP-EBNA1-YACs in mouse cells. *Human molecular genetics*, 9(4), pp. 617-29.

Hyde, S. C., Gill, D. R., Higgins, C. F., Trezise, A. E., MacVinish, L. J., Cuthbert, A. W., et al. (1993). Correction of the ion transport defect in cystic fibrosis transgenic mice by gene therapy. *Nature*, 362(6417), pp. 250-5.

Ivics, Z. % Izsvak, S. (2010). The expanding universe of transposon technologies for gene and cell engineering. *Mobile DNA*, 1:25

Jackson, D. A., Juranek, S., & Lipps, Hans J. (2006). Designing nonviral vectors for efficient gene transfer and long-term gene expression. *Molecular therapy*, 14(5), pp. 613-26.

Kakeda, M., Hiratsuka, M., Nagata, K., Kuroiwa, Y., Kakitani, M., Katoh, M., et al. (2005). Human artificial chromosome (HAC) vector provides long-term therapeutic transgene expression in normal human primary fibroblasts. *Gene therapy*, 12(10), pp. 852-6.

Keravala, A., Lee, S., Thyagarajan, Bhaskar, Olivares, Eric C, Gabrovsky, V. E., Woodard, L. E., et al. (2009). Mutational derivatives of PhiC31 integrase with increased efficiency and specificity. *Molecular therapy*, 17(1), pp. 112-20.

Koppel, I., Aid-Pavlidis, T., Jaanson, K., Sepp, M., Pruunsild, P., Palm, K., et al. (2009). Tissue-specific and neural activity-regulated expression of human BDNF gene in BAC transgenic mice. *BMC neuroscience*, 10, 68.

Kotin, R. M., Linden, R. M., & Berns, K I. (1992). Characterization of a preferred site on human chromosome 19q for integration of adeno-associated virus DNA by non- homologous recombination. *EMBO journal*, 11(13), pp. 5071-8.

Kotzamanis, G., Abdulrazzak, H., Gifford-Garner, J., Haussecker, PL., Cheung, W., Grillot- Courvalin, C., et al. (2009). CFTR expression from a BAC carrying the complete

human gene and associated regulatory elements. *Journal of cellular and molecular medicine*, 13(9A), pp. 2938-48.

Kotzamanis, George, Cheung, W., Abdulrazzak, H., Perez-Luz, S., Howe, Steven, Cooke, Howard, et al. (2005). Construction of human artificial chromosome vectors by recombineering. *Gene*, pp. 351, 29-38.

Kotzamanis, G., & Huxley, C.. (2004). Recombining overlapping BACs into a single larger BAC. *BMC biotechnology*, 4, 1.

Laner, A., Goussard, S., Ramalho, A. S., Schwarz, T., Amaral, M. D., Courvalin, P., et al. (2005). Bacterial transfer of large functional genomic DNA into human cells. *Gene therapy*, 12(21), pp. 1559-72.

Larsen, M. D. B., Griesenbach, U, Goussard, S., Gruenert, D. C., Geddes, D M, Scheule, R K, et al. (2008). Bactofection of lung epithelial cells in vitro and in vivo using a genetically modified Escherichia coli. *Gene therapy*, 15(6), pp. 434-42.

Li, Q., Harju, S., & Peterson, K. R. (1999). Locus control regions: coming of age at a decade plus. Trends in genetics : TIG, 15(10), 403-8.

Lipps, H J, Jenke, A. C. W., Nehlsen, K., Scinteie, M. F., Stehle, I M, & Bode, J. (2003).Chromosome-based vectors for gene therapy. *Gene*, 304, pp. 23-33.

Liu, F., Song, Y., & Liu, D. (1999). Hydrodynamics-based transfection in animals by systemic administration of plasmid DNA. *Gene therapy*, 6(7), pp. 1258-66.

Liu, Q., Perez, C. F., & Wang, Yaming. (2006). Efficient site-specific integration of large transgenes by an enhanced herpes simplex virus/adeno-associated virus hybrid amplicon vector. *Journal of virology*, 80(4), pp. 1672-9.

Liu, R., Li, Yiming, Hu, R., Jin, T., Deng, S., Liang, W., et al. (2010). A site-specific genomic integration strategy for sustained expression of glucagon-like peptide-1 in mouse muscle for controlling energy homeostasis. *Biochemical and biophysical research communications*, 403(2), pp. 172-7.

Lufino, M. M. P., Manservigi, R., & Wade-Martins, Richard. (2007). An S/MAR-based infectious episomal genomic DNA expression vector provides long-term regulated functional complementation of LDLR deficiency. *Nucleic acids research*, 35(15), e98.

Magin-Lachmann, C, Kotzamanis, G, D'Aiuto, L, Cooke, H, Huxley, C, & Wagner, E. (2004). In vitro and in vivo delivery of intact BAC DNA -- comparison of different methods. *Journal of gene medicine*, 6(2), pp. 195-209.

Magin-Lachmann, C., Kotzamanis, G., D'Aiuto, L., Wagner, E., & Huxley, C. (2003). Retrofitting BACs with G418 resistance, luciferase, and oriP and EBNA-1 - new vectors for in vitro and in vivo delivery. *BMC biotechnology*, 3(1), 2.

Manson, A. L., Trezise, A. E., MacVinish, L. J., Kasschau, K. D., Birchall, N., Episkopou, V., et al. (1997). Complementation of null CF mice with a human CFTR YAC transgene. *EMBO journal*, 16(14), pp. 4238-49.

Manzini, S., Vargiolu, A., Stehle, Isa M, Bacci, M. L., Cerrito, M. G., Giovannoni, R., et al. (2006). Genetically modified pigs produced with a nonviral episomal vector. *Proceedings of the National Academy of Sciences USA*, 103(47), pp. 17672-7.

Martin, D. I., & Whitelaw, E. (1996). The vagaries of variegating transgenes. *BioEssays*, 18(11), pp. 919-23.

Maruyama, H., Higuchi, N., Nishikawa, Y., Hirahara, H., Iino, N., Kameda, S., et al. (2002). Kidney-targeted naked DNA transfer by retrograde renal vein injection in rats. *Human gene therapy*, 13(3), pp. 455-68..

Masucci, M. G., & Ernberg, I. (1994). Epstein-Barr virus: adaptation to a life within the immune system. *Trends in microbiology*, 2(4), pp. 125-30.

McCarthy, V. A., & Harris, Ann. (2005). The CFTR gene and regulation of its expression. *Pediatric pulmonology*, 40(1), pp. 1-8.

McCarty, D. M., Young, Samuel M, & Samulski, R Jude. (2004). Integration of adeno- associated virus (AAV) and recombinant AAV vectors. *Annual review of genetics*, 38, pp. 819-45.

Mejía, J. E., & Larin, Z. (2000). The assembly of large BACs by in vivo recombination.*Genomics*, 70(2), pp. 165-70.

Melman, A., Bar-Chama, N., McCullough, A., Davies, K., & Christ, G. (2007). Plasmid-based gene transfer for treatment of erectile dysfunction and overactive bladder: results of a phase I trial. *The Israel Medical Association journal*, 9(3), pp. 143-6.

Meyer, K. B., Thompson, M. M., Levy, M. Y., Barron, L. G., & Szoka, F. C. (1995). Intratracheal gene delivery to the mouse airway: characterization of plasmid DNA expression and pharmacokinetics. *Gene therapy*, 2(7), pp. 450-60.

Morishita, R., Makino, H., Aoki, M., Hashiya, N., Yamasaki, K., Azuma, J., et al. (2010). Phase I/IIa Clinical Trial of Therapeutic Angiogenesis Using Hepatocyte Growth Factor Gene Transfer to Treat Critical Limb Ischemia. *Arteriosclerosis, thrombosis, and vascular biology.*

Murisier, F., Guichard, S., & Beermann, F. (2006). A conserved transcriptional enhancer that specifies Tyrp1 expression to melanocytes. *Developmental biology*, 298(2), pp. 644-55.

Muyrers, J. P., Zhang, Y., & Stewart, A. F. (2001). Techniques: Recombinogenic engineering-- new options for cloning and manipulating DNA. *Trends in biochemical sciences*, 26(5), pp. 325-31.

Oehmig, A, Cortés, M L, Perry, K F, Sena-Esteves, M., Fraefel, C., & Breakefield, X O. (2007). Integration of active human beta-galactosidase gene (100 kb) into genome using HSV/AAV amplicon vector. *Gene therapy*, 14(14), pp. 1078-91.

Osoegawa, K., Mammoser, A. G., Wu, C., Frengen, E., Zeng, C., Catanese, J. J., et al. (2001). A bacterial artificial chromosome library for sequencing the complete human genome. *Genome research*, 11(3), pp. 483-96.

Papapetrou, E. P., Ziros, P. G., Micheva, I. D., Zoumbos, N. C., & Athanassiadou, A. (2006). Gene transfer into human hematopoietic progenitor cells with an episomal vector carrying an S/MAR element. *Gene therapy*, 13(1), pp. 40-51.

Piechaczek, C., Fetzer, C., Baiker, A., Bode, J, & Lipps, H J. (1999). A vector based on the SV40 origin of replication and chromosomal S/MARs replicates episomally in CHO cells. *Nucleic acids research*, 27(2), pp. 426-8.

Pérez-Luz, S., Abdulrazzak, H., Grillot-Courvalin, Catherine, & Huxley, Clare. (2007). Factor VIII mRNA expression from a BAC carrying the intact locus made by homologous recombination. *Genomics*, 90(5), pp. 610-9.

Pérez-Luz, S., & Díaz-Nido, J. (2010). Prospects for the use of artificial chromosomes and minichromosome-like episomes in gene therapy. *Journal of biomedicine & biotechnology*, 2010.

Quiñones, H. I., Savage, T. K., Battiste, J., & Johnson, J. E. (2010). Neurogenin 1 (Neurog1) expression in the ventral neural tube is mediated by a distinct enhancer and preferentially marks ventral interneuron lineages. *Developmental biology*, 340(2), pp. 283-92.

Raper, S. E., Chirmule, N., Lee, F. S., Wivel, N. A., Bagg, A., Gao, G.-ping, et al. (n.d.). Fatal systemic inflammatory response syndrome in a ornithine transcarbamylase deficient patient following adenoviral gene transfer. *Molecular genetics and metabolism*, 80(1-2), pp. 148-58.

Raviprakash, K., & Porter, K. R. (2006). Needle-free injection of DNA vaccines: a brief overview and methodology. *Methods in molecular medicine*, 127, pp. 83-9.

Recchia, A., & Mavilio, F. (2006). Site-specific integration into the human genome: ready for clinical application? *Rejuvenation research*, 9(4), pp. 446-9.

Riordan, J. R., Rommens, J. M., Kerem, B., Alon, N., Rozmahel, R., Grzelczak, Z., et al. (1989). Identification of the cystic fibrosis gene: cloning and characterization of complementary DNA. *Science*, 245(4922), pp. 1066-73.

Romero, N. B., Braun, S., Benveniste, O., Leturcq, F., Hogrel, J.-Y., Morris, G. E., et al. (2004). Phase I study of dystrophin plasmid-based gene therapy in Duchenne/Becker muscular dystrophy. *Human gene therapy*, 15(11), pp. 1065-76.

Rommens, J. M., Iannuzzi, M. C., Kerem, B., Drumm, M. L., Melmer, G., Dean, M., et al. (1989). Identification of the cystic fibrosis gene: chromosome walking and jumping. *Science* ,245(4922), pp. 1059-65.

Sarsero, J. P., Li, L., Holloway, T. P., Voullaire, Lucille, Gazeas, S., Fowler, K. J., et al. (2004). Human BAC-mediated rescue of the Friedreich ataxia knockout mutation in transgenic mice. *Mammalian genome*, 15(5), pp. 370-82.

Schedl, A., Montoliu, L., Kelsey, G., & Schütz, G. (1993). A yeast artificial chromosome covering the tyrosinase gene confers copy number-dependent expression in transgenic mice. *Nature*, 362(6417), pp. 258-61.

Scherer, F., Anton, M., Schillinger, U., Henke, J., Bergemann, C., Krüger, A., et al. (2002). Magnetofection: enhancing and targeting gene delivery by magnetic force in vitro and in vivo. *Gene therapy*, 9(2), pp. 102-9.

Sclimenti, C. R., & Calos, M P. (1998). Epstein-Barr virus vectors for gene expression and transfer. *Current opinion in biotechnology*, 9(5), pp. 476-9.

Shen, M. H., Yang, J. W., Yang, J., Pendon, C., & Brown, W. R. (2001). The accuracy of segregation of human mini-chromosomes varies in different vertebrate cell lines, correlates with the extent of centromere formation and provides evidence for a trans-acting centromere maintenance activity. *Chromosoma*, 109(8), pp. 524-35.

Sheyn, D., Kimelman-Bleich, N., Pelled, G., Zilberman, Y., Gazit, D., & Gazit, Z. (2008). Ultrasound-based nonviral gene delivery induces bone formation in vivo. *Gene therapy*, 15(4), pp. 257-66.

Smith, A. N., Wardle, C. J., & Harris, A. (1995). Characterization of DNASE I hypersensitive sites in the 120kb 5' to the CFTR gene. *Biochemical and biophysical research communications*, 211(1), pp. 274-81.

Song, Y. K., Liu, F., Chu, S., & Liu, D. (1997). Characterization of cationic liposome-mediated gene transfer *in vivo* by intravenous administration. *Human gene therapy*, 8(13), pp. 1585-94.

Srivastava, A., Lusby, E. W., & Berns, K I. (1983). Nucleotide sequence and organization of the adeno-associated virus 2 genome. *Journal of virology*, 45(2), pp. 555-64.

Takahashi, K., Tanabe, K., Ohnuki, M., Narita, M., Ichisaka, T., Tomoda, K., et al. (2007). Induction of pluripotent stem cells from adult human fibroblasts by defined factors. *Cell*, 131(5), pp. 861-72.

Thyagarajan, Bhaskar, Liu, Y., Shin, S., Lakshmipathy, U., Scheyhing, K., Xue, H., et al. (2008). Creation of engineered human embryonic stem cell lines using phiC31 integrase. *Stem cells*, 26(1), pp. 119-26.

Tischer, B. K., Schumacher, D., Beer, M., Beyer, J., Teifke, J. P., Osterrieder, K., et al. (2002). A DNA vaccine containing an infectious Marek's disease virus genome can confer protection against tumorigenic Marek's disease in chickens. *Journal of general virology*, 83(Pt 10), pp. 2367-76.

Tong, J., Lillico, S. G., Bi, M. J., Qing, T., Liu, X. F., Wang, Yuanyuan, et al. (2010). Tissue- specific and expression of porcine growth hormone gene in BAC transgenic mice. *Transgenic research.*

Tros de Ilarduya, C., Sun, Y., & Düzgüneş, N. (2010). Gene delivery by lipoplexes and polyplexes. *European journal of pharmaceutical sciences*, 40(3), pp. 159-70.

Venken, K. J. T., He, Y., Hoskins, R. A., & Bellen, H. J. (2006). P[acman]: a BAC transgenic platform for targeted insertion of large DNA fragments in D. melanogaster. *Science*, N.Y.), 314(5806), pp. 1747-51.

Verma, I. M., & Weitzman, M. D. (2005). Gene therapy: twenty-first century medicine. *Annual review of biochemistry*, 74, pp. 711-38.

Voet, T., Vermeesch, J., Carens, A., Dürr, J., Labaere, C., Duhamel, H., et al. (2001). Efficient male and female germline transmission of a human chromosomal vector in mice. *Genome research*, 11(1), pp. 124-36.

Voigt, K., Izsvák, Z., & Ivics, Z. (2008). Targeted gene insertion for molecular medicine. *Journal of molecular medicine*, 86(11), pp. 1205-19.

Wade-Martins, R, White, R. E., Kimura, H., Cook, P. R., & James, M. R. (2000). Stable correction of a genetic deficiency in human cells by an episome carrying a 115 kb genomic transgene. *Nature biotechnology*, 18(12), pp. 1311-4.

Wade-Martins, R., Saeki, Y., & Chiocca, E. A. (2003). Infectious delivery of a 135-kb LDLR genomic locus leads to regulated complementation of low-density lipoprotein receptor deficiency in human cells. *Molecular therapy*, 7(5 Pt 1), pp. 604-12.

Walther, W., Siegel, R., Kobelt, D., Knösel, T., Dietel, M., Bembenek, A., et al. (2008). Novel jet-injection technology for nonviral intratumoral gene transfer in patients with melanoma and breast cancer. *Clinical cancer research*, 14(22), pp. 7545-53.

Wang, G., Bunnell, B. A., Painter, R. G., Quiniones, B. C., Tom, S., Lanson, N. A., et al. (2005). Adult stem cells from bone marrow stroma differentiate into airway epithelial cells: potential therapy for cystic fibrosis. *Proceedings of the National Academy of Sciences USA*, 102(1), pp. 186-91.

Wells, D J. (2004). Gene therapy progress and prospects: electroporation and other physical methods. *Gene therapy*, 11(18), pp. 1363-9.

Wolff, J A, Malone, R. W., Williams, P., Chong, W., Acsadi, G., Jani, A., et al. (1990). Direct gene transfer into mouse muscle in vivo. *Science*, 247(4949 Pt 1), pp. 1465-8.

Xenariou, S, Griesenbach, U, Ferrari, S., Dean, P., Scheule, R K, Cheng, S H, et al. (2006). Using magnetic forces to enhance non-viral gene transfer to airway epithelium in vivo. *Gene therapy*, 13(21), pp. 1545-52.

Xenariou, Stefania, Liang, H.-D., Griesenbach, Uta, Zhu, J., Farley, Raymond, Somerton, L., et al. (2010). Low-frequency ultrasound increases non-viral gene transfer to the mouse lung. *Acta biochimica et biophysica Sinica*, 42(1), pp. 45-51.

Xing, L., Salas, M., Lin, C.-S., Zigman, W., Silverman, W., Subramaniyam, S., et al. (2007). Faithful tissue-specific expression of the human chromosome 21-linked COL6A1 gene in BAC-transgenic mice. *Mammalian genome*, 18(2), pp. 113-22.

Yamashita, Y.-ichi, Shimada, M., Tachibana, K., Harimoto, N., Tsujita, E., Shirabe, K., et al. (2002). In vivo gene transfer into muscle via electro-sonoporation. *Human gene therapy*, 13(17), pp. 2079-84.

Yang, N. S., Burkholder, J., Roberts, B., Martinell, B., & McCabe, D. (1990). In vivo and in vitro gene transfer to mammalian somatic cells by particle bombardment. *Proceedings of the National Academy of Sciences USA*, 87(24), pp. 9568-72.

Yates, J. L., Warren, N., & Sugden, B. (1985.). Stable replication of plasmids derived from Epstein-Barr virus in various mammalian cells. *Nature*, 313(6005), pp. 812-5.

Yu, J., Vodyanik, M. A., Smuga-Otto, K., Antosiewicz-Bourget, J., Frane, J. L., Tian, S., et al. (2007). Induced pluripotent stem cell lines derived from human somatic cells. *Science*, 318(5858), pp. 1917-20.

Zeira, E., Manevitch, A., Khatchatouriants, A., Pappo, O., Hyam, E., Darash-Yahana, M., et al. (2003). Femtosecond infrared laser-an efficient and safe in vivo gene delivery system for prolonged expression. *Molecular therapy*, 8(2), pp. 342-50.

Zhang, G., Budker, V., & Wolff, JA. (1999). High levels of foreign gene expression in hepatocytes after tail vein injections of naked plasmid DNA. *Human gene therapy*, 10(10), pp. 1735-7.

Zhang, XM., & Huang, JD. (2003). Combination of overlapping bacterial artificial chromosomes by a two-step recombinogenic engineering method. *Nucleic acids research*, 31(15), e81.

2 Chapter

GENE THERAPY AND VACCINE DEVELOPMENT: NON-VIRAL DELIVERY SYSTEMS

1. INTRODUCTION: AN OVERVIEW

Gene therapy is the procedure of treating a specific disease by introducing genetic material to induce a therapeutic effect [Stone, 2010]. A variety of strategies, including "gene replacement," "gene correction," and "gene augmentation," can be employed to correct the defective gene of a diagnosed patient [Katare and Aeri, 2010]. In replacement therapy, a normal gene is inserted into the genome to allow its product to supplant that of a defective gene. This method may be appropriate for recessive disorders, which are characterized by a deficiency of an enzyme or other proteins. The approach may not be effective in treating dominant disorders that are associated with the production of an anomalous gene product, which interferes with the product of a normal gene, despite the fact that the gene functions in the genome and provides an appropriate regulatory sequence [Katare and Aeri, 2010]. The replacement of a mutant gene or a portion of it with a normal sequence is necessary for corrective gene therapy. This can be accomplished through the utilization of recombinant technology. The suppression of a specific mutation by a transfer RNA that is introduced into a cell is another form of corrective therapy [Katare and Aeri, 2010]. In gene augmentation, the introduction of a normal genetic sequence into a host genome alters the expression of the mutant gene in the defective cell, while the defective host gene remains unaffected. In general, the recipient cells of gene therapy may be either somatic or germline cells. Germline cell therapy entails the modification of genes in germ cells, which will be transmitted to subsequent generations. Somatic cell therapy entails the insertion of DNA into specific somatic cells, such as bone marrow stem cells, fibroblasts, hepatocytes, or mycocytes [Katare and Aeri, 2010]. The majority of genetic engineering laboratories worldwide are employing this type of gene therapy.

Gene therapy offers significant potential for the treatment of a variety of diseases, including cancer, infections, and genetic disorders [Park et al., 2006]. Although the genetic mutations that underlie a variety of diseases are well-understood, the delivery of a corrective gene to the diseased organs/tissues continues to be a remarkable challenge: [Stone, 2010]. The development of appropriate gene delivery systems may be one of the most critical factors in the successful implementation of gene therapy. Gene delivery systems should be engineered to ensure that the genetic materials are not prematurely degraded in the systemic bloodstream and that the therapeutic genes are transferred to the target cells in an efficient manner. All molecules that possess intracellular functionality will necessitate intracellular delivery systems. RNA molecules, oligonucleotides, and encoding genes, for instance, must enter cells and target the nucleus when transcription is the objective. A common requirement is the prevention of endosomal uptake, which may result in degradation and denaturation, irrespective of the molecules used for delivery [Gould and Chernajovsky, 2007]. A variety of methods are currently being developed to facilitate the delivery of all of these categories of molecules to disease sites. In order to attain the objective in its entirety, additional advancements in cell-targeting strategies are necessary.

At present, a variety of techniques, both older and more recently discovered, have been developed for therapeutic gene transfer. Various viral and non-viral options are accessible for clinical and fundamental research [Gardlik et al., 2005]. DNA-based vaccines are gaining popularity among these investigations. They activate the CD4+ T cells of the Th1 subset, thereby facilitating a cellular immune response that is efficacious against pathogens. Conversely, the recombinant protein vaccines induce a humoral response by activating the Th2 subset of T cells. The studies demonstrated that DNA vaccines have been effective in safeguarding animals from influenza, herpes, rabies, malaria, and leishmaniasis [Katare and Aeri, 2010]. Nevertheless, the approach's value has been diminished by the potential drawbacks of DNA vaccines. In order to enhance the efficiency of antigen delivery and vaccine efficacy, the non-viral vector has demonstrated specific advantages in overcoming the challenges that both peptide/protein and gene-based vaccines have faced [Chen and Huang, 2005. For instance, clinical trials have confirmed the efficacy of the liposome-based vaccine, and additional human trials are currently underway. This chapter provides a concise overview of the non-viral delivery routes and methods for gene transfer that are employed in the development of vaccines and gene therapy.

2. GENE DELIVERY SYSTEMS

The most straightforward method of gene delivery involves injecting naked DNA that encodes the therapeutic protein. However, due to its low efficiency, it is necessary to employ specialized molecules and methods to enhance gene delivery. A vector is a system that serves multiple functions, such as (a) providing protection from gene degradation, (b) facilitating the delivery of genes into the target cells and their nucleus, and (c) ensuring gene transcription within the cell. Additionally, the optimal DNA vehicle should be appropriate for clinical use. It must be both cost-effective and straightforward to manufacture and purify in large quantities [Gardlik et al., 2005]. Two types of vectors have been used as vehicles for gene transfer: 1) viral vectors for gene transduction (e.g., retroviral, adenoviral, adeno-associated viral, and lentiviral vectors) and 2) non-viral vectors for gene transfection based on lipids, water-soluble polycations, non-condensing polymers, and nano/micro-particles [Gardlik et al., 2005; Katare, 2010]. Nevertheless, each vector possesses its own set of advantages and disadvantages.

2.1 Viral vectors (biological delivery systems)

Viral techniques use various classes of viruses as a tool for gene delivery [Gardlik et al., 2005; Stone, 2010]. Viruses introduce their DNA into the cells with high efficiency. Therefore, it is possible to take advantage of this system by introducing a foreign gene into the virus and then using the properties of the virus to deliver this gene with high efficiency into the target cells [Gardlik et al., 2005]. Gene therapy vectors are being developed by genetic modification of *retroviruses*, *adenoviruses*, *poxviruses*, *parvoviruses* (*adeno-associated viruses*), *herpesviruses* etc. [Gardlik et al., 2005; Stone, 2010]. Unlike wild type viruses, these vectors are used to transfer therapeutic genes into target cells and thus are engineered by deleting the essential genes which allow replication, assembling or infection. Replication deficiency ensures the safety of viral vectors, but on the other hand, vectors need to be produced in large amounts of virus particles. For this purpose, there are specialized cell lines called "packaging cell lines" (PCLs) engineered to replace a function of a deleted viral gene and for the production of recombinant viruses [Gardlik et al., 2005]. However, the interaction between a vector and a host-cell genome cannot be completely eliminated. Some disadvantages of viral delivery are addressed in Table 1.

Currently, developed Viruses as transfer vectors are divided into two classes, following their different strategies for replication and survival: a) Non-lytic viruses, including *retroviruses* and *lentiviruses*, produce virions from the cellular membrane of an infected cell, leaving the host cell relatively intact; b) Lytic viruses, including human *adenovirus* and *herpes simplex virus* families, destroy the infected cell after replication and virion production. This native nature of the original viruses determines the use of each recombinant replication- defective viral vector in clinical applications [Table 2]. Despite some limitations on the use of viral vectors regarding safety and reproducibility, they are still the most used gene transfer vehicles [Gardlik et al., 2005; Katare, 2010; Gupta et al., 2004].

Table 1. Viral delivery system disadvantages [Templeton and Lasic, 1999; Guptaet al., 2004]

No.	Disadvantage
1	Generation of an immune response to expressed viral proteins that subsequently kill the target cells producing a therapeutic gene product
2	Random integration of some viral vectors into the host chromosome
3	Clearance of viruses delivered systemically
4	Difficulties in engineering viral envelopes or capsids to achieve specific delivery to cells other than those with natural tropism for the virus
5	Possible recombination of the viral vector with DNA sequences in the host chromosome that generates a replication-competent, infectious virus
6	Inability to administer certain viral vectors more than once
7	High costs in producing large amounts of high-titer viral stocks for use in the clinic
8	Limited size of the nucleic acid that can be packaged and used for viral gene therapy

Table 2. Utilization of viral vectors for gene delivery [Katare and Aeri, 2010]

Vector	Genome	Structure	Properties
Adenoviruses	dsDNA	Capsid	Transient expression, strong immunogenicity
Alphaviruses	RNA	Envelope	Transient, but extreme, expression levels; low immunogenicity
HSV	dsDNA	Envelope	Latent infection, long-term expression, low toxicity (mutant)
Lentiviruses	RNA	Envelope	Genome integration, long term expression, safety concerns low titers, inefficient production
Retroviruses	RNA	Envelope	Genome integration, long-term expression
Adeno-associated viruses (AAV)	ssDNA	Capsid	Slow expression onset, genome integration, long term expression, inefficient large-scale virus production

In clinical studies, a recombinant *vaccinia* virus vector has been developed to express single or multiple T cell co-stimulatory molecules as a vector for local gene therapy in patients with malignant melanoma. This approach generated local and systemic tumor immunity and induced effective clinical responses in patients with metastatic disease [Kim-Schulze and Kaufman, 2009]. Furthermore, PSA-TRICOM vaccine (prostate-specific antigen plus a TRIad of co-stimulatory molecules; PROSTVAC) includes a priming vaccination with recombinant *vaccinia* (rV)-PSA-TRICOM and booster vaccinations with recombinant fowlpox (rF)-PSA-TRICOM. Each vaccine consists of the transgenes for PSA, including an agonist epitope, and three immune co-stimulatory molecules (B7.1, ICAM-1, and LFA3; designated TRICOM) [Kaufman, 2002]. The efficacy of PSA-TRICOM has been evaluated in phase II clinical trials in patients with metastatic hormone-refractory prostate cancer (mHRPC). PANVAC-VF, another poxviral-based vaccine, consists of a priming vaccination with rV encoding CEA (6D), MUC1 (L93), and TRICOM plus booster vaccinations with rF expressing the identical transgenes. CEA (6D) and MUC1 (L93) represent carcinoembryonic antigen and mucin 1 glycoprotein, respectively, with a single amino acid substitution designed to enhance their immunogenicity. This vaccine is currently under evaluation in several different types of CEA or MUC1-expressing carcinomas and in patients with a life expectancy more than three months [Vergati et al., 2010].

2.2 Non-viral vectors (Non-biological gene delivery systems)

In comparison with virus-derived vectors, non-viral vectors have several advantages, such as the safety of administration without immunogenicity, almost unlimited transgene size and the possibility of repeated administration [Gardlik et al., 2005]. Non-viral gene delivery systems generally consist of three categories: (a) naked DNA delivery, (b) lipid-based and (c) polymer-based delivery [Park et al., 2006]. Therapeutic gene can be introduced into the target cell either as an insert in plasmid with regulation sequences, what enables the regulation control of expression (inducible promoter) or as a PCR product. The simplest way of gene introduction is an injection of naked DNA into target cells. Such a naked plasmid DNA was used in several pre-clinical and clinical trials [Gardlik et al., 2005]. For example, some positive results were gained in cancer therapy by intra-tumoral injection of tumor suppressor genes or cytokines. However, this approach does not have the transfection efficiency of viral vectors. Low transfection efficacy and short-term expression still remain the main disadvantages of naked DNA gene transfer compared with viral vectors. Thus, many techniques have been developed to improve the introduction of therapeutic gene. Physical methods like electroporation and gene gun increase the entry of transgene into target cells. In addition, chemical methods like lipoplexes (DNA-liposomes complexes) and polyplexes (DNA-polymers complexes) improve the stability of DNA and also facilitate the entry into the cell [Gardlik et al., 2005]. Main methods of gene delivering systems are summarized in figure 1.

3. PHYSICAL DELIVERY SYSTEMS

A number of methods utilizing various physical techniques have been developed to facilitate the transfer of foreign genes into the host cells [Katare and Aeri, 2010]. Among them, electroporation and gene gun are further involved in preclinical and clinical trials.

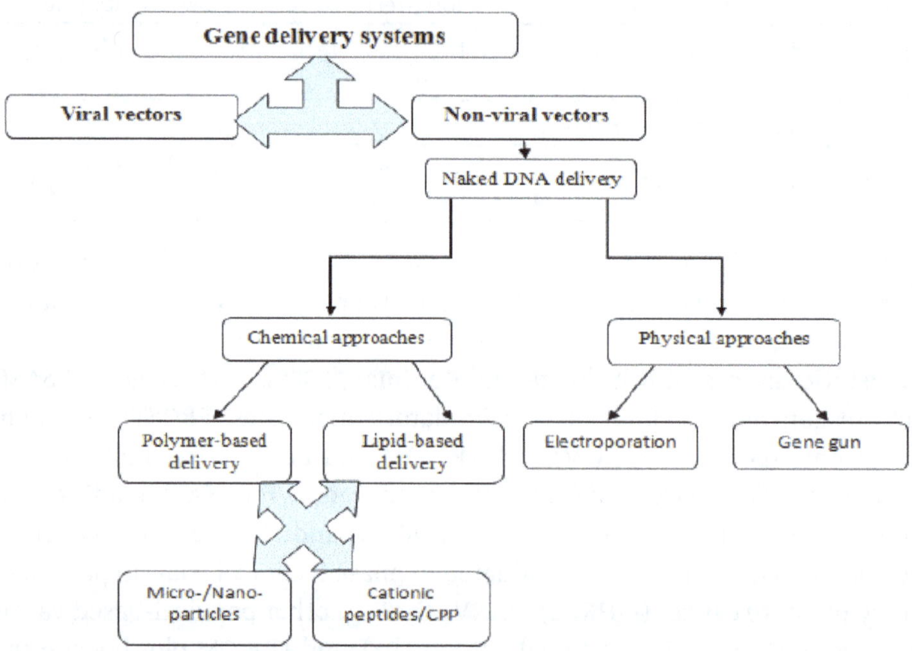

Fig. 1. Summary of the main methods of gene delivery systems

3.1 Electroporation

One of the methods that improve DNA penetration of the cell is electroporation [Lee et al., 2009; , Harrison et al., 1998; Rossini et al., 2002; Ahmad et al., 2009; Collins et al., 2006; Kang et al., 2011;]. *In vivo* use of electroporation is done by injecting naked DNA followed by electric pulses from electrodes that are located *in situ* in the target tissues. Successful use of electroporation was observed in transfecting muscles, brain, skin, liver, and tumors [Gardlik et al., 2005; Garcia-Frigola et al., 2007; Umeda et al., 2004; Babiuk et al., 2006; Harrison et al., 1998; Kang et al., 2011]. Since every tissue is specific and has its own characteristics, there are no generally accepted optimal conditions of electroporation that are suitable for effective transfection. These are dependent both on the amplitude and duration of the electric pulses and on the amount and concentration of DNA [Gardlik et al., 2005]. The generated pulse may be either a high voltage (1.5 kV) rectangular wave pulse for a short duration or a low voltage (350 V) pulse for a longer duration [Katare and Aeri, 2010].

Up to now, several clinical trials have been planned using the electroporation with DNA vaccines for cancer therapy such as: a) Intra-tumoral IL-12 DNA plasmid (pDNA) [ID: NCT00323206, phase I clinical trials in patients with malignant melanoma]; 2) Intratumoral VCL-IM01 (encoding IL-2) [ID: NCT00223899; phase I clinical trials in patients with metastatic melanoma]; 3) Xenogeneic tyrosinase DNA vaccine [ID: NCT00471133, phase I clinical trials in patients with melanoma]; 4) VGX-3100 [ID: NCT00685412, phase I clinical trials for HPV infections], and 5) IM injection prostate-specific membrane antigen (PSMA)/ pDOM fusion gene [ID: UK-112, phase I/II clinical trials for prostate cancer] [Bodles- Brakhop and Draghia-Akli, 2008; Bodles-Brakhop et al., 2009].

Furthermore, Hepatitis C virus DNA vaccine showed acceptable safety when delivered by Inovio Biomedical's electroporation delivery system in phase I/II clinical study at Karolinska University Hospital. ChronVac-C is a therapeutic DNA vaccine being given to individuals already infected with hepatitis C virus with the aim to clear the infection by boosting a cell-mediated immune response against the virus. This vaccination was among the first infectious disease DNA vaccine to be delivered in humans using electroporation- based DNA delivery [Bodles-Brakhop et al., 2009].

3.2 Gold bullet/gene gun

Similar to electroporation, another method, called "gene-gun", does not require the presence of complicated and potentially toxic delivery systems [Gardlik et al., 2005]. The gene transfer is mediated by small particles of gold on which the DNA is bounded. These particles are then shot into the cell under great pressure and speed (with the help of compressed helium) and so pass the membrane barrier [Katare and Aeri, 2010]. At first, the gene-gun was developed for gene transfer into plant cells; then, its use has expanded to gene transfer into the mammalian cells. Effective development of the gene-gun was also achieved in the field of DNA vaccination. The latest clinical experiments focus on cancer vaccines against various human tumors [Gardlik et al., 2005]. This method has been successfully used to deliver DNA *in vivo* into liver, skin, pancreas, muscle, spleen and tumors. Expression of reporter genes (e.g. firefly luciferase and β- galactosidase) or therapeutic genes (human growth hormone) have also been reported by this method [Gardlik et al., 2005]. Recently, gene gun-mediated transgene delivery system has been used for skin vaccination against melanoma using tumor-associated antigen (TAA) human gpl00 and reporter gene assays as experimental systems [Aravindaram and Yang, 2009 S]. In addition, the delivery of HPV DNA vaccines using intradermal administration through gene gun was shown to be the most efficient method of vaccine administration in comparison with routine intramuscular injection [Ogris and Wagner, 2002].

Currently, a HPV16 DNA vaccine encoding a signal sequence linked to an attenuated form of HPV16 E7 (E7 detox) and fused to heat shock protein 70 [(Sig/E7detox/HSP70)] has been used in clinical trials. In a previous study, the immunologic and anti-tumor responses have been evaluated by the pNGVL4a-Sig/ E7 (detox)/ HSP70 vaccine administered using three different delivery methods including needle intramuscular, biojector and gene gun. According to obtained results, DNA vaccine administered via gene gun generated the highest number of E7-speci⇒c CD8+ T cells as compared to needle intramuscular and biojector administrations in mice model [Trimble et al., 2003].

4. CHEMICAL DELIVERY SYSTEMS

In order to facilitate the effective transfer of non-viral DNA into the cells, synthetic vectors improving the DNA admission into the cell and protecting it from undesirable degradation were designed. The most chemical delivery systems were derived from lipids or synthetic polymers [Templeton and Lasic, 1999].

4.1 Lipoplexes (cationic lipids/liposomes)

Plasmid DNA can be covered by lipids into organized structures such as liposomes or micelles [Templeton and Lasic, 1999]. The complex of DNA with lipids is called lipoplex. Lipoplexes can be divided into two types: 1) Anionic and neutral liposomes: At first, these kinds of lipids were used for the construction of synthetic vectors. Although, they were characterized by safety, compatibility with body fluids and the possibility of tissue-specific gene transfer, but the level of transduced cell expression was relatively low. At present, new neutral and anionic liposomes suitable for *in vivo* gene therapy are being constructed [Gardlik et al., 2005; Gupta et al., 2004]; 2) Cationic liposomes: These lipids are naturally produced complexes with negatively charged DNA. Moreover, their positive charge allows interactions with the negatively charged cell membrane and thus penetration into the cell is permitted [Gardlik et al., 2005]. Cationic liposomes ensure effective protection against the degradation of the foreign DNA by the cell. The interactions of liposomes with DNA and the subsequent lipoplex formation are dependent on several physical conditions (pH, charge) as well as structural characteristics of the liposomes. The most frequent use of DNA-liposome complexes is in gene transfer into cancer cells, where the applied genes stimulate anti-tumor immune responses or genes decreasing the activity of oncogenes [Gardlik et al., 2005]. Recent studies revealed the ability of lipoplex gene transfer into the epithelial cells of the respiratory tract, which supports their usage in the therapy of respiratory diseases and cystic fibrosis. Their expression in all main organs, mostly in lungs, was observed after intravenous administration of lipoplexes. Targeted transfection can be gained, to some extent, by the addition of tissue-specific target ligand. It is suggested that the transfection is based on endocytosis of the host cell [Gardlik et al., 2005].

The advantages of using liposomes for gene therapy are included as: 1) lack of immunogenicity; 2) lack of clearance by complement system using improved formulations; 3) unlimited size of nucleic acids that can be delivered, from single nucleotides up to large mammalian artificial chromosomes containing several thousand kilobases; 4) ability to perform repeated administrations *in vivo* without adverse consequences; 5) low cost and relative ease of generating nucleic acid: liposome complexes that deliver therapeutic gene products in large scale; 6) safety, because plasmids used for non-viral delivery contain no- viral sequences, thereby precluding generation of an infectious virus; 7) naked DNA carried by liposome increases its uptake by antigen-presenting cells (APCs); 8) naked DNA carried by liposome enhances both humoral and cellular immunity; 9) naked DNA carried by liposome induces cytotoxic T lymphocyte response [Trimble et al., 2003].

For vaccine development, a general overview of different lipid-based particulate delivery systems, their composition, preparation methods, typical size, route of administration and model antigens has been listed by Myschik J. *et al.*, 2009 [Myschik et al., 2009]. Stimuvax (BLP25 liposome vaccine, L-BLP25, Oncothyreon partnered with Merck KGaA) is a cancer vaccine designed to induce an immune response against the extracellular core peptide of MUC1, a type I membrane glycoprotein widely expressed on many tumors (i.e., lung cancer, breast cancer, prostate cancer and colorectal cancer) [Vergati et al., 2010]. Stimuvax consists of MUC1 lipopeptide BLP25 [STAPPAHGVTSAPDTRPAPGSTAPPK (Pal) G], an immunoadjuvant monophosphoryl lipid A, and three lipids (cholesterol, dimyristoyl phosphatidylglycerol, and dipalmitoyl phosphatidylcholine), capable of enhancing the delivery of the vaccine to APCs. A randomized phase II B clinical trial evaluated the effect of Stimuvax on survival and toxicity in 171 patients with stage III B and IV non-small cell lung cancer (NSCLC), after stable disease or response to first-line chemotherapy. Based on these data, Merck is currently conducting three large phase III clinical trials of Stimuvax. This study will involve more than 1300 patients [Vergati et al., 2010].

Furthermore, a cationic lipid DNA complex (CLDC) consisting of DOTIM/cholesterol liposomes and plasmid DNA, containing immunostimulatory CpG and non-CpG motifs has been designed, with potential immunostimulating and anti-neoplastic activities. Upon systemic administration, TLR-directed cationic lipid-DNA complex JVRS-100 enters dendritic cells (DCs) and macrophages; immunostimulatory DNA binds to and activates

Toll-like receptors (TLRs), which may result in the generation of anti-tumor natural killer (NK) cell and T-cell responses by the innate immune system. In addition, as a vaccine adjuvant, this agent may induce a strong cytotoxic T-lymphocyte (CTL) response to co-administered antigen. The efficacy of JVRS-100 has been evaluated in phase I clinical trials for the treatment of patients with Relapsed or Refractory Leukemia [ID: NCT00860522].

4.2 Polyplex (polysaccharides/cationic polymers)

A wide range of polymeric vectors have been utilized to deliver therapeutic genes *in vivo*. The modification of polymeric vectors has also shown successful improvements in achieving target-specific delivery and in promoting intracellular gene transfer efficiency [Park et al., 2006, Ogris and Wagner, 2002]. Various systemic and cellular barriers, including serum proteins in blood stream, cell membrane, endosomal compartment and nuclear membrane, were successfully avoided by designing polymer carriers having a smart molecular structure [Park et al., 2006; Ogris and Wagner, 2002]. Vectors based on a complex of polymers with DNA are called polyplexes. Most of them consist of cationic polymers and their production is regulated by ionic interactions [Gardlik et al., 2005].

In contrast to lipoplexes, some polyplexes (polylysin) are not able to release intracellular DNA into the cytoplasm [Gardlik et al., 2005]. For this purpose, co-transfection with endosome-lytic agents (inactivated *adenovirus*) is needed. On the other hand, polymers such as polyethylenimine have a mechanism of endosome disruption and there is thus no need for transfection with endosome-lytic agents. Polyethylenimine is used as a vector in aerosol inhalation gene therapy. It is a non-invasive and relatively effective gene transfer, especially into the respiratory tract, with permanent gene expression in this target region without undesirable expression in other tissues [Gardlik et al., 2005]. The size of the polymer determines the transfection efficiency and is specific for each individual gene transfer. Furthermore, the size of the aerosol particles determines the place of action, and thus the specificity of inhalation gene therapy [Gardlik et al., 2005]. An alternative to polyplexes can also be the use of polymer nanoparticles. Two types of such complexes have been characterized as gelatin-DNA and chitosan-DNA. In comparison with naked DNA, transfection using nanoparticles shows increased expression *in vivo* when administrated intratracheally or intramuscularly. Some polymeric vectors are summarized in the following sections:

Poly (L-lysine) (PLL)-based gene delivery systems

PLL has been widely used as a non-viral gene carrier since the formation of polyelectrolyte complexes between PLL and DNA was identified [Park et al., 2006]. Although, PLLs with high molecular weight have some properties suitable for a gene carrier, the PLL/DNA complexes showed a relatively high cytotoxicity and a tendency to aggregate and precipitate depending on the ionic strength. PEGylation [PEG: polyethylene glycol] of cationic polymers is known to greatly improve the problems of cytotoxicity, aggregation and non- specific protein adsorption *in vivo* [Park et al., 2006]. Some examples are mentioned as following:

- **Sugar-conjugated PLL:** Lactose and galactose have been used as conjugation partners with polymeric gene carriers for targeting asialoglycoprotein of hepatocytes. Lac-PEG- PLL showed much less cytotoxicity and higher stability and solubility in a physiological condition compared to PLL [Park et al., 2006].
- **Arterial-wall binding peptide (AWBP)-conjugated PLL:** Arterial-wall binding peptide (AWBP) is a peptide containing the arterial-wall binding domain (1000-1016 amino acids) of apoB-100 protein, a major protein component of LDL. The AWBP was conjugated to PLL via a PEG linkage (AWBP-PEG-g-PLL). When interacted with a plasmid DNA, AWBP-PEG-g-PLL could form spherical shaped complexes with a size of
- 100 nm and showed dramatic increase of transfection efficiency (150-180-fold), compared to PLL and PEG-g-PLL, in bovine aorta endothelial cells and smooth muscle cells. The presence of free AWBP in the transfection medium reduced the transfection efficiency of AWBP-PEG-g-PLL, suggesting that AWBP-PEG-g-PLL could be used as a tissue-selective gene carrier [Park et al., 2006].
- **Antibody-PLL conjugates:** Antibody-antigen interaction is one of the most specific interactions in biological systems. A monoclonal antibody against leukemia-specific JL- 1 antigen (anti-JL-1-Ab) was conjugated with PLL by periodate-mediated oxidation of carbohydrate moiety in the Fc domain of the antibody, followed by reaction with PLL. The anti-JL-1-Ab-PLL conjugate demonstrated significantly higher transfection efficiency than PLL or lipofectin in leukemia (Molt 4) cells [Park et al., 2006].
- **Folate-conjugated PLL:** Folate receptor has been identified as a potential target molecule of various cancer cells. The receptor is up-regulated and over-expressed in a number of rapidly growing malignant tumor cells, resulting in a dramatic promotion of the cellular uptake of folate. Therefore, the conjugation of folate to a variety of polymeric carriers has been chosen as a popular strategy for the target-specific delivery of anti-cancer therapeutics to the folate receptor-bearing tumor cells. A folate-PLL conjugate that incorporates a PEG spacer between folate and PLL (Fol-PEG-PLL) was also synthesized. The Fol-PEG-PLL was coated onto the complexes of PEI/DNA for a receptor-mediated gene transfer. The formulated complexes exhibited much higher transfection efficiency than PEI/DNA or lipofectamine/DNA complexes in the presence of 10% serum, suggesting that the PEG segment of Fol-PEG-PLL could increase the cellular uptake by receptor-mediated endocytosis and efficiently stabilize the complexes by increasing their solubility as well as by reducing the non-specific adsorption of serum proteins. The formulation also showed much lower cytotoxicity than PEI/DNA complexes [Park et al., 2006].
- **The Terplex system:** The Terplex system is a low-density lipoprotein (LDL)-mediated targeting system, where the LDL specifically interacts with the LDL receptors on the cell surface. LDL receptors are membrane-anchored proteins present in many cell types including hepatocytes,

endothelial cells and myocytes. The stearyl-PLL conjugate synthesized by N-alkylation of PLL with stearyl bromide interacts with a plasmid DNA to form complexes. The stearyl group could then bind to LDL via hydrophobic interaction to form the supramolecular gene carrier, the terplex. The Terplex system showed efficient *in vitro* transfection in a variety of cells including smooth muscle cells (A7R5), and human lung fibroblasts (CCD-32 Lu). The systemic administration of the Terplex system demonstrated prolonged circulation time compared to naked DNA [Park et al., 2006].

Polyethylenimine (PEI)-based gene carriers

- PEI has been one of the most popularly employed cationic gene carriers due to its superior transfection efficiency in many different types of cells. The buffering property of PEI leads to protect the DNA from degradation in the endosomal compartment during the maturation of the endosome to lysosome, facilitating intracellular trafficking of DNA. High cation density of PEI also contributes to the formation of highly condensed particles by interacting with DNA. However, the property may confer significant cytotoxicity. Studies with linear PEIs showed even higher transfection efficiency and lower cytotoxicity compared to branched PEI [Park et al., 2006]. There are different spectra utilizing PEI as described in following sections:
- **PEI-PEG:** PEI-grafted PEGs (PEI-g-PEG) with different PEG grafting ratios were synthesized to address the cytotoxicity and aggregation problems of PEI. Cell cytotoxicity of PEI-g-PEG was greatly reduced, while the transfection efficiency of PEI- g-PEG was still comparable to that of PEI. Cytotoxicity was independent of molecular weight of PEG but affected by the degree of PEG substitution [Park et al., 2006].
- **PEI conjugates with targeting moieties:** a) PEI-g-PEG-RGD: An angiogenic endothelial cell targeted gene delivery system (PEI-g-PEG-RGD) was developed by incorporating the $\alpha v\beta 3/\alpha v\beta 5$ integrin binding RGD peptide. b) Antibody-conjugated PEIs: A monoclonal antibody against human epidermal growth factor receptor-2 (HER-2) was conjugated to linear PEI for targeted gene transfer to cancer cells. The HER-2 antibody- PEI conjugate showed enhanced transfection efficiency in HER-2 over-expressing human breast adenocarcinoma cells (Sk-Br-3) compared to unmodified PEI [Park et al., 2006].
- **Folate-conjugated PEIs:** Folate-polyethylene glycol-folate-grafted-polyethylenimine (FPF-g-PEI) was synthesized by grafting folate-PEG-folate to PEI. A PEI-PEG-folate (PEI-PEG-FOL) conjugate was used as a carrier for a plasmid encoding small interfering RNA (siRNA) targeting green fluorescence protein (GFP). The complexes between the PEI-PEG-FOL and the siRNA-expressing plasmid showed an efficient suppression of GFP expression compared to unmodified PEI complexes in folate receptor over- expressing cells (KB), which stably expressed GFP [Park et al., 2006].

Water-soluble lipopolymer (WSLP): The water-soluble lipopolymer (WSLP) was synthesized by conjugating chlosteryl chloroformate to a low molecular weight PEI (1.8 kDa). WSLP interacts with DNA to form stable colloidal particles (70 nm). The PEI moiety of WSLP confers a buffering effect, which could facilitate endosomal escape of the WSLP/DNA complex. It was also reported that the dodecylation of PEI enhanced the cellular uptake and transfection efficiency. In a similar way, the hydrophobic cholesterol moiety of WSLP would give a chance to form small and stable complexes, resulting in enhanced cellular uptake and transfection efficiency. WSLP showed higher transfection efficiency and much lower cytotoxicity than 25 kDa PEI, suggesting that WSLP has the advantages from PEI as well as from chlosterol. Intratumoral injection of WSLP/p2CMVmIL-12 complexes to tumor-bearing mice showed a significant improvement in the retardation of tumor growth and survival rate [Park et al., 2006].

Biodegradable polycations

The backbone linkages of most polymeric gene carriers consist of a –C-C- bond or amide bond, which are not degraded in physiological solutions. The non-degradable non-viral carriers are not easily removed by physiological clearance systems and therefore, can possibly accumulate within cells or tissues to elicit further cytotoxicity [Park et al., 2006]. To solve the problems, several biodegradable polycations have been synthesized and evaluated as potential gene carriers. Generally, the biodegradable polycations showed much less cytotoxicity and higher transfection efficiency compared to an unmodified polycations, such as PLL or PEI [Park et al., 2006]. Some examples include: Poly (α-[4-aminobutyl]-L-glycolic acid) (PAGA), a biodegradable PLL analogue; Poly(β-amino ester)s; Poly(2-aminoethyl propylene phosphate) (PPE-EA); Degradable PEIs; The biodegradable PEIs were synthesized by crosslinking low molecular weight PEI (0.8 kDa) with either PEG-bis- succinimidyl succinate or disulfide-containing cross-linkers, such as dithiobis (succinimidylpropionate) (DSP) and dimethyl 3,3'-dithiobispropionimidate (DTBP) [Park et al., 2006].

Neutral and non-condensing polymer-based gene delivery systems

Although, cationic polymers, which electrostatically interact with DNA to neutralize its negative charge and condense DNA into nanosized particles, are generally considered as gene carriers, neutral polymers such as polyvinyl alcohol (PVA) and polyvinyl-pyrolidone (PVP) can also be used for gene transfer. They can protect DNA from enzymatic degradation and facilitate cellular uptake of DNA. These polymers may interact with DNA via hydrogen bonding and/or van der Waals interactions; van der Waals interaction of the hydrophobic vinyl backbones may cover around DNA to make its surface more hydrophobic. In a study, intramuscularly administered PVP/plasmid DNA formulation resulted in a significant increase in the number and distribution of the reporter-gene expressing cells in rat tibialis, compared to naked plasmid [Park et al., 2006].

5. MICRO-/NANO-PARTICLES

Another approach to DNA-vaccine delivery involves microparticle-based technologies to target APCs [Ulmer et al., 2006]. Microencapsulation of DNA, or association of DNA with microcapsules, has led to enhancement of CTL responses to encoded proteins [Doria-Rose and Haigwood, 2003]. Biodegradable, non-antigenic poly-lactide polyglycolide (PLGA or PLG) microspheres offer many advantages as a vaccine delivery system. Both cellular and humoral immune responses can be elicited to antigens encapsulated in, or conjugated onto PLG microspheres. Particles used typically range in size from 1 to 10 μm in diameter, a size that is readily phagocytosed by dendritic cells and other antigen-presenting cells (APCs). Microspheres elicit both $CD8^+$ and $CD4^+$ T cell responses by releasing antigen intracellularly [Doria-Rose and Haigwood, 2003]. Biodegradable PLGA nanoparticles (NPs) have been investigated for sustained and targeted/ localized delivery of different agents, including drugs, proteins and peptides and recently, plasmid DNA owing to their ability to protect DNA from degradation in endolysosomes. PLGA-based nanotechnology has been widely used in diagnosis and treatment of cancer. These NPs have been shown to stimulate the immune response as measured by an increase in IL-2 and IFN-γ in spleen homogenates [Lu et al., 2009]. The majority of the existing literature involving PLGA polymers has tended to be focused on PLGA microspheres. In the last 10 years, microspheres have been used extensively for the injectable delivery of vaccine antigens, both for viral and bacterial antigens. Similar to microspheres, PLGA NPs have been shown to effectively enhance immune responses. The major obstacle is providing delivery vehicles with the adequate surface molecules for recognition by the immune system and for more effective targeting. It is likely, therefore, that future studies of PLGA NPs as vaccine candidates will focus on improving these features, as recently tested by grafting RGD peptides (arginine–glycine– aspartic acid-containing synthetic peptides) covalently onto PEG moieties on the surface of PLGA NPs [Lu et al., 2009].

These polymers have been designated as feasible candidates for drug delivery systems, anti- cancer agents and vaccine immunotherapy. For example, DNA vaccine delivery to APCs has been facilitated by microencapsulation of plasmid DNA, which encodes HPV E6/E7 antigenic proteins. The capsule is formed from polymeric PGLA microparticles. These resulting microparticles have a greater propensity toward APC uptake compared to naked DNA. This technique allows HPV DNA plasmid to be condensed inside the microparticle. The physical and chemical properties of the PGLA scaffold make DNA inaccessible to nuclease and preventing degradation, allowing for a sustained release of DNA and enhancing transfection efficiency *in vitro* [Lin et al., 2010]. In mice, microspheres containing HPV plasmid encoding HPV E6/E7 antigens have been shown to elicit a strong antigen- specific cytotoxic T cell response. Using this technology, microencapsulated DNA vaccine termed ZYC-101 encoding multiple HLA-A2 restricted HPV E7 epitopes has undergone Phase I trials in patients with CIN2/3 lesions and high-grade anal intraepithelial neoplasia. In both trials, intramuscularly administered vaccine was well tolerated, and in some patients had resulted in histological regression of the lesions as well as generation of E7-specific IFN- γ expressing T cells. A newer version of the DNA vaccine, ZYC-101a, which encodes HPV16 and HPV18 E6- and E7-derived epitopes has been used in phase II clinical trial in patients with CIN 2/3 lesions [Lin et al., 2010].

The multi-functional nano-devices based on the dendritic polymer or dendrimers are also being applied to a variety of cancer therapies to improve their safety and efficacy. Technical advances have been focused on the development of a linking strategy that allows the dendrimer molecules to be linked via complementary oligonucleotides. At present, further applications of dendrimers in photodynamic therapy, boron neutron capture therapy, and gene therapy for cancer are being examined [Baker, 2009].

Recently, the modified fluorescent nanoparticles have been synthesized as a targeting and delivery system, by conjugating both tumor targeting agent and chemokines to the nanoparticles, in order to attract immune cells toward tumor cells. Biodegradable chitosan nanoparticles encapsulating quantum dots were prepared, with suitable surface modification to immobilize both tumor targeting agent and chemokine on their surfaces [Chatterjee and Zhang, 2007]. Fluorescent chitosan coated quantum dots (QDs) were used to act as bi-functional bridging units between cancer and immune cells. This nanoparticulate form of delivery promises the advantages of enhanced tumor selectivity and longer half- lives, thereby enhancing effectiveness of the immune response and reduction in systemic toxicity [Chatterjee and Zhang, 2007]. Furthermore, the rapid development of Quantum Dots (QDs) technology has already fulfilled some of the hopes of developing new, more effective cancer-imaging probes. First, stable encapsulation of QDs with amphiphilic polymers has prevented the quenching of QD fluorescence in the aqueous *in vivo* environment. Second, QDs are relatively inert and stable. Finally, successful conjugation of QDs with biomolecules has probably made active targeting them to tumors. Despite their success so far in cancer imaging, there are challenges in enhancing sensitivity, maximizing specificity and minimizing toxicity of QDs, which must be undertaken before clinical applications can proceed [Zhang et al., 2008]. A major parameter limiting immune responses to vaccination is the number of activated APCs that capture antigen and migrate to draining lymph nodes. The use of cellular magnetic resonance imaging (MRI) is a promising approach for this purpose [Long et al., 2009]. In a study, an *in vivo* labeling method was described, which relies upon cell-to-cell transfer of super-paramagnetic iron oxide (SPIO) from tumor cells to endogenous APCs, *in situ*, for quantification of APC delivery to lymph nodes in a tumor vaccine model. Mice were immunized with a tumor cell–based vaccine that was labeled with SPIO. APCs that had captured SPIO were imaged over time as they accumulated in lymph nodes. It was indicated that MRI is capable of monitoring, *in vivo*, the trafficking of magnetically labeled APCs inducing a tumor-specific immune response, and that these cells can be magnetically recovered *ex vivo*. Excellent correlation was observed between *in vivo* and *ex vivo* quantification of APCs, with resolution sufficient to detect increased APC trafficking elicited by an adjuvant [Long et al., 2009].

6. CATIONIC PEPTIDES/CELL-PENETRATING PEPTIDES (CPP)/TROJAN PEPTIDES

The studies have shown that a number of peptides and proteins are able to penetrate the cell membrane and enter the cell. It has been observed that many cargo molecules that are covalently attached to these peptides will be translocated into the cell. Recently, various natural and/ or synthetic cell-penetrating peptides (CPP) have known as efficient tools in vaccine design as they are capable of delivering therapeutic targets into cellular compartments. In fact, the cell membrane is impermeable to hydrophilic substances and delivery into cells could be facilitated by linking to CPP. Different cargos such as drugs, peptide/ protein, oligonucleotide/ DNA/ RNA, nanoparticles, liposomes, bacteriophages, fluorescent dyes and quantum dots have been linked to CPPs for intracellular delivery with possible use in future vaccine design [Brooks et al., 2010]. Two applications of CPP already validated in vaccine studies are delivery of tumor-associated antigens into antigen- presenting cells (APCs) and use as a non-viral gene delivery vehicle in DNA vaccines. There are two methods for designing CPP incorporating immunogenic antigens: A) chemical linking via covalent bonds B) coupling via recombinant fusion constructs produced by bacterial expression vectors. The orientation of the peptide and cargo and the type of linkage are likely important [Brooks et al., 2010]. In addition, the utilized CPP, attached cargo, concentration and cell type, all significantly affect the mechanism of internalization. The mechanism of cellular uptake and subsequent processing still remains controversial. It is now apparent that CPP mediate intracellular delivery via both endocytic and non-endocytic pathways [Brooks et al., 2010; Jarver and Langel, 2004; Wagstaff and Jans, 2006]. An attractive feature of using polypeptides as gene delivery vectors is incorporating multiple functional domains into one polypeptide chain, such as a DNA-binding domain linked with a receptor-targeting domain. This kind of polypeptides will recognize and bind to cell surface receptors that are unique to target cells and deliver the bound DNA into the cells through receptor-mediated endocytosis. Therefore, this process may ensure the therapeutic effect in desired cells and limit the potential side effects caused by transgene expression into the non-targeted cells [Zeng and Wang, 2005].

Several studies have shown that oligo-deoxynucleotides (ODN) with immune-stimulating sequences (ISS) containing CpG motifs facilitate the priming of MHC class I- restricted $CD8^+$ T cell responses to proteins or peptides. Therefore, ODN/cationic peptide complexes are potent tools for priming $CD8^+$ T cell immunity [Schirmbeck et al., 2003]. The complex formation required electrostatic linkage of the positively charged peptide to the negatively charged ODN. Conjugation of immunostimulatory DNA or ODN to protein antigens facilitates the rapid, long-lasting, and potent induction of cell-mediated immunity. It was shown that ODN (with or without CpG-containing sequences) are potent Th1-promoting adjuvants when bound to cationic peptides covalently linked to antigenic epitopes, a mode of antigen delivery existing in many viral nucleocapsids [Schirmbeck et al., 2003]. Table 3 contains a list of peptides that have been investigated for their ability to penetrate the cell [Futaki, 2005; Brooks et al., 2010].

Table 3. Protein transduction domains (PTD)

Class	Protein	Functional peptide	Sequence	Translocation efficiency	Reference
Basic	TAT [HIV]	**Tat and related peptides**			Vives et al., 1997
		HIV-1 Tat (48-60)	GRKKRRQRRRPPQ	+++	Futaki, 2005; Brooks et al., 2010
		R9-Tat	GRRRRRRRRRPPQ	+++	Futaki, 2005
Basic		**Arginine-rich RNA binding peptides**			Futaki, 2005; Brooks et al., 2010
		HIV-1 Rev-(34-50)	TRQARRNRRRRWRERQR	+++	Futaki, 2005
		R7W	Fluo-RRRRRRRW-NH2	+++	Futaki, 2005
		TatP59W	Fluo- GRKKRRQRRRPWQ-NH2	+++	Futaki, 2005
		FHV Coat-(35-49)	RRRRNRTRRNRRRVR	+++	Futaki, 2005
		BMV Gag-(7-25)	KMTRAQRRAAARRNR WTAR	+++	Futaki, 2005
		HTLV-II REX-(4- 16)	TRRQRTRRARRNR	+++	Futaki, 2005
		CCMV Gag-(7-25)	KLTRAQRRAAARKNKR NTR	++	Futaki, 2005
		P22 N-(14-30)	NAKTRRHERRRKLAIER	++	Futaki, 2005
Basic	VP22 [HSV]	VP22	DAATATRGRSAASRPTE RPRAPARSASRPRRPVE	+++	Cashman et al., 2002
	Human Calcitonin	hCT (9-32)	LGTYTQDFNKFHTFPQT AIGVGAP		Trehin et al., 2004
	Vascular endothelial cadherin [Mouse]	pVEC	LLIILRRRIRKQAHAHSK		Elmquist et al., 2001
	Related to γ- Zein [Maize]	Sweet arrow peptide (SAP)	(VRLPPP)3		Fernandez-Carneado et al., 2004
	Related to γ- Zein [Maize]		(Tyr-ZnDPA)n		Johnson et al., 2008
Basic/ amphiphilic	**Antennapedia homeodomain [*Drosophila*]**	Antennapedia (43-58) [penetratin]	RQIKIWFQNRRMKWKK	+++	Derossi et al., 1994
		Fluoro-Penetratin	Fluo- RQIKIWFQNRRMKWKK -NH2		Futaki, 2005
		Pen2W2F	KLALKLALKALKAALKL A-NH2		Futaki, 2005
		PenArg	Fluo- RQIRIWFQNRRMRWRR-NH2		Futaki, 2005
		PenLys	Fluo- KQIKIWFQNKKMKWKK -NH2	+	Futaki, 2005
		E N-(1-22)	MDAQTRRRERRAEKQA QWKAAN	+	Futaki, 2005

Class	Protein	Functional peptide	Sequence	Translocation efficiency	Reference
		B21 N-(12-29)	TAKTRYKARRAELIAER R	+	Futaki, 2005
		Yeast PRP6-(129- 144)	TRRNKRNRIQEQLNRK	+	Futaki, 2005
		Hum U2AF-(142- 153)	SQMTRQARRLYV	-	Futaki, 2005
Chimera (synthetic)	**Galanin/ Mastoparan)**	Transportan	GWTLNSAGYLLGKINLK ALAALAKKIL		Pooga et al., 1998
	Galanin/ Mastoparan)	Model amphipathic peptide (MAP)	KLALKLALKALKAALKL A amide		Oehlke et al., 1998
	Hydrophobic- NLS	Pep-1	KETWWETWWTEWSQP KKKRKV		Morris et al., 2001
	Hydrophobic- NLS	CADY	Ac- GLWRALWRLLRSLWRL LWRA-cysteamide		Crombez et al., 2009
	Hydrophobic- NLS	Membrane translocating sequence peptide (MTS)	AAVALLPAVLLALLP		Rojas et al., 2008
	HIVGp41-NLS of SV40 T-antigen	MPG	GALFLGWLGAAGSTMG APKKKRKV		Morris et al., 2008

Furthermore, the HIV Tat derived peptide is a small basic peptide that has been successfully shown to deliver a large variety of cargoes, from small particles to proteins, peptides and nucleic acids. The "transduction domain" or region conveying the cell penetrating properties is clearly confined to a small stretch of basic amino acids, with the sequence RKKRRQRRR (residues 49–57) [Riedl et al., 2004; Brooks et al., 2005]. This polycationic nanopeptide is known to be a transfection enhancer of plasmid DNA. The conditions of DNA-peptide complex formation and DNA/Tat ratio have significant impact on the level of transgene expression and degree of DNA protection from nuclease attack [Hellgren et al., 2004]. The conjugation of this peptide to ovalbumin (OVA) resulted in efficient stimulation of MHC class I-restricted T cell responses *in vitro* and, more importantly, the generation of CTLs *in vivo* [Kim et al., 1997]. Also, soluble Tat-antigen conjugates can deliver the antigen directly to the MHC class I processing pathway and thereby increase the generation of antigen-specific CD8+ T cells *in vitro* [Kim et al., 1997; Riedl et al., 2004]. A fusion protein containing the carboxy-terminal end of Tat (amino acids: 49–86) linked to the HPV16 E7 oncoprotein enhanced tumor specific immune responses *in vivo* [Giannouli et al., 2003]. In C57BL/6 mice, E7-Tat mixed with Quil A generated efficient prophylactic and therapeutic suppression of HPV16-positive C3 tumor outgrowth. This study offers a new strategy for improving subunit cancer vaccines [Giannouli et al., 2003]. Particularly, a Tat-derived peptide in combination with a PEG-PEI copolymer could be a promising candidate as gene delivery vehicle intended for pulmonary administration. Tat-PEG-PEI represents a new approach to non-viral gene carrier for lung therapy, comprising protection for plasmid DNA, low toxicity and significantly enhanced transfection efficiency under *in vivo* conditions [Kleemann et al., 2005].

It has been shown that covalent attachment of low molecular weight polyethyleneimine (PEI) improves Tat peptide mediated gene delivery *in vitro* [Alexis et al., 2006; Putnam et al., 2001; Wang, 2006]. In our recent study, two delivery systems including polymer PEI 25 kDa and polymer peptide hybrid as PEI600-Tat conjugate were used to compare their efficiency for HPV16 E7 DNA transfection *in vitro*. Our data indicated that both delivery systems including PEI 25 kDa and PEI600-Tat conjugate are efficient tools for E7 gene transfection. In fact, PEI potency for E7 gene transfection is higher than PEI600-Tat *in vitro*, but its toxicity is obstacle *in vivo* [Bolhassani et al., 2008]. Using HPV16 E7 as a model antigen, the effect of PEI600-Tat conjugate has been evaluated on the potency of antigen-specific immunity in mice model. Assessment of lymphoproliferative and cytokine responses against recombinant E7 protein (rE7) showed that PEI600-Tat/E7DNA complex at certain ratio induces Th1 response. This study has demonstrated that PEI600-Tat conjugate is efficient to improve immune responses *in vivo* [Bolhassani et al., 2009].

Moreover, synthetic peptides containing a nuclear localization signal (NLS) can be bound to the DNA and the resulting DNA-NLS complexes can be recognized as a nuclear import substrate by specific intracellular receptor proteins [53]. For example, conjugation of an NLS to a Minimalistic Immunogenically Defined Gene Expression (MIDGE) vector encoding a truncated and secreted form of BHV-1 glycoprotein D (tgD) improved the tgD expression *in vitro* and induced both humoral and cellular immune responses in mice [Zheng et al., 2006]. This strategy could be applied as an efficient pathway in enhancement of DNA vaccine potency against cancer.

On the other hand, one of the CPPs that have currently received extensive attention in the field of DNA vaccination is the herpes simplex virus (HSV-1) protein VP22 [Brooks et al., 2010]. VP22 can form compacted complexes with short oligonucleotides and form particles of spherical nature with a size range of 0.3 to 1 μm in diameter. These particles entered cells efficiently within 2 to 4 hours. Furthermore, VP22 enables spreading of the antigenic peptide to the cells surrounding the transfected cells [Brooks et al., 2010]. Efforts have been made to increase the potency of DNA vaccines by exploiting the cell-to-cell spreading capabilities of the HSV-1 VP22 protein or the analogous protein from bovin herpesvirus 1 [Ulmer et al., 2006]. The significance of VP22 in intercellular spreading has been demonstrated through *in vitro* studies linking VP22 to p53, thymidine kinase, cytosine deaminase and Green Fluorescent Protein (GFP). These proteins were observed to be distributed to nuclei of surrounding cells [Lin et al., 2010]. Furthermore, vaccination with DNA encoding HPV16E7 linked to the HSV type 1 VP22 elicited the enhanced E7- specific memory $CD8^+$ T lymphocytes and anti-tumor effects against E7-expressing tumor cells [Michel et al., 2002]. Also, VP22 has been used for HPV DNA vaccines targeting the E6 protein [Lin et al., 2010]. Various groups have demonstrated that DNA constructs which encode fusion proteins of VP22 linked to an antigen increase the immune responses in mice and cattle. Bovine herpesvirus VP22 (BVP22) and Marek's disease virus VP22 (MVP-1) are both closely related by their structural homology to HSV-1 VP22, and can also have a significant role in intercellular spreading. Hung *et al.* has demonstrated that mice vaccinated with DNA encoding MVP22/E7 significantly increased numbers of IFN- γ-secreting, E7-specific $CD8^+$ T cell precursors compared to mice vaccinated with wild- type E7 DNA alone, which directly lead to a stronger tumor prevention response. Similarly, immunization of mice and cattle with DNA vaccine coding for BVP22 linked to truncated glycoprotein D (BVP-tgD) was shown to generate a stronger tgD-specific immune response compared to animals vaccinated with tgD alone. Taken together, DNA vaccine encoding VP22 linked to antigens represents a promising approach to enhance DNA vaccine potency [Lin et al., 2010].

To evaluate the VP22 role in gene therapy of hepatocellular carcinomas (HCCs), the expression vectors were constructed for N- and C-terminal fragments of VP22-p53 fusion proteins and investigated the VP22-mediated shuttle effect in hepatoma cells by co- transfection experiments. VP22-mediated trafficking was not detectable in hepatoma cells *in vitro* by fluorescence microscopy [Zender et al., 2002]. For *in vivo* experiments, the recombinant adenoviruses Ad5CMVp53 and Ad5CMVp53-VP22 were constructed. In contrast to the *in vitro* experiments, intercellular trafficking of VP22-p53 could be observed in subcutaneous tumors of hepatoma cells by fluorescence microscopy, indicating a stronger shuttle effect in solid tumors compared to cell culture experiments [Zender et al., 2002]. In our current study, Herpes simplex virus type 1 (HSV-1) VP22 protein was employed to enhance DNA vaccine potency of *Leishmania major* amastin antigen in BALB/c mice model. Vaccination with the VP22-amastin-EGFP fusion construct elicited significantly higher IFN-gamma response upon antigen stimulation of splenocytes from immunized mice compared to amastin as a sole antigen. These results suggest that the development of DNA vaccines encoding VP22 fused to a target *Leishmania* antigen would be a promising strategy to improve immunogenicity and DNA vaccine potency [Bolhassani et al., 2011].

7. HYBRID VECTORS

A promising approach to overcome the limitations and develop the advantages of the individual types of vectors is their combination. Several types of hybrid vectors have been known: A) Virosomes are produced by the fusion of lipoplexes (liposomes with DNA) with inactivated HVJ virus (hemagglutinating virus of Japan) or influenza virus [Gardlik et al., 2005]. It was shown that the efficiency of gene transfer into the respiratory tract is higher than cationic liposomes or viral vectors. In addition, they are very well tolerated from the immunological view, so even repeated injection does not influence the efficiency and safety of transfer; B) The second type is represented by hybrids that were generated by mixing cationic liposomes or polymers with adenoviral vector. These are effective mainly in cells which do not have viral receptors. In addition, it was proved that an inactivated adenovirus attachment improves the efficiency of the transfer mediated by cationic liposomes or polymers [Gardlik et al., 2005]; C) Hybrid viruses can be produced by a combination of various types of viral vectors, and they represent a system which employs the main advantages of both viruses [Gardlik et al., 2005].

8. BACTERIAL DELIVERY SYSTEMS

The most recent approach in targeting the gene therapy is the use of bacterial systems as vectors for transfer and gene expression in tumor cells. The principle lies in the ability of some anaerobic bacteria to selectively colonize hypoxic areas (e.g., tumors) and replicate there. Therefore, it is possible to achieve selective expression of therapeutic genes specifically in tumors. Currently known and tested bacterial vectors have been divided into two groups: A) Strictly anaerobic bacteria (the species *Clostridium* and *Bifi dobacterium*) are used in *in vivo* experiments. *Clostridium* is the most important bacterial species for use as a vector. On the other hand, the non-pathogenic *Bifi dobacterium*, is naturally present in the human gastrointestinal tract and provides higher safety [Gardlik et al., 2005]. B) The second group consists of attenuated auxotrophic strains of *Salmonella typhimurium* that require the presence of tumor specific nutrition factors for selective replication. They use these factors for their own metabolism, thus prohibiting the tumor cells from utilizing them and growing. In *in vivo* experiments, high levels of *Salmonella* (109 bacteria/g tissue) were obtained. A similar tumor inhibitory effect was shown by application of *Salmonella* producing tymidin-kinase [Gardlik et al., 2005]. Although the above bacteria are characterized by high selectivity in tissue colonization, it is necessary to ensure the maximum level of specificity and therapeutic efficiency of bacterial vectors. For this purpose, promoters inducible by radiation were constructed. The transcription of such promoters is conditioned by irradiation with visible light. Bacteria are known for their resistance to irradiation (having a relatively small genome and effective DNA repair mechanisms). Therefore the localization of a therapeutic gene under the control of a radiation-inducible promoter ensures that cytotoxic proteins are expressed only in bacteria colonizing currently irradiated tissues. This strategy allows eliminating the possibility of expression in non-tumor and hypoxic tissues. The latest approach using prokaryotes in gene therapy is a system of transformed bacteria producing a therapeutic protein *in situ* under exogenous induction regulation [Gardlik et al., 2005].

9. EUKARYOTIC DELIVERY SYSTEMS

9.1 *Leishmania tarentolae* as a novel live vector

Although live recombinant vectors (bacterial or viral recombinant vectors) have been known to develop new vaccine strategies against pathogens (e.g., HIV-1), their use as vaccine candidates in human is delayed due to problems related to pre-existing immunity, inefficient antigen delivery or presentation and toxicity issues. Therefore, it is necessary to develop new live-vaccine vectors that are able to enhance antigen presentation and elicit potent immune responses without the risk of developing disease in humans. Recently, a lizard parasitic protozoan that is not pathogenic to humans, *Leishmania tarentolae* (*L. tarentolae*), has been used as a candidate vaccine against visceral leishmaniasis [Breton et al., 2005], and HIV-1 [Breton et al., 2007]. *L. tarentolae* can elicit T-cell proliferation and the production of gamma interferon (IFN-γ), skewing the T-cell response towards a Th1-cell phenotype, and it provides inflammatory responses for the APC and acts as an immunostimulatory adjuvant. Unlike other pathogenic *Leishmania* strains, *L.tarentolae* lacks the potential to replicate within the targeted APCs and is eliminated after several days from the infected murine host [Breton et al., 2005; Breton et al., 2007]. It has been shown that a single intraperitoneal injection of *L. tarentolae* could elicit a protective immune response against infectious challenge with *L. donovani* in susceptible BALB/c mice [Breton et al., 2005]. Similarly, a single intraperitoneal administration of the A2-recombinant *L. tarentolae* strain could induce high levels of IFN-gamma and protect BALB/c mice against *L. infantum* challenge [Mizbani et al., 2009]. Interestingly, a recombinant *L. tarentolae* vaccine expressing high levels of full-length HIV-1 Gag elicited cell-mediated immunity in mice model and decreased HIV-1 replication in human tonsillar tissue following exposure to HIV-1 infection [Breton et al., 2007]. These data suggest that the use of *L. tarentolae* as a live vaccine vector may represent a promising approach for improving immunity and safety of candidate live vaccines against *Leishmania* infections and likely other intracellular pathogens for which T- cell mediated responses are critical for the development of protective immunity [Breitling et al, 2002; Breton et al., 2005].

9.2 Yeast as an efficient tool in vaccine development

Recent research has suggested that yeast cell wall components exhibit a variety of adjuvant properties. DC maturation is the consequence of interactions between yeast and DCs. When whole recombinant yeast is internalized by DCs, it can deliver heterologous antigens to both MHC class I and class II pathways and elicit potent cell-mediated immunity [Capilla et al., 2009; Bian et al., 2010; Haller et al., 2007]. Vaccination with Saccharomyces cerevisiae (S.cerevisiae) that expresses tumor-associated antigens can induce antigen-specific T-cell responses and protect animals from tumor challenge. In addition, *S. cerevisiae* is inherently nonpathogenic, and clinical studies have not demonstrated any toxicity in heat-killed recombinant *S. cerevisiae*. The inherent adjuvant properties of *S. cerevisiae* eliminate the necessity for additional adjuvants, and yeast can be effortlessly engineered to express multiple antigens. These attributes render *S. cerevisiae* a potential vaccine vehicle for infectious diseases and cancer [Capilla et al., 2009; Bian et al., 2010; Haller et al., 2007]. *S. cerevisiae* expression systems are subject to certain constraints and drawbacks. For instance, *S. cerevisiae* has a propensity to hyperglycosylate recombinant proteins, which results in the termination of N-linked carbohydrate chains with alpha-1, 3-linked mannose residues, which is considered allergenic. Another limitation is that this species is restricted in its ability to utilize a variety of carbon sources [Bian et al., 2010]. At present, the limitations of *S. cerevisiae* may be circumvented by two additional species, Hansenula polymorpha and Pichia pastoris, both of which are members of the Saccharomycetaceae family [Bazan et al., 2009; Bian et al., 2010]. Conversely, these two species are extensively employed as industrial platforms for the production of heterologous proteins [Maleki et al., 2010; Bian et al., 2010].

10. CONCISE SUMMARY

A variety of techniques have been developed and are currently being developed to ensure the safe and efficient delivery of therapeutic DNA. Research in the field of molecular biology is bolstered by the perspectives and aspirations that are linked to gene therapy. Despite the commencement of clinical trials, there are still a number of constraints that must be resolved prior to the implementation of routine clinical use. The primary objective of gene therapy is to create non-toxic, efficient gene carriers that can encapsulate and deliver foreign genetic materials to specific cell types, including malignant cells. In order to deliver therapeutic genes into cancer cells, both viral and non-viral vectors were devised and assessed. Numerous viruses, including retrovirus, adenovirus, herpes simplex virus, adeno-associated virus, and poxvirus, have been modified to preserve their high gene transfer capability and eliminate their toxicity. Non-viral vectors have been further emphasized as an alternative in delivery systems due to the limitations associated with viral vectors. Cationic polymers, cationic peptides, and cationic liposomes comprise the primary non-viral vectors. Currently, numerous modifications to the existing delivery systems and novel carrier systems have been developed to enhance the efficiency of transfection. In addition, the immune response's outcome can be influenced by the route of immunization, which can alter the interaction between the vaccine and various APCs at the injection site. Therefore, the therapeutic response is evidently influenced by the routes of administration and formulation of DNA, which alter the immune pathway. In vivo electroporation and gene gun delivery were the most effective methods of DNA vaccination. Nevertheless, it is imperative to conduct a more thorough examination of the results of ongoing clinical trials, particularly in conjunction with the success or failure of specific gene therapy delivery methodologies.

REFERENCES

Ahmad, S.; Casey, G.; Sweeney, P.; Tangney, M. & O'Sullivan G.C. (2009). Prostate stem cell antigen DNA vaccination breaks tolerance to self-antigen and inhibits prostate cancer growth. *Molecular Therapy*, Vol. 17, No. 6, pp. 1101-1108

Alexis, F.; Lo, S.L. & Wang, S. (2006). Covalent attachment of low molecular weight poly (ethyleneimine) improves Tat peptide mediated gene delivery. *Adv. Mater.*, Vol. 18, pp. 2174-2178

Aravindaram, K. & Yang, N.S. (2009). Gene gun delivery systems for cancer vaccine approaches. *Methods Mol. Biol.*, Vol. 542, pp. 167-178

Babiuk, S.; Drunen Little, S. & Babiuk, L.A. (2006). Delivery of DNA vaccines using electroporation. *Methods in Molecular Medicine*, Vol. 127, DNA vaccines: Methods and Protocols

Baker, J.R. (2009). Dendrimer-based nanoparticles for cancer therapy. *Nanotechnology for Hematology*, pp. 708-719

Bazan, S.B.; Chaves, A.A.M.; Aires, K.A.; Cianciarullo, A.M.; Garcea, R.L. & Ho, P.L. (2009). Expression and characterization of HPV16 L1 capsid protein in *Pichia pastoris*. *Arch Virol.*,Vol. 154, pp. 1609-1617

Bian, G.; Cheng, Y.; Wang, Z.; Hu, Y.; Zhang, X.; Wu, M.; Chen, Z.; Shi, B.; Sun, S.; Shen, Y.; Chen, E.; Yao, X.; Wen, Y. & Yuan Z. (2010). Whole recombinant *Hansenula polymorpha* expressing hepatitis B virus surface antigen (yeast-HBsAg) induces potent HBsAg-specific Th1 and Th2 immune responses. *Vaccine*, Vol. 28, pp. 187-194

Bodles-Brakhop, A.M. & Draghia-Akli, R. (2008). DNA vaccination and gene therapy: optimization and delivery for cancer therapy. *Expert Rev. Vaccines*, Vol. 7, pp. 1085-1101

Bodles-Brakhop, A.M.; Heller, R. & Draghia-Akli, R. (2009). Electroporation for the delivery of DNA-based vaccines and immunotherapeutics: current clinical developments. *Molecular Therapy*, Vol. 17, No. 4, pp. 585-592

Bolhassani, A.; Ghasemi, N.; Servis, C.; Taghikhani, M. & Rafati, S. (2008). Comparison of two delivery systems efficiency by using polyethylenimine (PEI) for plasmid HPV16 E7 DNA transfection into COS-7 cells. *Modarres J. Med. Sci.* Vol. 11, pp. 15-19

Bolhassani, A.; Ghasemi, N.; Servis, C.; Taghikhani, M. & Rafati, S. (2009). The efficiency of a novel delivery system (PEI600-Tat) in development of potent DNA vaccine using HPV16 E7 as a model antigen. *Drug Deliv.*, Vol. 16, pp. 196-204

Bolhassani, A.; Gholami, E.; Zahedifard, F.; Moradin, N.; Parsi, P.; Doustdari, F.; Seyed, N.; Papadopoulou, B. & Rafati, S. (2011). *Leishmania major*: Protective capacity of DNA vaccine using amastin fused to HSV-1 VP22 and EGFP in BALB/c mice model. *Experimental Parasitology*, pp. 1-9.

Breitling, R.; Klingner, S.; Callewaert, N.; Pietrucha, R.; Geyer, A.; Ehrlich, G.; Hartung, R.; Muller, A.; Contreras, R.; Beverley, S.M. & Alexandrov, K. (2002). Non-pathogenic trypanosomatid protozoa as a platform for protein research and production. *Protein Expression and Purification*, Vol. 25, pp. 209-218

Breton, M.; Tremblay, M.J.; Ouellette, M. & Papadopoulou B. (2005). Live nonpathogenic parasitic vector as a candidate vaccine against visceral leishmaniasis. *Infection and Immunity*, Vol. 73, No. 10, pp. 6372-6382

Breton, M.; Zhao, C.; Ouellette, M.; Tremblay, M.J. & Papadopoulou, B. (2007). A recombinant non-pathogenic *Leishmania* vaccine expressing human immunodeficiency virus 1 (HIV-1) Gag elicits cell-mediated immunity in mice and decreases HIV-1 replication in human tonsillar tissue following exposure to HIV-1 infection. *Journal of General Virology*, Vol. 88, pp. 217-225

Brooks, N.A.; Pouniotis, D.S.; Tang, C.K.; Apostolopoulos, V. & Pietersz, G.A. (2010). Cell- penetrating peptides: application in vaccine delivery. *Biochimica et biophysica Acta*, Vol. 1805, pp. 25-34

Brooks, H.; Lebleu, B. & Vives, E. (2005). Tat peptide-mediated cellular delivery: back to basics. *Adv. Drug Deliv. Rev.* Vol. 57, pp. 559-577

Brooks, N.A.; Pouniotis, D.S.; Tang, C.K.; Apostolopoulos, V. & Pietersz, G.A. (2010). Cell- penetrating peptides: application in vaccine delivery. *Biochimica et biophysica Acta*, Vol. 1805, pp. 25-34

Capilla, J.; Clemons, K.V.; Liu, M.; Levine, H.B. & Stevens, D.A. (2009). *Saccharomyces cerevisiae* as a vaccine against coccidioidomycosis. *Vaccine*, Vol. 27, pp. 3662-3668

Cashman, S.M.; Sadowski, S.L.; Morris, D.J.; Frederick, J. & Kumar-Singh, R. (2002). Intercellular trafficking of *adenovirus*-delivered HSV VP22 from the retinal pigment epithelium to the photoreceptors-implications for gene therapy. *Mol. Ther.*, Vol. 6, pp. 813-823

Chatterjee, D.K. & Zhang Y. (2007). Multi-functional nanoparticles for cancer therapy. *Science and Technology of Advanced Materials*, Vol. 8, pp. 131-133

Chen, W.C. & Huang, L. (2005). Non-viral vector as vaccine carrier. *Advances in Genetics*, Vol. 54, pp. 315-337

Collins, C.G.; Tangney, M.; Larkin, J.O.; Casey, G.; Whelan, M.C.; Cashman, J.; Murphy, J.; Soden, D.; Vejda, S.; McKenna, S.; Kiely, B.; Collins, J.K.; Barrett, J.; Aarons, S. & O'Sullivan, G.C. (2006). Local gene therapy of solid tumors with GM-CSF and B7-1 eradicates both treated and distal tumors. *Cancer gene Therapy*, Vol. 13, pp. 1061-1071

Crombez, L.; Aldrian-Herrada, G.; Konate, K.; Nguyen, Q.N.; McMaster, G.K.; Brasseur, R.; Heitz, F. & Divita G. (2009). A new potent secondary amphipathic cell-penetrating peptide for siRNA delivery into mammalian cells. *Mol. Ther.*, Vol. 17, pp. 95-103

Doria-Rose, N.A. & Haigwood, N.L. (2003). DNA vaccine strategies: candidates for immune modulation and immunization regimens. *Methods*, Vol. 31, pp. 207-216

Derossi, D.; Joliot, A.H.; Chassaing, G. & Prochiantz, A. (1994). The third helix of the *Antennapedia* homeodomain translocates through biological membranes. *J. Biol.Chem.*, Vol. 269, pp. 10444-10450

Elmquist, A.; Lindgren, M.; Bartfai, T. & Langel, U. (2001). VE-cadherin-derived cell-penetrating peptide, pVEC, with carrier functions. *Exp. Cell. Res.*, Vol. 269, pp. 237-244

Fernandez-Carneado, J.; Kogan, M.J.; Castel, S. & Giralt, E. (2004). Potential peptide carriers: amphipathic proline-rich peptides derived from the N-terminal domain of gamma- zein, Angew. *Chem. Int. Ed. Engl.*, Vol. 43, pp. 1811-1814

Futaki, S. Membrane-permeable arginine-rich peptides and the translocation mechanisms. (2005). *Adv. Drug Deliv. Rev.* Vol. 57, No. 4, pp. 547-58

Garcia-Frigola, C.; Carreres, M.I.; Vegar, C. & Herrera, E. (2007). Gene delivery into mouse retinal ganglion cells by in utero electroporation. *B.M.C. Developmental Biology*, Vol. 7, No. 103, pp. 1-10

Gardlik, R.; Palffy, R.; Hodosy, J.; Lukacs, J.; Turna, J. & Celec P. (2005). Vectors and delivery systems in gene therapy. *Med. Sci. Monit.* Vol. 11, No. 4, pp. RA 110-121

Giannouli, C.; Brulet, J.M.; Gesché, F.; Rappaport, J.; Burny, A.; Leo, O. & Hallez, S. (2003). Fusion of a tumor-associated antigen to HIV-1 Tat improves protein-based immunotherapy of cancer. *Anticancer Res.* Vol. 23, pp. 3523-3532.

Gould, D.J. & Chernajovsky, Y. (2007). Novel delivery methods to achieve immunomodulation. *Curr. Opin. Pharmacol.* Vol. 7, No. 4, pp. 445-450

Gupta, P.N.; Singh, P.; Mishra, V.; Jain, S.; Dubey, P.K. & Vyas, S.P. (2004). Topical immunization: Mechanistic insight and novel delivery systems. *Indian Journal of Biotechnology*, Vol. 3, pp. 9-21

Haller, A.A.; Lauer, G.M.; King, T.H.; Kemmler, C.; Fiolkoski, V.; Lu, Y., Bellgrau, D.; Rodell, T.C.; Apelian, D.; Franzusoff A. & Duke, R.C. (2007). Whole recombinant yeast- based immunotherapy induces potent T cell responses targeting HCV NS3 and Core proteins. *Vaccine*, Vol. 25, pp. 1452-1463

Harrison, R.L.; Byrne, B.J. & Tung, L. (1998). Electroporation-mediated gene transfer in cardiac tissue. *FEBS Letters*, Vol. 435, pp. 1-5

Hellgren, I.; Gorman, J. & Sylven, C. (2004). Factors controlling the efficiency of Tat- mediated plasmid DNA transfer. *J. Drug Target*, Vol. 12; pp. 39-47

Jarver, P. & Langel, U. (2004). The use of cell-penetrating peptides as a tool for gene regulation. D.D.T., Vol. 9, pp. 395-402

Johnson, J.R.; Jiang, H. & Smith, B.D. (2008). Zinc(II)-coordinated oligotyrosine: a new class of cell penetrating peptide. *Bioconjug. Chem.* Vol. 19, pp. 1033-1099

Kang, T.H.; Monie, A.; Wu, L.S.; Pang, X.; Hung, C.F. & Wu, T.C. (2011). Enhancement of protein vaccine potency by *in vivo* electroporation mediated intramuscular injection. *Vaccine*, Vol. 29, No. 5, pp. 1082-9

Katare, D.P. & Aeri, V. (2010). Progress in gene therapy: A Review. *I.J.T.P.R.* Vol. 1, No. 2, pp. 33-41

Kaufman, H.L.; Flanagan, K.; Lee, C.S.; Perretta, D.J. & Horig H. (2002). Insertion of IL-2 and IL-12 genes into *vaccinia* virus results in effective anti-tumor responses without toxicity. *Vaccine*, Vol. 20, pp. 1862-1869

Kim, D.T.; Mitchell, D.J.; Brockstedt, D.G.; Fong, L.; Nolan, G.P.; Fathman, C.G.; Engleman, E.G. & Rothbard, J.B. (1997). Introduction of soluble proteins into the MHC class I pathway by conjugation to an HIV tat peptide. *J. Immunol.*, Vol. 159, pp. 1666-1668

Kim-Schulze, S. & Kaufman, H.L. (2009). Gene therapy for anti-tumor vaccination. Methods in Molecular Biology, *Gene Therapy of Cancer*, Walther, W; Stein, U.S. (eds.), Vol. 542, pp. 515-527

Kleemann, E.; Neu, M.; Jekel, N.; Fink, L.; Schmehl, T.; Gessler, T.; Seeger, W. & Kissel, T. (2005). Nano-carriers for DNA delivery to the lung based upon a TAT-derived peptide covalently coupled to PEG-PEI. *Journal of Controlled Release*, Vol. 109, pp. 299-316

Lee, W.G.; Demirci, U. & Khademhosseini, A. (2009). Microscale electroporation: challenges and perspectives for clinical applications. *Integr. Biol.*, Vol. 1, pp. 242-251

Lin, K.; Roosinovich, E.; Ma, B.; Hung, C.F. and Wu, T.C. (2010). Therapeutic HPV DNA vaccines. *Immunol Res*, pp.1-27

Long, C.M.; van Laarhoven, H.W.M.; Bulte, J.W.M. & Levitsky, H.I. (2009). Magnetovaccination as a novel method to assess and quantify dendritic cell tumor antigen capture and delivery to lymph nodes. *Cancer Res.*, Vol. 69, pp. 3180–3187

Lu, J.M.; Wang, X.; Marin-Muller, C.; Wang, H., Lin, P.H., Yao, Q. & Chen C. (2009). Current advances in research and clinical applications of PLGA-based nanotechnology. *Expert Rev. Mol. Diagn.*, Vol. 9, pp. 325-341

Maleki, A.; Roohvand, F.; Tajerzadeh, H.; Khanahmad, H.; Nobari, M.B.; Beiruti, A. & Rouholamini Najafabadi A. (2010). High expression of methylotrophic yeast- derived recombinant human erythropoietin in a pH-controlled batch system. *Avicenna Journal of Medical Biotechnology*, Vol. 2, No. 4, pp. 197-206.

Michel, N.; Osen, W.; Gissmann, L.; Schumacher, T.N.; Zentgraf, H. & Müller M. (2002). Enhanced immunogenicity of HPV16 E7 fusion proteins in DNA vaccination. *Virology*, Vol. 294, pp. 47-59

Mizbani, A.; Taheri, T.; Zahedifard, F., Taslimi, Y., Azizi, H.; Azadmanesh, K.; Papadopoulou, B. & Rafati, S. (2009). Recombinant *Leishmania tarentolae* expressing the A2 virulence gene as a novel candidate vaccine against visceral leishmaniasis. *Vaccine*, Vol. 28, No. 1, pp. 53-62

Morris, M.C.; Vidal, P.; Chaloin, L.; Heitz, F. & Divita, G. (1997). A new peptide vector for efficient delivery of oligonucleotides into mammalian cells. *Nucleic Acids Res.*, Vol. 25, pp. 2730-2736

Morris, M.C.; Depollier, J.; Mery, J.; Heitz, F. & Divita G. (2001). A peptide carrier for the delivery of biologically active proteins into mammalian cells. *Nat. Biotechnol.* Vol. 19, pp. 1173-1176.

Myschik, J.; Rades, T. & Hook, S. (2009). Advances in lipid-based subunit vaccine formulations. *Current Immunology Reviews*, Vol. 5, pp. 42-48

Oehlke, J.; Scheller, A.; Wiesner, B.; Krause, E.; Beyermann, M.; Klauschenz, E.; Melzig, M. & Bienert, M. (1998). Cellular uptake of an alpha-helical amphipathic model peptide with the potential to deliver polar compounds into the cell interior non- endocytically. *Biochem.Biophys.Acta*, Vol. 1414, pp. 127-139

Ogris, M. & Wagner E. (2002). Targeting tumors with non-viral gene delivery systems. *D.D.T.*, Vol. 7, No. 8, pp. 479-485

Park, T.G.; Jeong, J.H. & Kim, S.W. (2006).Current status of polymeric gene delivery systems. *Advanced Drug Delivery Reviews*, Vol. 58, pp. 467-486

Pooga, M.; Hallbrink, M.; Zorko, M. & Langel, U. (1998). Cell penetration by transportan. F.A.S.E.B. J., Vol. 12, pp. 67-77.

Putnam, D.; Gentry, C.A.; Pack, D.W. & Langer, R. (2001). Polymer based gene delivery with low cytotoxicity by a unique balance of side-chain termini. *Proc. Natl. Acad. Sci.*, Vol. 98, pp. 1200-1205

Riedl, P.; Reimann, J. & Schirmbeck R. (2004). Peptides containing antigenic and cationic domains have enhanced, multivalent immunogenicity when bound to DNA vaccines. *J. Mol. Med.*, Vol. 82, pp. 144-152

Riedl, P.; Reimann, J. & Schirmbeck, R. (2004). Complexes of DNA vaccines with cationic, antigenic peptides are potent, polyvalent CD8+ T cell-stimulating immunogens. *Methods Mol. Med.*,Vol. 127, pp. 159-169

Rojas, M.; Donahue, J.P.; Tan, Z. & Lin, Y.Z. (1998). Genetic engineering of proteins with cell membrane permeability. *Nat. Biotechnol.*, Vol. 16, pp. 370-375

Rossini, K.; Sandri, M. & Dona, M. (2002). High level of gene transfer into adult skeletal muscle by *in vivo* direct electroporation. *Basic Appl. Mol.*, Vol. 12, No. 3, pp. 97-100

Schirmbeck, R.; Riedl, P.; Zurbriggen, R.; Akira, S. & Reimann, J. (2003). Antigenic epitopes fused to cationic peptide bound to oligonucleotides facilitate toll-like receptor 9- dependent, but CD4+ T cell help-independent, priming of CD8+ T cells. *The Journal of Immunology*, Vol. 171, pp. 5198-5207

Stone, D. (2010). Novel viral vector systems for gene therapy. *Viruses*, Vol. 2, pp. 1002-1007 Templeton, N.S. & Lasic D.D. (1999). New directions in liposome gene delivery. *Molecular Biotechnology*, Vol. 11, pp. 175-180

Trehin, R.; Krauss, U.; Muff, R.; Meinecke, M.; Beck-Sickinger, A.G. & Merkle, H.P. (2004). Cellular internalization of human calcitonin derived peptides in MDCK monolayers: a comparative study with Tat (47-57) and penetratin (43-58). *Pharm. Res.*, Vol. 21, pp. 33-42

Trimble, C.; Lin, C.T.; Hung, C.F.; Pai, S.; Juang, J.; He, L.; Gillison, M.; Pardoll, D.; Wu, L. & Wu, T.C. (2003). Comparison of the CD8+ T cell responses and antitumor effects generated by DNA vaccine administered through gene gun, biojector and syringe. *Vaccine*, Vol. 21, pp. 4036–4042.

Ulmer, J.B.; Wahren, B. & Liu, M.A. (2006). Gene-based vaccines: recent technical and clinical advances. *Trends in Molecular Medicine*, Vol. 12, pp. 216-222

Umeda, Y.; Marui, T.; Matsuno, Y.; Shirahashi, K.; Iwata, H.; Takagi, H.; Matsumoto, K.; Nakamura, T.; Kosugi, A.; Mori, Y. & Takemura, H. (2004). Skeletal muscle targeting *in vivo* electroporation-mediated HGF gene therapy of bleomycin-induced pulmonary fibrosis in mice. *Laboratory Investigation*, Vol. 84, pp. 836-844

Vergati, M.; Intrivici, C.; Huen, N.Y.; Schlom, J. & Tsang, K.Y. (2010). Strategies for cancer vaccine development. *Journal of Biomedicine and Biotechnology*, 1-13.

Vives, E.; Brodin, P. & Lebleu, B. (1997). A truncated HIV-1 Tat protein basic domain rapidly translocates through the plasma membrane and accumulates in the cell nucleus. *J. Biol. Chem.*, Vol. 272, pp. 16010-16017

Wagstaff, K.M. & Jans, D.A. (2006). Protein transduction: cell penetrating peptides and their therapeutic applications. *Current Medicinal Chemistry*, Vol. 13, pp. 1371-1387.

Wang, S. (2006). Tat peptide conjugates of low molecular weight polyethylenimine as effective non-viral gene delivery vectors. *Mol. Ther.*, Vol. 13, pp. 76

Zender, L.; Kuhnel, F.; Kock, R.; Manns, M. & Kubicka S. (2002). VP22-mediated intercellular transport of p53 in hepatoma cells *in vitro* and *in vivo*. *Cancer Gene Therapy*, Vol. 9, pp. 489-496.

Zhang, H.; Yee, D. & Wang, C. (2008). Quantum Dots for cancer diagnosis and therapy: biological and clinical perspectives. *Nanomedicine*, Vol. 3, pp. 83-91

Zheng, C.; Juhls, C.; Oswald, D.; Sack, F.; Westfehling, I.; Wittig, B.; Babiuk, L.A. & Hurk

S.D.L. (2006). Effect of different nuclear localization sequences on the immune responses induced by a MIDGE vector encoding bovine herpesvirus-1 glycoprotein D. *Vaccine*, Vol. 24, pp. 4625-4629

3 Chapter

DESIGNING VECTORS TO TARGET SPECIFIC TISSUES: GENE DELIVERY SYSTEMS

1. INTRODUCTION: AN OVERVIEW

The concept of gene therapy was established over forty years ago. In a historic lecture, Edward Tatum declared, "We can anticipate that viruses will be utilized effectively for the benefit of humanity, in theoretical studies regarding somatic cell genetics and the potential for genetic therapy." We can even be somewhat optimistic about the long-term potential of therapy that is based on the isolation, design, synthesis, and introduction of new genes into defective cells of specific organs. Marshall Niremberg predicted two years later that it would be feasible to program cells with synthetic messages within 25 years. However, he cautioned that this should be postponed until the necessary wisdom is acquired to apply this knowledge for the betterment of humanity.

Rogers and Pfudere made the first attempts to construct vectors in 1968. However, the initial gene therapy clinical trial was not conducted in a patient with adenosine deaminase (ADA) deficiency until 1990 (Friedmann, 1992). In the past few years, there have been substantial advancements in the development of therapies for a variety of conditions, including cancer, AIDS, Parkinson's disease, X-linked severe combined immunodeficiency (X-SCID), hemophilias, cystic fibrosis, Leber's congenital amaurosi, and β-thalassemia. However, an adolescent patient with a urea cycle defect passed away due to an anaphylactic reaction a few hours following the injection of an adenoviral vector. Five children who were enrolled in a clinical trial to correct an X-SCID developed a leukemia-like condition shortly thereafter as a result of the activation of an oncogene (Edelstein, 2007). Initially, the tragic demise of a volunteer in a clinical trial for rheumatoid arthritis that utilized an adenoassociated virus vector was attributed to a side effect of the vector (Williams, 2007). Nevertheless, subsequent investigations revealed that this was the result of an infection with Histoplasma capsulatum, which was generated as a result of an immunosuppressive state that was provoked by concurrent systemic anti-TNF-alpha therapy administered via the drug adalimumab (Williams, 2007). The progression of gene therapy was halted as a result of these failures, which raised significant concerns regarding the procedure's efficacy. It also prompted the necessity for additional fundamental research on the immunogenic aspects and the development of more efficient and secure vectors.

In the past 35 years, a plethora of viral and non-viral vectors have been created to create a vector that is capable of delivering genes to specific cell types, has a high efficiency of gene transfer, produces therapeutic levels of gene expression over extended periods, and minimizes the generation of side effects (Verma & Weitzman, 2005). However, the discipline has made significant progress in the development of an ideal vector; however, the primary obstacle remains the pursuit of a cell-specific vector. It is highly probable that a vector that is not specifically targeted and remains in the general circulation will be sequestered by the liver or may wind up in cells that do not require the transfected gene product. Furthermore, gene therapy continues to face two primary challenges: (i) the natural obstruction of the gene or its product by the blood-brain barrier in the central nervous system (CNS) and (ii) the delivery of the gene to poorly circulated tissues, such as bone, which has only been accomplished with low transfection efficiencies to date.

This chapter will investigate the modifications that are implemented in viral and non-viral vectors to improve their natural tropism. Modification objectives can be summarized as follows, regardless of the vector: (i) tropism expansion to enable gene delivery to cells that are difficult to transfect with a non-modified vector, (ii) cell-type-specific transfection to mitigate side effects, and (iii) stealth vector enhancement to reduce protein interactions or cells that restrict its therapeutic activity, such as the immune system (Yu & Schaffer, 2005).

2. RETROVIRAL VECTORS

Retroviruses are a group of positive-stranded RNA viruses that are enveloped, diploid, and RNA. This family comprises alpharetrovirus, betaretrovirus, deltaretrovirus, gammaretrovirus, lentivirus, and spumavirus (foamy virus). Only the latter three have been employed as gene therapy vectors (Verma & Weitzman, 2005). The genome is primarily composed of three genes: (i) gag, which encodes for viral matrix, capsid, and nucleocapsid proteins; (ii) pol, which encodes for protease, reverse transcriptase, and integrase; and (iii) env, which encodes a bipartite membrane-anchored surface (SF) protein (Escors & Breckpot, 2010). Furthermore, ancillary genes are present in complex retroviruses, such as lentivirus and foamy virus, to regulate gene expression, assembly, and replication (Escors & Breckpot, 2010). Glycoproteins (GPs) are crucial for viral tropism, receptor recognition, and cell entry during the viral life cycle (Liang et al., 2009). The fusion and insertion of the capsid into the cytoplasm are induced by the conformational change of the viral GP that results from its binding to the cellular surface protein (Yu & Schaffer, 2005). Most of the advantages and limitations of retrovirus targeting are associated with this cell entry mechanism.

The properties of retroviral vectors utilized in gene therapy enable the transduction of a wide range of cells (Cronin et al., 2005; Frecha et al., 2008). Nevertheless, the utilization of targeted retroviral vectors would be advantageous, as it would enable the delivery of the transgene to cells that are not typically transduced or restrict the delivery to cells that should not be transduced (Cronin et al., 2005; Frecha et al., 2008; Lavillette et al., 2001). In general, retroviral targeting can be accomplished by: (i) engineering the GP to produce a vector that recognizes a specific cell surface protein (direct targeting), (ii) pseudotyping with GPs derived from other retrovirus and that recognize specific receptors, (iii) using ligands that block the cell entry of the vector to the cells expressing the receptor for the ligand, while allowing the transduction of other cell types (indirect targeting), or (iv) using biospecific antibodies, chemically crosslinking ligands, and grafting polymers (Fig. 1) (Lavillette et al., 2001; Verhoeyen & Cosset, 2004; Yu & Schaffer, 2005).

2.1 Retroviral targeting by engineering the glycoproteins

Direct targeting consists of the insertion of ligands or antibody fragments into the viral GP (Fig. 1), which allows recognition of cell surface proteins different from those naturally used by the vector (Lavillette et al., 2001). Direct modification of a GP was first reported in 1990, with the successful fusion of CD4 with the GP of a gammaretroviral vector (Young et al., 1990). Later, a functional murine leukemia virus (MLV) vector bearing a GP protein fused with an antibody fragment against a hapten (4-hidroxy-5-iodo-3-nitrophenacetyl caproate) demonstrated binding of the modified retroviral vector to hapten, and the ability to package a transgene and transduce culture cells (Russell et al., 1993).

After these pioneer works revealed the feasibility of GP engineering, different authors studied the real potential of this targeting approach. The fusion of the erythropoietin polypeptide with the GP of a MLV vector showed for the first time that a chimeric GP was not only correctly incorporated into the virus envelope but also allowed the specific transduction of cells expressing the EPO receptor (Kasahara et al., 1994). A similar result was reported with a gammaretrovirus bearing a GP fused with a single-chain antigen-binding site against a cell surface expressed on human carcinoma cells (Kasahara et al., 1994).

Nevertheless, different studies have exposed the limitations of this approach. The first evidence of this limitation was observed with vectors carrying the Ram-1 phosphate transporter or the epidermal growth factor (EFG) into the GP, which showed a reduced transduction efficiency although the chimeric GPs were correctly expressed, processed, and exposed in the viral envelope (Cosset et al., 1995). A similar observation was reported for a gammaretroviral vector with a GP bearing a peptide binding to the αvβ3 integrin (Wu et al., 2000a), or with the stromal cell derived factor 1-alpha (Katane et al., 2002). These low transductions efficiencies were associated with loss or impairment of viral and cell surface proteins binding and subsequent membrane fusion and penetration of the viral core into the cytosol (Frecha et al., 2008; Lavillette et al., 2001; Yu & Schaffer, 2005). In addition, these results strongly suggest that only few set of peptides sequences can be inserted into the GPs without altering the mechanism of retrovirus cell entry (Verhoeyen & Cosset, 2004). The insertion of spacer sequences between the inserted peptide and the GP has proved to overcome this issue, allowing a significant improvement in the transduction efficiency of the modified vectors (Kayman et al., 1999).

2.2 Retroviral targeting by pseudotyping

Pseudotyping involves the production of viral particles bearing GP derived from other enveloped viruses (Fig. 1). This strategy was developed due to the narrow cellular transduction profile observed in vectors with native GPs (Cronin et al., 2005; Yu & Schaffer, 2005). Production of pseudotyped retroviral vectors is a well-established protocol (Bischof & Cornetta, 2010), which in most cases is used to allow virus purification, improve the cellular transduction, extend the range of transduced cells, and reduce cell toxicity (Cronin et al., 2005). However, it should be possible to employ natural tropism in GPs used for pseudotyping to target vectors bearing these proteins.

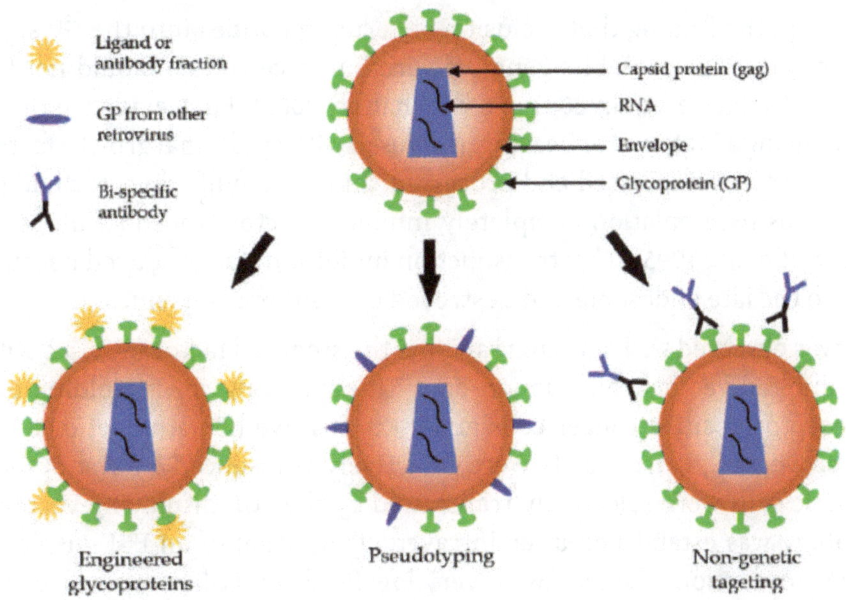

Fig. 1. Strategies for retroviral targeting. Retroviral vectors can be modified by ligand or antibody fraction insertion (engineered glycoproteins), use of glycoprotein (GP) from other retrovirus (pseudotyping), or the use of bi-specific antibodies (non-genetic targeting).

A large set of GPs has been evaluated for vector pseudotyping, among which vesicular stomatitis virus glycoprotein (VSV-G) is the most used [reviewed in (Bischof & Cornetta, 2010; Cronin et al., 2005)]. However, only a few GPs have allowed the transduction of different cell types compared to those observed VSV-G pseudotyped vectors. The use of a GP derived from rabies virus (RV) for pseudotyping an HIV-1 lentiviral vector allowed the transduction of neurons usually non-transduced with VSV-G vectors, since the RV glycoprotein induced a retrograde transport along axons (Mazarakis et al., 2001). The infusion of an animal model of familial amyotrophic lateral sclerosis (ALS) with a RV-HIV-1 vector carrying the gene of vascular endothelial growth factor (VEGF) delayed the onset of the disease and slowed progression in treated animals. Although treatment was initiated at the onset of paralysis CNS tropism of the vector was still significantly improved (Azzouz et al., 2004).

An interesting tropism was also observed for an HIV-1 vector pseudotyped with a GP from the lymphocytic choriomeningitis virus (LCMV) (Miletic et al., 2004). *In-vitro* and *in-vivo* evaluation of a vector pseudotyped with a LCMV GP showed that it transduced, almost exclusively, astrocytes; while VSV-G pseudotypes vectors infected neurons as well as astrocytes. In addition, LCMV-HIV-1 vector presented a specific transduction of infiltrating tumor cells, while VSV-G-HIV-1 vectors transduced mostly normal brain cells in infiltrating tumor areas. An HIV-1 vector pseudotyped with an Ebola Zaire virus-derived GP appeared useful in the treatment of airway diseases (e.g. cystic fibrosis). This vector allowed *in-vitro* and *in-vivo* transduction of airway epithelial cells, which was not observed with a VSV-G vector (Kobinger et al., 2001).

A recent modification of this approach used an HIV-1 vector pseudotyped with the GP and the fusion protein from a measles virus. The GP had an epidermal growth factor (EGF) or a single chain antibody against CD20 (Funke et al., 2008). The CD20-bearing vector was able to transduce primary and CD20-positive B cells both alone or within a cell mixture, while a VSV-G vector did not transduce these cell types.

1.3 Indirect retroviral targeting

This strategy is based on the finding that inclusion of certain peptides into the GP significantly impairs transduction of some cells. This prohibits gene delivering into cells that should not be transduced and extends viral tropism (Lavillette et al., 2001; Yu & Schaffer, 2005). First evidence of this approach was discovered while producing a MLV vector bearing a GP fused with epidermal growth factor (EGF). Although the chimeric GP was correctly expressed and processed, and the modified vector had the ability to bind to the EGF receptor, this manipulation completely inhibited vector transduction into cells expressing the EGF receptor (Cosset et al., 1995). This transduction inhibition was produced because of the modified vector was directed to the late endosome and destroyed by lysosomal enzymes.

A similar result was observed with a vector carrying the stem cell factor (SCF, a c-Kit receptor ligand), which selectively inhibited vector transduction on c-Kit-expressing cells (Fielding et al., 1998). When a mixture of EGF receptor-positive cancer cells and c-Kit-positive hematopoietic cells were exposed to the above mentioned vectors, cancer cells were selectively transduced by the SCF-displaying vector, whereas hematopoietic cells were selectively transduced by the EGF-displaying vector. *In-vivo* proof-of-concept of this strategy was established after intravenous infusion of an EGF-displaying HIV-1 vector, showing preferable transduction of spleen with very low levels on EGF-receptor rich tissues (e.g. liver). Furthermore, the VSV-G vector transduced heart, skeletal muscle, lung, brain, kidney, ovaries and bone-marrow (Peng et al., 2001). Transduction inhibition of human and canine cells was also observed for a spleen necrosis virus bearing EFG, which was restored after cleavage via factor Xa at a site located between EFG and GP (Merten et al., 2003).

A variant of this strategy takes advantage of protease-activatable gene delivery vehicles. Matrix metalloproteinases (MMPs) are commonly overexpressed in angiogenesis, inflammation, and cancer invasion. The inclusion of a MMPs-cleavable peptide within an EGF-retroviral vector permitted that in the presence of exogenous MMPs, the infectivity of the MMP-EGF vector but not of the EGF-vector, could be restored. This MMPss-sensitive vector could efficiently discriminate between two different cell types, infecting only MMP- positive cells (Peng et al., 1997).

2.4 Non-genetic retroviral targeting

This strategy involves the use of antibodies or adapter molecules bound to the GP (Fig. 1) (Lavillette et al., 2001; Yu & Schaffer, 2005). This was the first approach used for cell targeting of a gammaretrovirus by using two biotinilated antibodies bridged by streptavidin: one against a GP and the other one against a specific cell membrane marker (Roux et al., 1989). This strategy allowed the specific transduction of cells expressing class I or II major histocompatibility complex (MHC) molecules by using monoclonal antibodies against these MHC molecules and against GP from the retroviral vector (Roux et al., 1989). For targeting of a retroviral vector to folate receptor-expressing epidermoid carcinoma cells, a myeloproliferative sarcoma retrovirus was modified with folic acid (Reddy et al., 2001). Although receptor binding was observed, this vector was not able to induce gene expression. Similar results were observed for a MLV vector bearing a single chain anti-folate receptor antibody to produce vectors targeting ovarian cancer cells (Pizzato et al., 2001), manifesting the limitations of folate receptor targeting. A modification of this approach combines genetic modification of the GP by inclusion of an IgG-binding domain and the conjugation with an antibody that reacts with specific cell surface molecules expressing the antigen (Ohno et al., 1997). By changing the monoclonal antibody it was possible to transduce efficiently and specifically a variety of cell lines. The potential of this approach was confirmed by *in-vivo* evaluation of modified vectors conjugated with antibodies against surface proteins expressed in cancer cells, indicating that after intravenous infusion the vector preferentially transduced these cells (Liang et al., 2009). This strategy has been used both with gammaretroviral and lentiviral vectors. It represents a powerful tool that can be easily modified to obtain a site-specific gene expression for the treatment of cancers, genetic, infectious and immune diseases.

3. ADENOVIRAL VECTORS

An adenovirus (AdV) is a non-enveloped virus formed by an icosahedral protein capsid surrounding a lineal double-stranded DNA of 36 kb (Douglas, 2007). The AdV capsid is characterized by the presence of 252 different capsomers and long fibers protruding from each of the twelve vertices. Seven polypeptides form this complex capsid: hexon (II), penton base (III), fiber (IV), IIIa, VI, VIII, and IX (Campos & Barry, 2007; Sharma et al., 2009). The cell receptor depends on the virus subgroup: A, E and F subgroups use the cell surface coxsackievirus B and adenovirus receptor (CAR); B1 and B2 subgroups use CD46, CD80/86, receptor X, or heparan sulfate proteoglycan (HSPG); C subgroup uses CAR, HSPG, MHC-I, vascular cell adhesion molecule-I (VCAM-I), or integrins; and D subgroup uses CAR, sialic acid, or CD46 (Arnberg, 2009; Campos & Barry, 2007; Sharma et al., 2009).

Adenoviral vectors have been used as tools for gene therapy since the late 80s (Friedmann, 1992), and the first clinical trial was started in 1993 (Douglas, 2007). Since then, over 392 clinical trials using AdV vectors have been carried out. Currently AdV are the most used vectors for gene therapy and represent 24% of clinical trials total (Edelstein, 2007). In addition, the first commercially approved gene therapy product, Gendicine, is based on an AdV vector engineered to express p53 for treatment of patients with head and neck squamous cell carcinoma (Peng, 2005).

AdV vectors are mainly derived from serotypes 2 (AdV2) and 5 (AdV5). They have the advantage of inducing high levels of gene expression, are able to infect both diving and non- dividing cells, and can be purified to high titers. Furthermore, AdV2 and AdV5 have high bloodstream stability with a reduced risk of insertional mutagenesis, since their genome remains extrachromosomal (Douglas, 2007; Edelstein, 2007; Volpers & Kochanek, 2004). Nonetheless, several drawbacks have been identified during *in-vivo* evaluation of these vectors: (i) presence of pre-existing antibodies that rapidly neutralize the vector, (ii) after intravenous administration the vector is mainly taken up by liver cells, limiting the vector to reach its target tissue in adequate concentrations, and (iii) use of high doses in an effort to overcome these problems has not proven to be an adequate and safe approach (Arnberg, 2009). Three strategies can be used to solve these issues: (i) use of vectors from other subgroups different from AdV2 and AdV5 (subgroup C), (ii) use of non-human serotypes, and (iii) retargeting of the vector to cells or tissues of interest (Arnberg, 2009). In this section we will explore different strategies for AdV retargeting that include genetic modification of the capsid, use of molecular adaptors and chemical modification of the capsid (Fig. 2) (Campos & Barry, 2007).

3.1 Genetic modification of the AdV capsid

Genetic modification of the viral capsid involves incorporation of foreign peptides into exposed regions of the capsid. In this manner the gene encoding for the peptide is inserted into the vector's genome (Fig. 2) (Campos & Barry, 2007). The possibility of altering the AdV vector capsid was first evaluated by the insertion of octapeptides into the hexon (Crompton et al., 1994) or the fiber knob (Krasnykh et al., 1998). This established the possibility of producing viable vectors with foreign proteins present on the vector capsid surface. Further experiments revealed the possibility of expanding the vector tropism by inserting a heparin- binding domain (Wickham et al., 1996), or an Arg-Gly-Asp (RGD)-containing peptide into the fiber knob (Dmitriev et al., 1998).

Although different studies indicated the feasibility to target vectors to specific cell types, genetic modification often resulted in failure to rescue viable viruses, or in an impaired virus packing, peptide exposure, and vector transduction (Leissner et al., 2001; Wu et al., 2005). Because the insertion of a pre-selected peptide into a fiber knob often fails to generate an adenovirus vector, use of random peptide libraries displayed directly on the AdV capsid allows isolation of viable vectors with high affinity for specific tissues or cells (Miura et al., 2007). A similar approach involves the pre-selection of peptides by using a phage display peptide library, from which peptides with high affinity for a certain cell type are selected and then inserted into the AdV vector capsid. This has allowed the targeting of AdV vectors to epithelium cells (Nicklin et al., 2001b), tumor cells (Rittner et al., 2007), and neurons (Schmidt et al., 2007). Most efforts to modify the AdV vector capsid have been on vector targeting to tumor cells (Bachtarzi et al., 2008). Some examples of this approach include: (i) production of an AdV vector bearing a CAR protein fused with EGF producing an enhanced gene transfer efficiency in pancreatic carcinoma cells (Wesseling et al., 2001); (ii) a vector with a polylysine motif in the fiber for a CAR-independent binding to HSPG for specific transduction of different breast cancer cells (Ranki et al., 2007); and (iii) vectors with high affinity for prostate adenocarcinoma (Wu et al., 2010), and human colon carcinoma cells (Rittner et al., 2007).

In addition to cancer cell, genetically modified AdV vectors have been used as tools for gene delivery to other cell types. The insertion of an RGD-modified AdV allowed the production of a vector that mediated cell entry via RGD binding to integrins. Administration into the synovial lining improved the outcome of gene therapy for arthritis (Bakker et al., 2001). Targeting to endothelial cells, an important target in vascular gene therapy, was possible after insertion of the SIGYPLP peptide into the fiber protein in combination with fiber mutations that block natural CAR binding (detargeting) (Nicklin et al., 2001b). For specific kidney gene delivery, the intravenous administration of HTTHREP- and HITSLLS-bearing AdV vectors resulted in selective renal targeting, specifically of tubular epithelium and glomeruli (Denby et al., 2007).

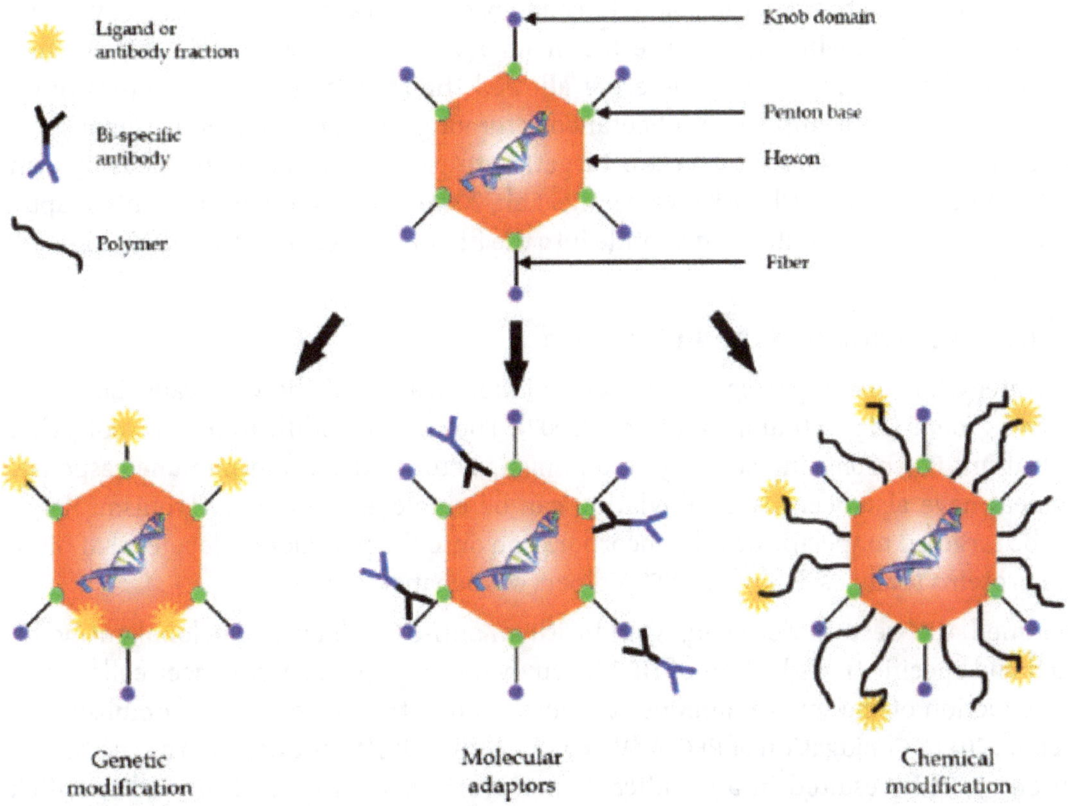

Fig. 2. Strategies for Adenoviral vectors targeting. Adv vectors can be modified by ligand or antibody fraction insertion (genetic modification), use of bi-specific antibodies (molecular adaptors), or conjugation with polymers bearing or not a ligand (chemical modification).

Although genetically modified AdV vectors have been widely studied, most of these modifications have been carried out only into fiber or hexo. Recently it was determined that protein IX in the capsid is a viable platform for the insertion of single-chain variable- fragment antibodies (scFv) or single-domain antibodies (sdAb) for AdV vector retargeting. Even so, only sbAb enhanced virus infection of cells expressing the targeted receptor. Thus, proving that the nature of the ligand can significantly affect vector targeting as already had been observed with fiber and hexo (Poulin et al., 2010).

3.2 Molecular adaptors for AdV targeting

Although genetic modification is the most direct form of vector targeting it has the disadvantage of not all peptides can be inserted into the vector capsid. An approach to overcome this difficulty is to use molecular adaptors consisting of a capsid-binding domain fused with a cell-binding domain (Fig. 2) (Campos & Barry, 2007). This strategy has the advantages of not requiring a correct processing of the cell-binding peptide, since it is not translated with the viral capsid proteins; and that virtually any cell-binding peptide can be fused with the capsid.

The primary adaptor molecules are bi-specific antibodies against a capsid protein and a cellular receptor. The use of a single-chain bi-specific antibody directed against human EGFR and fiber knob significantly increased gene transfer into primary glioma cells and organotypic glioma spheroids, while reducing natural tropism dramatically (Grill et al., 2001). The human epithelial cell adhesion molecule

(EpCAM) is highly expressed in malignant lesions of the stomach and esophagus. By using bi-specific antibodies against adenovirus fiber knob protein and EpCAM a specific transduction of gastric and esophageal cancer cell lines was observed with reduced transduction in normal cells (Heideman et al., 2001). For other epithelial tumors, e.g. colon, lung and breast, it was possible to target vectors by using a bi-specific adapter protein, which fused the ectodomain of CAR with a single-chain anti-carcinoembryonic antigen (CEA) antibody. This adaptor molecule allowed the targeting of AdV vectors to CEA-positive epithelial tumor cells in cell culture, in subcutaneous tumor grafts, and in hepatic tumor grafts (Li et al., 2007). A similar strategy was used for vector targeting to pancreatic carcinoma cells by using a vector bearing an EGF peptide fused with CAR (Wesseling et al., 2001). The combination of this adaptor molecule with the insertion of an RGD-containing peptide into the fiber knob resulted in a significant enhancement of gene transfer.

3.3 Chemical modification of the AdV capsid

The third strategy for AdV targeting is the chemical modification of the viral capsid by using primary-amine reactive groups (Fig. 2) (Campos & Barry, 2007). The capsid modification with polyethylene glycol (PEG - PEGylation) is a promising strategy to diminish vector toxicity and immune response. Although other polymers have also been used PEGylation can be employed to modify tropism. Polymer-coated vectors can be targeted to specific cells by incorporation onto the polymer molecules that bind to specific cell-surface proteins (Campos & Barry, 2007; Volpers & Kochanek, 2004).

PEG-modified AdV vectors were conjugated with an anti-HER2/neu monoclonal antibody leading to an enhanced and specific transduction of HER2/neu over-expressed breast cancer cells. Furthermore, a significant reduction of the innate immune response against the vector was accomplished (Jung et al., 2007; Kim et al., 2011). Conjugation of PEG-AdV vectors with an RGD peptide led to a specific $\alpha v\beta 3$ integrin vector cell entry. This resulted in a significant improvement in transduction and specificity of gene delivery into endothelial cells. These events have implications on the treatment of rheumatoid arthritis, inflammatory bowel disease and epithelial tumors (Eto et al., 2005; Ogawara et al., 2006). Transduction of bone marrow derived human mesenchymal stem cells (MSC) can be significantly improved by using a PEGylated AdV conjugated with a blocked poly-L-lysine. However, modification on vector tropism was not clearly reported in this study (Park et al., 2010).

Although a great part of systemically administrated vector particles is scavenged by Kupffer cells in the liver, the use of a PEG- or a dextran-coated vectors allowed the specific transduction of hepatocytes independent of the presence of Kupffer cells, emphasizing the potential for therapeutic liver-directed gene transfer (Prill et al., 2010). Similar result can be observed by using a multivalent hydrophilic polymer based on poly-[N-(2- hydroxypropyl) methacrylamide] conjugated with EGF or VCAM, producing a CAR-independent binding and uptake into EGF- or VCAM-positive target cells selectively in mixed culture and also in xenografts *in-vivo* (Fisher et al., 2001).

4. ADENO-ASSOCIATED VIRAL VECTORS

Adeno-associated virus (AAV) are non-enveloped virus belonging to the *Parvoviridae* family that need a helper virus, such as AdV or herpes simplex, for efficient infection and replication (Flotte, 2004). AAV are formed by a single-stranded DNA genome of 4.7 kb that contains two open reading frames (ORFs): (i) *rep*, which encodes for proteins Rep78, Rep68, Rep52 and Rep40, involved in virus genome replication, packing and integration; and (ii) *cap*, which encodes for the capsid proteins VP1, VP2 and VP3 (Wu et al., 2006). These ORFs are flanked by two Inverted Terminal Repeats (ITRs), involved in complementary DNA synthesis, Rep binding proteins, and site-specific genome integration in human chromosome 19 (Wu et al., 2006).

AAV capsid is formed by 60 subunits of VP1, VP2 and VP3 in a 1:1:10 ratio (Michelfelder & Trepel, 2009). These proteins share the C-terminal sequence, while the N-terminal sequences differ according to the start codon, being VP1>VP2>VP3 (Wu et al., 2006). Capsid structure has been elucidated for AAV2, AAV3, AAV4, AAV8 and AAV9 serotypes (Govindasamy et al., 2006; Lerch et al., 2010; Mitchell et al., 2009; Nam et al., 2007; Xie et al., 2002). The difference within capsid proteins allows for each serotype to use a specific receptor: (i) AAV2 and AAV3 use the HSPG, (ii) AAV1, 4 and 5 use glycans with sialic acid ends, and AAV8 uses the 37/67 kDa laminin receptor (Nam et al., 2007; Summerford & Samulski, 1998; Wu et al., 2006). AAV can infect a wide number of tissues including liver, lung, central nervous system, muscle, and heart; although as a result of the differences in receptor used, each serotype has a characteristic tropism (Flotte, 2004; Verma & Weitzman, 2005).

From the first clinical trial for cystic fibrosis patients with AAV vectors in 1995, over 75 clinical trials have been conducted. Pathologies such as α1-antitrypsin deficiency, Alzheimer's disease, Canavan's disease, hemophilia B, Leber congenital amaurosis, Parkinson's disease and muscular dystrophy; have attained promising results and no direct side effects have been associated with the vector (Warrington & Herzog, 2006). Although AAV vectors have shown to be promising tools for therapeutic gene delivery, they cannot transduce all cell types and could be useful to restrict its transduction to specific cell types. Strategies to modify the natural tropism of AAV vectors include: (i) the insertion of ligands into the viral capsid, (ii) use of chimeric or mosaic capsids, and (iii) conjugation with ligands through non-genetic modifications (Fig. 3) (Choi et al., 2005; Kwon & Schaffer, 2008; Michelfelder & Trepel, 2009).

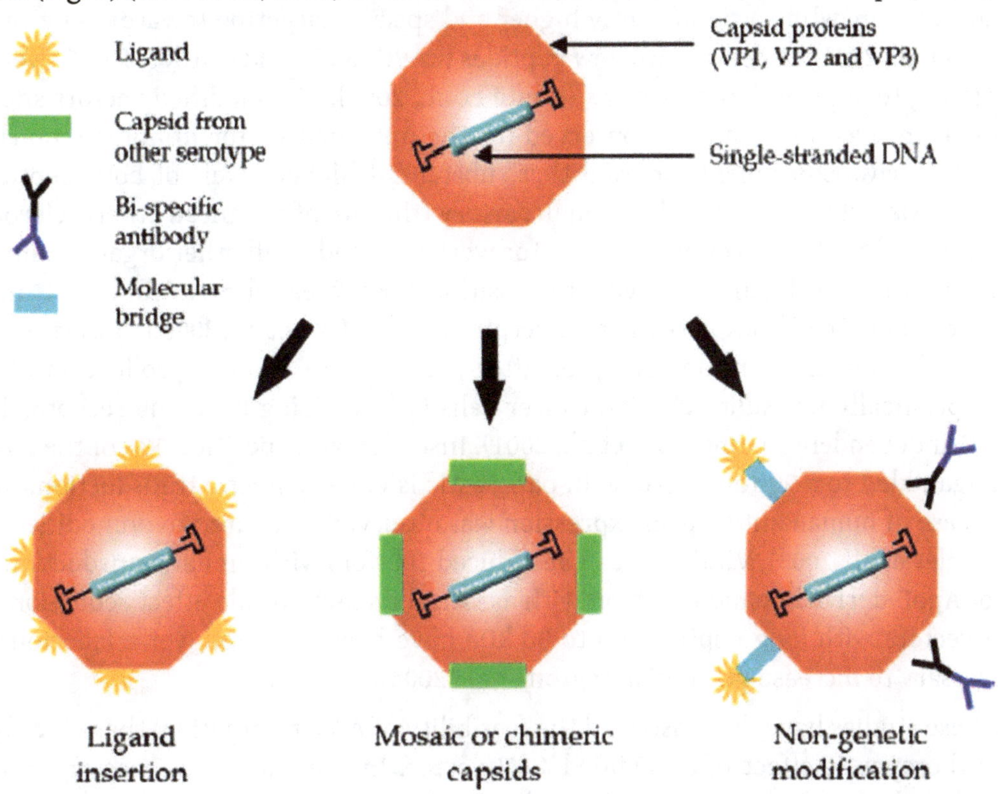

Fig. 3. Strategies for AAV vectors retargeting. AAV vectors can be modified by insertion of ligands into the capsid's proteins (ligand insertion), mixing capsid proteins from different serotypes (mosaic or chimeric capsids), or absorption of molecules onto the capsid using bi- specific antibodies or ligands fused with molecular bridges (non-genetic modification).

4.1 Targeting by ligand insertion

Mutagenesis analysis have permitted the identification of positions within the viral capsid that allow peptide insertion with little or no effect on DNA packaging and virus trafficking. For AAV2-derived vectors, peptides inserted in positions 1, 34, 138, 161, 459, 584, 587 and 588 (relative to VP1 start codon) are displayed on the vector surface, and allow production of vectors with similar viral titers to those observed for unmodified AAV2 vectors (Shi et al., 2001; Wu et al., 2000b). Most of the studies have used positions 138 (VP2 N-terminal), 587 and 588 (HSPG binding domain) to insert peptides ranging from 5 to 272 amino acids (Loiler et al., 2003; Michelfelder et al., 2009; Nicklin et al., 2001a; Perabo et al., 2006; Shi et al., 2001; White et al., 2007; White et al., 2004; Wu et al., 2000b; Yu et al., 2009). Capsid protein modifications have improved gene delivery to lung (Kwon & Schaffer, 2008), endothelial cells (Nicklin et al., 2001a), pancreatic islets (Loiler et al., 2003), vascular tissue (White et al., 2004), atherosclerotic lesions (White et al., 2007), muscle (Yu et al., 2009), myocardium (Yang et al., 2009), and cancer cells (Michelfelder et al., 2009).

Although the therapeutic effect of these vectors still remains to be seen, there is a long list of reports showing different advantages of peptide insertion within different positions of the viral capsid. Insertion after position 588 provided a complete AAV2 retargeting with inhibition of binding to HSPG (Büning et al., 2003; Perabo et al., 2006). AAV vectors designed for vascular tissue targeting were constructed by insertion after position 587 of peptides MSLTTPPAVARP and MTPFPTSNEANL, which showed increased transduction and specificity for venous endothelial cells and reduction of hepatocytes transduction (White et al., 2004). *In-vivo* experiments with these modified AAV vectors demostrated a reduction in liver and kidney transduction, while a significantly higher and specific targeting towards vena cava cells was observed (White et al., 2004). Recently, two new peptides (CAPGPSKSC and CNHRYMQMC) were evaluated for AAV retargeting to atherosclerotic lesions (White et al., 2007). The modified vectors showed higher levels of *in vitro* transduction than those observed for the untagged vector in human, murine, and rat endothelial cells. *In vivo* experiments showed that substantial higher levels of both peptide-modified AAV2 vectors were detected in the brachiocephalic artery (the site of advanced atherosclerotic plaques) and aorta, whereas reduced levels of modified vector were detected in all other organs examined (White et al., 2007). Serpin receptor ligand inserted after position 138 increased by 15-fold the transduction of vector on a lung epithelial cell line expressing the serpin receptor (Wu et al., 2000b). Insertion of a peptide that binds to the luteinizing hormone receptor after position 139 allowed production of a modified vector able to specifically transduce ovarian cancer cells (a luteinizing hormone receptor-bearing cell line) in an HSPG-independent manner (Shi et al., 2001). Insertion after position 138 of the 28-amino acid ApoE-derived ligand led to a 90-fold increase in pancreatic islet cells *in-vitro* transduction. Additionally, a four-fold increase of human antitrypsin expression was observed compared to unmodified AAV vector transduction (Loiler et al., 2003). Viral dose can be reduced 100-fold with similar transduction results as a consequence of ApoE-derived ligand insertion with a critical impact on side effect reduction. Functional AAV-modified vectors with long peptides up to 30 kDa have been reported, yet a laborious producing protocol is necessary to increase yield (Warrington et al., 2004).

Although these studies have demonstrated the feasibility of AAV retargeting, there is only one study evaluating the therapeutic effect of a modified AAV vector. In the study, AAV vectors were designed for brain-endothelial targeting, using peptides identified by phage display (Chen et al., 2009). After intravenous administration of modified vectors in MPS VII mice, there was extensive expression of enzyme in brain, leading to correction of lysosomal storage and reversal of established behavioral deficits (Chen et al., 2009).

During the last years we have worked in the development of a gene therapy strategy for the Mucopolysaccharidosis IV A (Morquio A disease) by using AAV vectors (Alméciga-Díaz et al., 2010; Alméciga-Díaz et al., 2009; Gutierrez et al., 2008). Morquio A disease, as well as other mucopolysaccharidosis, has a marked bone involvement, and an effective therapy should focus on the treatment in these manifestations. Recently, we designed a bone-targeting AAV vector bearing an aspartic acid octapeptide inserted immediately after the N-terminal of the VP2 capsid protein. The vector was designed to interact with hydroxyapatite (HA), a main component of bone (Tomatsu et al., 2010). We observed that the unmodified AAV vector had low bone affinity, while the bone-targeting vector had significantly higher HA affinity with up to 36.6-fold higher vector genome copies in cortical bone compared with the unmodified vector matrix (Alméciga-Díaz CJ, Montaño A, Barrera L, Tomatsu S., unpublished data).

4.2 Chimeric or mosaic capsids

A mosaic capsid AAV is a virion that is composed of a mixture of viral capsid proteins from different serotypes, which are mixed during viral assembly. On the other hand, a chimeric capsid AAV is a vector produced by the insertion of a sequence from another wild-type AAV into the ORF of the capsid gene (Fig. 3) (Choi et al., 2005; Michelfelder & Trepel, 2009). AAV vectors produced by the mixture of capsid proteins of AAV serotypes 1 to 5 led to high-titer viral particles with mixtures of serotype 1, 2, or 3; whereas intermediate titers were observed from AAV5 mixtures (Rabinowitz et al., 2004). Transduction levels varied depending on capsid mixture, producing a synergistic effect in transduction when AAV1 capsids were combined with AAV2 or AAV3 (Rabinowitz et al., 2004). Similar results were observed for a vector produced by the mixture of AAV1 and AAV2 capsid proteins, which exhibited similar titers to those observed for native vectors. The biodistribution profile combined transduction characteristics of both parent vectors (i.e. muscle and liver transduction) (Hauck et al., 2003).

A variant of this strategy uses capsid proteins bearing foreign peptides mixed with wild- type capsids. An AAV2 vector containing wild-type and immunoglobulin-binding Z34C fragment of protein A-bearing capsid proteins; made possible the production of a vector with high virus particle and transduction titers, and the capacity to transduce selectively and efficiently MO7e (human megakaryoblastic leukemia cell line) and Jurkat cells (T lymphocyte cells line) (Gigout et al., 2005). A mosaic vector produced by the mixture of wild-type and integrin-targeted capsid proteins resulted in a vector with a 50 - 100-fold enhancement in endothelial cell gene transfer. This is higher compared to the vector produced only with integrin-targeted capsid (Stachler & Bartlett, 2006).

4.3 Adsorption of receptor ligands

This strategy involves the binding of molecules onto the viral capsid to alter natural tropism of AAV vectors (Fig. 3). Initial approaches used bi-specific antibodies consisting of an anti-AAV antibody cross-linked with another antibody that binds specifically to a cellular receptor (Choi et al., 2005). In the first approach, an anti-AAV antibody was chemically cross-linked to Fab arms of the αIIβ3 integrin binding monoclonal AP-2 antibody, allowing the specific transduction of megakaryocyte cells expressing the integrin (Ponnazhagan et al., 1996).

In a second approach, a high-affinity biotin-avidin interaction was used as a molecular bridge for the adsorption of streptavidin-fused ligands (Ponnazhagan et al., 2002). This strategy did not affect virus production and allowed the vector to be specifically targeted to EGFR- or fibroblast growth factor receptor (FGFR)-positive cells after conjugation of EGF or FGF onto the biotinilated capsid. Recently the exogenous glycation of the viral capsid was presented as a potential alternative to redirect vectors from liver to skeletal and cardiac muscle after systemic administration in mice (Horowitz et al., 2011).

5. NON-VIRAL VECTORS

Non-viral vectors for gene delivery offer several advantages over viral vectors, since these include lack of an immune response, simple synthesis, and low scale-up production cost (Niidome & Huang, 2002). Non-viral vectors may include lipoplexes (DNA-liposome complex), polyplexes (DNA-polymer complex), lipopolyplexes (DNA-liposome-polymer complex) and peptide-based complexes (Douglas, 2008); which can be mixed with physical methods (e.g. electroporation, ultrasound, gene gun or hydrodynamic infusion) to improve the gene delivery (Niidome & Huang, 2002). In addition, most used non-viral vectors do not integrate into the host genome reducing the risk of mutagenic events in the transfected cells (Kim et al., 2010). However, transduction efficiencies are significantly lower than that of viral vectors (Douglas, 2008). The reader is referred to other chapters in this book edition for more thorough treatises of non-viral vectors. Although some physical methods allow gene delivery to a specific cell type, we will only consider some strategies developed to modify vector tropism. It is important to note that although most of these studies have been evaluated exclusively *in-vitro*, they constitute proof-of-concept.

1.1 Lipoplexes

Lipoplexes are the result of a complex formation between cationic liposomes (artificial closed vesicles of lipid bilayer membranes) and genetic material. They can be considered a synthetic means of encapsulating genetic material until it reaches its cellular target (Tros et al., 2010). The main strategies for targeting of lipoplexes involve use of chelator lipids that bind metal-bearing ligands, or the modification with polymers bearing a ligand (Fig. 4).

The insertion of the chelator lipid 3 (nitrilotriacetic acid)-ditetradecylamine (NTA(3)-DTDA) in a lipoplex preparation allowed the engraftment of a His-tagged vascular epidermal growth factor (VEGF) ligand and specific transfection of cells expressing the VEGF receptor (Fig. 4) (Herringson et al., 2009). A similar approach was used to target small iRNAs to B lymphocytes by using a lipoplex bearing the NTA(3)-DTDA chelator lipid engrafted with a His-tagged CD4 or to cells expressing the receptor for tumor necrosis factor alpha (TNFα) by engrafting with a His-tagged TNFα ligand (Herringson & Altin, 2009).

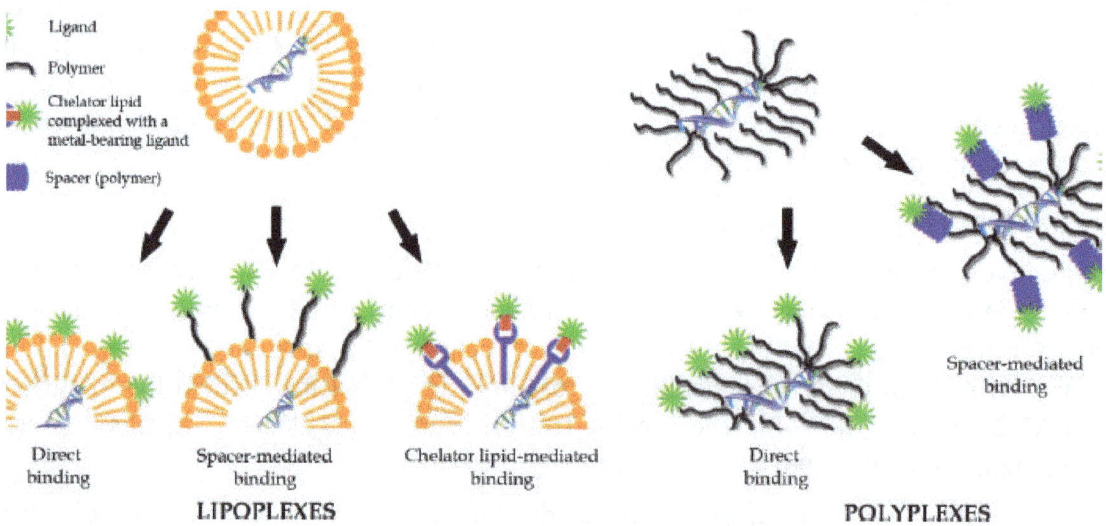

Fig. 4. Modification of lipoplexes and polyplexes. Ligands can be bound to non-viral vectors through direct binding, spacer (polymer)-mediated binding, or chelator lipid-mediated binding (ligand should be fused with a metal).

To improve hepatic specific delivery of a lipoplex after systemic administration, PEG molecules were added to the complex and galactose molecules were covalently affixed at the distal end of the PEG chain to allow an active targeting to the asialoglycoprotein-receptor present on hepatocytes. This resulted in an 18-fold increase in hepatic gene expression (Morille et al., 2009). Inclusion of a PEG conjugated with a nuclear localization signal (NLS) and added to a lipoplex preparation permitted transfection of endothelial cells in the presence of serum. This method was used as a model for transfection of blood-brain barrier cells considered difficult targets for non-modified viral and non-viral vectors (Zhang et al., 2009). A similar strategy was used for targeting of a lipoplex to tumor cells expressing the tumor-associated glycoprotein (TAG)-72 (Kim et al., 2008). In this work, a Fab' fragment from an anti-TAG-72 monoclonal antibody was conjugated with PEG and included within the lipoplex, providing specific gene delivery to TAG-expressing tumor cells both *in-vitro* and *in-vivo*, while non-targeted lipoploxes did not produce the transfection of those cells.

Ligands can be also directly conjugated with liposome molecules (Fig. 4). A hyaluronic acid- liposome conjugate resulted in an increased gene delivery to breast cancer cells with high CD44 levels. Conversely, cells with low CD44 levels were not transfected (Surace et al., 2009). Conjugation of the peptide 4-fluorobenzoyl-RR-(L-3-(2-naphthyl)alanine)-CYEK-(L- citrulline)-PYR-(L-citrulline)-CR with a liposome permitted targeting of a lipoplex to CXCR4-expressing cells, with transfection efficiency depending upon CXCR4 expression levels (Driessen et al., 2008). Conjugation of a liposome with transferrin presented a significant increment in iRNA delivery to hepatocarcinoma cells, with reduced toxicity and enhanced specific gene silencing compared to non-modified lipoplexes (Cardoso et al., 2007). Transferin-conjugated lipoplexes were also used to evaluate specific gene delivery to a metastatic mammary carcinoma cell line, although the results did not demonstrate a significant increase in cytotoxicity with modified lipoplexes in comparison with non- modified lipoplex (Lopez-Barcons et al., 2005).

Another strategy based on *ex-vivo* gene therapy assessed the challenge of nuclear entry by using plasmids transfected with a cationic lipid vector in MSC (Hoare et al., 2010). To enhance gene expression in MSC, they used NLS peptides to direct the plasmid to the nucleus. A significant increase was observed in reporter gene delivery compared to non- modified plasmid. They also noticed the lipid had a protective impact on the plasmid-NLS complex. Differentiation potential was not affected by NLS peptides.

5.2 Polyplexes

Polyplexes are non-viral vectors formed by genetic material coupled with cationic polymers including polylysine (PLL), polyarginine, PEG, polyethylenimine (PEI), polyamidoamine (PAMAM), dendrimers, and chitosan. The mechanism involves interaction between negatively charged DNA or iRNA molecules with positively charged molecules within the polymer (Tros et al., 2010). To target polyplexes to specific cells, they can be modified by incorporation of ligands as small molecules, vitamins, carbohydrates, peptides, or proteins (Wagner et al., 2005).

PEG-derived vectors are widely used as gene therapy tools as a result of their well-known capacity to enhance gene delivery and induce a longer DNA half-life time (Midoux et al., 2008). Due to the possibility to bind ligands in the terminal ends of the polymer (Fig. 4), PEG-derived vectors are commonly used for cell specific gene delivery using non-viral vectors (Hughes & Rao, 2005). PEG polyplexes have been conjugated with a transferrin peptide to target vectors to receptor-expressing cells in two prostate cancer cell lines (Nie et al., 2011). In addition they have been conjugated with EGFR ligand or lactose for specific transfection of human hepatocarcinoma cell lines (Klutz et al., 2011; Oishi et al., 2007).

PEG has also been used as a spacer to bind ligands to polyplexex (Fig. 4), although it has been associated in some cases with reduced transfection efficiencies. A complex of DNA and the HIV-derived TAT protein (which has important translocation abilities due to its strong cell surface adherence) was conjugated with an ICAM-1 ligand by using a PEG spacer allowing an efficient and specific transfection of ICAM 1 expressing cells (Khondee et al., 2011). Similarly, a PEI-derived polyplex was engineered for enhanced and specific delivery of VEGF iRNA in prostate cancer cells. The conjugation with prostate cancer-binding peptide via a PEG spacer, showed a significantly higher VEGF inhibition than with the unmodified PEI polyplex in human prostate carcinoma cells (Kim et al., 2009). Anionic liposomes and cationic polymers do not display an efficient targeting to liver or tumors cells. Thus, a lipopolyplex was constructed with a liposome. A PEI polymer and a PEG polymer bearing a monoclonal antibody permitted an *in-vitro* and *in-vivo* transfection of tumor cells (Hu et al., 2010). A similar strategy was used to target an iRNA to tumor cells expressing the EGFR by using this lipopolyplex bearing the EGFR ligand (Hu et al., 2011).

Additionally, the direct binding of ligands to other polyplexes has been assessed. The transfection of cells that overexpress the folic acid receptor in certain human tumor cells was facilitated by the conjugation of folic acid with the aminomethacrylate-phosphoryl-choline-based copolymer (Lam et al., 2009). The efficient transfection of HeLa cells with low and high densities of $\alpha v\beta 3$ integrin was achieved by a PEI-derived polyplex conjugated with an Arg-Gly-Asp peptide, which binds to the $\alpha v\beta 3$ integrin receptor (Ng et al., 2009). Antibodies can be conjugated with polyplexes to target genetic material to specific cell types, as observed in viral vectors and lipoplexes. A breast cancer cell line has been successfully overexpressed of the human epidermal growth factor receptor-2 (HER2) through transfection with a polyplex conjugated with an anti-HER2 monoclonal antibody (Chiu et al., 2004). An additional method for targeting bone marrow stromal cells in adult MSCs involved the modification of a PLL polymer with a palmitic acid substitution. The transfection efficiency was comparable to that of adenoviral vectors, which paves the way for the efficient transfection of these cells without the adverse effects and cell toxicity that are commonly associated with viral vectors (Incani et al., 2007).

The in-vivo evaluations of these modified polyplexes involve the use of a PEI-derived complex that contains a FGFR ligand for the transfection of tumor cell lines. It demonstrated the in-vitro and in-vivo specific transduction of tumor cells overexpressing the FGFR (Li & Huang, 2006) and the use of a poly[N-(2-hydroxypropyl)methacrylamide]-derived polyplex modified with an anti-P-selectin antibody. Newman et al. (2009) observed in-vivo specific and enhanced gene delivery to inflamed cremasteric venules. A PEI polyplex that contained an antibody against a tumor-specific protein via a PEG spacer also exhibited a substantial retargeting to tumor cells, with low expression in non-tumor cells (Duan et al., 2010).

6. CONCISE SUMMARY

The consolidation of gene therapy as a viable treatment, vector tropism limitations, and adverse effects can be significantly influenced by the modification of gene therapy vectors for specific cell type gene delivery. Regardless of the form of vector, vector modifications can be classified into three fundamental strategies: genetic modification (ligand insertion within the capsid), chemical modification (polymers or ligand cross-linking), or the utilization of adaptor molecules (bi-specific antibodies). The majority of studies have concentrated on the vector design and proof-of-concept of specific gene delivery, while only a small number have assessed the therapeutic impact of these modifications. In that regard, future research should concentrate on evaluating the potential of these modified vectors to enhance gene therapy trials in order to transition from laboratory science to clinical trials. The development of modified vectors for gene delivery to the central nervous system (CNS) and bone should receive particular attention. Given that they exhibit a restricted transfection profile with the vectors that are currently available under normal circumstances. Lastly, parallel research should be conducted in tissue-specific promoters to enhance the processes of vector production, reduce the immune response, and design safer integrating vectors.

REFERENCES

Alméciga-Díaz, C., Montano, A.M., Tomatsu, S. & Barrera, L. (2010). Adeno-associated virus gene transfer on Morquio A: effect of promoters and sulfatase-modifying Factor 1. *FEBS Journal*. Vol. 277, No. 17, (September 2010), pp. 3608-3619, ISSN 1742-464X.

Alméciga-Díaz, C., Rueda-Paramo, M., Espejo, A., Echeverri, O., Montano, A., Tomatsu, S. & Barrera, L. (2009). Effect of Elongation Factor 1α promoter and SUMF1 over in-vitro expression of N-acetylgalactosamine-6-sulfate sulfatase. *Molecular Biology Reports*. Vol. 36, No. 7, (September 2009), pp. 1863-1870, ISSN 0301-4851.

Arnberg, N. (2009). Adenovirus receptors: implications for tropism, treatment and targeting. *Reviews in Medical Virology*. Vol. 19, No. 3, (May 2009), pp. 165-178, ISSN 1052-9276.

Azzouz, M., Ralph, G.S., Storkebaum, E., Walmsley, L.E., Mitrophanous, K.A., Kingsman, S.M., Carmeliet, P. & Mazarakis, N.D. (2004). VEGF delivery with retrogradely transported lentivector prolongs survival in a mouse ALS model. *Nature*. Vol. 429, No. 6990, (May 2004), pp. 413-417, ISSN 0028-0836.

Bachtarzi, H., Stevenson, M. & Fisher, K. (2008). Cancer gene therapy with targeted adenoviruses. *Expert Opinion in Drug Delivery*. Vol. 5, No. 11, (November 2008), pp. 1231-1240, ISSN 1742-5247.

Bakker, A.C., Van de Loo, F.A., Joosten, L.A., Bennink, M.B., Arntz, O.J., Dmitriev, I.P., Kashentsera, E.A., Curiel, D.T. & van den Berg, W.B. (2001). A tropism-modified adenoviral vector increased the effectiveness of gene therapy for arthritis. *Gene Therapy*. Vol. 8, No. 23, (December 2001), pp. 1785-1793, ISSN 0969-7128.

Bischof, D. & Cornetta, K. (2010). Flexibility in cell targeting by pseudotyping lentiviral vectors. *Methods in Molecular Biology*. Vol. 614, (March 2010), pp. 53-68, ISSN 1940-6029.

Büning, H., Ried, M., Perabo, L., Gerner, F., Huttner, N., Enssle, J. & Hallek, M. (2003). Receptor targeting of adeno-associated virus vectors. *Gene Therapy*. Vol. 10, (July 2003), pp. 1142-1151, ISSN 0969-7128.

Campos, S.K. & Barry, M.A. (2007). Current advances and future challenges in Adenoviral vector biology and targeting. *Current Gene Therapy*. Vol. 7, No. 3, (Jun 2007), pp. 189-204, ISSN 1566-5232.

Cardoso, A.L., Simoes, S., de, A.L.P., Pelisek, J., Culmsee, C., Wagner, E. & Pedroso, d.L.M.C. (2007). siRNA delivery by a transferrin-associated lipid-based vector: a non-viral strategy to mediate gene silencing. *Journal of Gene Medicine*. Vol. 9, No. 3, (March 2007), pp. 170-183, ISSN 1099-498X.

Chen, Y.H., Chang, M. & Davidson, B.L. (2009). Molecular signatures of disease brain endothelia provide new sites for CNS-directed enzyme therapy. *Nature Medicine*. Vol. 15, No. 10, (October 2009), pp. 1215-1218, ISSN 1546-170X.

Chiu, S.J., Ueno, N.T. & Lee, R.J. (2004). Tumor-targeted gene delivery via anti-HER2 antibody (trastuzumab, Herceptin) conjugated polyethylenimine. *Journal of Controlled Release*. Vol. 97, No. 2, (June 2004), pp. 357-369, ISSN 0168-3659.

Choi, V., McCarty, D. & Samulski, R.J. (2005). AAV Hybrid Serotypes: Improved Vectors for Gene Delivery. *Current Gene Therapy*. Vol. 5, (June 2005), pp. 299-310, ISSN 1566-5232.

Cosset, F.L., Morling, F.J., Takeuchi, Y., Weiss, R.A., Collins, M.K. & Russell, S.J. (1995). Retroviral retargeting by envelopes expressing an N-terminal binding domain. *Journal of Virology*. Vol. 69, No. 10, (October 1995), pp. 6314-6322, ISSN 0022-538X.

Crompton, J., Toogood, C.I., Wallis, N. & Hay, R.T. (1994). Expression of a foreign epitope on the surface of the adenovirus hexon. *Journal of Gene Medicine*. Vol. 75 (Pt 1), (January 1994), pp. 133-139, ISSN 0022-1317.

Cronin, J., Zhang, X.Y. & Reiser, J. (2005). Altering the tropism of lentiviral vectors through pseudotyping. *Current Gene Therapy*. Vol. 5, No. 4, (August 2005), pp. 387-398, ISSN 1566-5232.

Denby, L., Work, L.M., Von Seggern, D.J., Wu, E., Mcvey, J.H., Nicklin, S.A. & Baker, A.H. (2007). Development of renal-targeted vectors through combined in vivo phage display and capsid engineering of adenoviral fibers from serotype 19p. *Molecular Therapy*. Vol. 15, No. 9, (September 2007), pp. 1647-1654, ISSN 1525-0016.

Dmitriev, I., Krasnykh, V., Miller, C.R., Wang, M., Kashentseva, E., Mikheeva, G., Belousova, N. & Curiel, D.T. (1998). An adenovirus vector with genetically modified fibers demonstrates expanded tropism via utilization of a coxsackievirus and adenovirus receptor-independent cell entry mechanism. *Journal of Virology*. Vol. 72, No. 12, (December 1998), pp. 9706-9713, ISSN 0022-538X.

Douglas, J.T. (2007). Adenoviral vectors for gene therapy. *Molecular Biotechnology*. Vol. 36, No. 1, (May 2007), pp. 71-80, ISSN 1073-6085.

Douglas, K.L. (2008). Toward development of artificial viruses for gene therapy: a comparative evaluation of viral and non-viral transfection. *Biotechnology Progress*. Vol. 24, No. 4, (July 2008), pp. 871-883, ISSN 1520-6033.

Driessen, W.H., Fujii, N., Tamamura, H. & Sullivan, S.M. (2008). Development of peptide- targeted lipoplexes to CXCR4-expressing rat glioma cells and rat proliferating endothelial cells. *Molecular Therapy*. Vol. 16, No. 3, (March 2008), pp. 516-524, ISSN 1525-0024.

Duan, Y., Yang, C., Zhang, Z., Liu, J., Zheng, J. & Kong, D. (2010). Poly(ethylene glycol)- grafted polyethylenimine modified with G250 monoclonal antibody for tumor gene therapy. *Human Gene Therapy*. Vol. 21, No. 2, (February 2010), pp. 191-198, ISSN 1557-7422.

Edelstein, M. (June 2010). Vectors used in gene therapy clinical trials, In: *Gene Therapy Clinical Trials Worldwide*, Febrary 2011, Available from: http://www.wiley.com.uk/genetherapy/clinical/.

Escors, D. & Breckpot, K. (2010). Lentiviral Vectors in Gene Therapy: Their Current Status and Future Potential. *Archivum Immunologiae et Therapiae Experimentalis*, (Feb 2010), ISSN 1661-4917.

Eto, Y., Gao, J.Q., Sekiguchi, F., Kurachi, S., Katayama, K., Maeda, M., Kawasaki, K., Mizuguchi, H., Hayakawa, T., Tsutsumi, Y., Mayumi, T. & Nakagawa, S. (2005). PEGylated adenovirus vectors containing RGD peptides on the tip of PEG show high transduction efficiency and antibody evasion ability. *Journal of Gene Medicine*. Vol. 7, No. 5, (May 2005), pp. 604-612, ISSN 1099-498X.

Fielding, A.K., Maurice, M., Morling, F.J., Cosset, F.L. & Russell, S.J. (1998). Inverse targeting of retroviral vectors: selective gene transfer in a mixed population of hematopoietic and nonhematopoietic cells. *Blood*. Vol. 91, No. 5, (March 1998), pp. 1802-1809, ISSN 0006-4971.

Fisher, K.D., Stallwood, Y., Green, N.K., Ulbrich, K., Mautner, V. & Seymour, L.W. (2001). Polymer-coated adenovirus permits efficient retargeting and evades neutralising antibodies. *Gene Therapy*. Vol. 8, No. 5, (March 2001), pp. 341-348, ISSN 0969-7128.

Flotte, T. (2004). Gene Therapy Progress and Prospects: Recombinant adeno-associated virus (rAAV) vectors. *Gene Therapy*. Vol. 11, (May 2004), pp. 805-810, ISSN 0969-7128.

Frecha, C., Szecsi, J., Cosset, F.L. & Verhoeyen, E. (2008). Strategies for targeting lentiviral vectors. *Current Gene Therapy*. Vol. 8, No. 6, (December 2008), pp. 449-460, ISSN 1566-5232.

Friedmann, T. (1992). A brief history of gene therapy. *Nature Genetics*. Vol. 2, (October 1992), pp. 93-98, ISSN 1061-4036.

Funke, S., Maisner, A., Muhlebach, M.D., Koehl, U., Grez, M., Cattaneo, R., Cichutek, K. & Buchholz, C.J. (2008). Targeted cell entry of lentiviral vectors. *Molecular Therapy*. Vol. 16, No. 8, (August 2008), pp. 1427-1436, ISSN 1525-0016.

Gigout, L., Rebollo, P., Clement, N., Jr., K.H.W., Muzyczka, N., Linden, R.M. & Weber, T. (2005). Altering AAV Tropism with Mosaic Viral Capsids. *Molecular Therapy*. Vol. 11, No. 6, (June 2005), pp. 856-865, ISSN 1525-0016.

Govindasamy, L., Padron, E., McKenna, R., Muzyczka, N., Kaludov, N., Chiorini, J.A. & Agbandje-McKenna, M. (2006). Structurally mapping the diverse phenotype of adeno-associated virus serotype 4. *Journal of Virology*. Vol. 80, No. 23, (December 2006), pp. 11556-11570, ISSN 0022-538X.

Grill, J., Van Beusechem, V.W., Van Der Valk, P., Dirven, C.M., Leonhart, A., Pherai, D.S., Haisma, H.J., Pinedo, H.M., Curiel, D.T. & Gerritsen, W.R. (2001). Combined targeting of adenoviruses to integrins and epidermal growth factor receptors increases gene transfer into primary glioma cells and spheroids. *Clinical Cancer Research*. Vol. 7, No. 3, (March 2001), pp. 641-650, ISSN 1078-0432.

Gutierrez, M., Garcia-Vallejo, F., Tomatsu, S., Ceron, F., Alméciga-Díaz, C., Domínguez, M. & Barrera, L. (2008). Construction of an Adenoassociated virus-derived vector for the treatment of Morquio A disease. *Biomedica*. Vol. 28, No. 3, (September 2008), pp. 448-459, ISSN 0120-5147.

Hauck, B., Chen, L. & Xiao, W. (2003). Generation and characterization of chimeric recombinant AAV vectors. *Molecular Therapy*. Vol. 7, No. 3, (March 2003), pp. 419-425, ISSN 1525-0016.

Heideman, D.A., Snijders, P.J., Craanen, M.E., Bloemena, E., Meijer, C.J., Meuwissen, S.G., van Beusechem, V.W., Pinedo, H.M., Curiel, D.T., Haisma, H.J. & Gerritsen, W.R. (2001). Selective gene delivery toward gastric and esophageal adenocarcinoma cells via EpCAM-targeted adenoviral vectors. *Cancer Gene Therapy*. Vol. 8, No. 5, (May 2001), pp. 342-351, ISSN 0929-1903.

Herringson, T.P. & Altin, J.G. (2009). Convenient targeting of stealth siRNA-lipoplexes to cells with chelator lipid-anchored molecules. *Journal of Controlled Release*. Vol. 139, No. 3, (November 2009), pp. 229-238, ISSN 1873-4995.

Herringson, T.P., Patlolla, R.R. & Altin, J.G. (2009). Targeting of plasmid DNA-lipoplexes to cells with molecules anchored via a metal chelator lipid. *The Journal of Gene Medicine*. Vol. 11, No. 11, (November 2009), pp. 1048-1063, ISSN 1521-2254.

Hoare, M., Greiser, U., Schu, S., Mashayekhi, K., Aydogan, E., Murphy, M., Barry, F., Ritter, T. & O'Brien, T. (2010). Enhanced lipoplex-mediated gene expression in mesenchymal stem cells using reiterated nuclear localization sequence peptides. *Journal of Gene Medicine*. Vol. 12, No. 2, (February 2010), pp. 207-218, ISSN 1521- 2254.

Horowitz, E.D., Weinberg, M.S. & Asokan, A. (2011). Glycated AAV Vectors: Chemical Redirection of Viral Tissue Tropism. *Bioconjugate Chemistry*, (March 2011), ISSN 1520-4812.

Hu, Y., Li, K., Wang, L., Yin, S., Zhang, Z. & Zhang, Y. (2010). Pegylated immuno- lipopolyplexes: A novel non-viral gene delivery system for liver cancer therapy. *Journal of Controlled Release*. Vol. 144, No. 1, (May 2010), pp. 75-81, ISSN 1873-4995.

Hu, Y., Shen, Y., Ji, B., Wang, L., Zhang, Z. & Zhang, Y. (2011). Combinational RNAi gene therapy of hepatocellular carcinoma by targeting human EGFR and TERT. *European Journal of Pharmaceutical Sciences*. Vol. 42, No. 4, (March 2011), pp. 387-391, ISSN 1879-0720.

Hughes, J.A. & Rao, G.A. (2005). Targeted polymers for gene delivery. *Expert opinion on drug delivery*. Vol. 2, No. 1, (January 2005), pp. 145-157, ISSN 1742-5247.

Incani, V., Tunis, E., Clements, B.A., Olson, C., Kucharski, C., Lavasanifar, A. & Uludag, H. (2007). Palmitic acid substitution on cationic polymers for effective delivery of plasmid DNA to bone marrow stromal cells. *Journal of Biomedical Materials Research A*. Vol. 81, No. 2, (May 2007), pp. 493-504, ISSN 1549-3296.

Jung, Y., Park, H.J., Kim, P.H., Lee, J., Hyung, W., Yang, J., Ko, H., Sohn, J.H., Kim, J.H., Huh, Y.M., Yun, C.O. & Haam, S. (2007). Retargeting of adenoviral gene delivery via Herceptin-PEG-adenovirus conjugates to breast cancer cells. *Journal of Controlled Release*. Vol. 123, No. 2, (November 2007), pp. 164-171, ISSN 1873-4995.

Kasahara, N., Dozy, A.M. & Kan, Y.W. (1994). Tissue-specific targeting of retroviral vectors through ligand-receptor interactions. *Science*. Vol. 266, No. 5189, (November 1994), pp. 1373-1376, ISSN 0036-8075.

Katane, M., Takao, E., Kubo, Y., Fujita, R. & Amanuma, H. (2002). Factors affecting the direct targeting of murine leukemia virus vectors containing peptide ligands in the envelope protein. *EMBO Reports*. Vol. 3, No. 9, (September 2002), pp. 899-904, ISSN 1469-221X.

Kayman, S.C., Park, H., Saxon, M. & Pinter, A. (1999). The hypervariable domain of the murine leukemia virus surface protein tolerates large insertions and deletions, enabling development of a retroviral particle display system. *Journal of Virology*. Vol. 73, No. 3, (March 1999), pp. 1802-1808, ISSN 0022-538X.

Khondee, S., Baoum, A., Siahaan, T.J. & Berkland, C.J. (2011). Calcium-crosslinked LABL- TAT complexes effectively target gene delivery to ICAM-1 expressing cells. *Molecular Pharmaceutics*. Vol. In Press, (April 2011), ISSN 1543-8392.

Kim, K.S., Lee, Y.K., Kim, J.S., Koo, K.H., Hong, H.J. & Park, Y.S. (2008). Targeted gene therapy of LS174 T human colon carcinoma by anti-TAG-72 immunoliposomes. *Cancer Gene Therapy*. Vol. 15, No. 5, (May 2008), pp. 331-340, ISSN 1476-5500.

Kim, N.Y., Choi, Y.B., Kang, C.I., Kim, H.H., Yang, J.M. & Shin, S. (2010). An hydrophobically modified arginine peptide vector system effectively delivers DNA into human mesenchymal stem cells and maintains transgene expression with differentiation. *Journal of Gene Medicine*. Vol. 12, No. 9, (September 2010), pp. 779- 789, ISSN 1521-2254.

Kim, P.H., Sohn, J.H., Choi, J.W., Jung, Y., Kim, S.W., Haam, S. & Yun, C.O. (2011). Active targeting and safety profile of PEG-modified adenovirus conjugated with herceptin. *Biomaterials*. Vol. 32, No. 9, (March 2011), pp. 2314-2326, ISSN 1878-5905

Kim, S.H., Lee, S.H., Tian, H., Chen, X. & Park, T.G. (2009). Prostate cancer cell specific VEGF siRNA delivery system using cell targeting peptide conjugated polyplexes. *Journal of Drug Targeting*. Vol. 17, No. 4, (May 2009), pp. 311-317, ISSN 1029-2330.

Klutz, K., Schaffert, D., Willhauck, M.J., Grunwald, G.K., Haase, R., Wunderlich, N., Zach, C., Gildehaus, F.J., Senekowitsch-Schmidtke, R., Goke, B., Wagner, E., Ogris, M. & Spitzweg, C. (2011). Epidermal Growth Factor Receptor-targeted (131)I-therapy of Liver Cancer Following Systemic Delivery of the Sodium Iodide Symporter Gene. *Molecular Therapy*. Vol. 19, No. 4, (April 2011), pp. 676-685, ISSN 1525-0024.

Kobinger, G.P., Weiner, D.J., Yu, Q.C. & Wilson, J.M. (2001). Filovirus-pseudotyped lentiviral vector can efficiently and stably transduce airway epithelia in vivo. *Nature Biotechnology*. Vol. 19, No. 3, (March 2001), pp. 225-230, ISSN 1087-0156.

Krasnykh, V., Dmitriev, I., Mikheeva, G., Miller, C.R., Belousova, N. & Curiel, D.T. (1998). Characterization of an adenovirus vector containing a heterologous peptide epitope in the HI loop of the fiber knob. *Journal of Virology*. Vol. 72, No. 3, (March 1998), pp. 1844-1852, ISSN 0022-538X.

Kwon, I. & Schaffer, D.V. (2008). Designer gene delivery vectors: molecular engineering and evolution of adeno-associated viral vectors for enhanced gene transfer. *Pharmaceutical Research*. Vol. 25, No. 3, (March 2008), pp. 489-499, ISSN 0724-8741.

Lam, J.K., Armes, S.P., Lewis, A.L. & Stolnik, S. (2009). Folate conjugated phosphorylcholine-based polycations for specific targeting in nucleic acids delivery. *Journal of Drug Targeting*. Vol. 17, No. 7, (August 2009), pp. 512-523, ISSN 1029-2330.

Lavillette, D., Russell, S.J. & Cosset, F.L. (2001). Retargeting gene delivery using surface- engineered retroviral vector particles. *Current Opinion in Biotechnology*. Vol. 12, No. 5, (October 2001), pp. 461-466, ISSN 0958-1669.

Leissner, P., Legrand, V., Schlesinger, Y., Hadji, D.A., van Raaij, M., Cusack, S., Pavirani, A. & Mehtali, M. (2001). Influence of adenoviral fiber mutations on viral encapsidation, infectivity and in vivo tropism. *Gene Therapy*. Vol. 8, No. 1, (Jannuary 2001), pp. 49-57, ISSN 0969-7128.

Lerch, T.F., Xie, Q. & Chapman, M.S. (2010). The structure of adeno-associated virus serotype 3B (AAV-3B): Insights into receptor binding and immune evasion. *Virology*. Vol. 03, No. 1, (July 2010), pp. 26-36, ISSN 1096-0341.

Li, H.J., Everts, M., Pereboeva, L., Komarova, S., Idan, A., Curiel, D.T. & Herschman, H.R. (2007). Adenovirus tumor targeting and hepatic untargeting by a coxsackie/adenovirus receptor ectodomain anti-carcinoembryonic antigen bispecific adapter. *Cancer Research*. Vol. 67, No. 11, (Jun 2007), pp. 5354-5361, ISSN 0008-5472.

Li, S.D. & Huang, L. (2006). Gene therapy progress and prospects: non-viral gene therapy by systemic delivery. *Gene Therapy*. Vol. 13, No. 18, (September 2006), pp. 1313-1319, ISSN 0969-7128.

Liang, M., Morizono, K., Pariente, N., Kamata, M., Lee, B. & Chen, I.S. (2009). Targeted transduction via CD4 by a lentiviral vector uses a clathrin-mediated entry pathway. *Journal of Virology*. Vol. 83, No. 24, (December 2009), pp. 13026-13031, ISSN 1098-5514.

Loiler, S., Conlon, T., Song, S., Tang, Q., Jr, K.W., Agarwal, A., Kapturczak, M., Li, C., Ricordi, C., Atkinson, M., Muzyczka, N. & Flotte, T. (2003). Targeting recombinant adeno-associated virus vectors to enhance gene transfer to pancreatic islets and liver. *Gene Therapy*. Vol. 10, (September 2003), pp. 1551-1558, ISSN 0969-7128.

Lopez-Barcons, L.A., Polo, D., Llorens, A., Reig, F. & Fabra, A. (2005). Targeted adriamycin delivery to MXT-B2 metastatic mammary carcinoma cells by transferrin liposomes: effect of adriamycin ADR-to-lipid ratio. *Oncology Reports*. Vol. 14, No. 5, (November 2005), pp. 1337-1343, ISSN 1021-335X.

Mazarakis, N.D., Azzouz, M., Rohll, J.B., Ellard, F.M., Wilkes, F.J., Olsen, A.L., Carter, E.E., Barber, R.D., Baban, D.F., Kingsman, S.M., Kingsman, A.J., O'Malley, K. & Mitrophanous, K.A. (2001). Rabies virus glycoprotein pseudotyping of lentiviral vectors enables retrograde axonal transport and access to the nervous system after peripheral delivery. *Human Molecular Genetics*. Vol. 10, No. 19, (September 2001), pp. 2109-2121, ISSN 0964-6906.

Merten, C.A., Engelstaedter, M., Buchholz, C.J. & Cichutek, K. (2003). Displaying epidermal growth factor on spleen necrosis virus-derived targeting vectors. *Virology*. Vol. 305, No. 1, (January 2003), pp. 106-114, ISSN 0042-6822.

Michelfelder, S., Kohlschutter, J., Skorupa, A., Pfennings, S., Muller, O., Kleinschmidt, J.A. & Trepel, M. (2009). Successful expansion but not complete restriction of tropism of adeno-associated virus by *in vivo* biopanning of random virus display peptide libraries. *PLoS One*. Vol. 4, No. 4, (April 2009), pp. e5122, ISSN 1932-6203.

Michelfelder, S. & Trepel, M. (2009). Adeno-associated viral vectors and their redirection to cell-type specific receptors. *Advances in Genetics*. Vol. 67, No. Pt A, (November 2009), pp. 29-60, ISSN 0065-2660.

Midoux, P., Breuzard, G., Gomez, J.P. & Pichon, C. (2008). Polymer-based gene delivery: a current review on the uptake and intracellular trafficking of polyplexes. *Current Gene Therapy*. Vol. 8, No. 5, (Oct), pp. 335-352, ISSN 1566-5232.

Miletic, H., Fischer, Y.H., Neumann, H., Hans, V., Stenzel, W., Giroglou, T., Hermann, M., Deckert, M. & Von Laer, D. (2004). Selective transduction of malignant glioma by lentiviral vectors pseudotyped with lymphocytic choriomeningitis virus glycoproteins. *Human Gene Therapy*. Vol. 15, No. 11, (November 2004), pp. 1091-1100, ISSN 1043-0342.

Mitchell, M., Nam, H.J., Carter, A., McCall, A., Rence, C., Bennett, A., Gurda, B., McKenna, R., Porter, M., Sakai, Y., Byrne, B.J., Muzyczka, N., Aslanidi, G., Zolotukhin, S. & Agbandje-McKenna, M. (2009). Production, purification and preliminary X-ray crystallographic studies of adeno-associated virus serotype 9. *Acta Crystallographica Section F: Structural Biology and Crystallization Communications*. Vol. 65, No. Pt 7, (July 2009), pp. 715-718, ISSN 1744-3091.

Miura, Y., Yoshida, K., Nishimoto, T., Hatanaka, K., Ohnami, S., Asaka, M., Douglas, J.T., Curiel, D.T., Yoshida, T. & Aoki, K. (2007). Direct selection of targeted adenovirus vectors by random peptide display on the fiber knob. *Gene Therapy*. Vol. 14, No. 20, (October 2007), pp. 1448-1460, ISSN 0969-7128.

Morille, M., Passirani, C., Letrou-Bonneval, E., Benoit, J.P. & Pitard, B. (2009). Galactosylated DNA lipid nanocapsules for efficient hepatocyte targeting. *International Journal of Pharmaceutics*. Vol. 379, No. 2, (September 2009), pp. 293-300, ISSN 1873-3476.

Nam, H.-J., Douglas, M., Padron, E., Gurda, B., McKenna, R., Kohlbrenner, E., Aslanidi, G., Byrne, B., Muzyczka, N., Zolotukhin, S. & Agbandje-McKenna, M. (2007). Structure of Adeno-Associated virus serotype 8, a gene therapy vector. *Journal of Virology*. Vol. 81, No. 22, (August 2007), pp. 12260-12271, ISSN 0022-538X.

Newman, C.M., Crosdale, D.J., Fisher, K.D., Briggs, S.S., Norman, K.E., Seymour, L.W. & Hellewell, P.G. (2009). P-selectin dependent targeting to inflamed endothelium of recombinant P-selectin glycoprotein ligand-1 immunoglobulin chimera-coated poly[N-(2-hydroxypropyl) methacrylamide]-DNA polyplexes in vivo visualised by intravital microscopy. *Journal of Gene Medicine*. Vol. 11, No. 4, (April 2009), pp. 326- 334, ISSN 1521-2254.

Ng, Q.K., Sutton, M.K., Soonsawad, P., Xing, L., Cheng, H. & Segura, T. (2009). Engineering clustered ligand binding into nonviral vectors: alphavbeta3 targeting as an example. *Molecular Therapy*. Vol. 17, No. 5, (May 2009), pp. 828-836, ISSN 1525-0024.

Nicklin, S.A., Buening, H., Dishart, K.L., Alwis, M.d., Girod, A., Hacker, U., Thrasher, A.J., Ali, R.R., Hallek, M. & Baker1, A.H. (2001a). Efficient and Selective AAV2- Mediated Gene Transfer Directed to Human Vascular Endothelial Cells. *Molecular Therapy*. Vol. 4, (September 2001), pp. 174-181, ISSN 1525-0016.

Nicklin, S.A., Von Seggern, D.J., Work, L.M., Pek, D.C., Dominiczak, A.F., Nemerow, G.R. & Baker, A.H. (2001b). Ablating adenovirus type 5 fiber-CAR binding and HI loop insertion of the SIGYPLP peptide generate an endothelial cell-selective adenovirus. *Molecular Therapy*. Vol. 4, No. 6, (December 2001), pp. 534-542, ISSN 1525-0016.

Nie, Y., Schaffert, D., Rodl, W., Ogris, M., Wagner, E. & Gunther, M. (2011). Dual-targeted polyplexes: One step towards a synthetic virus for cancer gene therapy. *Journal of Controlled Release*. Vol. In press, (March 2011), ISSN 1873-4995.

Niidome, T. & Huang, L. (2002). Gene therapy progress and prospects: nonviral vectors. *Gene Therapy*. Vol. 9, No. 24, (December 2002), pp. 1647-1652, ISSN 0969-7128.

Ogawara, K., Kuldo, J.M., Oosterhuis, K., Kroesen, B.J., Rots, M.G., Trautwein, C., Kimura, T., Haisma, H.J. & Molema, G. (2006). Functional inhibition of NF-kappaB signal transduction in alphavbeta3 integrin expressing endothelial cells by using RGD- PEG-modified adenovirus with a mutant IkappaB gene. *Arthritis Research & Therapy*. Vol. 8, No. 1, (January 2006), pp. R32, ISSN 1478-6362.

Ohno, K., Sawai, K., Iijima, Y., Levin, B. & Meruelo, D. (1997). Cell-specific targeting of Sindbis virus vectors displaying IgG-binding domains of protein A. *Nature Biotechnology*. Vol. 15, No. 8, (August 1997), pp. 763-767, ISSN 1087-0156.

Oishi, M., Nagasaki, Y., Nishiyama, N., Itaka, K., Takagi, M., Shimamoto, A., Furuichi, Y. & Kataoka, K. (2007). Enhanced growth inhibition of hepatic multicellular tumor spheroids by lactosylated poly(ethylene glycol)-siRNA conjugate formulated in PEGylated polyplexes. *ChemMedChem*. Vol. 2, No. 9, (September 2007), pp. 1290-1297, ISSN 1860-7187.

Park, J.W., Mok, H. & Park, T.G. (2010). Physical adsorption of PEG grafted and blocked poly-L-lysine copolymers on adenovirus surface for enhanced gene transduction. *Journal of Controlled Release*. Vol. 142, No. 2, (March 2010), pp. 238-244, ISSN 1873-4995.

Peng, K.W., Morling, F.J., Cosset, F.L., Murphy, G. & Russell, S.J. (1997). A gene delivery system activatable by disease-associated matrix metalloproteinases. *Human Gene Therapy*. Vol. 8, No. 6, (April 1997), pp. 729-738, ISSN 1043-0342.

Peng, K.W., Pham, L., Ye, H., Zufferey, R., Trono, D., Cosset, F.L. & Russell, S.J. (2001). Organ distribution of gene expression after intravenous infusion of targeted and untargeted lentiviral vectors. *Gene Therapy*. Vol. 8, No. 19, (October 2001), pp. 1456-1463, ISSN 0969-7128.

Peng, Z. (2005). Current status of gendicine in China: recombinant human Ad-p53 agent for treatment of cancers. *Human Gene Therapy*. Vol. 16, No. 9, (September 2005), pp. 1016-1027, ISSN 1043-0342.

Perabo, L., Goldnau, D., White, K., Endell, J., Boucas, J., Humme, S., Work, L.M., Janicki, H., Hallek, M., Baker, A.H. & Büning, H. (2006). Heparan Sulfate Proteoglycan Binding Properties of Adeno-Associated Virus Retargeting Mutants and Consequences for Their *In Vivo* Tropism. *Journal of Virology*. Vol. 80, No. 14, (July 2006), pp. 7265- 7269, ISSN 0022-538X.

Pizzato, M., Blair, E.D., Fling, M., Kopf, J., Tomassetti, A., Weiss, R.A. & Takeuchi, Y. (2001). Evidence for nonspecific adsorption of targeted retrovirus vector particles to cells. *Gene Therapy*. Vol. 8, No. 14, (July 2001), pp. 1088-1096, ISSN 0969-7128.

Ponnazhagan, S., Mahendra, G., Kumar, S., Thompson, J.A. & Castillas, M. (2002). Conjugate-based targeting of recombinant adeno-associated virus type 2 vectors by using avidin-linked ligands. *Journal of Virology*. Vol. 76, No. 24, (December 2002), pp. 12900-12907, ISSN 0022-538X.

Ponnazhagan, S., Wang, X.S., Woody, M.J., Luo, F., Kang, L.Y., Nallari, M.L., Munshi, N.C., Zhou, S.Z. & Srivastava, A. (1996). Differential expression in human cells from the p6 promoter of human parvovirus B19 following plasmid transfection and recombinant adeno-associated virus 2 (AAV) infection: human megakaryocytic leukaemia cells are non-permissive for AAV infection. *The Journal of general virology*. Vol. 77 (Pt 6), (Jun 1996), pp. 1111-1122, ISSN 0022-1317.

Poulin, K.L., Lanthier, R.M., Smith, A.C., Christou, C., Risco Quiroz, M., Powell, K.L., O'Meara, R.W., Kothary, R., Lorimer, I.A. & Parks, R.J. (2010). Retargeting of adenovirus vectors through genetic fusion of a single-chain or single-domain antibody to capsid protein IX. *Journal of Virology*. Vol. 84, No. 19, (October 2010), pp. 10074-10086, ISSN 1098-5514.

Prill, J.M., Espenlaub, S., Samen, U., Engler, T., Schmidt, E., Vetrini, F., Rosewell, A., Grove, N., Palmer, D., Ng, P., Kochanek, S. & Kreppel, F. (2010). Modifications of Adenovirus Hexon Allow for Either Hepatocyte Detargeting or Targeting With Potential Evasion From Kupffer Cells. *Molecular Therapy*, (October 2010), ISSN 1525- 0024.

Rabinowitz, J.E., Bowles, D.E., Faust, S.M., Ledford, J.G., Cunningham, S.E. & Samulski, R.J. (2004). Cross-dressing the virion: the transcapsidation of adeno-associated virus serotypes functionally defines subgroups. *Journal of Virology*. Vol. 78, No. 9, (May 2004), pp. 4421-4432, ISSN 0022-538X.

Ranki, T., Kanerva, A., Ristimaki, A., Hakkarainen, T., Sarkioja, M., Kangasniemi, L., Raki, M., Laakkonen, P., Goodison, S. & Hemminki, A. (2007). A heparan sulfate-targeted conditionally replicative adenovirus, Ad5.pk7-Delta24, for the treatment of advanced breast cancer. *Gene Therapy*. Vol. 14, No. 1, (January 2006), pp. 58-67, ISSN 0969-7128.

Reddy, J.A., Clapp, D.W. & Low, P.S. (2001). Retargeting of viral vectors to the folate receptor endocytic pathway. *Journal of Controlled Release*. Vol. 74, No. 1-3, (July 2001), pp. 77-82, ISSN 0168-3659.

Rittner, K., Schreiber, V., Erbs, P. & Lusky, M. (2007). Targeting of adenovirus vectors carrying a tumor cell-specific peptide: *in vitro* and *in vivo* studies. *Cancer Gene Therapy*. Vol. 14, No. 5, (May 2007), pp. 509-518, ISSN 0929-1903.

Roux, P., Jeanteur, P. & Piechaczyk, M. (1989). A versatile and potentially general approach to the targeting of specific cell types by retroviruses: application to the infection of human cells by means of major histocompatibility complex class I and class II antigens by mouse ecotropic murine leukemia virus-derived viruses. *Proceedings of the National Academy of Sciences of the United States of America*. Vol. 86, No. 23, (December 1989), pp. 9079-9083, ISSN 0027-8424.

Russell, S.J., Hawkins, R.E. & Winter, G. (1993). Retroviral vectors displaying functional antibody fragments. *Nucleic Acids Research*. Vol. 21, No. 5, (March 1993), pp. 1081-1085, ISSN 0305-1048.

Schmidt, A., Haas, S.J., Hildebrandt, S., Scheibe, J., Eckhoff, B., Racek, T., Kempermann, G., Wree, A. & Putzer, B.M. (2007). Selective targeting of adenoviral vectors to neural precursor cells in the hippocampus of adult mice: new prospects for in situ gene therapy. *Stem Cells*. Vol. 25, No. 11, (November 2007), pp. 2910-2918, ISSN 1549-4918.

Sharma, A., Li, X.X., Bangari, D.S. & Mittal, S.K. (2009). Adenovirus receptors and their implications in gene delivery. *Virus Research*. Vol. 143, No. 2, (August 2009), pp. 184-194, ISSN 0168-1702.

Shi, W., Arnold, G. & Bartlett, J. (2001). Insertional Mutagenesis of the Adeno-Associated Virus Type 2 (AAV2) Capsid Gene and Generation of AAV2 Vectors Targeted to Alternative Cell-Surface Receptors. *Human Gene Therapy*. Vol. 12, (September 2001), pp. 1697-1711, ISSN 1043-0342.

Stachler, M.D. & Bartlett, J.S. (2006). Mosaic vectors comprised of modified AAV1 capsid proteins for efficient vector purification and targeting to vascular endothelial cells. *Gene Therapy*. Vol. 13, No. 11, (June 2006), pp. 926-931, ISSN 0969-7128.

Summerford, C. & Samulski, R.J. (1998). Membrane-associated heparan sulfate proteoglycan is a receptor for adeno-associated virus type 2 virions. *Journal of Virology*. Vol. 72, No. 2, (Febrary 1998), pp. 1438-1445, ISSN 0022-538X.

Surace, C., Arpicco, S., Dufay-Wojcicki, A., Marsaud, V., Bouclier, C., Clay, D., Cattel, L., Renoir, J.M. & Fattal, E. (2009). Lipoplexes targeting the CD44 hyaluronic acid receptor for efficient transfection of breast cancer cells. *Molecular Pharmaceutics*. Vol. 6, No. 4, (July 2009), pp. 1062-1073, ISSN 1543-8384.

Tomatsu, S., Montano, A.M., Alméciga-Díaz, C. & Barrera, L. (2010). Delivery of therapeutic agents to the bone. US20100008979. USA Tros, d.I.C., Sun, Y. & Duzgunes, N. (2010). Gene delivery by lipoplexes and polyplexes. *European Journal of Pharmaceutical Sciences*. Vol. 40, No. 3, (June 2010), pp. 159-170, ISSN 1879-0720.

Verhoeyen, E. & Cosset, F.L. (2004). Surface-engineering of lentiviral vectors. *Journal of Gene Medicine*. Vol. 6 (Suppl 1), (Febrary 2004), pp. S83-94, ISSN 1099-498X.

Verma, I.M. & Weitzman, M.D. (2005). Gene Therapy: Twenty-First Century Medicine. *Annual Reviews in Biochemistry*. Vol. 74, (July 2005), pp. 711-738, ISSN 1545-4509.

Volpers, C. & Kochanek, S. (2004). Adenoviral vectors for gene transfer and therapy. *Journal of Gene Medicine*. Vol. 6 (Suppl 1), (Febrary 2004), pp. S164-171, ISSN 1099-498X.

Wagner, E., Culmsee, C. & Boeckle, S. (2005). Targeting of polyplexes: toward synthetic virus vector systems. *Advances in Genetics*. Vol. 53, pp. 333-354, ISSN 0065-2660.

Warrington, K. & Herzog, R. (2006). Treatmeant of human disease by adeno-associated viral gene transfer. *Human Genetics*. Vol. 119, (April 2006), pp. 571-603, ISSN 0340-6717.

Warrington, K.H., Gorbatyuk, O.S., Harrison, J.K., Opie, S.R., Zolotukhin, S. & Muzyczka, N. (2004). Adeno-Associated Virus Type 2 VP2 Capsid Protein Is Nonessential and Can Tolerate Large Peptide Insertions at Its N Terminus. *Journal of Virology*. Vol. 78, No. 12, (January 2004), pp. 6595-6609, ISSN 0022-538X.

Wesseling, J.G., Bosma, P.J., Krasnykh, V., Kashentseva, E.A., Blackwell, J.L., Reynolds, P.N., Li, H., Parameshwar, M., Vickers, S.M., Jaffee, E.M., Huibregtse, K., Curiel, D.T. & Dmitriev, I. (2001). Improved gene transfer efficiency to primary and established human pancreatic carcinoma target cells via epidermal growth factor receptor and integrin-targeted adenoviral vectors. *Gene Therapy*. Vol. 8, No. 13, (July 2001), pp. 969-976, ISSN 0969-7128.

White, K., Büning, H., Kritz, A., Janicki, H., McVey, J., Perabo, L., Murphy, G., Odenthal, M., Work, L., Hallek, M., Nicklin, S. & Baker, A. (2007). Engineering adeno-associated viures 2 vectors for targeted gene delivery to atherosclerotic lessions. *Gene Therapy*. Vol. 15, No. 6, (April 2007), pp. 443-451, ISSN 0969-7128.

White, S.J., Nicklin, S.A., Büning, H., Brosnan, M.J., Leike, K., Papadakis, E.D., Hallek, M. & Baker, A.H. (2004). Targeted Gene Delivery to Vascular Tissue In Vivo by Tropism- Modified Adeno-Associated Virus Vectors. *Circulation*. Vol. 109, (January 2004), pp. 513-519, ISSN 0009-7322.

Wickham, T.J., Roelvink, P.W., Brough, D.E. & Kovesdi, I. (1996). Adenovirus targeted to heparan-containing receptors increases its gene delivery efficiency to multiple cell types. *Nature Biotechnology*. Vol. 14, No. 11, (November 1996), pp. 1570-1573, ISSN 1087-0156.

Wu, B.W., Lu, J., Gallaher, T.K., Anderson, W.F. & Cannon, P.M. (2000a). Identification of regions in the Moloney murine leukemia virus SU protein that tolerate the insertion of an integrin-binding peptide. *Virology*. Vol. 269, No. 1, (March 2000), pp. 7-17, ISSN 0042-6822.

Wu, H., Han, T., Belousova, N., Krasnykh, V., Kashentseva, E., Dmitriev, I., Kataram, M., Mahasreshti, P.J. & Curiel, D.T. (2005). Identification of sites in adenovirus hexon for foreign peptide incorporation. *Journal of Virology*. Vol. 79, No. 6, (March 2005), pp. 3382-3390, ISSN 0022-538X.

Wu, P., Kudrolli, T.A., Chowdhury, W.H., Liu, M.M., Rodriguez, R. & Lupold, S.E. (2010). Adenovirus Targeting to Prostate-specific Membrane Antigen through Virus- displayed Semi-random Peptide Library Screening. *Cancer Research*, (July 2010), ISSN 1538-7445.

Wu, P., Xiao, W., Conlon, T., Hughes, J., Agbandje-McKenna, M., Ferkol, T., Flotte, T. & Muzyczka, N. (2000b). Mutational Analysis of the Adeno-Associated Virus Type 2 (AAV2) Capsid Gene and Construction of AAV2 Vectors with Altered Tropism. *Journal of Virology*. Vol. 74, No. 18, (September 2000), pp. 8635-8647, ISSN 0022- 538X.

Wu, Z., Asokan, A. & Samulski, R. (2006). Adeno-associated virus serotypes: vector toolkit for human gene therapy. *Molecular Therapy*. Vol. 14, No. 3, (September 2006), pp. 316-327, ISSN 1525-0016.

Xie, Q., Bu, W., Bhatia, S., Hare, J., Somasundaram, T., Azzi, A. & Chapman, M.S. (2002). The atomic structure of adeno-associated virus (AAV-2), a vector for human gene therapy. *Proceedings of the National Academy of Sciences of the United States of America*. Vol. 99, No. 16, (August 2002), pp. 10405-10410, ISSN 0027-8424.

Yang, L., Jiang, J., Drouin, L.M., Agbandje-McKenna, M., Chen, C., Qiao, C., Pu, D., Hu, X., Wang, D.Z., Li, J. & Xiao, X. (2009). A myocardium tropic adeno-associated virus (AAV) evolved by DNA shuffling and in vivo selection. *Proceedings of the National Academy of Sciences of the United States of America*. Vol. 106, No. 10, (March 2009), pp. 3946-3951, ISSN 1091-6490.

Young, J.A., Bates, P., Willert, K. & Varmus, H.E. (1990). Efficient incorporation of human CD4 protein into avian leukosis virus particles. *Science*. Vol. 250, No. 4986, (December 1990), pp. 1421-1423, ISSN 0036-8075.

Yu, C.Y., Yuan, Z., Cao, Z., Wang, B., Qiao, C., Li, J. & Xiao, X. (2009). A muscle-targeting peptide displayed on AAV2 improves muscle tropism on systemic delivery. *Gene Therapy*. Vol. 16, No. 8, (August 2009), pp. 953-962, ISSN 1476-5462.

Yu, J.H. & Schaffer, D.V. (2005). Advanced targeting strategies for murine retroviral and adeno-associated viral vectors, In: *Advances in Biochemical Engineering/Biotechnology*, Scheper, T. (ed). 147-167, Springer, ISBN 103-540-28404-4, Berlin.

Williams, D. (2007). RAC Reviews Serious Adverse Event Associated with AAV Therapy Trial. *Molecular Therapy*, Vol. 15, No. 12, (December 2007), pp. 2053-2054, ISSN 1525-0016.

Zhang, H., Mitin, A. & Vinogradov, S.V. (2009). Efficient transfection of blood-brain barrier endothelial cells by lipoplexes and polyplexes in the presence of nuclear targeting NLS-PEG-acridine conjugates. *Bioconjugate Chemistry*. Vol. 20, No. 1, (January 2009), pp. 120-128, ISSN 1520-4812.

4 Chapter

EXAMINATION OF THE TRANSFECTION BARRIERS INVOLVED IN THE DELIVERY OF NON-VIRAL NANOPARTICULATE GENES IN VARIOUS CELL LINES

1. INTRODUCTION: AN OVERVIEW

The development of efficient non-viral vectors for optimized gene transfer into target cells has become a critical objective in order to enhance the delivery efficiency of genetic material into cells in vitro and *in vivo.* In particular, the feasibility of non-viral vector systems, including cationic lipids and polymers, as a delivery system has been extensively studied [1-4]. In cell lines, non-viral gene transfer has been efficiently achieved through the use of cationic lipid/DNA aggregates, resulting in low cytotoxicity and no immunogenicity [5]. Regrettably, the efficiency of non-viral gene transfer into primary cells remains inefficient, resulting in low transgene expression in vivo [6]. The introduction of foreign DNA into primary cells necessitates the meticulous selection of the gene transfer technique, in contrast to the effective transfection of most cell lines, which can be achieved through a variety of methods. In contrast to viral strategies, which are associated with immunogenic risks, non-viral approaches have been demonstrated to be ineffective for the majority of primary cell types. This may be attributable to the necessity of surmounting biological obstacles in order to achieve successful gene delivery.

Consequently, the development of efficient gene delivery systems is contingent upon the understanding of the ingestion mechanism and the subsequent intracellular processing of non-viral gene delivery systems. Additionally, distinct pathways may be targeted in the context of particle internalization into cells.

Clathrin-dependent uptake is the most extensively studied of the numerous mechanisms that are believed to be involved in the cellular internalization of particles [7]. Nevertheless, other internalization pathways, including the caveolae-dependent pathway, macropinocytosis, phagocytosis, and the non-clathrin-non-caveolae dependent pathway, are also viable methods for gene delivery and subsequent processing in the cell [8].

The kinetics of intracellular processing and the efficiency of transfection may be influenced by the mode of internalization. A variety of obstacles could be the cause of limited transfection efficiency, depending on the uptake mechanism. Additionally, internalization may result in lysosomal degradation and digestion, recycling back to the membrane, transcytotic transport across the epithelial barrier, or delivery to other compartments, contingent upon the mode of cellular uptake.

After the complex has been released into the cytosol, it must overcome additional obstacles, including insufficient desaggregation, poor cytoplasmic transport, cytosolic digestion by nucleases, and low intra-nuclear DNA delivery. Targeting specific uptake pathways that enhance particle delivery efficiency is a promising approach to enhance the efficacy of non-viral vectors. A comprehensive examination of the various internalization pathways and the subsequent intracellular events associated with each case is necessary for the implementation of such a strategy.

We have focused on the first and second primary barriers to enhanced transfection efficiency in human primary cells in this study. In light of their significance in cardiovascular diseases, two distinct cell types were selected: human aortic endothelial cells (HAEC) and human aortic smooth muscle cells (HASMC). The principal targets in a cardiovascular gene-therapy strategy are both cells, as they are involved in the unwanted re-narrowing (restenosis) of cardiac vessels after angioplasty.

Clarification of the unique uptake mechanism of cationic lipid/DNA aggregates was necessary to obtain insight into additional intracellular processing and other obstacles that needed to be surmounted. In order to differentiate between the various endocytic pathways that are involved in the internalization of lipoplexes of the Rhodamine-labeled DC-30 lipoplexes (Rh-DC-30), both general and specific inhibitors of endocytic routes were analyzed using flow cytometry in the presence of lipoplexes. The inhibitors employed are summarized in Table 1. In order to gain a more comprehensive understanding of the intracellular fate of lipoplexes, spectral bio-imaging was employed to investigate their co-localization with a diverse array of molecules. A marker for clathrin-mediated uptake was employed with transferrin alexa fluor 488 (tf) [9-16], a marker for internalization via caveolae and related membrane structures with cholera toxin B alexa fluor 488 (chltx-B) [8, 17], and a marker for macropinocytosis with FITC-dextran [8, 17].

Subsequently, it was determined whether the low transfection efficacy was due to the lipoplex's endosomal release failure or the complex's desaggregation. Consequently, electroporation was employed to introduce plasmid DNA into the cytosol an additional time.

Table 1. Inhibitors for the specific endocytic routes

Inhibitor	Uptake route	Reference
Chlorpromazine (chlp)	Clathrin	[11]
Methyl-ß-cyclodextrin (mbCD)	Clathrin and caveolae	[22-24]
Filipin (fil)	Caveolae	[25-27]
Genistein (gen)	Caveolae	[28]
LY 29004 (Ly)	Macropinocytosis	[29, 30]
Wortmannin (wm)	Macropinocytosis	[29, 30]
Nocodazole (noco)	Microtubule depolymerization	[31]
Cytochalasin-D (cch-D)	Actin skeleton disruptor	[8]

2. MATERIALS

DC-30 was acquired from Avanti Polar Lipids (Birmingham, AL, USA). The Cy-5- and Rh- DNA Labeling Kits, Transferrin Alexa Fluor conjugate 488, Cholera Toxin Subunit B Alexa Fluor conjugate 488, and DAPI were acquired from Molecular Probes (Leiden, The Netherlands). pEGFP was obtained from BD Clontech Germany, Heidelberg, and cells were acquired from ATCC (American Type Culture Collection), Manassas, USA. All other chemicals, unless otherwise specified, were of analytical grade and acquired from Sigma/Fluka (Deisenhofen, Germany). Further information is provided in the relevant method section.

3. METHODS

3.1 Cell culture

Human vascular smooth muscle cells (HASMC) were cultivated in smooth muscle cell medium 2 (Promocell, Heidelberg, Germany) supplemented with 5 % (v/v) FCS and human aortic endothelial (HAEC) were maintained in endothelial cell medium MV (Promocell, Heidelberg, Germany) on 100 mm culture plates. For HAEC, tissue culture plates were first coated with sterile-filtered 2 % gelatine solution for 10 min and washed twice with PBS w/o calcium and magnesium before seeding. Cells were cultivated at 37 °C in a humidified atmosphere with 5 % (v/v) CO2.

3.2 Plasmid preparation

The plasmid pEGFP carries the green fluorescent protein coding region under the control of the cytomegalovirus immediate-early promoter region. It was isolated from *Escherichia coli* (Stratagene, Amsterdam, NL) with the Maxi-Prep Kit from Qiagen (Hilden, Germany). Isolated DNA was stored in TE buffer (100 mM NaCl, 10 mM Tris-HCl) at a concentration of 1 mg/mL at -20 °C after its purity was verified by determining the ratio of absorbance at 260/280 nm and by gel electrophoresis.

3.3 DNA labeling

Cy-5-labeled and Rh-labeled DNA was prepared as described for the Mirus Labeling Kit (Molecular Probes, Leiden, Netherlands).

3.4 Lipoplex formation

Enhanced green fluorescent plasmid pEGFP (either labeled for uptake studies or unlabeled for transfection studies) was mixed with DC-30® in a lipid/DNA ratio (w/w) of 5:1 according to the following protocol: lyophilized DC-30® was redispersed in sterile transfection medium (TM; 250 mM saCcharose, 25 mM NaCl) at a concentration of 1 mg/ml and incubated for 30 min at room temperature. Plasmid DNA and the respective amount of DC-30®-dispersion were diluted separately into equal volumes of TM in order to achieve the desired lipoplex concentration in 200 µL of lipoplex preparation. The dilutions were combined discontinuously by pipetting the plasmid into the liposome solution, gently mixing and incubating for at least 30 min at room temperature to allow the formation of the lipoplexes.

3.5 In vitro transfection assay

Cells were seeded in a 24-well cluster dish at a density of 104 cells per well 24 h prior to the experiments and cultivated in the appropriate growth medium with serum. After 24 h in culture the cells were washed with 1 ml PBS and then 400 µL growth medium containing serum was added to the cells. 200 µL of freshly prepared lipoplexes were added to the cells. After incubating for 5 h at 37 °C (5 % (v/v) CO2) the supernatants were removed and 1 ml of the appropriate growth medium was added to each well. Thereafter, the cells were cultured further for a total of 48 h at 37 °C, 5 % (v/v) CO2. In the control experiments cells were incubated with 200 µL culture medium or 200 µL TM and treated the same way as the lipoplex samples. Analysis was carried out by means of flow cytometry (see uptake and FACS analysis).

3.6 Electroporation experiments

Cells were cultivated to 80 % confluency as described above. Every 100 mm plate was trypsinated with 1 mL Trypsin/EDTA for 30 s (HAEC), and 90 s (HASMC) to detach the cells. Trypsinization was stopped by adding 3 ml culture medium with serum. After washing twice with ice-cold PBS, cells were redispersed in ice-cold electroporation buffer (100 mM Hepes, 137 mM NaCl, 4 mM Na_2HPO_4, 6 mM Dextrose) at a concentration of 107 cells/ ml. Electroporation experiments were carried out using a Gene Pulser II (BioRad Laboratories, Muenchen, Germany). 500 µL of cell suspension was poured in a chilled 0,4 cm gap cuvette. After adding 20 µg DNA (1 mg/mL TE-buffer pH 7,0; Tris 10 mM, EDTA 1 mM) the cuvette was placed on ice and electroporated with the following settings: for HAEC: 350 V, 750 µF; for HASMC: 500 V, 950 µF. After the shock, the cuvette stayed on ice for 10 min. Then, cells were transferred into a 100 mm culture plate, (gelatine-coated plates for HAEC, see section "cell culture") and incubated for 48 h at 37 °C and 5 % CO_2. Fluorescent pictures were taken with an Axiovert S 20 (Zeiss, Jena Germany, 20x).

As a positive control experiment, C3-Toxin plasmid was electroporated into the cells to see if DNA was delivered into the nucleus. Effective delivery was achieved when cells were rounded up and dead.

As cells were not vital enough for FACS analysis, fluorescent and non-fluorescent cells were counted via "Neubauer Zählkammer". Counting was repeated twice for every culture plate for at least 3 independent experiments.

3.7 Uptake studies and FACS analysis

Two days before the uptake experiment 10 000 HAEC and HASMC respectively were seeded onto a 24-well tissue culture plate. One hour before the experiment the culture medium was refreshed with 400 µL of cell specific medium containing 5 % FCS. Prior to the inhibition experiments the cells were incubated with the inhibitors (cytochalasin D (Cch- D, 10 µM) for 120 min, chlorpromazine (chlp, 56 µM), LY29004 (ly, 50 µM), wortmannin (wm, 50 nM) genistein (gen, 200 µM) and nocodazole (noco, 10 µM) for 60 min, filipin (fil, 5 µg/ml for HASMC, 10 µg/mL for HAEC) for 30 min and methyl-β-cyclodextrin (mbCD, 164 µM) for 15 min). Subsequently, the DC-30 based lipoplexes containing Cy5-labeled plasmid DNA were added and the cells were incubated at 37 °C for another 60 min. Then, cells were washed with PBS and detached with 200 µL trypsin/EDTA (0,5 mg/mL trypsin, 0,2 mg/mL EDTA) for 30 s (HAEC) or 90 s (HASMC) and harvested by centrifugation. The cell pellet was washed in one ml ice-cold PBS and resuspended again in 200 µL ice-cold PBS. Cell toxicity was measured by adding 2 µL 7-AAD (BD Biosciences), a fluorescent DNA- binding dye which only penetrates dead cells. Fluorescent positive cells were analyzed with a FACS Calibur using Cell Quest Pro software (BD Biosciences). 10 000 cells were measured for each sample. GFP was measured using fluorescence channel 1 (530+/-15 nm), Rh in fluorescence channel 2 (585+/-21 nm), 7-AAD in fluorescence channel 3 (661+/-8 nm) and Cy5 in fluorescence channel 4 (> 670 nm).

In order to determine the amount of lipoplex absorbed to the outside of the cell membrane, experiments were also performed at 4 °C. Furthermore, control experiments such as incubation only with Cy5-labeled DNA with buffer alone were performed.

3.8 Spectral bio-imaging

Cells were seeded onto 12-mm coverslips in 24-well plates 2 days prior to the experiment. Lipoplexes were prepared using DC-30® and Rh-labeled plasmid. Cells were then washed twice with PBS w/o calcium and magnesium and then incubated with 400 µL medium containing the lipoplexes. Cells were incubated simultaneously with the markers for the different pathways for 30 min to 2 h (tf alexa fluor 488 (10 µg/mL), chltx-B alexa fluor 488 (2 µg/mL) and FITC dextran 70 000, (1 µg/mL)). The amount of DNA was the same as that used in the uptake experiments. Prior to microscopic examination, the cells were fixed with 4 % paraformaldehyde (500 µL, 10 min, RT) and the coverslips were mounted on glass slides with 3 µL MobiGlow (MoBiTec, Goettingen, Germany), an antifading substance to reduce photobleaching effects.

Spectral bio-imaging was performed as described previously by Huth et al. [8] with a SpectraCube SD-200H system (Applied Spectral Imaging, Migdal HaEmek, Israel). An inverted fluorescence microscope (Axiovert S 100, Zeiss, Jena Germany) equipped with a high-pressure mercury lamp (HBO 100) for excitation and a triple bandpass filter set was used. All images were taken using a 100x/1.3 oil-immersion objective lens (Plan Neofluar, Zeiss, Jena, Germany). In the spectral range from 400 to 700 nm the objective lens shows only minimal fluctuations of transmission (85-90 %). The optical head attached to the microscope is composed of a Sagnac common-path interferometer and imaging optics including a cooled CCD camera (Hamamatsu C4880-85, Japan). Microscopic images were obtained with spectral bio-imaging 2.5 software

(Applied Spectral Imaging). The acquisition time of a desired image varied from 30 to 90 s depending on the brightness of the fluorescence and the image size. Cells were first incubated with only one dye to get single- colored images. For further analysis images were then transferred to the SpectraView 1.6 software (Applied Spectral Imaging).

4. RESULTS

4.1 Uptake experiments with inhibitors

For uptake experiments studied by means of flow cytometry, DNA was labeled with Cy-5. Cell experiments were performed at both 37 °C and 4 °C in order to distinguish cellular up- take from adsorption of lipoplexes to the outside of the cell membrane or their fusion with the cell membrane. It has been described that no energy dependent internalization process can take place at 4 °C [18]. As the size of the DC-30® lipoplexes was determined to be in the range of 300 and 500 nm [5], we assumed internalization to be an active process. The detection of positive fluorescent signal at 4 °C therefore refers to fusion or adsorption of lipoplex with the cell membrane. Fig. 1 represents the amount of fluorescent positive cells after one hour of incubation with lipoplexes at 37 °C.

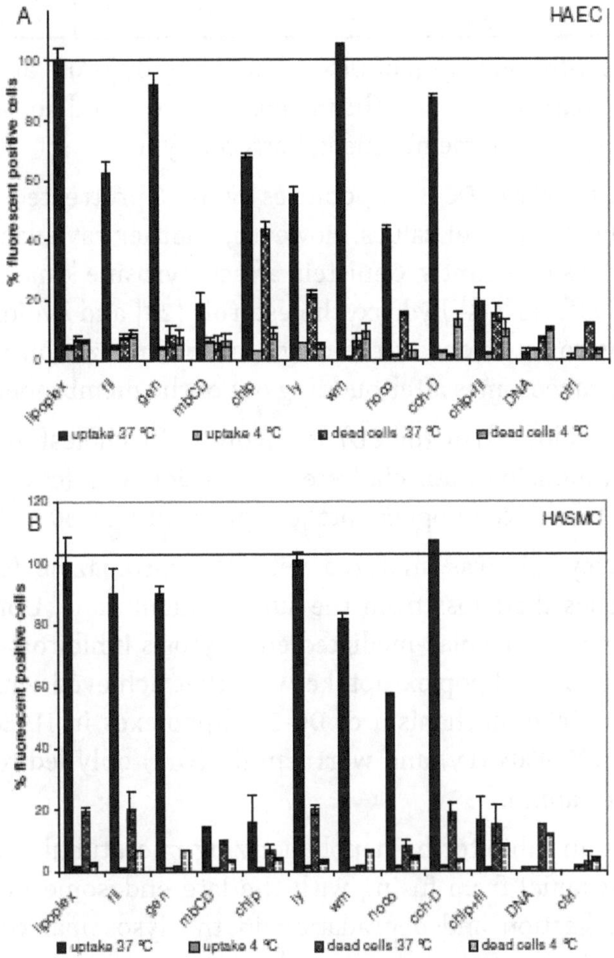

Fig. 1. Cells were incubated for one hour with Cy-5 labeled lipoplexes and analyzed by means of flow cytometry (Fl-4 channel). Prior to the measurement, every sample was additionally incubated with 7-AAD to determine the amount of dead cells (Fl-3 channel). Error bars indicate standard deviation (n=3 in 3 independent experiments).

The corresponding transfection efficiency is listed in Table 2 and shows that despite the fact that uptake of lipoplexes was successful in 24 % of HAEC, less than 1 % of HAEC were transfected. In HASMC, uptake reached as high as 80 %, whereas transfection was only 2%.

Only 1 % of fluorescent cells were detected after incubation at 4 °C and therefore adsorption to the outside of the cell membrane or fusion with the cell membrane is negligible and the fluorescent positive cells as seen in Fig. 1 can be considered as resulting from cellular uptake (internalization) of the lipoplexes. Analysis of cytotoxicity confirmed results which were published previously [5].

Table 2. Influence of transfection method on transfection efficiency

Transfection method	Transfection agent	Transfection efficiency [%] HAEC	Transfection efficiency [%] HASMC
Non-viral cationic lipid-mediated DNA-	DC-30 lipoplex	0,40 ± 0,02	2,01 ± 0,08
	DC-30 lipoplex + 100μM Chloroquine	0,70 ± 0,02	2,74 ± 0,14
delivery	DC-30 lipoplex + 100 μM Chloroquine	0,11 ± 0,03	2,47 ± 0,22
Electroporation	DNA	43 ± 7	51 ± 8

Fig. 2 and Fig. 3 show the internalization of DC-30 lipoplexes by HAEC and HASMC in the presence of inhibitors of different uptake pathways. Again the incubation at 4 °C did not show any passive adsorption or fusion of DC-30® lipoplexes with the membrane in both cell types.

As depicted in Fig. 2 A the uptake of DC-30® lipoplexes by HAEC pretreated with filipin (fil) was reduced by about 40 % to 60 % relative to control values. However, another caveolae- inhibitor, genistein (gen), did not reduce lipoplex uptake significantly. Genistein affects tyrosine kinase [19], which associates with caveolae. In contrast, filipin affects 3-ß- hydroxycholesterols [20] and avoids the formation of caveolin coats ab initio. Filipin seems to intervene at the origin of internalization whereas genistein seems to inhibit further processing of caveosomes after budding out of the membrane.

The inhibitor methyl-ß-cyclodextrin (mbCD) is involved in cholesterol depletion of the plasma membrane and therefore influencing both cholesterol-rich domains (caveolae and clathrin-mediated endocytosis) and was shown to reduce lipoplex uptake significantly by 80 %.

Clathrin-dependent endocytosis was analyzed with chlorpromazine (chlp) which interacts with clathrin-coated pits and causes their loss from the surface membrane. Uptake was reduced by about 25 %. Combining the clathrin and caveolae-mediated endocytosis inhibitors chlorpromazine and filipin leads to a comparable reduction of lipoplex uptake with that achieved with mbCD. Macropinocytosis is possibly involved in the uptake mechanism of DC-30® lipoplexes in HAEC. Pre-incubation with the macropinocytosis inhibitors LY29004 (Ly) and wortmannin (wm) only led to a significant reduction of uptake using Ly but not wortmannin.

Nocodazol (noco) acts as an inhibitor by depolymerizing microtubules and therefore prevents the transport vesicle (early endosome) from fusing with the late endosome in order to protect the early endosome´s content from digestion and degradation in the lysosomal compartment. Applying this inhibitor showed a 50 % reduction in fluorescent signal.

Incubation with cytochalasin D (Cch-D), a disrupter of the actin cytoskeleton, resulted in a reduction of about 10 % and did not seem to be strongly involved in the uptake or processing of DC-30® lipoplexes. In this context, the involvement of macropinocytosis, which is strongly actin dependent, and the different inhibition rates caused by LY and wortmannin does not clarify whether or not lipoplexes are internalized via macropinocytosis.

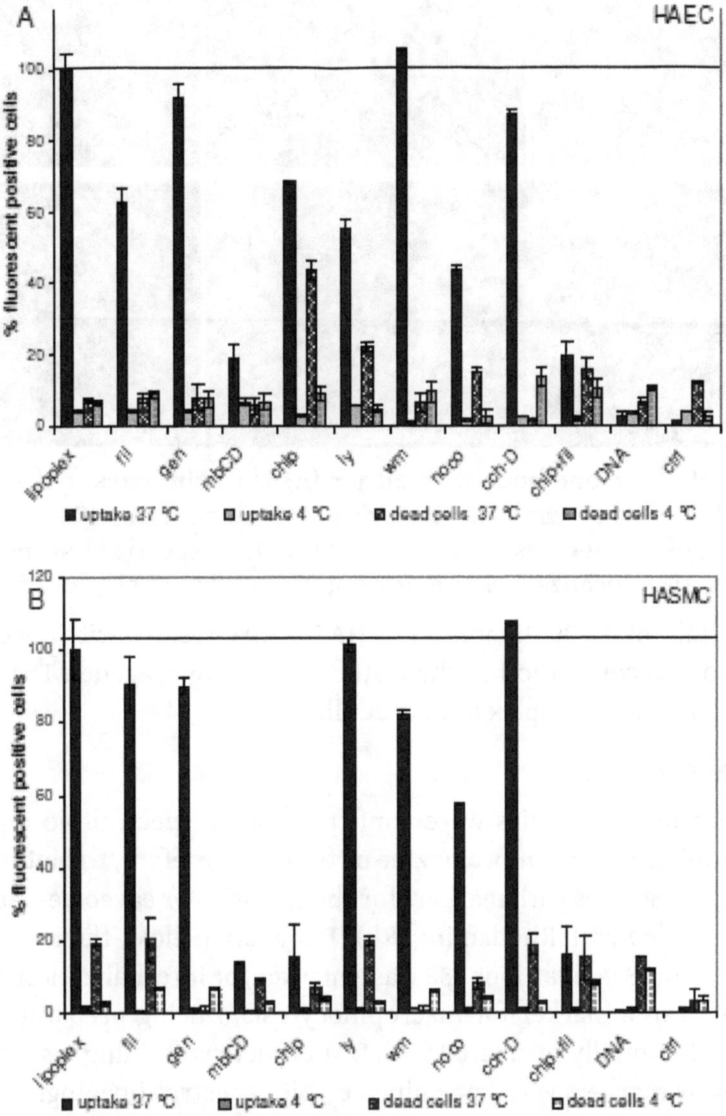

Fig. 2. After pre-incubation with different inhibitors (see Table 1) (x-range), HAEC (A) and HASMC (B) were incubated for one hour with Cy-5 labeled DC-30® lipoplexes and analyzed by means of flow cytometry (Fl-4 channel). Prior to measurement, every sample was also incubated with 7-AAD to determine the amount of dead cells (Fl-3 channel). The fluorescence intensity of lipoplexes without inhibitors was set to 100%. Error bars indicate standard deviation (n=3 in 3 independent experiments).

Fig. 2 B presents data on the internalization of DC-30® lipoplexes by HASMC. Cells pretreated with the caveolae-blocking reagents filipin and genistein showed only a slightly decreased uptake of lipoplexes. In contrast, upon incubation with the clathrin inhibitor chlorpromazine the highest reduction of uptake was observed (about 85 %). The same result could be achieved with mbCD, which is involved in both clathrin and caveolae-mediated pathways. Again, when combining the specific caveolae and clathrin inhibitors (chlp and fil), lipoplex uptake could not be further reduced. Macropinocytosis is also supposed to be involved in lipoplex uptake in these cells, but pre-treatment with wortmannin alone caused only a minor reduction (about 10 %) whereas Ly did not seem to effect the uptake mechanism.

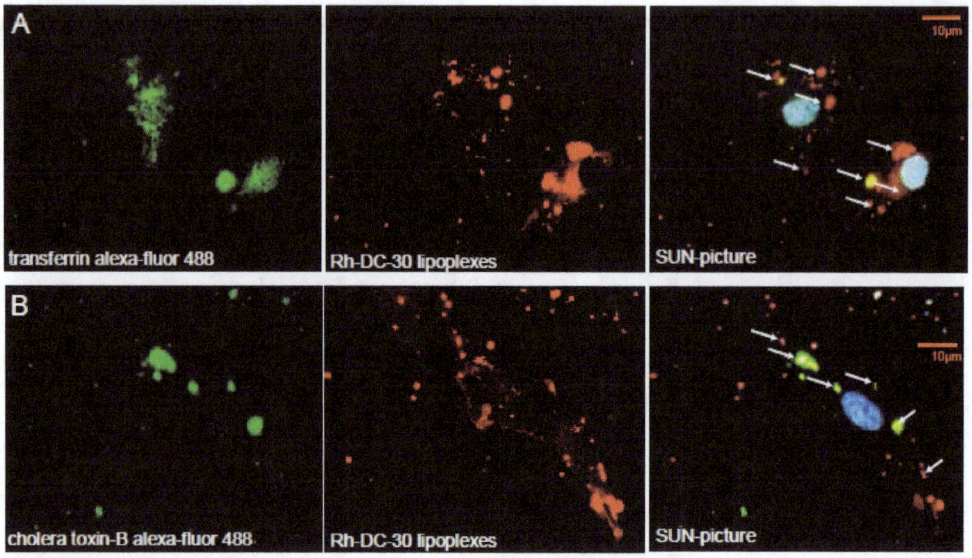

Fig. 3. HAEC were incubated for one hour with either (A) clathrin marker tf alexa-fluor 488 and Rh-lipoplexes or (B) caveolae marker choleratoxin-subunit B alexa-fluor 488 and Rh- lipoplexes. After subsequent DAPI staining of the nucleus, pictures were taken by spectral bio-imaging. I: localization of Rho-lipoplexes; II: localization of markers; III: overlay of single colour images.

On the other hand, uptake of DC-30® lipoplexes by HASMC pre-treated with nocodazole was diminished by about 45 %, which shows a correlation to the clathrin-blocking reagents. The actin-disrupting agent Cch-D did not cause a reduction of lipoplex uptake at all.

4.2 Microscopic uptake studies

The results obtained with uptake studies were confirmed using spectral-bio-imaging. Cells were first pre-incubated with the inhibitors chlormpromazine or filipin. Therefore, the internalization routes were examined via co-localization studies with markers for the clathrin- or caveolae-dependent pathways. The DNA of the lipoplex was labeled with Rhodamine (Rh). Transferrin alexa-fluor 488 was used as a clathrin marker, Choleratoxin- subunit B-Alexa Fluor 488 was a marker for internalization via caveolae-dependent endocytosis and FITC-dextran a marker for macropinocytosis. During preparation for the microscopic experiments, cells were additionally incubated with DAPI (nucleus staining) as a control. Co-localization of the lipoplex and pathway markers was determined by bio- spectral-imaging.

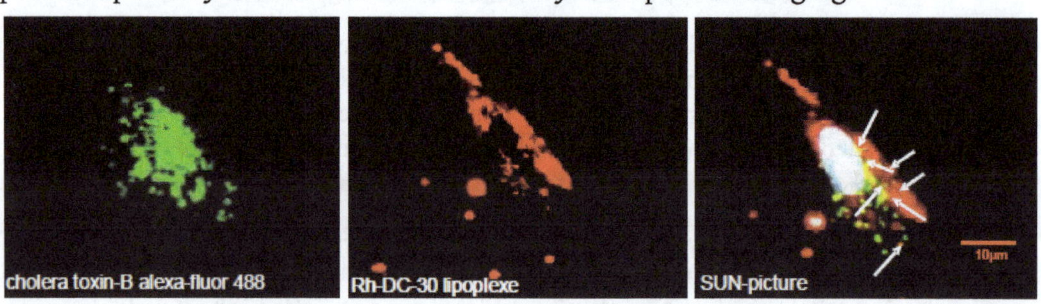

Fig. 4. HASMC were incubated for one hour with clathrin marker tf alexa-fluor 488 and Rh- lipoplexes. After subsequent DAPI staining of the nucleus pictures were taken by spectral bio-imaging. I: localization of Rh-lipoplexes; II: localization of tf alexa-fluor 488; III: overlay of single colour images.

In HAEC after one h incubation time, co-localization of lipoplexes with tf-488 (Fig. 3A) as well as with chltx-B-488 (Fig. 3B) was detected which indicates uptake via clathrin- and caveolae-dependent pathways. No co-localization with the macropinocytosis marker FITC- dextran 70,000 could be shown.

In HASMC, Rh-labeled lipoplexes only showed co-localization with tf-488 (Fig. 4), the marker for clathrin-dependent uptake.

In both cases, spectral bio-imaging studies confirmed the results achieved by flow cytometry.

4.3 Effects on transfection efficiency

As cellular internalization of lipoplexes was successfully achieved in both HAEC and HASMC by clathrin-dependent endocytosis, which results in the transport of lipoplexes via endosomes (Fig. 5), further investigations examined possible barriers after uptake.

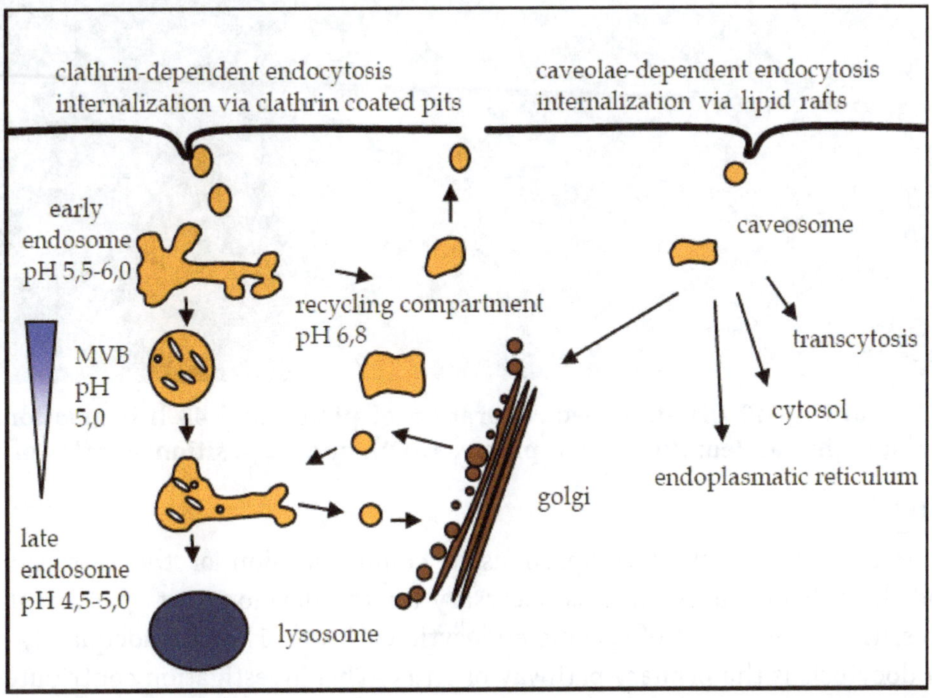

Fig. 5. Cartoon of the main cellular internalization routes of lipoplexes. On the left hand the clathrin-dependent pathway and on the right hand the caveolae-mediated uptake mechanisms are shown. Whereas the first pathway has been well investigated, caveolin- mediated internalization has not been examined in detail to date. The first step in the clathrin-dependent pathway is the budding of clathrin-coated vesicles resulting in early or sorting endosomes. Recycling back to the membrane happens in recycling endosomes. Early endosomes transform in MVB and subsequently late endosomes are formed which end up as lysosomes. Active exchange between the golgi apparatus and late endosomes is known. Particles, internalized via caveolae from so-called caveosomes offering various processing possibilities such as transcytosis, cytosolic release, delivery to the ER or transport to the golgi apparatus.

For these experiments DNA was introduced into the cell by electroporation. With this method, endosomal release and disintegration of the DNA from the lipoplex are no longer involved. Basically, it can be determined that the introduced DNA is not degraded in the cytosol and is able to cross the nuclear membrane to enter the nucleus. Indeed, electroporation resulted in transfection efficiencies of 43 % in HAEC and 51 % in HASMC (Table 2, Fig. 6), indicating that transfection is restricted either by insufficient endosomal release or insufficient release of DNA from the lipoplex.

To investigate insufficient endosomal release, transfection experiments with DC-30® lipoplexes were accomplished with HAEC and HASMC as described before [5]. Additionally, prior to lipoplex incubation the cells were incubated with different concentrations of chloroquine, which represses lysosomal degradation by increasing the pH [21] to facilitate endosomal escape and therefore improve transfection efficiency. Addition of chloroquine did not lead to an increase in transfection efficiency (Table 2). This shows no insufficient endosomal escape of lipoplex.

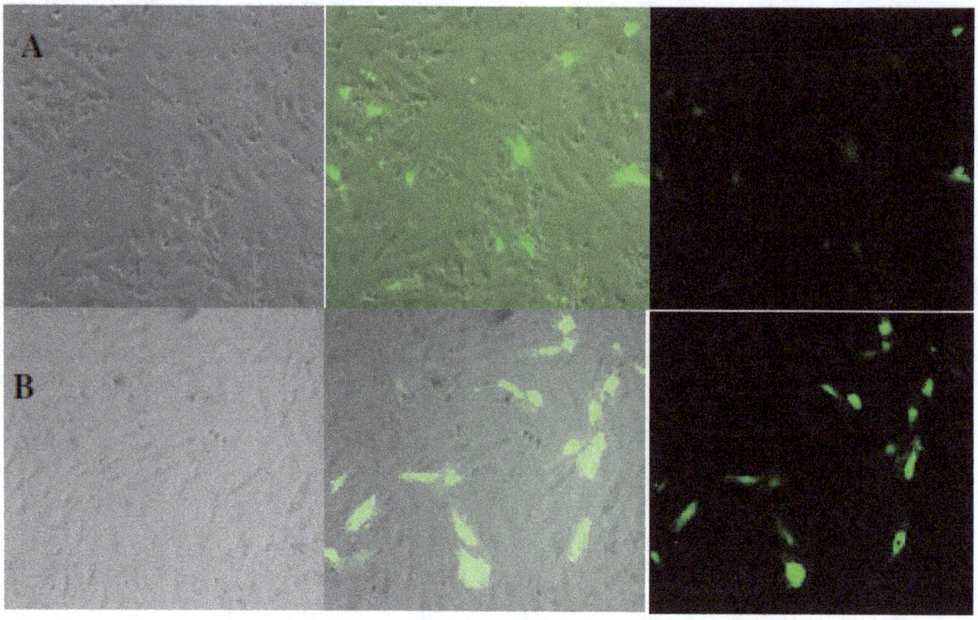

Fig. 6. HAEC (A) and HASMC (B) after electroporation of pEGFP and 48 h incubation. Left- handed: brightfield; Right handed: fluorescent picture (GFP), middle position: overlay of pictures.

5. DISCUSSION

This study illustrates that a more comprehensive comprehension of the intracellular processing mechanism and the delivery mechanism is necessary for the development of efficient non-viral gene delivery systems. The involvement of specific endocytic pathways is still inadequately defined, despite the fact that endocytosis is the primary pathway of entry. This investigation contributes to the further elucidation of the mechanisms and obstacles involved in the absorption of genes in primary cells.

It demonstrated that the insufficient delivery into the cell is not the cause of the poor transfection efficiency of primary human aorta and smooth muscle cells (HAEC and HASMC). The uptake of lipoplex was 20% in HAEC and 80% in HASMC, while transfection was less than 3% in both cases. In addition, it was demonstrated that the direct transfer of plasmid DNA into the cytosol of the cell through electroporation effectively transfected 40 to 50% of the cells. These data indicate that DNA is capable of crossing the nucleolar membrane and exhibiting activity after being present in the cytosol.

In the present study, it was demonstrated that lipoplexes are partially internalized in HAEC through caveolae-mediated endocytosis. The caveolae route is recognized for its ability to facilitate the transport of nanoparticulate drug delivery systems into and/or through the cell without the involvement of lysosomal compartments. It is crucial to avoid cellular routes that transport material to lysosomes in the case of medications that are sensitive to acidic hydrolysis or enzymatic degradation, such as biomacromolecules such as polypetides or nucleic acids. Conversely, the study demonstrated that effective transfection does not necessarily follow uptake via caveolae. Transfection did not occur, despite the caveolae route partially achieving uptake in HAEC. This is an important factor to take into account, as it is evident that it is insufficient to concentrate the research on the modification of particulate drug carriers, such as the attachment of specific ligands, the introduction of surface charge or pH-sensitivity, or the alteration of particle size or elasticity, in order to initiate caveolae-mediated endocytosis.

The evidence that transfection of primary human aorta endothelium and smooth muscle cells was not effective was bolstered by the use of a variety of methods (flow cytometry, spectral bio-imaging, electroporation) and materials (markers and inhibitors of endocytosis pathways). This was facilitated by the insufficient dissociation of plasmid DNA from the lipid/DNA delivery system. This discovery is of paramount significance for the advancement of gene therapy drug delivery.

REFERENCES

1. K. Lundstrom, T. Boulikas, Viral and non-viral vectors in gene therapy: technology development and clinical trials, Technol Cancer Res Treat 2 (2003) 471-486.
2. M. Thomas, A.M. Klibanov, Non-viral gene therapy: polycation-mediated DNA delivery, Appl Microbiol Biotechnol 62 (2003) 27-34.
3. R.I. Zhdanov, O.V. Podobed, V.V. Vlassov, Cationic lipid-DNA complexes (lipoplexes) for gene transfer and therapy, Bioelectrochemistry 58 (2002) 53-64.
4. H. Boulaiz, J.A. Marchal, J. Prados, C. Melguizo, A. Aranega, Non-viral and viral vectors for gene therapy, Cell Mol Biol (Noisy-le-grand) 51 (2005) 3-22.
5. K. Kiefer, J. Clement, P. Garidel, R. Peschka-Süss, Transfection efficiency and cytotoxicity of nonviral gene transfer reagents in human smooth muscle and endothelial cells, Pharm Res 21 (2004) 1009-1017.
6. O. Gresch, F.B. Engel, D. Nesic, T.T. Tran, H.M. England, E.S. Hickman, I. Korner, L. Gan, S. Chen, S. Castro-Obregon, R. Hammermann, J. Wolf, H. Muller-Hartmann,
7. M. Nix, G. Siebenkotten, G. Kraus, K. Lun, New non-viral method for gene transfer into primary cells, Methods 33 (2004) 151-163.
8. M.B. Bally, P. Harvie, F.M. Wong, S. Kong, E.K. Wasan, D.L. Reimer, Biological barriers to cellular delivery of lipid-based DNA carriers, Adv Drug Deliv Rev 38 (1999) 291- 315.
9. U.S. Huth, R. Schubert, R. Peschka-Süss, Investigating the uptake and intracellular fate of pH-sensitive liposomes by flow cytometry and spectral bio-imaging, J Control Release 110 (2006) 490-504.
10. S. Rothenberger, B.J. Iacopetta, L.C. Kuhn, Endocytosis of the transferrin receptor requires the cytoplasmic domain but not its phosphorylation site, Cell 49 (1987) 423-431.
11. L. Killisch, P. Steinlein, K. Romisch, R. Hollinshead, H. Beug, G. Griffiths, Characterization of early and late endocytic compartments of the transferrin cycle. Transferrin receptor antibody blocks erythroid differentiation by trapping the receptor in the early endosome, J Cell Sci 103 (1992) 211-232.
12. L.H. Wang, K.G. Rothberg, R.G. Anderson, Mis-assembly of clathrin lattices on endosomes reveals a regulatory switch for coated pit formation, J Cell Biol 123 (1993) 1107-1117.
13. Y. Shitara, Y. Kato, Y. Sugiyama, Effect of brefeldin A and lysosomotropic reagents on intracellular trafficking of epidermal growth factor and transferrin in Madin-Darby canine kidney epithelial cells, J Control Release 55 (1998) 35-43.
14. L. Pelkmans, J. Kartenbeck, A. Helenius, Caveolar endocytosis of simian virus 40 reveals a new two-step vesicular-transport pathway to the ER, Nat Cell Biol 3 (2001) 473-483.
15. A. Lakkaraju, Y.E. Rahman, J.M. Dubinsky, Low-density lipoprotein receptor-related protein mediates the endocytosis of anionic liposomes in neurons, J Biol Chem 277 (2002) 15085-15092.
16. M. Karin, B. Mintz, Receptor-mediated endocytosis of transferrin in developmentally totipotent mouse teratocarcinoma stem cells, J Biol Chem 256 (1981) 3245-3252.
17. I. Killisch, P. Steinlein, K. Romisch, R. Hollinshead, H. Beug, G. Griffiths, Characterization of early and late endocytic compartments of the transferrin cycle. Transferrin receptor antibody blocks erythroid differentiation by trapping the receptor in the early endosome, J Cell Sci 103 (Pt 1) (1992) 211-232.

18. K. Sandvig, B. van Deurs, Membrane traffic exploited by protein toxins, Annu Rev Cell Dev Biol 18 (2002) 1-24.

19. S. Kessner, A. Krause, U. Rothe, G. Bendas, Investigation of the cellular uptake of E- Selectin-targeted immunoliposomes by activated human endothelial cells, Biochim Biophys Acta 1514 (2001) 177-190.

20. R.G. Parton, B. Joggerst, K. Simons, Regulated internalization of caveolae, J Cell Biol 127 (1994) 1199-1215.

21. K. Simons, E. Ikonen, Functional rafts in cell membranes, Nature 434 (1997) 569-572. [21] W.T. Godbey, K.K. Wu, A.G. Mikos, Tracking the intracellular path of poly(ethylenimine)/DNA complexes for gene delivery, Proc Natl Acad Sci U S A 96 (1999) 5177-5181.

22. S.K. Rodal, G. Skretting, O. Garred, F. Vilhardt, B. van Deurs, K. Sandvig, Extraction of cholesterol with methyl-beta-cyclodextrin perturbs formation of clathrin-coated endocytic vesicles, Mol Biol Cell 10 (1999) 961-974.

23. A. Subtil, I. Gaidarov, K. Kobylarz, M.A. Lampson, J.H. Keen, T.E. McGraw, Acute cholesterol depletion inhibits clathrin-coated pit budding, Proc Natl Acad Sci U S A 96 (1999) 6775-6780.

24. P. Thomsen, K. Roepstorff, M. Stahlhut, B. van Deurs, Caveolae are highly immobile plasma membrane microdomains, which are not involved in constitutive endocytic trafficking, Mol Biol Cell 13 (2002) 238-250.

25. R.M. Pelletier, M.L. Vitale, Filipin vs enzymatic localization of cholesterol in guinea pig, mink, and mallard duck testicular cells, J Histochem Cytochem 42 (1994) 1539- 1554.

26. P.A. Orlandi, P.H. Fishman, Filipin-dependent inhibition of cholera toxin: evidence for toxin internalization and activation through caveolae-like domains, J Cell Biol 141 (1998) 905-915.

27. J.E. Schnitzer, P. Oh, E. Pinney, J. Allard, Filipin-sensitive caveolae-mediated transport in endothelium: reduced transcytosis, scavenger endocytosis, and capillary permeability of select macromolecules, J Cell Biol 127 (1994) 1217-1232.

28. L. Pelkmans, D. Puntener, A. Helenius, Local actin polymerization and dynamin recruitment in SV40-induced internalization of caveolae, Science 296 (2002) 535-539.

29. U. Huth, A. Wieschollek, Y. Garini, R. Schubert, R. Peschka-Süss, Fourier transformed spectral bio-imaging for studying the intracellular fate of liposomes, Cytometry A 57 (2004) 10-21.

30. N. Araki, M.T. Johnson, J.A. Swanson, A role for phosphoinositide 3-kinase in the completion of macropinocytosis and phagocytosis by macrophages, J Cell Biol 135 (1996) 1249-1260.

31. R.B. Kelly, Microtubules, membrane traffic, and cell organization, Cell 61 (1990) 5-7.

5 Chapter

THE MECHANICAL AGITATION METHOD FOR GENE TRANSFER IN EX-VIVO GENE THERAPY

1. INTRODUCTION: AN OVERVIEW

Gene therapy is a medical technique that modifies the genes in a person's cells to treat disorders. The concept of gene therapy arose when molecular biology progressed from the study of DNA molecules to the manipulation of genes within cells. A surge of global research was undertaken following the initial introduction of the notion of gene therapy to the scientific community. The initial objective was to directly alter the genes of patients using *in vivo* gene therapy methods. Nevertheless, the introduction of DNA molecules into the body by in vivo gene therapy occurs at a significantly low rate in terminally differentiated tissue cells, which generally lack the ability to self-renew (Tenenbaum et al., 2003). In vivo gene therapy has a transient effect on correcting a faulty gene in patients due to its brief duration (Kaloss et al., 1999). There has been significant pursuit in the development of gene therapy approaches that aim to permanently retain the corrected patient gene.

Ex vivo gene therapeutic approaches are being studied as alternative solutions to gene therapy in order to address the transient nature of repaired genes in in vivo gene therapy. Ex vivo gene therapy involves the cultivation of surgically extracted adult stem cells, such as mesenchymal stem cells or hematopoietic stem cells, in a laboratory system. The therapeutic DNA molecules are inserted into the separated cells, and these transfected cells are subsequently delivered into the patients. When adult stem cells are used in ex vivo gene therapy, the corrected genes that are introduced are usually expressed permanently if they are corrected properly. This is because adult stem cells have the ability to renew themselves (Dube & Denis, 1995; Muller-Sieburg & Sieburg, 2006; Tseng et al., 2006; Nehlin & Barington 2009). *Ex vivo* gene therapy refers to the process of genetically modifying cells in a laboratory setting, separate from the human body. Nevertheless, typical somatic cells, such as adult stem cells, have a limited capacity for replication and are susceptible to epigenetic alterations. As a result, maintaining long-term cultures of somatic cells that have been separated from the body is extremely challenging (Beyer & Da sliva, 2006; Tonti & Mannello, 2008). It is advisable to minimize the cultivation of adult stem cells in ex vivo gene therapy over a long period of time. Hence, it is imperative to promptly and effectively introduce therapeutic DNA molecules into isolated cells.

Current gene transfer techniques involve the use of gene delivery vehicles that contain therapeutic DNA molecules. However, these vehicles only have limited interaction with their target cells through passive diffusion. As a result, the chances of successful gene delivery are restricted. Our laboratory has devised a highly effective technique for delivering therapeutic genes to adult stem cells via mechanical agitation (Park et al., 2009).

This method involves the use of mechanical agitation to enhance the movement of gene delivery vehicles and target cells in cell suspensions. As a result, there is an increased interaction between the vehicles and cells, facilitating gene transfer. Our mechanical agitation method has significantly improved the efficiency of gene transfer in ex vivo gene therapy, specifically in transfection and transduction, surpassing all other known gene transfer protocols.

2. BASIC PRINCIPLES OF CURRENT GENE THERAPEUTIC APPROACHES

Gene therapy is classified as somatic gene therapy or germ line genc therapy. The application of current molecular genetic techniques used during the manipulation of transgenic or knock-out animals would definitely make gene therapy possible in virtually any type of germ line. However, all civilized societies in the world currently legally prohibit any attempts to genetically modify embryos. Thus far, gene therapy essentially implies somatic gene therapy. Compared to the easy genetic manipulation of embryonic stem cells, the genetic manipulation of somatic cells, including adult stem cells, is limited such that none of the somatic gene therapies are used practically thus far. Therefore, it is not surprising that the main quest of current gene therapy is to improve the efficiency of genetic manipulation in gene therapy, and the future success of gene therapy depends on the efficiency of genetic manipulation.

Genetic manipulation in gene therapy can be achieved by two different approaches: direct genetic manipulation of somatic cells in the body and genetic manipulation of autologous cells outside of the body. These two different strategies for gene delivery are termed *in vivo* and *ex vivo*, respectively. In the *in vivo* strategy, therapeutic genes are delivered into cells *in situ* using a variety of vectors to produce therapeutic proteins in specific sites in the body. In *ex vivo* gene therapy, genetically modified autologous cells are surgically implemented into the body. The *ex vivo* and *in vivo* gene therapies both have positive and negative aspects. Although gene therapy has been a very hot topic in biomedical science for several decades, it is still in its infancy, and a number of hurdles must be overcome to achieve the practical application of gene therapy to patients.

2.1 *In vivo* gene therapy

In vivo gene therapy is a process in which a therapeutic gene is delivered through a vector directly into thc target cells of patients to produce a therapeutic effect that prevents or treats diseases (Fig. 1). Theoretically, once an ideal gene delivery vehicle for a therapeutic gene transfer is developed, the *in vivo* gene therapy should involve a very simple procedure: the injection of a solution containing the gene delivery vehicle into the body. Because of this potentially easy treatment procedure in clinics, *in vivo* gene therapy is considered the preferred gene therapeutic method than *ex vivo* gene therapy. However, *in vivo* gene therapy has a basic and fundamental problem in the delivery of therapeutic genes to target cells: the low efficiency of gene transfer.

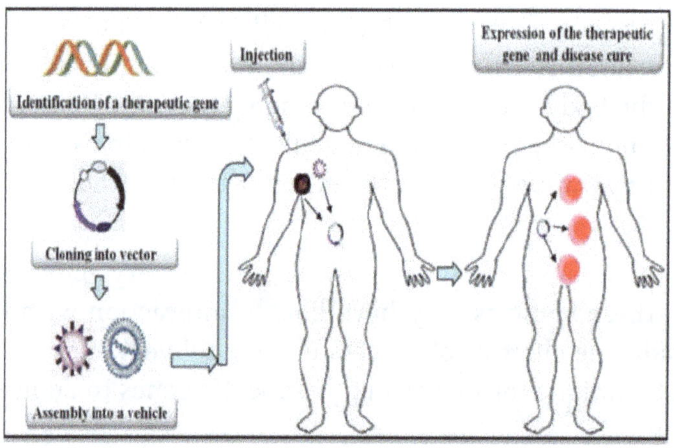

Fig. 1. Strategies for *in vivo* gene therapy. *In Vivo* gene therapy involves introduction of therapeutic DNA directly into the patient body. The DNA is introduced by cell-specific direct injection into tissue in need. Once inside the body and in contact with the specifically targeted cells, the inserted DNA is incorporated into the tissue cells where it encodes the production of the needed protein.

The bottleneck in development of *in vivo* gene therapeutic methods has been the development of an efficient method for delivery of a therapeutic gene into the target cells of the body. The main reason for poor gene delivery efficiency in *in vivo* gene therapy is rooted to the nature of the body. The cells in the body are typically surrounded by an extracellular matrix that usually provides structural support to the cells in addition to performing various other important functions (Fig. 2). The main constituent of the body is the extracellular matrix, not cells (Suki & Jason, 2008). For example, collagen proteins, which are one of the components of the extracellular matrix, constitute approximately 25-35% of the protein content of the entire body, implying that the extracellular matrix occupies the main volume of the body (Khan et al., 2009). The injected gene delivery vehicles must pass through the extracellular matrix to deliver therapeutic proteins into target cells in *in vivo* gene therapy. However, because the extracellular matrix spatially occupies such a large portion of the body, there is an unsolvable limitation for efficient gene transfer in *in vivo* gene therapy.

In addition to the low efficiency of gene delivery, *in vivo* gene therapy has another problem. The gene transfer vector is obligatorily exposed to the immune system of the body. This exposure causes an immune response that blocks gene delivery entirely. Overall, the potential immune response is another factor contributing to the low efficiency of gene delivery in *in vivo* gene therapy. Therefore, development of an ideal gene delivery vehicle for *in vivo* gene therapy is so extremely challenging that, until now, none of the *in vivo* gene therapeutic methods have not a satisfactory result.

2.2 *Ex vivo* gene therapy

In *ex vivo* gene therapy, cells are removed from a patient, maintained in culture to introduce a therapeutic gene into the cells, and then transplanted into the patient (Fig. 3). The role of the transplanted cells, which are genetically modified, is to deliver a recombinant gene product into the patient's body.

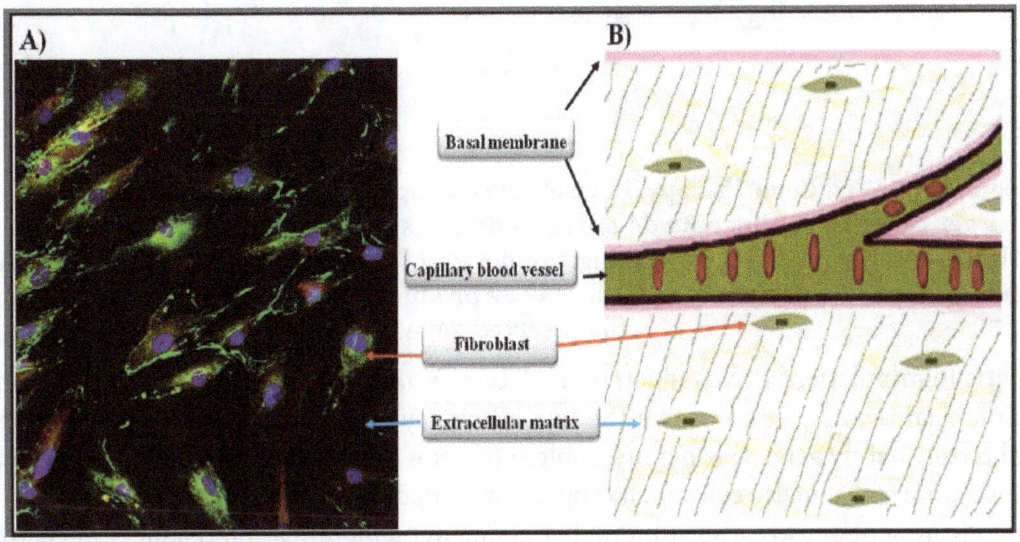

Fig. 2. Typical Anatomical Structure of Connective Tissue. A) The Confocal Microscopic Image of a Mouse Connective Tissue. B) Schematic Illustration Depicting Extracellular Matrix. The extracellular matrix (ECM) is the extracellular part of animal tissue that usually provides structural support to the animal cells in addition to performing various other important functions. The extracellular matrix is the defining feature of connective tissue in animals.

The main disadvantage of *ex vivo* gene therapy is that it requires the surgical removal of cells from the body and transplantation of the cells back to the body. These surgical steps are very painful. However, *ex vivo* therapy has several advantages over *in vivo* gene therapy. First, the efficiency of gene transfer into the targeted cells is very high compared to *in vivo* gene therapy because gene delivery is performed under controlled, optimized conditions. Second, the transduced cells can be enriched if the vector has a selectable gene marker. Third, the immunological side effects that are caused by gene delivery vehicles in *in vivo* gene therapy are usually minimized in *ex vivo* gene therapy.

2.3 *Ex-vivo* gene therapy as a practical option to correct a defective gene permanently

In vivo gene therapy introduces the therapeutic genes directly into the patient by intravascular injection. Because this approach is much simpler and less technically demanding than *ex vivo* gene therapy, which requires two surgical steps, the science of *in vivo* gene therapy has been preferentially developed. However, as discussed above, the nature of the mammalian body imposes innate, unsolved problems for *in vivo* gene therapy to achieve gene expression at therapeutically effective levels. The limitation of *in vivo* gene therapy for practical application is mainly due to low efficiency gene transfer.

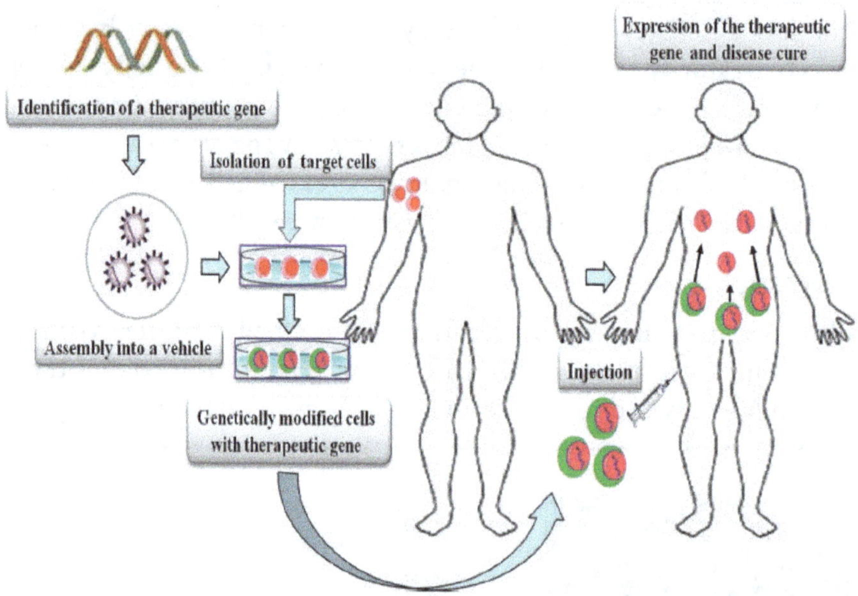

Fig. 3. Strategies for *Ex Vivo* Gene Therapy. *Ex vivo* gene therapy is performed with the genetic alterations of patients target cells happening outside of the body in as culture. Target cells from the patient are infected with a recombinant virus containing the desired therapeutic gene. These modified cells are then reintroduced into the patient body, where they produce the needed proteins that correspond to the inserted gene.

Because genetic manipulations are conducted in a lab in *ex vivo* gene therapy, the application of current molecular biological techniques can solve the low efficiency of gene transfer. *Ex vivo* gene therapy has the potential to ultimately solve this major problem of gene therapy, and it could be practically used in clinics. However, normal somatic cells, including adult stem cells, do not only propagate well in a typical cell culture environment and are also vulnerable to epigenetic modification (Islam et al., 2007; Martinez-Climent et al., 2006), requiring that the transfer of therapeutic genes to the isolated cells be performed as soon as possible. Therefore, one of the key factors for the success of *ex vivo* gene therapy is to deliver therapeutic DNA molecules into isolated cells promptly with high efficiency. If these problems could be solved successfully, the *ex vivo* technique could be practically applied to patients in the near future.

3. CURRENT METHODS FOR GENE DELIVERY IN *EX VIVO* GENE THERAPY

As the name implies, the success of gene therapy depends on introducing therapeutic genes into target cells with high efficiency. Since Friedmann and Roblin formulated the concept of gene therapy in 1972

(Friedmann & Roblin, 1972), the biggest challenge in gene therapy has been the development of a method to deliver therapeutic genes to target cells with high efficiency. Although gene delivery in *in vivo* gene therapy is much easier than in *ex vivo* gene therapy, gene delivery into primary cells of *in vitro* cell cultures is also quite difficult.

Typical efficiencies of gene delivery to primary cells are 5-10% in most current methods (Cai et al., 2002; Ding et al., 1999; Eiges et al., 2001; Lakshmipathy et al., 2004; Peister et al., 2004), which is not high enough for satisfactory *ex vivo* gene therapy. Because of this, many different methods of gene delivery have been developed using primary cells for *ex vivo* gene therapy. Generally, gene delivery methods can be divided into two categories, viral and non-viral.

3.1 Viral methods for gene delivery in *ex vivo* gene therapy

Viruses have evolved specialized molecular mechanisms to efficiently transport their genomes into cells. Viral vectors have developed by taking advantage of the molecular mechanisms of the virus to deliver exogenous DNA into target cells. Currently, viral vectors are frequently used molecular biology tools for delivery of genetic material into cells of a living organism (*in vivo*) or in cell culture (*in vitro*). Viral vectors are tailored to their specific applications but share a key property. Because viral vectors are essentially created from pathogenic viruses, they are modified to minimize the pathogenic properties of the original viruses. This usually involves the deletion of a portion of the viral genome that is critical for viral replication. Such viral vectors can efficiently infect cells but once the infection has taken place, they cannot replicate. The viral vectors require helper genes to provide the missing proteins for production of new virions. Replication of viral vectors is usually conducted in packaging cells that were engineered with helper genes. Therefore, viral vectors can only replicate in packaging cells and exist solely to deliver exogenous DNA to target cells where the viral vector cannot replicate.

Table 1 summarizes the types of viral vectors currently developed. In general, viral vectors are very efficient in terms of gene delivery into target cells. However, viral vectors have common problems including the following: (i) a limited DNA capacity, (ii) expression of viral genes, (iii) initiation of the antiviral immune response, (iv) reversion to a replication competent state and (v) decreasing expression over time.

Viral vectors can be classified as DNA or RNA viral vectors. DNA viral vectors are derived from viruses such as adenovirus or herpes virus, which carry their genetic material in the form of DNA. Because these viral vectors persist as an extrachromosomal element after delivery into target cells, the viral vectors remain only temporarily. One advantage of the episomal presence of the vectors is that the gene expression level is high. Because exogenous genes cannot stay indefinitely, these vectors are not suitable for *ex vivo* gene therapy. Currently, these vectors are mostly used for *in vivo* gene therapy.

Retroviral and lentiviral vectors are examples of RNA viral vectors that are replicated in target cells via reverse transcriptase to produce DNA from their RNA genomes. The DNA is then incorporated into the host's genome by an integrase. Thereafter, the virus replicates as part of the host cell's DNA, permitting long-term expression of the exogenous gene and ensuring transmission of the exogenous gene to the progeny of transduced cells. Therefore, these vectors are suitable for *ex vivo* gene therapy in which permanent gene expression is required in *in vitro* cell culture. These vectors are also widely used for *in vivo* gene therapy.

3.2 Non-viral methods for gene delivery in *ex vivo* gene therapy

Gene therapy was originally devised for the treatment of inherited genetic diseases, such as hemophilia and cystic fibrosis. However, the realm of gene therapy has been expanding to develop strategies for cancer, infectious diseases like HIV, and various complex diseases, such as diabetes, dementia and hypertension. Genetic manipulations for these diseases are more complicated than genetic manipulations for the treatment of inherited genetic diseases. This means that current gene therapies need to deliver DNA, RNA, siRNA, or antisense sequences that alter gene expression within a specific cell population to manipulate cellular processes and responses. Viral vector-mediated gene deliveries are by far the most effective means of DNA delivery. However, the recombinant vector containing the therapeutic gene has to be packaged with viral coat proteins to make gene delivery possible, meaning that viral vector-mediated gene deliveries are limited to a DNA molecule of a certain size because the viral coat proteins have a limited DNA carrying capacity. Other than the physical limitation of viral vector-mediated gene deliveries, there are more limitations, such as immunotoxicity caused by viral coat proteins, restricted targeting of specific cell types, and recombination. Therefore, non-viral gene deliveries have been a very popular research topic, and many interesting and creative methods have been developed. The efficiency of gene delivery (*i.e.* transfection efficiency) is crucial to the success of non-viral gene deliveries. Various non-viral gene delivery methods currently developed could be classified into two groups: physical gene delivery methods and chemical gene delivery methods.

Table 1. A comparison of different viral vectors used for gene therapy

Vector	Genetic material	Packaging Capacity	Tropism	Inflam-matory Potential	Vector genome forms	Main) Advantages
Retrovirus	RNA	8 kb	Dividing Cells	Low	Integrated	Persistent gene transfer in dividing cells
Adenovirus	dsDNA	7.5 kb	Dividingand nondividing cells	High	Non-integrated	Extremely efficient transduction of most tissues
Adeno-associated Virus	ssDNA	< 5kb	Dividing and nondividing cells	Low	Nonintegrated (90% Integrated (>10%)	Noninflammatory; Non-pathogenic
Lentivirus	RNA	8 kb	Dividing and nondividing cells	Low	Integrated	Persistent gene transfer in most tissues
Herpes virus	dsDNA	> 30 kb	Dividing and nondividing cells	High	Non-integrated	Large packaging capacity; strong tropism for neurons

3.2.1 Physical gene delivery methods

Physical gene delivery methods are methods for transferring DNA molecules from the surrounding medium into cells. Naked DNA (*i.e.*, an uncomplexed form of DNA) is used in physical gene delivery methods. The easiest method to deliver genes into cells is to draw naked DNA into a microneedle and then inject the microneedle into cells to transfer the naked DNA directly to the cells. Though gene transfer efficiency by this method is very efficient, the method is very slow and laborious. The main drawback of this method is that microinjection can be only performed on one cell at a time, which means that this approach cannot be used for typical gene therapeutic approaches. The approach is limited to use for gene delivery into germ-line cells to produce transgenic organisms.

Currently, the most popular physical methods for gene delivery into cells are electroporation and sonoporation. The cellular membrane is punctured by an electric pulse (electroporation) (Neumann et al., 1982) or ultrasonic wave (sonoporation) (Yizhi et al., 2007). The pores in the cellular membrane are only temporarily formed, and DNA molecules pass through during the short period of time when the pores open. These methods are generally efficient and work well across a broad range of cell types. However, a high rate of cell death limits their use, especially in gene therapy. These methods are widely used for gene delivery of immortal cells in which cell viability is not a critical issue during gene transfer.

Another popular method for physical gene delivery is the use of particle bombardment. In this method, gold particles (gene gun) (Gan et al., 2000) or magnetic particles (magnetofection) (Scherer et al., 2002) are coated with naked DNA. In the gene gun method, the DNA-coated gold particles are shot into the cell using high pressure gas, and the particles pass through the cellular membrane to introduce the particles inside the cells. In the magnetofection method, a magnet is placed underneath the tissue culture dish to attract DNA-coated magnetic particles. Then, the DNA-coated magnetic particles come into contact with a cell monolayer to introduce the particles inside the cells. These methods yield reasonably high efficiency gene transfers, but do not yield better efficiencies compared to other non-viral gene transfer methods, despite the requirement for expensive equipment. Also, it is quite difficult to control the DNA entry pathway, and the metal particles in the cells following gene transfer could negatively affect cells. Therefore, these methods are not widely used.

3.2.2 Chemical gene delivery methods

Because DNA cannot pass through cellular membranes alone, various chemicals have been designed to aid the transfer of therapeutic genes into cells. The chemicals used in chemical gene delivery function to enhance the stability of the DNA molecule, to increase the efficiency of cellular uptake and intracellular trafficking, or to alter the distribution of the transferred DNA in the cells. These methods are very successful in terms of transferring genes into cells and are currently the most widely used methods. Also, chemical gene delivery methods are the easiest and most effective among various non-viral gene delivery methods developed thus far.

The most well-studied and effective approach for non-viral gene delivery is the use of cationic lipids. Positively charged cationic lipids naturally bind to negatively charged DNA in solution to condense DNA so that the DNA molecules and cationic lipids form complexes called lipoplexes. After lipoplexes are formed, the positively charged cationic lipids of the lipoplexes interact with cell membranes to allow cells to take up the lipoplexes by endocytosis. In typical cell physiology, endosomes that are formed as the result of endocytosis will fuse with lysosomes to degrade the lipoplexes containing the DNA. An exogenous gene in the lipoplexes would not have a chance to be released into the cytoplasm for gene expression if the endosomes are stable. Therefore, helper lipids are added to form lipoplexes to facilitate the endosomal escape of the exogenous gene (Herringson et al., 2009a, 2009b; Savva et al., 2005). This approach is very successful because it increases the transfection efficiency dramatically. There are various combinations of cationic lipids and helper lipids available. More than 40 products are commercially available for cationic lipid- based gene delivery, including LipoTAXI (Agilent Technologies), LipofectaminTM (Invitrogen), NanoJuice® (Merck), Transfectam® (Promega), and LipoJet TM (SignaGen Laboratories). The cationic lipid-based gene delivery shows a very high transfection efficiency of up to 90 in *in vitro* cell culture. Because the cationic lipid-based lipoplexes are not stably maintained in the blood, these methods are best for *ex vivo* gene therapy. However, cationic lipid-based lipoplexes show a very poor transfection efficiency with primary cells, such as stem cells, indicating that new methodological developments are required for the practical application of *ex vivo* gene therapy.

Other than cationic lipids, several different positively charged materials are used as a base material for non-viral DNA delivery, such as cationic polymers (Segura & Shea, 2001), cationic peptides consisting of poly-L-Lysine (D'Haeze et al., 2007; Mullen et al., 2000; Niidome et al., 1997), or other types of cationic proteins (De Lima et al., 1999; Jean et al., 2009; Lam et al., 2004; Lee et al., 2003; Oliveira et al., 2009;

Vighi et al., 2007). These approaches produce DNA carrying complexes that are more stable. However, the transfection efficiency of this method is not better than cationic lipid-based lipoplexes. Therefore, most of these methods are designed for *in vivo* gene therapy.

4. APPLICATION OF THE MECHANICAL AGITATION METHOD TO *EX VIVO* GENE THERAPY

One of the main obstacles for the application of adult stem cells in *ex vivo* gene therapy is the low efficiency of gene transfer to these cells. For example, electroporation or transfection in mesenchymal stem cells yields 5-10% gene delivery efficiency (Cai et al., 2002; Ding et al., 1999; Eiges et al., 2001; Lakshmipathy et al., 2004; Peister et al., 2004). Therefore, improved gene delivery methods would potentially be very beneficial for the practical application of *ex vivo* gene therapy in patient care. In current gene transfer protocols, virus particles or lipoplexes passively diffuse through the liquid culture medium to reach their target cells, which are layered on the bottom of a culture dish (Chuck & Palsson, 1996). Because the virus particles or lipoplexes contact the target cells by passive diffusion, increasing the chance of contact between virus particles or lipoplexes and their target cells would increase the chance of gene transfer and to promote higher transfer efficiencies. One simple way to increase contact between viruses or lipoplexes and target cells is through mechanical agitation.

A

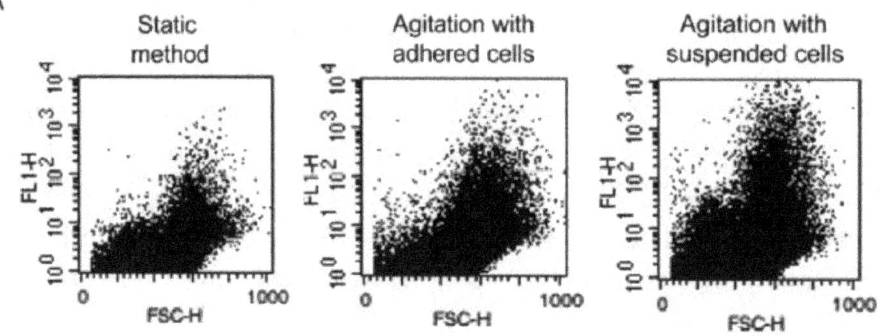

B

		Simultaneous mechanical agitation	
	Static method	Adhered cells	Suspended cells
% EGFP (+) cells	17 ± 2.7	30 ± 1.4	43 ± 2.5

Fig. 4. A Typical Example of Application of Machanical Agitation Method to Transduction of EGFP-Carrying Retrovirus (Park et al., 2009). (A) The representative FACS plots of rat mesenchymal stem cells after transduction with Retro-EGFP using the static method (left panel), mechanical agitation of viruses with adhered cells (middle panel) and simultaneous mechanical agitation of retroviruses with suspended cells (right panel). (B) Numerical representation of the transduction efficiencies of EGFP retrovirus under the static protocol versus the new agitation protocol. The transduction efficiency is defined as the percentage of cells expressing EGFP as measured using FACSCalibur. The mean percentage of GFP- positive cells is presented as the average of three independent transduction experiments (+/- SEM, n=3).

Based on this hypothesis, we developed a mechanical agitation method for retroviral transduction of primary cells or transfection by lipoplexes (Park et al., 2009). In this method, we simply implemented a step in which virus-containing or lipoplexes- containing cell suspensions are agitated to increase the movement of viruses or lipoplexes and target cells with the purpose of generating more frequent contact between them. Suspended target cells have a better chance of making physical contact with virus particles or lipoplexes than adherent target cells because of the possibility for three-dimensional contact between the cells and viruses or lipoplexes. The simple addition of the mechanical agitation step to the conventional transduction or transfection protocol increased gene transfer efficiency two-fold above the current rates these protocols (Fig. 4). In the following section, we describe one example of retroviral transduction using our mechanical agitation protocol.

4.1 Materials

1. Packaging cell line PT67 (Clontech).
2. PCAG-EGFP expression vector (Addgene, USA) and pMSCVneo retroviral vector (Clontech, USA).
3. Lipofectamine 2,000 (Invitrogen).
4. Rat mesenchymal stem cells (isolated by flushing the femurs of two-month-old female SD rats (Da-mool Bioscience Co., Korea).
5. DMEM/ High glucose (Hyclone, USA) with 10% FBS (Hyclone), 2 mM L-glutamin (Invitrogen, USA), 100 μg/ml streptomycin, and 100 U/ml penicillin (Sigma, USA).
6. Selection drug, G418 (Sigma) and Polybrene (Calbiochem, USA).
7. Rat mesenchymal stem cell characterization kit (Millipore,USA).
8. TrypLE Express (Invitrogen).
9. Carl Zeiss LSM510 Meta microscope.
10. 0.45 μm cellulose acetate filter (Millipore).
11. Rocker (SLS4, Seoulin, Korea).
12. Incubator at 37°C under 5% CO_2.
13. 24-well plates and 96-well plates.
14. E-Max micro-well reader (Molecular Devices, USA).
15. Fluorescence microscope (TE2000-S, Nikon, Japan).
16. FACSCalibur instrument (Becton Dickinson, USA).

4.2 One example protocol of retroviral transduction into mesenchymal stem cells

1. 1. Trypsinize pure rat mesenchymal stem cells (0.1 ml/cm2) for 3 min at 37°C.
2. Adjust the cell suspension to contain 5 × 105 cells/ml.
3. Mix a 1 ml aliquot of the trypsinized rat mesenchymal stem cells directly with 1 ml virus stock in the presence of 6 μg/ml polybrene (Calbiochem, USA).
4. Seed the mixing solution in a six-well plate (Falcon, USA).
5. Mechanically agitate the plate containing the mixture of rat mesenchymal stem cells and virus on a rocker (SLS4, Seoulin, Korea) at 20 rpm for 50 min while incubating at 37°C under 5% CO_2.
6. Incubate the plate at 37°C under 5% CO_2 for 24 h.
7. Replace the supernatant containing virus particles with fresh growth medium.
8. Observe EGFP fluorescence in the transduced rat mesenchymal stem cells with a fluorescence microscope (TE2000-S, Nikon, Japan) after an 86 h incubation.

5. CONCISE SUMMARY

As of March 2011, a total of 1703 gene therapy clinical trials have been completed since the first experiment in 1990 (source: http://www.wiley.com/legacy/wileychi/genmed/clinical/). Despite the initial gene therapy clinical study, there has been minimal advancement, resulting in the United States Food and Drug Administration withholding approval for any human gene therapy for genuine patient treatments.

While gene therapy is currently in the experimental phase, ex vivo gene therapeutic approaches have demonstrated significant potential in treating monogenic genetic diseases, as evidenced by the clinical trial results of adenosine deaminase deficiency and familial hypercholesterolemia (Cappelli et al., 2010 & Kassim et al., 2010). The positive outcomes of the clinical trials for these disorders provided strong motivation to persistently pursue ex vivo gene therapy.

Primary cells, such as adult stem cells, have a restricted capacity for self-renewal and are susceptible to epigenetic alterations (Dube & Denis, 1995; Muller-Sieburg & Sieburg, 2006; Tseng et al., 2006; Nehlin & Barington, 2009). It is not feasible to cultivate primary cells over an extended period of time. Hence, it is imperative to expeditiously and effectively introduce therapeutic DNA molecules into individual cells. Nevertheless, the rate at which exogenous DNA is successfully transferred into primary cells is extremely low, as indicated by studies conducted by Beyer and Da Silva in 2006, and Tonti and Mannello in 2008. The gene transfer efficiency in primary cells is significantly lower than that in cell lines using existing gene transfer methods. This means that improvement of the transfer efficiency of exogenous DNA into primary cells is the first obstacle for the practical use of for ex vivo gene therapy.

The mechanical agitation method described here resulted in a greater effectiveness of gene transfer, whether through transfection utilizing lipoplexes or transduction, compared to any existing static transduction method. This approach has the potential to be utilized in several existing transduction or transfection techniques by making minor modifications to the duration and intensity of agitation. This methodology is expected to enhance *ex vivo* gene treatments and in vitro gene transfer investigations using primary cells.

REFERENCES

Beyer, NN., & Da Silva, ML. (2006). Mesenchynal stem cell: isolation, in vitro expansion and characterization. *Handb. Exp. Phamacol.*, Vol.174, pp.249-282.

Cai, S., Ma, Q., Yu, X., Dang, G., & Ma, D. (2002). Expression of human VEGF(121) cDNA in mouse bone marrow stromal cells. *Chin. Med. J., (Engl)* Vol.115, pp.914-918.

Cappelli, B., & Aiuti, A. (2010). Gene therapy for adenosine deaminase deficiency. *Immunol Allergy Clin. North Am.*, Vol.30, No.2, pp.249-260.

Chuck, AS., & Palsson, BO. (1996). Consistent and high rates of gene transfer can be obtained using flow-through transduction over a wide range of retroviral titers. *Hum. Gene Ther.*, Vol.7, pp.743-750.

De Lima, MC., Simões, S., Pires, P., Gaspar, R., Slepushkin, V., & Düzgüneş, N. (1999). Gene delivery mediated by cationic liposomes: from biophysical aspects to enhancement of transfection. *Mol. Membr. Biol.*, Vol.16, No.1, pp.103-109.

D'Haeze, W., Leoff, C., Freshour, G., Noel, KD., & Carlson, RW. (2007). Rhizobium etli CE3 bacteroid lipopolysaccharides are structurally similar but not identical to those produced by cultured CE3 bacteria. *J. Biol. Chem.*, Vol.282, No.23, pp.17101-17113.

Eiges, R., Schuldiner, M., Drukker, M., Yanuka, O., Itskovitz-Eldor, J., & Benvenisty, N. (2001). Establishment of human embryonic stem cell-transfected clones carrying a marker for undifferentiated cells. *Curr. Biol.*, Vol. 11, pp.514-518.

Friedmann, T., & Roblin, R. (1972). Gene Therapy for Human Genetic Disease? *Science*, Vol.175, No.25, pp.949-955.

Gan, W., Grutzendler, J., Wong, WT., Wong, ROL., & Lichtman, JW. (2000). Multicolor Diolistic Labeling of the Nervous System Using Lipophilic Dye Combinations. *Neuron*, Vol. 27, Iss. 2, pp.219-225.

Herringson, TP., Patlolla, RR., & Altin, JG. (2009). Targeting of plasmid DNA-lipoplexes to cells with molecules anchored via a metal chelator lipid. *J. Gene. Med.*, Vol.11, No.11, pp.1048-1063.

Herringson, TP., & Altin, JG. (2009). Convenient targeting of stealth siRNA-lipoplexes to cells with chelator lipid-anchored molecules. *J. Control Release.*, Vol.139, No.3, pp.229-238.

Islam, MQ., Panduri, V., & Islam, K. (2007). Generation of somatic cell hybrids for the production of biologically active factors that stimulate proliferation of other cells. *Cell Prolif.*, Vol.40, No.1, pp.91-105.

Jean, M., Smaoui, F., Lavertu, M., Méthot, S., Bouhdoud, L., Buschmann, MD., & Merzouki, (2009). Chitosan-plasmid nanoparticle formulations for IM and SC delivery of recombinant FGF-2 and PDGF-BB or generation of antibodies. *Gene Ther.*, Vol.16, No.9, pp.1097-1110.

Kaloss, M., Linscott, M., Wey, C., Lu, P., Long, Z., McGarrity, GJ., Otto, E., & Lyons, RM. (1999). Distribution of retroviral vectors and vector producer cells using two routes of administration in rats. *Gene therapy*, Vol.6, pp.1389-1396.

Kassim, SH., Li, H., Vandenberghe, LH., Hinderer, C., Bell, P., Marchadier, D., Wilson, A., Cromley, D., Redon, V., Yu, H., Wilson, JM., & Rader, DJ. (2010). Gene therapy in a humanized mouse model of familial hypercholesterolemia leads to marked regression of atherosclerosis. *PLoS One*, Vol.5, No.10, e13424.

Khan, T., Muise, ES., Iyengar, P., Wang, ZV., Chandalia, M., Abate, N., Zhang, BB., Bonaldo, P., Chua, S., & Scherer, PE. (2009). Metabolic Dysregulation and Adipose Tissue Fibrosis: Role of Collagen VI. *Molecular and Cellular Biology*, Vol.29, No. 6, pp.1575– 1591.

Lakshmipathy, U., Pelacho, B., Sudo, K., Linehan, JL., Coucouvanis, E., Kaufman, DS., & Verfaillie, CM. (2004). Efficient transfection of embryonic and adult stem cells. *Stem Cells*, Vol. 22, pp.531-543.

Lam, JK., Ma, Y., Armes, SP., Lewis, AL., Baldwin, T., & Stolnik, S. (2004). Phosphorylcholine-polycation diblock copolymers as synthetic vectors for gene delivery. *J. Control Release.*, vol.100, No.2, pp.293-312.

Lee, LK., Siapati, EK., Jenkins, RG., McAnulty, RJ., Hart, SL., & Shamlou ,PA. (2003). Biophysical characterization of an integrin-targeted non-viral vector. *Med. Sci. Monit.*, Vol.9, No.1, pp.54-61.

Martinez-Climent , JA., Andreu, EJ., & Prosper, F. (2006). Somatic stem cells and the origin of cancer. *Clin. Transl. Oncol.*, Vol.8, No. 9, pp.647-63.

Mullen, PM., Lollo, CP., Phan, QC., Amini, A., Banaszczyk, MG., Fabrycki ,JM., Wu, D., Carlo, AT., Pezzoli , P., Coffin, CC., & Carlo, DJ. (2000). Strength of conjugate binding to plasmid DNA affects degradation rate and expression level *in vivo*. *Biochim. Biophys. Acta.*, Vol.1523, No. 1, pp.103-110.

Muller-Sieburg, CE., & Sieburg, HB. (2006). Clonal diversitiy of the stem cell compartment. *Cur. Opin. Hematol.*, Vol.13, No. 4, pp.243-248.

Nehlin, JO., & Barington, T. (2009). Strategies for future histocompatible stem cell therapy. *Biogerontology*, Vol.10, No.4, pp.339-376.

Neumann, E., Schaefer-Ridder, M., Wang, Y., & Hofschneider, PH. (1982). Gene transfer into mouse lyoma cells by electroporation in high electric fields. *EMBO J.*, Vol.1, No. 7, pp. 841–845.

Niidome, T., Ohmori, N., Ichinose, A., Wada, A., Mihara, H., Hirayama, T., & Aoyagi, H. (1997). Binding of cationic alpha-helical peptides to plasmid DNA and their gene transfer abilities into cells. *J. Biol. Chem.*, Vol.272, No.24, pp.15307-12.

Oliveira, AC., Ferraz, MP., Monteiro, FJ., & Simões, S. (2009). Cationic liposome-DNA complexes as gene delivery vectors: Development and behaviour towards bone- like cells. *Acta. Biomater.*, Vol.5, No.6, PP.2142-2151.

Park, JO., Park, SH. & Hong, ST. (2009). A new method for transduction of mesenchymal stem cells using mechanical agitation. *Mol. Cell*, Vol.28, No. 6, pp.515-520.

Peister, A., Mellad, JA., Wang, M., Tucker, HA., & Prockop, DJ. (2004). Stable transfection of MSCs by electroporation. *GeneTher.*, Vol. 11, pp.224-228.

Savva, M., Aljaberi, A., Feig, J., & Stolz, DB. (2005). Correlation of the physicochemical properties of symmetric 1,3-dialkoylamidopropane-based cationic lipids containing single primary and tertiary amine polar head groups with in vitro transfection activity. *Colloids Surf B Biointerfaces.*, Vol.43, No.1, pp.43-56.

Scherer, F., Anton, M., Schillinger, U., Henke, J., Bergemann, C., Krüger, A., Gänsbacher, B., & Plank, C. (2002). Magnetofection: enhancing and targeting gene delivery by magnetic force *in vitro* and *in vivo*. *Gene Ther.*, Vol. 9, No.2, pp. 102–109.

Segura, T., & Shea, LD. (2001). Materials for non-viral gene delivery. *Annual Review of Materials Research*, Vol. 31, pp. 25–46.

Suki, B., & Jason, HT. (2008). Bates Extracellular matrix mechanics in lung parenchymal diseases. *Respir.Physiol. Neurobiol.*, Vol.30, No.163, pp.33–43.

Tenenbaum, L., Lehtonen, E., & Monahan, PE. (2003). Evaluation of risks related to the use of adeno-associated virus-based vectors. *Curr. Gene Ther.*, Vol.3, No. 6, pp.545-565.

Tonti, GA., & Mannello, F. (2008). From bone marrow to therapeutic applications:different behaviour and genetic/epigenetic stability during mesenchymal stem cell expansion in autologous and foetal bovine sera? *Int. J. Dev. Biol.* Vol.52, pp.1023- 1032.

Tseng, AS., Engel, FB., & Keating, MT. (2006). The GSK-3 inhibitor BIO promotes proliferation in mammalian cardiomyocytes. *Chem. Biol.*, Vol.13, No.9, pp.957-963.

Vighi, E., Ruozi, B., Montanari, M., Battini, R., & Leo, E. (2007). Re-dispersible cationic solid lipid nanoparticles (SLNs) freeze-dried without cryoprotectors: characterization and ability to bind the pEGFP-plasmid. *Eur. J. Pharm. Biopharm.*, Vol.67, No.2, pp.320-328.

Yizhi, Song., Garaventa, G., Bottero, S., Leprini, AE., Pallestrini, E., Castello, E., Pallestrini, EA., & Huang, WE. (2007). Ultrasound-mediated DNA transfer for bacteria. *Nucleic Acids Res.*, Vol.35, No.19, pp. 1073–1080.

6 Chapter

GENE THERAPY FOR CERTAIN GENETIC DISEASES INVOLVES THE TRANSFER OF NORMAL HUMAN GENOMIC DNA INTO SOMATIC CELLS AND STEM CELLS TAKEN FROM PATIENTS

1. INTRODUCTION: AN OVERVIEW

1.1 Viral vectors for gene therapy

Gene therapy is a method of rectifying faulty genes in living organisms by transferring healthy genes into the cells of individuals with genetic disorders or cancer. It can also involve introducing novel genes into cells to produce therapeutic proteins. Various viruses, such as adenoviruses, alphaviruses, retroviruses, lentiviruses, adeno-associated viruses (AAV), and herpes simplex viruses type 1 (HSV-I), have been utilized as carriers to introduce healthy genes into patients' cells for gene therapy (Nayak & Herzog, 2010; Raper et al., 2003; Lundstrom, 2001, 2005; Aiuti et al., 2009; Bordignon et al., 1989, 1995; Cavazzana-Calvo et al., 2000; Ferrari et al., 1991; Halatsch et al., 2000; Dupré et al., 2004; Mortellaro et al., 2006; Jayandharan et al., 2011; Terzi & Zachariou, 2008; Epstein, 2009). Nevertheless, the utilization of these vectors was impeded by several constraints and obstacles. Retroviruses and lentiviruses have the ability to integrate into the DNA of humans and may lead to the development of cancer (Dave et al., 2004; Du et al., 2005; Hacein-Bey-Abina et al., 2003a, 2003b; Z. Li et al., 2002; Modlich et al., 2005; Seggewiss et al., 2006). Viruses have the ability to infect both healthy cells and defective/cancer cells in patients, as long as the cells possess receptors that can bind to the viruses. This can result in severe infections, inflammatory responses, and immune reactions.

1.2 Highly pathogenic (virulent) viruses, moderately pathogenic viruses, and lowly or mildly pathogenic viruses

Davies et al., 1987; Jones et al., 1987). The presence of many serotypes and subserotypes is mostly attributed to this significant factor. Due to the presence of 12 separate segments of double-stranded RNA in the seadornavirus genome, the process of reassortment allows for the creation of new genotypes by combining diverse strains of the virus in nature. In China, there are a minimum of 6 distinct genotypes of seadornavirus, along with multiple serotypes and subserotypes within the Chinese isolates (Q. Li et al., 1992; L. Song et al., 1995;L . Song & Chen, 1995, 1996; Tao et al., 1999; L. Xu et al., 2003; P. Xu et al., 1990; You et al., 1990). Similar seadornaviruses were isolated from mosquitoes collected in Indonesia (Brown et al., 1993) and Vietnam (Nabeshima et al., 2008). The virus was classified as a probable member of the genus Coltivirus previously, and later it was renamed as a member of a novel genus-Seadornavirus within the family of Reoviridae (Mohd Jaafar et al., 2005). Some viruses like M14-a nonpathogenic twelve-segmented double-stranded RNA virus isolated from mosquitoes in China (C. Huang et al., 1985, Liang et al., 1985) are lowly or mildly virulent viruses. The majority of viruses like hepatitis A, B, C, D, and E virus, polio virus, measles virus, mumps virus, West Nile virus, influenza virus, Coxsackie A virus, enterovirus 71, rhinoviruses, coronaviruses, norovirus, rubella virus, and the newly isolated member of bunyavirus which caused severe fever and thrombocytopenia syndrome in China (X-J. Yu et al., 2011), have moderate pathogenicity. There are very few human viruses are truly nonpathogenic viruses in nature, except some animal viruses that mainly infect animals but not humans. These mildly virulent viruses cannot cause obvious infections in humans (Csatary et al., 1985).

It is important to recognize that many mild viruses, although they may not cause severe infections in healthy individuals, can nonetheless pose a threat to individuals with weaker immune systems, such as late-stage cancer patients, elderly individuals, critically ill patients, and those with immunodeficiency diseases. The majority of viruses undergo modification and attenuation before to their use as vectors in gene therapy. However, in extremely rare cases, even these modified viral vectors can give rise to complications. An 18-year-old male with a partial deficit of ornithine transcarbamylase died following a clinical trial of gene therapy, despite the use of a modified human adenovirus type 5 virus as the vector in that experiment (Raper et al., 2003).

1.3 Reovirus is not an oncolytic virus

There exist A number of specialists worldwide have been endeavoring to utilize specific oncolytic viruses for the purpose of treating cancer (Pennisi, 1998). A significant challenge of these medicines is the difficulty in identifying an optimal wild-type oncolytic virus that exclusively targets cancer cells without affecting normal cells.

The LLC-MK2 cell line, derived from a normal rhesus monkey kidney, was first established in 1955 by Evans et al. (1959) and further studied by Hull et al. in 1956 and 1962. The mouse L929 cell line was initially characterized by Stanford et al. in 1948. The L929 cell line is a genetically identical strain derived from its parental mouse cell strains through cloning. The L. L cell strain was derived from the normal subcutaneous areolar and adipose tissue of a male mouse, as described by Earle in 1943 and Stanford et al. in 1948. These two cell lines were extensively utilized for the isolation, cultivation, and amplification of various types of viruses, including reoviruses.

Reovirus (Respiratory Enteric Orphan Virus) is a member of the family Reoviradae. It got the name originally because it was often isolated from human respiratory and enteric systems but no obvious human disease was associated with it. Reovirus can cause cytopathic effect (CPE) in many normal cell lines like rhesus monkey kidney LLC-MK2 and MA-104E, African green embryonic monkey kidney Vero, baby hamster kidney BHK-21, Buffalo green monkey kidney BGM, African green monkey kidney BS-C-1, Madin-Darby bovine kidney (MDBK), Madin-Darby canine kidney (MDCK), human embryonic intestinal (intestinal 407), human embryonic lung (HEL), and mouse L929 cells. After a few days of cell culture, like most other viruses, reovirus will destroy and lyse the cells it infected eventually in vitro (McClain et al., 1967; Nibert et al., 1991; Ridinger et al., 1982; Rozee & Easterbrook, 1970; Schiff et al, 2007; L. Song et al., 1995, 1999b, 2000, 2009).

Reovirus can infect and kill both normal cells and human tumor cells in vitro, as long as the cells have reovirus receptor-junctional adhesion molecule (Antar et al., 2009; L. Song et al., 1999b, 2000, 2009; van den Wollenberg et al., 2008; van Houdt et al., 2008). If a small number of reoviruses are injected into tumor tissues directly, the virus will infect and kill some tumor cells locally. In the meantime, the human immune system will fight with the virus, a lot of immune cells, such as T cells, B cells, natural killer cells, neutrophils, and macrophages will be recruited to the infection site, and the immune cells will produce antibodies, chemokines, and cytokines like interferons, interleukins; and after a few days, before the virus spreading to other parts of the body, the virus will be killed, and be cleared from the human body. If a large number of reoviruses are injected into the tumor body, the viruses will infect tumor cells and nearby normal cells, and spread to other organs of the body and cause systemic infection. This could be fatal for some cancer patients, as we know that many cancer patients have unbalanced, weakened, and dysfunctional immune systems. A great number of cancer patients are treated with radiation and immunosuppressive anticancer drugs, these anticancer therapies can damage immune cells further. Cancer patients with weakened immune systems have more chances to have opportunistic infections (Baggiolini et al., 1997; Bodey, 1986; Dunn et al., 2002, 2004; Locati & Murphy, 1999; Lodish et al., 2008; Luster, 1998; Murdoch & Finn, 2000; Nibert et al., 1991; Pitisuttithum et al., 2001; Schiff et al, 2007; Sutlu & Alici, 2009; Swann & Smyth, 2007; Virgin, 2007). This is the same problem we are facing when a patient has chemotherapies nowadays; there are rare drugs that only specifically and selectively target cancer cells but not normal cells. Over doses of anticancer drugs will kill both normal cells and tumor cells of patients, and lead to serious side effects and deaths; normal doses or small doses of anticancer drugs will not kill all the cancer cells, and the remained cancer cells will overexpress a membrane protein-P-glycoprotein, and be able to resistant to the cell kill effects of multi-anticancer drugs (Arkin et al., 1989; Croop et al., 1988; De Rosa et al., 2008; Debenham et al., 1982; Deuchars et al., 1987; Endicott & Ling, 1989; Goldstein et al., 1989; Juliano & Ling, 1976; Kobayashi et al., 1994, 1998; Moscow & Cowan, 1988; Pastan & Gottesman, 1987; Riordan et al., 1985; L. Song et al., 1999a).

In 2003, L-H. Song et al. discovered a reovirus in the throat swabs of a patient with severe acute respiratory syndrome (SARS) in Beijing. This reovirus was found to cause similar clinical symptoms to SARS in guinea pigs and macaques. Antarasena et al. identified and separated some avian reoviruses from chickens that had abrupt mortality in Thailand (Antarasena et al., 2002). Chua et al. (2007) revealed that a bat-derived reovirus has the capability to induce acute respiratory illness in people. It is important to note that more than 50% of individuals have been infected with reovirus at some point in their lives, with many of these infections occurring during early childhood (Selb & Weber, 1994; Tai et al., 2005). Furthermore, there is growing evidence suggesting that reovirus can infect normal human cells in the body and lead to various mild to severe illnesses such as upper respiratory illnesses and meningitis in humans (Johansson et al., 1996; Schiff et al., 2007; Tyler et al., 2004). Therefore, the previous belief that reoviruses were "orphan" viruses not associated with any human diseases is no longer accurate and should be revised. According to L. Song (2010), it is not appropriate to classify wild-type reovirus as an oncolytic virus, and it is improbable that reovirus could be a successful and feasible anticancer treatment.

1.4 Non-viral vectors for gene therapy

Peptide and polymer mediated gene therapy, which are non-viral vectors, have limited ability to create long-lasting expression of genes. Additionally, their efficacy in transferring genes into cells is significantly lower when compared to viral vectors (Al-Dosari & Gao, 2009; Cartier & Reszka, 2002; X. Gao et al., 2007; Niidome & Huang, 2002).

2. A POSSIBLE APPROACH FOR GENE THERAPY OF SOME GENETIC DISEASES

It is well known that a bacterium can obtain foreign DNA from another bacterium through a process of bacterial conjugation (Lederberg & Tatum, 1946). In this process, DNA is directly transferred from one cell into another cell via direct cell-to-cell contact or via a bridge-like structure between two cells. By this way, a bacteium's genomic DNA could be changed by homologous recombination with another bacterium's genomic DNA.

As above pointed out, multisegmented RNA viruses like influenza virus can reassort their genomic segments in vivo, if an animal is infected by two or more different strains of a virus in the same time period. This can create new strains of the virus, and the new strains carry the gene segments of their parental strains. There are some triple reassorted influenza virus strains in nature (V. Gregory et al., 2001; Ma et al., 2010; Octaviani et al., 2011; Rambaut et al., 2008; Smith et at., 2004; Vincent et al., 2006; Webby et al., 2000; X. Xu et al., 2002, 2004; Zhou et al., 1999).

Complementation test was used to study genetic subtypes (complementation groups) of a genetic disease like Fanconi anemia. If cells from two patients with Fanconi anemia were from two different complementation groups, the defective genes could be repaired after fusion of the two genetic complementation cells; and the hybrid cells were able to resist the attack of DNA cross-linking agents such as mitomycin C and diepoxybutane. Whereas, if the cells of two patients were from the same complementation group, the hybrid cells would still be sensitive to the attack of DNA cross-linking reagents (Buchwald, 1995; Duckworth- Rysiecki et al., 1985; Fanconi Anaemia/Breast Cancer Consortium, 1996; Giampietro et al., 1997; Joenje et al., 1995; Joenje & Patel, 2001; Levitus et al., 2004; Tischkowitz & Hodgson, 2003; Whitney et al., 1995). 14 different complementation groups of Fanconi anemia have been discovered, and 14 distinct Fanconi anemia genes have been identified (FANCA, FANCB, FANCC, FANCD1, FANCD2, FANCE, FANCF, FANCG, FNACI, FANCJ, FANCL, FANCM, FANCN, FANCP) (Alderton, 2011; Kim et al., 2011; Levitus et al., 2006; Reid et al., 2007; Smogorzewska et al., 2007; L. Song, 2009; Stoepker et al., 2011; Taniguchi & D'Andrea, 2006; Tischkowitz et al., 2008).

It is very interesting that a number of patients with some genetic diseases were cured or improved naturally later in their lives. Somatic mosaicism was involved in this form of miracle "natural gene therapy" process. Somatic mosaicism means that there are genetically different somatic cells exist in a given organism resulting from in vivo reversion of a mutated allele to wild type. Somatic mosaicism has been found in several genetic disorders, such as hemophilia B, tyrosinemia type I, Bloom syndrome, adenosine deaminase deficiency, epidermolysis bullosa, Wiskott-Aldrich syndrome, androgen insensitivity syndrome, T-cell immunodeficiency, leukocyte adhesion deficiency type 1, Duchenne muscular dystrophy, atypical X-linked severe combined immunodeficiency, and Fanconi anemia. The mechanism of somatic mosaicism is complicated; it might be due to epigenetic alterations of DNA, copy number variations, back mutation, gene conversion, frame- restoring mutation, DNA polymerase slippage, and compensatory mutation in cis. Homologous genetic recombination between paternally and maternally derived chromosomes also plays a vital role in somatic mosaicism (Darling et al., 1999; J. Jr. Gregory et al., 2001; M. Gross et al., 2002; Hirschhorn, 2003; Lo Ten Foe et al., 1997; Mankad et al., 2006; Müller & Williams, 2009; Notini et al., 2008; Piotrowski et al., 2008; L. Song, 2009; Wada et al., 2001, 2003; Waisfisz et al., 1999; Youssoufian & Pyeritz, 2002).

Homologous genetic recombination was first discovered in bacterium *Escherichia coli* strain K-12 in 1946 by Joshua Lederberg (May 23, 1925 – February 2, 2008), winner of the 1958 Nobel Prize in Physiology or Medicine (Lederberg & Tatum, 1946; Lederberg, 1947, 1987a, 1987b; Tatum & Lederberg, 1947).

Homologous recombination happens when two homologous DNA molecules meet in vivo. They pair up and exchange some sequences. Homologous genetic recombination occurs in the processes of mitosis and meiosis. Gene recombination takes place between two nonsister chromatids of the two homologous chromosomes by crossover during meiosis. Homologous recombination happens much more often in distal regions of chromosomes and on shorter arms of chromosomes. Crossover occurs at least one time per chromosome in each of the process of meiosis. Gene sequences are exchanged during the process of meiosis by crossover (Creighton & McClintock, 1931, 1935; Holliday, 1974; International Human Genome Sequencing Consortium, 2001; Weil, 2002; Whitby, 2005).

Crossover during the process of meiosis is really a smart and fair way to let a female's eggs or a male's sperms to have genetic information from both of her/his parents; and when an egg and a sperm form a new life in the womb, the new baby carries the genetic information from both of his/her grandparents on his/her father's side and grandparents on his/her mother's side; so a baby's genetic traits are inherited from his/her four biological grandparents, this could make the baby more diversity, flexible, and fit.

Radioactive materials like radon gas in some basement rooms, ultraviolet (UV) light, toxic chemicals, reactive-oxygen compounds, polluted air, and smoking, some viral infections all can damage human genomic DNA, cause mutations, and lead to cancers. Human cells are able to cope with outside and inside challenges, and to repair the damaged DNA molecules by several ways. One of the DNA repair mechanisms is homologous recombination to repair DNA gaps, DNA double-stranded breaks, and DNA interstrand crosslinks. A damaged chromatid can be repaired by its undamaged sister chromatid or its homologous nonsister chromatid through homologous recombination during mitosis. Sister chromatids are the preferred templates over homologous or heterologous chromosomes for recombination repair in yeast and mammalian cells. Instead of crossover, gene conversion is the major result of homologous recombination. Cells seem reluctant to crossover their chromatids unnecessarily to maintain their genome's integrity and stability (Hope et al., 2007; Jackson & Bartek, 2009; Johnson & Jasin, 2001; Kadyk & Hartwell, 1992; X. Li & Heyer, 2008; Lorenz & Whitby, 2006; Willers et al., 2004).

Gene targeting (gene knockout) is a technique that a vector is used to deliver a fragment of mutated DNA into embryonic stem cells and to target its homologous DNA in the genome by homologous genetic recombination. Thousands of genes in mice were knocked out by using this method. Mario R. Capecchi, Martin J. Evans, and Oliver Smithies were awarded jointly the Nobel Prize in Physiology or Medicine in 2007 for their discoveries of principles for creating knockout mice (Bradley et al., 1984; Bronson & Smithies, 1994; Capecchi, 1989a, 1989b; Koller & Smithies, 1992; Kuehn et al., 1987; Robertson et al., 1986).

It was assumed that plant tissue grafts did not have gene exchange, but this is not true. A recent discovery found that even in plant tissue grafts, some of their genes were exchanged (Stegemann & Bock, 2009).

Based on the above observations and experiments, I proposed a possible approach for gene therapy of some genetic diseases as indicated in figure 1 (L. Song, 2009).

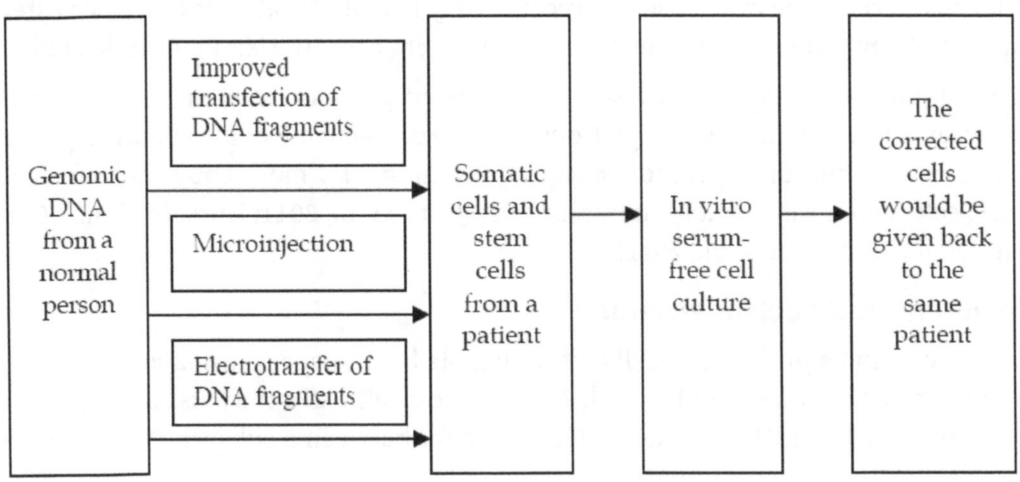

Fig. 1. Gene therapy of some genetic diseases by transferring normal human genomic DNA/DNA fragments into cells from patients.

Briefly, normal human genomic DNA or genomic DNA fragments from a healthy donor can be transferred into somatic cells and stem cells from a patient by microinjection, gene electrotransfer (electroporation), and improved transfection. After in vitro serum-free cell culture, the defective genes could be repaired by homologous genetic recombination, since the genomic DNAs of the two persons are considerably similar, although there are deletions, insertions, rearrangements, loss-of-function variants, and copy number variations (CNV) (Fujimoto et al, 2010; Levy et al., 2007; H. Park et al., 2010; The 1000 Genomes Consortium, 2010, 2011; J. Wang et al., 2008; Yim et al., 2010).

3. DISCUSSION

3.1 Stem cells, somatic cells

Stem cells were first discovered by Ernest Armstrong McCulloch (April 27, 1926–January 20, 2011) and James Edgar Till in 1963 (Becker et al., 1963; McCulloch et al., 1964; Siminovitch et al., 1963; Till et al., 1964; Weissman & Shizuru, 2008). Stem cells have the ability to self- renew and to differentiate into different cell types. Early stage embryonic stem cells can form all types of cells. Adult stem cells only can differentiate and generate specialized cells, like bone cells, liver cells, blood cells, skin cells. The stem cells that produce all the blood cell types are called hematopoietic stem cells (Bordignon, 2006; Spangrude et al., 1988; Thomson et al., 1998; Weissman & Shizuru, 2008).

Stem cells and their differentiated cells (somatic cells) maintain a balance-homeostasis. When under stress, stem cells are activated, and start to produce more differentiated cells (Jiang et al., 2009; Martinez-Agosto et al., 2007; Till et al., 1964; Wilson et al., 2004).

Somatic cells are the end products of stem cells, they are unable to self-renew.

3.2 Induced pluripotent stem cells

In recent years, a few transcription factors (genes) were introduced into somatic cells by lentiviral or retroviral vectors, to reprogram somatic cells into pluripotent stem cells (Liao et al., 2008; I. Park et al., 2008; Takahashi et al., 2007; J. Yu et al., 2007).

Some of the genes like c-Myc that used to form induced pluripotent stem cells are oncogenes, and as above revealed, retroviral or lentiviral vectors might integrate into genome DNA randomly, these risk factors might lead to producing cancer cells. A recent study revealed that human induced stem cells were easier and faster to form tumors than human embryonic stem cells (Gutierrez-Aranda et al., 2010).

Some recently published papers disclosed that induced pluripotent stem cells could induce more immune responses and immunological rejections in the recipient mice, and had more protein-coding point mutations, more abnormal epigenomic reprogramming, and more copy number variations than normal somatic cells and normal embryonic stem cells (Gore et al., 2011; Hussein et al., 2011; Laurent et al., 2011; Lister et al., 2011; Zhao et al., 2011).

3.3 Cell membrane and nuclear envelope

Both a eukaryotic cell and a prokaryotic cell have a flexible lipid bilayer plasma membrane that controls movement of molecules in and out of the cell. A eukaryotic cell has a nucleus, while a prokaryotic cell does not have a nucleus; this is the characteristic difference between a eukaryotic cell and a prokaryotic cell.

The eukaryotic cell nucleus is surrounded by a nuclear envelope with nuclear pores. The nuclear envelope has two layers: the out nuclear membrane which faces the cytoplasm, and the inner nucleic membrane which faces the nucleoplasm. The nuclear pores are formed by nuclear pore complexes (NPCs) that span the double lipid bilayer of the nuclear envelope. The NPCs are formed by about 30 proteins.

NPCs are gatekeepers of the nucleus (Alber et al., 2007; D'Angelo et al., 2006; D'Angelo & Hetzer, 2008; Devos et al., 2006; Fernandez- Martinez & Rout, 2009; Lam & Dean, 2010; Terry et al., 2007; Theerthagiri et al., 2010; E. Tran & Wente, 2006).

Ions and small molecules and DNA smaller than 200 bp can diffuse through the nuclear pore freely; while the transport of DNA molecules between 310 bp and 1500 bp from the cytosol to the nucleus is through an active transport process. DNA greater than 2 kb can rarely be seen in the nucleus (Cartier & Reszka, 2002; Hagstrom et al., 1997; Ludtke et al., 1999).

A foreign DNA molecule has to go through the human cell membrane, cytoplasm, and the nuclear envelope to reach the genomic DNA in the nucleus. This process can be performed and prompted by microinjection, electroporation, and transfection.

3.4 Microinjection

Microinjection technique has been used in transgenic animals for many years (Bishop & Smith, 1989; Chan & Yang, 2009; Charreau et al., 1996; Filipiak & Saunders, 2006; Ménoret et al., 2010; Tesson et al., 2005; Yang et al., 2008). Microinjection technique also has been used as a tool to clone animals-first, an unfertilized egg's nucleus is removed; then a nucleus of a somatic cell is microinjected into the denucleated egg; now the egg contains a whole copy of the diploid genomic DNA from the somatic cell and can be cultured in vitro to form a blastocyst; and the blastocyst is implanted into the womb of an animal; eventually a cloned animal is born (Campbell et al., 1996; Vajta & Gjerris, 2006; Willadsen, 1986; Wilmut et al., 1997).

Several different genes inserted into plasmids were microinjected into cultured mammalian somatic cells, and some genetic defective genes were corrected by homologous recombination (W. Anderson et al., 1980; Capecchi, 1980; Folger et al., 1982; Yamaizumi et al., 1983).

Feng et al. introduced a 110 kb whole human alpha globin gene cluster clone in a bacterial artificial chromosome (BAC) vector into fertilized eggs to generate transgenic mice by microinjection method. The human alpha globin gene cluster DNA was integrated into the mice genome, and human alpha globin mRNA was expressed in 3 transgenic mice (Feng et al., 2001). Similarly, Gao et al. generated transgenic mice carrying a BAC clone of a 116 kb human *apoAI/CIII/AIV/AV* gene cluster and a mutant in which the *apoCIII* enhancer was deleted from the 116 kb gene cluster by microinjection (J. Gao et al., 2005).

I assume that normal genomic DNA without a plasmid or an artificial chromosome can be directly microinjected into the nucleus of somatic cells and stem cells from a patient successfully. This method could have a higher homologous recombination rate and less immunological reactions. It is not a very convenient method, but I think it is worth the effort to try. It only needs 30 purified mouse hematopoietic stem cells to save 50% of lethally irradiated mice (Spangrude et al., 1988). Even one single stem cell transplant can significantly reconstruct the bone marrow function of some irradiated mice (Decker & Nyberg, 2001; Krause et al., 2001; Mankad et al., 2006; Osawa et al., 1996). Therefore, we might need to collect less than one hundred stem cells from a patient, and microinject normal genomic DNA into these cells. Hopefully, less than one hundred of these corrected cells are sufficient to improve a patient's physiological function significantly.

3.5 Electroporation

Electroporation or electropermeabilization has been used to transfer foreign plasmid DNA into bacteria, yeast, and mammalian cells (Escoffre et al., 2009; Favard et al., 2007; Golzio et al., 2010; Mir, 2009; Neumann et al., 1982; Somiari et al., 2000). It might be difficult to transfer large genomic DNA molecules into mammalian cells by this method. We might first digest the normal human genomic DNA by restriction enzymes, and then transfer the normal genomic DNA fragments into a patient's stem cells and somatic cells by electroporation in vitro. After a few days of in vitro cell culture in serum free media, the cells can be transplanted into the same patient. Of course, before conducting human clinical trials, this kind of experiment should be performed in animal models first.

3.6 Genomic DNA Transfection

Whole genomic DNA molecules are too big to be transfected into cells directly. Normal genomic DNA can be digested by a few restriction enzymes first, and then the purified genomic DNA fragments can be transfected into stem cells and somatic cells from patients.

Molecules commonly used for transfection are smaller than 10 kb; transfection efficiency is very low with plasmids of 12 kb or bigger (Campeau et al., 2001; Cartier & Reszka, 2002). Transfection is a relatively simple, easy, and convenient method to transfer a foreign DNA into a cell, but the current transfection methods cannot satisfy our needs when we want to transfer large DNA fragments. We have to improve the transfection efficiency, and new methods and advanced techniques are needed to transfer large genomic DNA fragments.

A cell culture medium with a little bit lower osmotic pressure can cause cell osmotic swelling, and the cells become bigger, cell membrane permeability is increased, the nuclear pores might become bigger also. Therefore, bigger size of DNA molecules might be easier to enter the swelling cells and reach the genomic DNA inside the nucleus. After transfection, the transfected cells are grown in a cell culture medium with normal osmotic pressure for a period of time, and let the cells to recover to normal. The recovered transfected cells can be transplanted into animal models of a genetic disease.

3.7 Genome sequencing and human genetic variation

Human somatic cells are diploid, each somatic cell has 23 homologous chromosome pairs (46 chromosomes), 23 of the chromosomes are from a sperm of the father and other 23 chromosomes are from an egg of the mother. The paired homologous chromosomes are similar in length, except the pair of X and Y chromosomes, an X chromosome is much longer than a Y chromosome. The human sperm cells and egg cells are haploid-each of them has 23 chromosomes.

Different species of animals or plants have different number of chromosomes. A chimpanzee has 48 chromosomes (Young et al., 1960), a dog has 78, a chicken has 78, a pig has 38, a cat has 38, a horse has 64, a cow has 60, a goat has 60, a sheep has 54, a mouse has 40, and a rat has 42 chromosomes separately (O'Brien et al., 1999); and a silkworm has 28 chromosomes (International Silkworm Genome Consortium, 2008; Xia et al., 2004). Wheat has three ploidy levels: diploid wheat (*Triticum urartu, Aegilops speltoides*, and *Ae. tauschii*) has 14 (2x), tetraploid wheat (*Triticum turgidum* ssp. *dicoccoides*) has 28 (4x), and hexaploid wheat (*Triticum aestivum*) has 42 (6x) chromosomes respectively; diploid wheat is the ancestor of the tetraploid and hexaploid wheat (Akhunov et al., 2005, 2010; Dvorak & Akhunov, 2005; S. Huang et al., 2002).

Each human gene has two alleles, one allele on each chromosome of the homologous pair. If the 2 alleles have the same sequence, they are called homozygotes; otherwise, they are called heterozygotes.

A specific phenotype (trait) might be determined by two alleles (recessive) or by one allele (dominant). In autosomal recessive genetic diseases like cystic fibrosis, sickle-cell anemia, and fanconi anemia (except FANCB), if a mutant gene appears on both of the paired homologous chromosomes of a person, this person has the genetic disease; if the mutant gene occurs on one of the homologous chromosomes of a person, this person is a carrier of the genetic disease. In X-linked recessive genetic diseases like Fanconi anemia subtype B, Duchenne muscular dystrophy, and Wiskott-Aldrich syndrome, if a male's X chromosome carries the mutated gene, this male has the genetic disease; if one of a female's X chromosomes carries the mutated gene, this female is a carrier; if both of a female's X chromosomes carry the mutated gene, this female has the genetic disease. On the other hand, in autosomal dominant genetic diseases like Huntington's disease, it only needs one mutated allele on any of the two homologous chromosomes to have the related genetic disease; there is no carrier of a dominant genetic disease, because every person who has the mutated allele gets the disease.

A diploid genome sequence showed that we are genetically more diverse than we have claimed before (International Human Genome Sequencing Consortium, 2001, 2004; Venter et al., 2001) based on the haploid genome sequences, and the difference between two homologous chromosomes of a pair of chromosomes inherited from one's parents is bigger than we thought before. There were more than 4.1 million DNA sequence variants in this new diploid genome. Single-base variations -single nucleotide polymorphisms (SNPs) are the major variants, small fragments insertions or deletions (indels), large fragments deletions and duplications- copy number variations also contribute to the genomic variation significantly (L. Gross, 2007; Levy et al., 2007).

J. Wang et al. sequenced a Chinese diploid genome sequence (named YH) and found about 3 million SNPs in YH's genome, of which 13.6% were new compared to the SNP database dbSNP. They compared the 3 known genome sequences and recognized that the genomes of YH, Venter, and Watson shared 1.2 million SNPs, and their unique SNPs were 31.8% (YH), 30.1% (Venter), and 33.0% (Watson) separately (J. Wang et al., 2008). Koreans and Chinese were historically related, and they might have the same ancestors. The diploid genome sequence of a Korean male (named SJK) was significantly different from the Chinese YH; there were 1.3 million different SNPs between the two persons; even though SJK shared more SNPs with YH than with Caucasians Venter and Watson, and the Nigerian male Yoruba. 420, 083 (12.2%) SNPs of SJK were not found in the dbSNP database before, and 39.87% of the SNPs were SJK-specific (S. Ahn et al., 2009).

More than 99% of the genomic DNA sequences of a Japanese male were same to the reference human genome, but there were still 3,132,608 single nucleotide variations (SNVs) compared to other six reported human genomes (Fujimoto et al., 2010).

3.8 Copy number variations (CNV)

We are in a new era of personalized genomic medicine. With the significantly advanced and simplified new DNA sequencing tools and methods available in a few years, we would be able to know the whole genomic DNA sequence of every person in a few hours at an affordable price (less than one thousand dollars) (L. Gross, 2007).

In addition to each person has his /her unique protein-coding sequences, deletions, insertions, and inversions, copy number variations is one of the main reasons that we are different from each other genetically. Copy number variations is a hot topic of research in recent years, the aim of the studies is to disclose some possible diseases caused/or influenced by copy number variations; and how copy number variations might determine, regulate, and affect our genetic traits and social behaviours. Park et al. discovered 5,177 CNVs in 30 individuals of Korean, Chinese and Japanese, of which 3,547 were putative Asian-specific CNVs (H. Park et al., 2010). Every genome has about 40.3 CNVs averagely; the median length of CNVs is 18.9 kb. About 8% regions of the human genome are occupied by CNV regions (Yim et al., 2010).

The actual number of protein-coding genes might be bigger than we have claimed. (A. Ahn & Kunkel, 1993; Black, 2003; Dwi Pramono et al., 2000; Muntoni et al., 2003; Nishio et al., 1994; L. Song et al., 2003; V. Tran et al., 2005; Zhang et al., 2007). We do not know the meaning and usefulness of these non-coding sequences clearly so far, only one thing we are almost certain is that: they must have meaning and usefulness. We read many books, newspapers, and journals; we watched hundreds of movies, TV shows; we travelled numerous places, and met a lot of people. We do not know why and how we can remember all these things, and why the childhood memories can be stored in our brains for many years, and the memories can be recalled after so many years. If we can transfer the information from one person's brain to a computer, it might take up millions of gigabyte DVD space. In a human brain cell, only genomic DNA molecules could have such big storage capabilities to store such huge quantities of information. The mechanism of memory is one of the biggest challenges of our human beings; we should be able to uncover the secret of our brains with our own brains if we are on the right track. One day we might be able to know all the secrets of the silkworm and other organisms including humans.

1.11 Graft-versus-host disease (GVHD)

It is often hard to find a human leukocyte antigen (HLA)-identical sibling or a well-matched HLA unrelated donor when a patient needs hematopoietic stem cell transplant (HSCT). Sometimes, a patient had to receive a mismatched or partially matched bone marrow transplant and cord-blood transplant, when there was no HLA-matched unrelated donor available, and when a transplant was needed urgently. Acute and chronic graft-versus-host disease is the most severe and common long-term side effect of allogeneic hematopoietic stem cell transplantation (HCT). Acute GVHD was more likely to occur after mismatched marrow transplantation. Chronic GVHD was the major cause of late death of HSCT patients (Eapen et al., 2010; Laughlin et al., 2001, 2004; Mastaglio et al., 2010; Rocha et al., 2004).

Cells seem to be able to tolerate foreign DNA without immunological reactions; this is proved by the animal cloning experiments, transgenic animal models, and human and animal replication phenomena. Therefore, the possible approach I described above might have great benefits and advantages. Hopefully some genetic diseases listed below could be cured or improved by using this gene therapy method.

1.12 Fanconi anemia

Fanconi anemia (FA) is a rare chromosomal recessive genetic disease. As above cited, there are at least 14 subtypes of Fanconi anemia, and 14 genes whose mutation can cause FA are cloned. FANCB gene is on the X chromosome, and it is the only one on sex chromosomes, the other 13 FA genes are on autosomes. FA was first described by the Swiss pediatrician Guido Fanconi (1892-1979) in 1927 (Joenje & Patel, 2001; Lobitz & Velleuer, 2006; L. Song, 2009; Tischkowitz & Hodgson, 2003).

There are mouse models of Fanconi anemia available currently; FancA, FancC, FancG, FancD1, and FancD2 genes have been deleted or mutated in the mice (Parmar et al., 2009).

These 5 mouse models of Fanconi anemia will be used to prompt the research of gene therapy of these 5 subtypes of Fanconi anemia, because we can use these mouse models to do animal experiments. Normal genomic DNA or normal DNA fragments can be microinjected/electrotransfered, and transfected into stem cells and somatic cells from a mutated mouse; the corrected stem cells and somatic cells can be transplanted back to the same mouse, to see if the Fanconi anemia mouse model's physiological function is improved by this kind of gene therapy.

1.13 Sickle-cell anemia

Sickle-cell anemia is an autosomal recessive genetic disease. It results from a mutation at the sixth codon of the β hemoglobin gene on chromosome 11 (the hydrophilic amino acid glutamic acid is replaced by the hydrophobic amino acid valine). This mutation causes red blood cells to become rigid and inflexible. The patient's red blood cells are difficult to go through small capillaries, leading to stroke, chronic pain, anemia, and infection. This disease affects more than 300, 000 people worldwide (Ataga, 2009; Chang et al., 2006; Ingram, 1956, 1957; Pawliuk et al., 2001; Wu et al., 2006).

1.14 Cystic fibrosis

Cystic fibrosis is a common autosomal recessive genetic disease caused by mutations of cystic fibrosis transmembrane conductance regulator (CFTR) gene on chromosome 7 in Caucasian population (Kerem et al., 1989; Riordan et al., 1989; Rommens et al., 1989). CFTR is a cAMP-regulated chloride channel; the CFTR gene mutations lead to the cAMP-induced chloride channel dysfunction, thereby alter the transport of chloride and associated liquid, cause problems in several organ systems including respiratory system, sweat glands, pancreas, intestine, liver and gallbladder. There are more than 1800 CFTR gene mutations in the world. Cystic fibrosis affacts more than 70, 000 individuals worldwide. In 2006, the median survival age for a person with cystic fibrosis was 37 (M. Anderson et al., 1991; Collaco & Cutting, 2008; Collins, 1992; Cutting, 2010; Lee et al., 2005; Rowntree & Harris, 2003; G. Wang et al., 2005; Zielenski, 2000).

1.15 Duchenne muscular dystrophy

Duchenne muscular dystrophy (DMD) is the most common form of muscular dystrophy, it affects about one of every 3500 males. DMD is an X-linked recessive muscle degenerative disease caused by the mutations of dystrophin. As the above stated, the DMD gene is the largest human gene (>2.4 million bp on chromosome X), its cDNA is 14 kb long. DMD gene encodes a single 427 kDa protein-dystrophin. Patient's muscle fibers do not have the 427 kDa dystrophin (Burghes et al., 1987; Campeau et al., 2001; Hoffmanet al., 1987; Koenig et al., 1987, 1988; Kunkel, 2004; Monaco et al., 1986; Nelson et al., 2009).

1.16 Huntington's disease

Huntington's disease (HD) was first described by George Huntington in 1872. HD is an autosomal dominant neurodegenerative disease caused by the mutation of the huntingtin (HTT) gene. HTT gene located at 4p16.3; it has longer CAG trinucletide repeats (more than 40 CAG repeats) in the first exon of the HTT gene than the normal gene. There are transgenic mouse, sheep and monkey models available for conducting animal experiments currently (Bates et al., 1997; Beilby, 2007; S. Davies & Ramsden, 2001; Jacobsen et al., 2010; MacDonald et al., 1993; Yang et al., 2008).

3.17 X-linked severe combined immunodeficiency (SCID-X1)

X-linked severe combined immunodeficiency (SCID-X1) is caused by the mutations of interleukin-2 receptor subunit gamma (IL2RG) gene. Patients with the disease lack of T cells and natural killer cells, their B cells are functionally impaired; therefore, they are extremely vulnerable to infections (Aiuti & Roncarolo, 2009; Cavazzana-Calvo et al., 2000; Gaspar et al., 2004; Hacein-Bey-Abina et al., 2002, 2010).

1.18 Adenosine deaminase deficiency (ADA)-SCID

ADA- SCID is a rare genetic disease caused by a mutation of a gene on chromosome 20; this gene encodes an enzyme called adenosine deaminase (ADA). The mutation can lead to lack of ADA enzyme, and the lack of ADA enzyme causes disorder of adenosine metabolism and severe combined immunodeficiency (Aiuti et al., 2002, 2009; Aiuti & Roncarolo, 2009; Bordignon et al., 1989, 1995; Ferrari et al., 1991; Gaspar et al., 2009; Mortellaro et al., 2006).

1.19 Wiskott-Aldrich syndrome (WAS)

Wiskott-Aldrich syndrome (WAS) is an X-linked recessive primary immunodeficiency disease caused by mutations of the WAS protein (WASP) gene. The WASP gene is located on chromosome Xp11.22–Xp11.23. It has 12 exons, and encodes 502 amino acids. Patients with Wiskott-Aldrich syndrome have smaller platelets and lymphocytes, and their platelet counts are decreased; they have bleeding problems, recurrent bacterial and viral infections, and higher risk of autoimmune diseases and cancers. This disease affects about 1-10 in 1 million of live births (Aiuti & Roncarolo, 2009; Bouma et al., 2009; Dupré et al., 2004; Jin et al., 2004; Qasim et al., 2009; Ramesh et al., 1997; Zhu et al., 1997).

1.20 Other diseases

This gene therapy method has the potential to treat various diseases, including Alzheimer's disease (Rogaev et al., 1995; Sherrington et al., 1995), Parkinson's disease (Terzi & Zachariou, 2008; Veeriah et al., 2010), X-chronic granulomatous disease (CGD) (Aiuti & Roncarolo, 2009; Kang et al., 2010), type I (insulin-dependent) (Efrat, 1998), and type II (non-insulin-dependent) (Freeman et al., 1999) diabetes.

4. CONCISE SUMMARY

Normal human genomic DNA can potentially be utilized as materials for in vivo homologous genetic recombination to correct faulty genes. Human genomic DNA or fragments of genomic DNA can be introduced into somatic cells or stem cells from a patient using microinjection, transfection, or electroporation. The rectified cells can be reimplanted into the identical patient. Cells exhibit a capacity to endure foreign DNA without triggering immune responses, suggesting that the aforementioned technique could serve as an efficient and straightforward gene therapy procedure with minimal or no immunological reactions or rejections. Undoubtedly, the implementation of gene therapy should only occur following rigorous and meticulously planned cellular and animal investigations, as well as human clinical trials.

REFERENCES

Ahn, A. & Kunkel, L. (1993). The Structural and Functional Diversity of Dystrophin. *Nature Genetics*, Vol.3, No.4, (April 1993), pp. 283-291, ISSN 1061-4036

Ahn, S., Kim, T., Lee, S., Kim, D., Ghang, H., Kim, D., Kim, B., Kim, S., Kim, W., Kim, C., Park, D., Lee, Y., Kim, S., Reja, R., Jho, S., Kim, C., Cha, J., Kim, K., Lee, B., Bhak, J. & Kim, S. (2009). The First Korean Genome Sequence and Analysis: Full Genome Sequencing for a Socio-ethnic Group. *Genome Research*, Vol.19, No.9, (September 2009), pp. 1622-1629, ISSN 1088-9051

Aiuti, A., Slavin, S., Aker, M., Ficara, F., Deola, S., Mortellaro, A., Morecki, S., Andolfi, G., Tabucchi, A., Carlucci, F., Marinello, E., Cattaneo, F., Vai, S., Servida, P., Miniero, R., Roncarolo, M. & Bordignon, C. (2002). Correction of ADA-SCID by Stem Cell Gene Therapy Combined with Nonmyeloablative Conditioning. *Science*, Vol.296, No.5577, (June 2002), pp. 2410-2413, ISSN 0036-8075

Aiuti, A., Cattaneo, F., Galimberti, S., Benninghoff, U., Cassani, B., Callegaro, L., Scaramuzza, S., Andolfi, G., Mirolo, M., Brigida, I., Tabucchi, A., Carlucci, F., Eibl, M., Aker, M., Slavin, S., Al-Mousa, H., Al Ghonaium, A., Ferster, A., Duppenthaler, A., Notarangelo, L., Wintergerst, U., Buckley, R., Bregni, M., Marktel, S., Valsecchi, M., Rossi, P., Ciceri, F., Miniero, R., Bordignon, C. & Roncarolo, M. (2009). Gene Therapy for Immunodeficiency due to Adenosine Deaminase Deficiency. *The New England Journal of Medicine*, Vol.360, No.5, (January 2009), pp. 447-458, ISSN 0028-4793

Aiuti, A. & Roncarolo, M. (2009). Ten Years of Gene Therapy for Primary Immune Deficiencies. *Hematology, American Society of Hematology Education Program Book*, Vol.2009, (2009), pp. 682-689, ISSN 1520-4391

Akhunov, E., Akhunova, A. & Dvorák, J. (2005). BAC Libraries of *Triticum urartu, Aegilops speltoides* and *Ae. tauschii*, the Diploid Ancestors of Polyploid Wheat. *TAG Theoretical and Applied Genetics*, Vol.111, No.8, (November 2005), pp. 1617-1622, ISSN 0040-5752

Akhunov, E., Akhunova, A., Anderson, O., Anderson, J., Blake, N., Clegg, M., Coleman- Derr, D., Conley, E., Crossman, C., Deal, K., Dubcovsky, J., Gill, B., Gu, Y., Hadam, J., Heo, H., Huo, N., Lazo, G., Luo, M., Ma, Y., Matthews, D., McGuire, P., Morrell, P., Qualset, C., Renfro, J., Tabanao, D., Talbert, L., Tian, C., Toleno, D., Warburton, M., You, F., Zhang, W. & Dvorak, J. (2010). Nucleotide Diversity Maps Reveal Variation in Diversity among Wheat Genomes and Chromosomes. *BMC Genomics*, Vol.11, (December 2010), pp. 702. doi: 10.1186/1471-2164-11-702, ISSN 1471-2164

Alber, F., Dokudovskaya, S., Veenhoff, L., Zhang, W., Kipper, J., Devos, D., Suprapto, A., Karni-Schmidt, O., Williams, R., Chait, B., Sali, A. & Rout, M. (2007). The Molecular Architecture of the Nuclear Pore Complex. *Nature*, Vol.450, No.7170, (November 2007), pp. 695-701, ISSN 1061-4036

Alderton, G. (2011). Genomic Instability: Expanding the Reach of Fanconi Anaemia. *Nature Reviews Cancer*, Vol.11, No.3, (March 2011), pp. 158-159, ISSN 1474-175X

Al-Dosari, M. & Gao, X. (2009) Nonviral Gene Delivery: Principle, Limitations, and Recent Progress. *The AAPS Journal*, Vol.11, No.4, (December 2009), pp. 671-681, ISSN 1550- 7416

Anderson, M., Gregory, R., Thompson, S., Souza, D., Paul, S., Mulligan, R., Smith, A. & Welsh, M. (1991). Demonstration that CFTR Is a Chloride Channel by Alteration of Its Anion Selectivity. *Science*, Vol.253, No.5016, (July 1991), pp. 202-205, ISSN 0036- 8075

Anderson, W., Killos, L., Sanders-Haigh, L., Kretschmer, P. & Diacumakos, E. (1980). Replication and Expression of Thymidine Kinase and Human Globin Gene Microinjected into Mouse Fibroblasts. *Proceedings of the National Academy of Sciences*, Vol.77, No.9, (September 1980), pp. 5399-5403, ISSN 0027-8424

Antar, A., Konopka, J., Campbell, J., Henry, R., Perdigoto, A., Carter, B., Pozzi, A., Abel, T. & Dermody, T. (2009). Junctional Adhesion Molecule-A Is Required for Hematogenous Dissemination of Reovirus. *Cell Host & Microbe*, Vol.5, No.1, (January 2009), pp. 59-71, ISSN 1931-3128

Antarasena, C., Prommuang, Po., Promkuntod, P., Trongwongsa, L., Prommuang, Pr. & Rujikwan, S. (2002). Isolation of Avian Reoviruses Associated with Diseases of Chickens in Southern Thailand. *Songklanakarin Journal of Science and Technology*, Vol.24, No.2, (April-June 2002), pp. 329-340, ISSN 0125-3395

Arkin, H., Ohnuma, T., Kamen, B., Holland, J., & Vallabhajosula, S. (1989). Multidrug Resistance in Human Leukemia Cell Line Selected for Resistance to Trimctrexate. *Cancer Research*, Vol.49, No.23, (December 1989), pp. 6556-6561, ISSN 0008-5472

Ataga, K. (2009). Novel Therapies in Sickle Cell Disease. *Hematology, American Society of Hematology Education Program Book*, Vol.2009, (2009), pp. 54-61, ISSN 1520-4391

Baggiolini, M., Dewald, B. & Moser, B. (1997). Human Chemokines: An Update. *Annual Review of Immunology*, Vol.15, (1997), pp. 675-705, ISSN 0732-0582

Balter, M. (2000). Emerging Diseases. On the Trail of Ebola and Marburg Viruses. *Science*, Vol.290, No.5493, (November 2000), pp. 923-925, ISSN 0036-8075

Barbour, V., Tufarelli, C., Sharpe, J., Smith, Z., Ayyub, H., Heinlein, C., Sloane-Stanley, J., Indrak, K., Wood, W. & Higgs, D. (2000). α -Thalassemia Resulting from a Negative Chromosomal Position Effect. *Blood, Journal of the American Society of Hematology*, Vol.96, No.3, (August 2000), pp. 800-807, ISSN 0006-4971

Bates, G., Mangiarini, L., Mahal, A. & Davies, S. (1997). Transgenic Models of Huntington's Disease. *Human Molecular Genetics*, Vol.6, No.10, (1997), pp. 1633-1637, ISSN 0964- 6906

Batten, C., Maan, S., Shaw, A., Maan, N. & Mertens, P. (2008). A European Field Strain of Bluetongue Virus Derived from Two Parental Vaccine Strains by Genome Segment Reassortment. *Virus Research*, Vol.137, No.1, (October 2008), pp. 56-63, ISSN 0168- 1702

Becker, A., McCulloch E. & Till J. (1963). Cytological Demonstration of the Clonal Nature of Spleen Colonies Derived from Transplanted Mouse Marrow Cells. *Nature*, Vol.197, No.4866, (February 1963), pp. 452-454, ISSN 0028-0836

Beilby, J. (2007). DNA: Where to Now? *The Clinical Biochemist Reviews*, Vol.28, No.2, (May 2007), pp. 52-59, ISSN 0159–8090

Berget, S., Moore, C. & Sharp, P. (1977). Spliced Segments at the 5' Terminus of Adenovirus 2 Late mRNA. *Proceedings of the National Academy of Sciences*, Vol.74, No.8, (August 1977), pp. 3171-3175, ISSN 0027 8424

Bishop, J. & Smith, P. (1989). Mechanism of Chromosomal Integration of Microinjected DNA. *Molecular Biology and Medicine*, Vol.6, No.4, (August 1989), pp. 283-298, ISSN 0735-1313

Black, D. (2003). Mechanisms of Alternative Pre-messenger RNA Splicing. *Annual Review of Biochemistry*, Vol.72, (July 2003), pp. 291-336, ISSN 0066-4154

Bodey, G. (1986). Infection in Cancer Patients. A Continuing Association. *The American Journal of Medicine*, Vol.81, No.1A, (July 1986), pp. 11-26, ISSN 0002-9343

Bordignon, C., Yu, S., Smith, C., Hantzopoulos, P., Ungers, G., Keever, C., O'Reilly, R. & Gilboa, E. (1989). Retroviral Vector-mediated High-efficiency Expression of Adenosine Deaminase (ADA) in Hematopoietic Long-term Cultures of ADA- deficient Marrow Cells. *Proceedings of the National Academy of Sciences*, Vol.86, No.17, (September 1989), pp. 6748-6752, ISSN 0027-8424

Bordignon, C., Notarangelo, L., Nobili, N., Ferrari, G., Casorati, G., Panina, P., Mazzolari, E., Maggioni, D., Rossi, C., Servida, P., Ugazio, A. & Mavilio, F. (1995). Gene Therapy in Peripheral Blood Lymphocytes and Bone Marrow for ADA-immunodeficient Patients. *Science*, Vol.270, No.5235, (October 1995), pp. 470-475, ISSN 0036-8075

Bordignon, C. (2006). Stem-cell Therapies for Blood Diseases. *Nature*, Vol.441, No.7097, (June 2006), pp. 1100-1102, ISSN 0028-0836

Bouma, G., Burns, S. & Thrasher, A. (2009). Wiskott-Aldrich Syndrome: Immunodeficiency Resulting from Defective Cell Migration and Impaired Immunostimulatory Activation. *Immunobiology*, Vol.214, No.9-10, (September 2009), pp. 778-790, ISSN 0171-2985

Bradley, A., Evans, M., Kaufman, M. & Robertson, E. (1984). Formation of Germ-line Chimaeras from Embryo-derived Teratocarcinoma Cell Lines. *Nature*, Vol.309, No.5965, (May 1984), pp. 255-256, ISSN 0028-0836

Bronson, S. & Smithies, O. (1994). Altering Mice by Homologous Recombination Using Embryonic Stem Cells. *The Journal of Biological Chemistry*, Vol.269, No.44, (November 1994), pp. 27155-27158, ISSN 0021-9258

Brown, S., Gorman, B., Tesh, R. & Knudson, D. (1993) Coltiviruses Isolated from Mosquitoes Collected in Indonesia. *Virology*, Vol.196, No.1, (September 1993), pp. 363-367, ISSN 0042-6822

Buchwald, M. (1995). Complementation Groups: One or More per Gene? *Nature Genetics,* Vol.11, No.3, (November 1995), pp. 228-230, ISSN 1061-4036

Burghes, A., Logan, C., Hu, X., Belfall, B., Worton, R. & Ray, P. (1987). A cDNA Clone from the Duchenne/Becker Muscular Dystrophy Gene. *Nature*, Vol.328, No.6129, (July- August), pp. 434-437, ISSN 0028-0836

Campbell, K., McWhir, J., Ritchie, W. & Wilmut, I. (1996). Sheep Cloned by Nuclear Transfer from a Cultured Cell Line. *Nature*, Vol.380, No.6569, (March 1996), pp. 64-66, ISSN 0028-0836

Campeau, P., Chapdelaine, P., Seigneurin-Venin, S., Massie, B. & Tremblay, J. (2001). Transfection of Large Plasmids in Primary Human Myoblasts. *Gene Therapy*, Vol.8, No.18, (September 2001), pp. 1387-1394, ISSN 0969-7128

Capecchi, M. (1980). High Efficiency Transformation by Direct Microinjection of DNA into Cultured Mammalian Cells. *Cell*, Vol.22, No.2 part 2, (November 1980), pp. 479– 488, ISSN 0092-8674

Capecchi, M. (1989a). The New Mouse Genetics: Altering the Genome by Gene Targeting. *Trends in Genetics*, Vol.5, No.3, (March 1989), pp. 70-76, ISSN 0168-9479

Capecchi, M. (1989b). Altering the Genome by Homologous Recombination. *Science*, Vol.244, No.4910, (June 1989), pp. 1288-1292, ISSN 0036-8075

Cartier, R. & Reszka, R. (2002). Utilization of Synthetic Peptides Containing Nuclear Localization Signals for Nonviral Gene Transfer Systems. *Gene Therapy*, Vol.9, No.3, (February 2002), pp. 157-167, ISSN 0969-7128

Cavazzana-Calvo, M., Hacein-Bey, S., de Saint Basile, G., Gross, F., Yvon, E., Nusbaum, P., Selz, F., Hue, C., Certain, S., Casanova, J., Bousso, P., Deist, F. & Fischer, A. (2000). Gene Therapy of Human Severe Combined Immunodeficiency (SCID)-X1 Disease. *Science,* Vol.288, No.5466, (April 2000), pp. 669-672, ISSN 0036-8075

Chan, A & Yang, S. (2009). Generation of Transgenic Monkeys with Human Inherited Genetic Disease. *Methods: A Companion to Methods in Enzymology*, Vol.49, No.1, (September 2009), pp. 78-84, ISSN 1046-2023

Chang, J., Ye, L. & Kan, Y. (2006). Correction of the Sickle Cell Mutation in Embryonic Stem Cells. *Proceedings of the National Academy of Sciences*, Vol.103, No.4, (January 2006), pp. 1036-1040, ISSN 0027-8424

Charreau, B., Tesson, L., Soulillou, J., Pourcel, C. & Anegon, I. (1996). Transgenesis in Rats: Technical Aspects and Models. *Transgenic Research*, Vol.5, No.4, (July 1996), pp. 223- 234, ISSN 0962-8819

Chow, L., Gelinas, R., Broker, T. & Roberts, R. (1977). An Amazing Sequence Arrangement at the 5' Ends of Adenovirus 1 Messenger RNA. *Cell*, Vol.12, No.1, (September 1977), pp. 1-8, ISSN 0092-8674

Chua, K., Crameri, G., Hyatt, A., Yu, M., Tompang, M., Rosli, J., McEachern, J., Crameri, S., Kumarasamy, V., Eaton, B. & Wang, L. (2007). A previously Unknown Reovirus of Bat Origin Is Associated with an Acute Respiratory Disease in Humans. *Proceedings of the National Academy of Sciences*, Vol.104, No.27, (July 2007), pp. 11424-11429, ISSN 0027-8424

Collaco, J. & Cutting, G. (2008). Update on Gene Modifiers in Cystic Fibrosis. *Current Opinion in Pulmonary Medicine*, Vol.14, No.6, (November 2008), pp. 559-566, ISSN 1070-5287

Collins, F. (1992). Cystic Fibrosis: Molecular Biology and Therapeutic Implications. *Science*, Vol.256, No.5058, (May 1992), pp. 774-779, ISSN 0036-8075

Creighton, H. & McClintock, B. (1931). Correlation of Cytological and Genetical Crossing- Over in *Zea Mays*. *Proceedings of the National Academy of Sciences*, Vol.17, No.8, (August 1931), pp. 492-497, ISSN 0027-8424

Creighton, H. & McClintock, B. (1935). The Correlation of Cytological and Genetical Crossing-Over in Zea Mays. A Corroboration. *Proceedings of the National Academy of Sciences*, Vol.21, No.3, (March 1935), pp. 148-150, ISSN 0027-8424

Croop, J., Gros, P. & Housman, D. (1988). Genetics of Multidrug Resistance. *The Journal of Clinical Investigation*, Vol.81, No.5, (May 1988), pp. 1303-1309, ISSN 0021-9738

Csatary, L., Romvary, J., Kasza, L., Schaff, S. & Massey, R. (1985). In vivo Interference between Pathogenic and Non-pathogenic Viruses. *Journal of Medicine*, Vol.16, No.5- 6, (1985), pp. 563-573, ISSN 1090-1221

Cutting, G. (2010). Modifier Genes in Mendelian Disorders: The Example of Cystic Fibrosis. *Annals of the New York Academy of Sciences*, Vol.1214, (December 2010), pp. 57-69, ISSN 0077-8923

D'Angelo, M., Anderson, D., Richard, E. & Hetzer, M. (2006). Nuclear Pores Form de novo from Both Sides of the Nuclear Envelope. *Science*, Vol.312, No.5772, (April 2006), pp. 440-443, ISSN 0036-8075

D'Angelo, M. & Hetzer, M. (2008). Structure, Dynamics and Function of Nuclear Pore Complexes. *Trends in Cell Biology*, Vol.18, No.10, (October 2008), pp. 456-466, ISSN 0962-8924

Darling, T., Yee, C., Bauer, J., Hintner, H. & Yancey, K. (1999). Revertant Mosaicism: Partial Correction of a Germ-line Mutation in COL17A1 by a Frame-restoring Mutation. *The Journal of Clinical Investigation*, Vol.103, No.10, (May 1999), pp. 1371-1377, ISSN 0021-9738

Dave, U., Jenkins, N. & Copeland, N. (2004). Gene Therapy Insertional Mutagenesis Insights. *Science*, Vol.303, No.5656, (January 2004), pp. 333, ISSN 0036-8075

Davies, C., Jones, L., Green, B. & Nuttall, P. (1987). *In vivo* Reassortment of Thogoto Virus (A Tick-borne Influenza-like Virus) following Oral Infection of Rhipicephalus appendiculatus Ticks. *Journal of General Virology*, Vol.68, No.part 9, (September 1987), pp. 2331-2338, ISSN 0022-1317

Davies, S. & Ramsden, D. (2001). Huntington's Disease. *Molecular Pathology*, Vol.54, No.6, (December 2001), pp. 409-413, ISSN 0021-9746

De Rosa, M., Ackerley, C., Wang, B., Ito, S., Clarke, D. & Lingwood, C. (2008). Inhibition of Multidrug Resistance by Adamantylgb3, a Globotriaosylceramide Analog. *The Journal of Biological Chemistry*, Vol.283, No.8, (February 2008), pp. 4501-4511, ISSN 0021-9258

Debenham, P., Kartner, N., Siminovitch, L., Riordan, J. & Ling, V. (1982). DNA-mediated Transfer of Multiple Drug Resistance and Plasma Membrane Glycoprotein Expression. *Molecular and Cellular Biology*, Vol.2, No.8 (August 1982), pp. 881-889, ISSN 1098-5549

Decker, G. & Nyberg, S. (2001). Evidence that a Single Stem Cell Can Lead to Multi-organ Engraftment. *Liver Transplantation,* Vol.7, No.12, (December 2001), pp. 1085-1087, ISSN 1527-6465

Deuchars, K., Du, R., Naik, M., Evernden-Porelle, D., Kartner, N., van der Bliek, A. & Ling, V. (1987). Expression of Hamster P-glycoprotein and Multidrug Resistance in DNA-mediated Transformants of Mouse LTA Cells. *Molecular and Cellular Biology*, Vol.7, No.2 (February 1987), pp. 718-724, ISSN 1098-5549

Devos, D., Dokudovskaya, S., Williams, R., Alber, F., Eswar, N., Chait, B., Rout, M. & Sali, A. (2006). Simple Fold Composition and Modular Architecture of the Nuclear Pore Complex. *Proceedings of the National Academy of Sciences*, Vol.103, No.7, (February 2006), pp. 2172-2177, ISSN 0027-8424

Du, Y., Spence, S., Jenkins, N., Copeland, N. (2005). Cooperating Cancer-gene Identification through Oncogenic-retrovirus-induced Insertional Mutagenesis. *Blood, Journal of the American Society of Hematology,* Vol.106, No.7, (October 2005), pp. 2498-2505, ISSN 0006-4971

Duckworth-Rysiecki, G., Cornish, K., Clarke, C. & Buchwald, M. (1985). Identification of Two Complementation Groups in Fanconi Anemia. *Somatic Cell and Molecular Genetics*, Vol.11, No.1, (January 1985), pp. 35-41, ISSN 0740-7750

Dunn, G., Bruce, A., Ikeda, H., Old, L. & Schreiber, R. (2002). Cancer Immunoediting: from Immunosurveillance to Tumor Escape. *Nature Immunology*, Vol.3, No.11, (November 2002), pp. 991-998, ISSN 1529-2908

Dunn, G., Old, L. & Schreiber, R. (2004). The Three Es of Cancer Immunoediting. *Annual Review of Immunology*, Vol.22, (2004), pp. 329-360, ISSN 0732-0582

Dupré, L., Trifari, S., Follenzi, A., Marangoni, F., Lain de Lera, T., Bernad, A., Martino, S., Tsuchiya, S., Bordignon, C., Naldini, L., Aiuti, A. & Roncarolo, M. (2004). Lentiviral Vector-mediated Gene Transfer in T Cells from Wiskott-Aldrich Syndrome Patients Leads to Functional Correction. *Molecular Therapy*, Vol.10, No.5, (November 2004), pp. 903-915, ISSN 1525-0016

Dvorak, J. & Akhunov, E. (2005). Tempos of Gene Locus Deletions and Duplications and Their Relationship to Recombination Rate during Diploid and Polyploid Evolution in the Aegilops-Triticum alliance. *Genetics*, Vol.171, No.1, (September 2005), pp. 323-332, ISSN 0016-6731

Dwi Pramono, Z., Takeshima, Y., Surono, A., Ishida, T. & Matsuo, M. (2000). A Novel Cryptic Exon in Intron 2 of the Human Dystrophin Gene Evolved from an Intron by Acquiring Consensus Sequences for Splicing at Different Stages of Anthropoid Evolution. *Biochemical and Biophysical Research Communications*, Vol.267, No.1, (January 2000), pp. 321-328, ISSN 0006-291X

Eapen, M., Rocha, V., Sanz, G., Scaradavou, A., Zhang, M., Arcese, W., Sirvent, A., Champlin, R., Chao, N., Gee, A., Isola, L., Laughlin, M., Marks, D., Nabhan, S., Ruggeri, A., Soiffer, R., Horowitz, M., Gluckman, E., Wagner, J., Center for International Blood and Marrow Transplant Research, Acute Leukemia Working Party Eurocord (the European Group for Blood Marrow Transplantation) & National Cord Blood Program of the New York Blood Center. (2010). Effect of Graft Source on Unrelated Donor Haemopoietic Stem-cell Transplantation in Adults with Acute Leukaemia: A Retrospective Analysis. *The Lancet Oncology*, Vol.11, No.7, (July 2010), pp. 653-660, ISSN 1470-2045

Earle, W. (1943). Production of Malignancy in vitro. IV. The Mouse Fibroblast Cultures and Changes Seen in the Living Cells. *The Journal of the National Cancer Institute*, Vol.4, (1943), pp. 165-212, ISSN 0027-8874

Efrat, S. (1998). Prospects for Gene Therapy of Insulin-dependent Diabetes Mellitus. *Diabetologia. Clinical and Experimental Diabetes and Metabolism.* Vol.41, No.12, (December 1998), pp. 1401-1409, ISSN 0012-186X

Endicott, J. & Ling, V. (1989). The Biochemistry of P-glycoprotein-mediated Multidrug Resistance. *Annual Review of Biochemistry*, Vol.58, (July 1989), pp. 137-171, ISSN 0066-4154

Entrez Gene: HBA1 Hemoglobin alpha 1. (Updated on 16-Jan-2011). *NCBI*, February 7, 2011, Available from: http://www.ncbi.nlm.nih.gov/sites/entrez?Db=gene&Cmd=ShowDetailView&Te rmToSearch=3039

Epstein, A. (2009). Progress and Prospects: Biological Properties and Technological Advances of Herpes Simplex Virus Type 1-based Amplicon Vectors. *Gene Therapy*, Vol.16, No.6, (June 2009), pp. 709–715, ISSN 0969-7128

Escoffre, J., Portet, T., Wasungu, L., Teissié, J., Dean, D. & Rols, M. (2009). What Is (Still Not) Known of the Mechanism by Which Electroporation Mediates Gene Transfer and Expression in Cells and Tissues. *Molecular Biotechnology,* Vol.41, No.3, (March 2009), pp. 286-295, ISSN 1073-6085

Evans, V., Kerr, H., McQuilkin, W., Earle, W. & Hull, R. (1959). Growth in vitro of a Long- term Strain of Monkey-kidney Cells in Medium NCTC 109 Free of Any Added

Protein. *The American Journal of Hygiene,* Vol.70, (November 1959), pp. 297-302, ISSN 0096-5294

Falls, J., Pulford, D., Wylie, A. & Jirtle, R. (1999). Genomic Imprinting: Implications for Human Disease. *The American Journal of Pathology,* Vol.154, No.3, (March 1999), pp. 635-647, ISSN 0002-9440

Fanconi Anaemia/Breast Cancer Consortium. (1996). Positional Cloning of the Fanconi Anaemia Group A Gene. *Nature Genetics,* Vol.14, No.3, (November 1996), pp. 324- 328, ISSN 1061-4036

Favard, C., Dean, D. & Rols, M. (2007). Electrotransfer as a Non Viral Method of Gene Delivery. *Current Gene Therapy,* Vol.7, No.1, (February 2007), pp. 67-77, ISSN 1566- 5232

Feng, D., Liu, D., Huang, Y., Wu, L., Li, T., Wu, M., Tang, X. & Liang, C. (2001). The Expression of Human alpha -like Globin Genes in Transgenic Mice Mediated by Bacterial Artificial Chromosome. *Proceedings of the National Academy of Sciences,* Vol.98, No.26, (December 2001), pp. 15073-15077, ISSN 0027-8424

Fernandez-Martinez, J. & Rout, M. (2009). Nuclear Pore Complex Biogenesis. *Current Opinion in Cell Biology,* Vol.21, No.4, (August 2009), pp. 603-612, ISSN 0955-0674

Ferrari, G., Rossini, S., Giavazzi, R., Maggioni, D., Nobili, N., Soldati, M., Ungers, G., Mavilio, F., Gilboa, E. & Bordignon, C. (1991). An *in vivo* Model of Somatic Cell Gene Therapy for Human Severe Combined Immunodeficiency. *Science,* Vol.251, No.4999, (March 1991), pp. 1363-1366, ISSN 0036-8075

Filipiak, W. & Saunders, T. (2006). Advances in Transgenic Rat Production.*Transgenic Research,* Vol.15, No.6, (December 2006), pp. 673-686, ISSN 0962-8819

Folger, K., Wong, E., Wahl, G. & Capecchi, M. (1982). Patterns of Integration of DNA Microinjected into Cultured Mammalian Cells: Evidence for Homologous Recombination between Injected Plasmid DNA Molecules. *Molecular and Cellular Biology,* Vol.2, No.11, (November 1982), pp. 1372-1387, ISSN 1098-5549

Freeman, D., Leclerc, I. & Rutter, G. (1999). Present and Potential Future Use of Gene Therapy for the Treatment of Non-insulin Dependent Diabetes Mellitus. *International Journal of Molecular Medicine,* Vol.4, No.6, (December 1999), pp. 585- 592, ISSN 1107-3756

Fujimoto, A., Nakagawa, H., Hosono, N., Nakano, K., Abe, T., Boroevich, K., Nagasaki, M., Yamaguchi, R., Shibuya, T., Kubo, M., Miyano, S., Nakamura, Y. & Tsunoda, T. (2010). Whole-genome Sequencing and Comprehensive Variant Analysis of a Japanese Individual Using Massively Parallel Sequencing. *Nature Genetics,* Vol.42, No.11, (November 2010), pp. 931–936, ISSN 1061-4036

Gao, J., Wei, Y., Huang, Y., Liu, D., Liu, G., Wu, M., Wu, L., Zhang, Q., Zhang, Z., Zhang, R. & Liang, C. (2005). The Expression of Intact and Mutant Human apoAI/CIII/AIV/AV Gene Cluster in Transgenic Mice. *The Journal of Biological Chemistry,* Vol.280, No.13, (April 2005), pp. 12559-12566, ISSN 0021-9258

Gao, X., Kim, K. & Liu, D. (2007). Nonviral Gene Delivery: What We Know and What Is Next. *The AAPS Journal,* Vol.9, No.1, (March 2007), pp. E92-E104, ISSN 1550-7416

Garten, R., Davis, C., Russell, C., Shu, B., Lindstrom, S., Balish, A., Sessions, W., Xu, X., Skepner, E., Deyde, V., Okomo-Adhiambo, M., Gubareva, L., Barnes, J., Smith, C., Emery, S., Hillman, M., Rivailler, P., Smagala, J., de Graaf, M., Burke, D., Fouchier, R., Pappas, C., Alpuche-Aranda, C., López-Gatell, H., Olivera, H., López, I., Myers,

C., Faix, D., Blair, P., Yu, C., Keene, K., Dotson, P. Jr., Boxrud, D., Sambol, A., Abid, S., St George, K., Bannerman, T., Moore, A., Stringer, D., Blevins, P., Demmler- Harrison, G., Ginsberg, M., Kriner, P., Waterman, S., Smole, S., Guevara, H., Belongia, E., Clark, P., Beatrice, S., Donis, R., Katz, J., Finelli, L., Bridges, C., Shaw, M., Jernigan, D., Uyeki, T., Smith, D., Klimov, A. & Cox, N. (2009). Antigenic and Genetic Characteristics of Swine-origin 2009 A (H1N1) Influenza Viruses Circulating in Humans. *Science,* Vol.325, No.5937, (July 2009), pp. 197-201, ISSN 0036-8075

Gaspar, H., Parsley, K., Howe, S., King, D., Gilmour, K., Sinclair, J., Brouns, G., Schmidt, M., Von Kalle, C., Barington, T., Jakobsen, M., Christensen, H., Al Ghonaium, A., White, H., Smith, J., Levinsky, R., Ali, R., Kinnon, C. & Thrasher, A. (2004). Gene Therapy of X-linked Severe Combined Immunodeficiency by Use of a Pseudotyped Gammaretroviral Vector. *The Lancet,* Vol.364, No.9452, (December 2004), pp. 2181- 2187, ISSN 0140-6736

Gaspar, H., Aiuti, A., Porta, F., Candotti, F., Hershfield, M. & Notarangelo, L. (2009). How I Treat ADA Deficiency. *Blood, Journal of the American Society of Hematology*, Vol.114, No.17, (October 2009), pp. 3524-3532, ISSN 0006-4971

Giampietro, P., Verlander, P., Davis, J. & Auerbach, A. (1997). Diagnosis of Fanconi Anemia in Patients without Congenital Malformations: An International Fanconi Anemia Registry Study. *The American Journal of Human Genetics*, Vol.68, No.1, (January 1997), pp. 58-61, ISSN 0148-7299

Goldman, E. (2008). Transfer RNA, In: *Encyclopedia of Life Sciences*, DOI: 10.1002/9780470015902.a0000878.pub2, John Wiley & Sons, Ltd, Retrieved from http://onlinelibrary.wiley.com/doi/10.1002/9780470015902.a0000878.pub2/full

Goldstein, L., Galski, H., Fojo, A., Willingham, M., Lai, S., Gazdar, A., Pirker, R., Green, A., Crist, W., Brodeur, G., Lieber, M., Cossman, J., Gottesman, M. & Pastan, I. (1989). Expression of a Multidrug Resistance Gene in Human Cancers. *Journal of the National Cancer Institute*, Vol.81, No.2, (January 1989), pp. 116-124, ISSN 0027-8874

Golubovsky, M. & Manton, K. (2005). Genome Organization and Three Kinds of Heritable Changes: General Description and Stochastic Factors. *Frontiers in Bioscience*, Vol.10, (January 2005), pp. 335-344, ISSN 1093-9946

Golzio, M., Escoffre, J., Portet, T., Mauroy, C., Teissié, J., Dean, D. & Rols, M. (2010). Observations of the Mechanisms of Electromediated DNA Uptake--from Vesicles to Tissues. *Current Gene Therapy*, Vol.10, No.4, (August 2010), pp. 256-266, ISSN 1566- 5232

Gore, A., Li, Z., Fung, H., Young, J., Agarwal, S., Antosiewicz-Bourget, J., Canto, I., Giorgetti, A., Israel, M., Kiskinis, E., Lee, J., Loh, Y., Manos, P., Montserrat, N., Panopoulos, A., Ruiz, S., Wilbert, M., Yu, J., Kirkness, E., Izpisua Belmonte, J., Rossi, D., Thomson, J., Eggan, K., Daley, G., Goldstein, L. & Zhang K. (2011). Somatic Coding Mutations in Human Induced Pluripotent Stem Cells. *Nature*, Vol.471, No.7336, (March 2011), pp. 63-67, ISSN 0028-0836

Gregory, J. Jr., Wagner, J., Verlander, P., Levran, O., Batish, S., Eide, C., Steffenhagen, A., Hirsch, B. & Auerbach, A. (2001). Somatic Mosaicism in Fanconi Anemia: Evidence of Genotypic Reversion in Lymphohematopoietic Stem Cells. *Proceedings of the National Academy of Sciences*, Vol.98, No.5, (February 2001), pp. 2532-2537, ISSN 0027-8424

Gregory, V., Bennett, M., Orkhan, M., Al Hajjar, S., Varsano, N., Mendelson, E., Zambon, M., Ellis, J., Hay, A. & Lin, Y. (2001). Emergence of Influenza A H1N2 Reassortant Viruses in the Human Population during 2001. *Virology*, Vol.300, No.1, (August 2002), pp. 1-7, ISSN 0042-6822

Gross, L. (2007). A New Human Genome Sequence Paves the Way for Individualized Genomics. *PLoS Biol*, Vol.5, No.10, (October 2007), pp. e266. doi:10.1371/journal.pbio.0050266, ISSN 1544-9173

Gross, M., Hanenberg, H., Lobitz, S., Friedl, R., Herterich, S., Dietrich, R., Gruhn, B., Schindler, D. & Hoehn, H. (2002). Reverse Mosaicism in Fanconi Anemia: Natural Gene Therapy via Molecular Self-correction. *Cytogenetic and Genome Research*, Vol.98, No.2-3, (2002), pp. 126-135, ISSN 1424–8581

Groudine, M., Kohwi-Shigematsu, T., Gelinas, R., Stamatoyannopoulos, G. & Papayannopoulou, T. (1983). Human Fetal to Adult Hemoglobin Switching: Changes in Chromatin Structure of the beta-globin Gene Locus. *Proceedings of the National Academy of Sciences*, Vol.80, No.24, (December 1983), pp. 7551-7555, ISSN 0027-8424

Gutierrez-Aranda, I., Ramos-Mejia, V., Bueno, C., Munoz-Lopez, M., Real, P., Mácia, A., Sanchez, L., Ligero, G., Garcia-Parez, J. & Menendez, P. (2010). Human Induced Pluripotent Stem Cells Develop Teratoma More Efficiently and Faster than Human Embryonic Stem Cells Regardless the Site of Injection. *Stem Cells*, Vol.28, No.9, (September 2010), pp. 1568-1570, ISSN 1066-5099

Hacein-Bey-Abina, S., Le Deist, F., Carlier, F., Bouneaud, C., Hue, C., De Villartay, J., Thrasher, A., Wulffraat, N., Sorensen, R., Dupuis-Girod, S., Fischer, A., Davies, E., Kuis, W., Leiva, L. & Cavazzana-Calvo, M. (2002). Sustained Correction of X-linked Severe Combined Immunodeficiency by ex vivo Gene Therapy. *The New England Journal of Medicine*, Vol.346, No.16, (April 2002), pp. 1185-1193, ISSN 0028-4793

Hacein-Bey-Abina, S., von Kalle, C., Schmidt, M., Le Deist, F., Wulffraat, N., McIntyre, E., Radford, I., Villeval, J., Fraser, C., Cavazzana-Calvo, M. & Fischer, A. (2003a). A Serious Adverse Event after Successful Gene Therapy for X-linked Severe Combined Immunodeficiency. *The New England Journal of Medicine*, Vol.348, No.3, (January 2003), pp. 255–256, ISSN 0028-4793

Hacein-Bey-Abina, S., von Kalle, C., Schmidt, M., McCormack, M., Wulffraat, N., Leboulch, P., Lim, A., Osborne, C., Pawliuk, R., Morillon, E., Sorensen, R., Forster, A., Fraser, P., Cohen, J., de Saint Basile, G., Alexander, I., Wintergerst, U.,

Frebourg, T., Aurias, A., Stoppa-Lyonnet, D., Romana, S., Radford-Weiss, I., Gross, F., Valensi, F., Delabesse, E., Macintyre, E., Sigaux, F., Soulier, J., Leiva, L., Wissler, M., Prinz, C., Rabbitts, T., Le Deist, F., Fischer, A. & Cavazzana-Calvo, M. (2003b). LMO2- associated Clonal T Cell Proliferation in Two Patients after Gene Therapy for SCID- X1. *Science,* Vol.302, No.5644, (October 2003), pp. 415–419, ISSN 0036-8075

Hacein-Bey-Abina, S., Hauer, J., Lim, A., Picard, C., Wang, G., Berry, C., Martinache, C., Rieux-Laucat, F., Latour, S., Belohradsky, B., Leiva, L., Sorensen, R., Debré, M., Casanova, J., Blanche, S., Durandy, A., Bushman, F., Fischer, A. & Cavazzana- Calvo, M. (2010). Efficacy of Gene Therapy for X-linked Severe Combined Immunodeficiency. *The New England Journal of Medicine*, Vol.363, No.4, (July 2010), pp. 355-364, ISSN 0028-4793

Hagstrom, J., Ludtke, J., Bassik, M., Sebestyén, M., Adam, S. & Wolff, J. (1997). Nuclear Import of DNA in Digitonin-permeabilized Cells. *Journal of Cell Science,* Vol.110, No.part 18, (September 1997), pp. 2323-2331, ISSN 0021-9533

Halatsch, M., Schmidt, U., Bötefür, I., Holland, J. & Ohnuma, T. (2000). Marked Inhibition of Glioblastoma Target Cell Tumorigenicity in vitro by Retrovirus-mediated Transfer of a Hairpin Ribozyme against Deletion-mutant Epidermal Growth Factor Receptor Messenger RNA. *Journal of Neurosurgery*, Vol.92, No.2, (February 2000), pp. 297- 305, ISSN 0022-3085

Hall, J. (1990). Genomic Imprinting: Review and Relevance to Human Diseases. *The American Society of Human Genetics,* Vol.46, No.5, (May 1990), pp. 857-873, ISSN 0002-9297

Hardison, R. (1996). A Brief History of Hemoglobins: Plant, Animal, Protist, and Bacteria. *Proceedings of the National Academy of Sciences*, Vol.93, No.12, (June 1996), pp. 5675- 5679, ISSN 0027-8424

Hawkins, J. (1988). A Survey on Intron and Exon Lengths. *Nucleic Acids Research*, Vol.16, No.21, (November 1988), pp. 9893-9908, ISSN 0305-1048

Higgs, D., Vickers, M., Wilkie, A., Pretorius, I., Jarman, A. & Weatherall, D. (1989). A Review of the Molecular Genetics of the Human alpha-globin Gene Cluster. *Blood, Journal of the American Society of Hematology,* Vol.73, No.5, (April 1989), pp. 1081-1104, ISSN 0006-4971

Hirschhorn, R. (2003). In vivo Reversion to Normal of Inherited Mutations in Humans. *Journal of Medical Genetics*, Vol.40, No.10, (October 2003), pp. 721-728, ISSN 0022- 2593

Hoffman, E., Brown, R. Jr. & Kunkel, L. (1987). Dystrophin: The Protein Product of the Duchenne Muscular Dystrophy Locus. *Cell*, Vol.51, No.6, (December 1987), pp. 919- 928, ISSN 0092-8674

Holliday R. (1974). Molecular Aspects of Genetic Exchange and Gene Conversion. *Genetics*, Vol.78, No.1, (September 1974), pp. 273-287, ISSN 0016-6731

Holmes, E., Ghedin, E., Miller, N., Taylor, J., Bao, Y., St George, K., Grenfell, B., Salzberg, S., Fraser, C., Lipman, D. & Taubenberger, J. (2005). Whole-genome Analysis of Human Influenza A Virus Reveals Multiple Persistent Lineages and Reassortment among Recent H3N2 Viruses. *PLoS Biology*, Vol.3, No.9, (September 2005), pp. e300, ISSN 1544-9173

Holmes, K., Tresnan, D. & Zelus, B. (1997). Virus-receptor Interactions in the Enteric Tract. Virus-receptor Interactions. *Advances in Experimental Medicine and Biology*, Vol.412, (1997), pp. 125-133, ISSN 0065-2598

Hope, J., Cruzata, L., Duvshani, A., Mitsumoto, J., Maftahi, M. & Freyer, G. (2007). Mus81- Eme1-dependent and -independent Crossovers Form in Mitotic Cells during Double-strand Break Repair in *Schizosaccharomyces pombe. Molecular and Cellular Biology*, Vol.27, No.10, (May 2007), pp. 3828-3838, ISSN 0270-7306

Huang, C. Liang, H. & Jia, F. (1985). Beneficial Role of a Non-pathogenic Orbi-like Virus: Studies on the Interfering Effect of M14 Virus in Mice and Mosquitoes Infected with Japanese Encephalitis Virus. *Intervirology*, Vol.24, No.3, (1985), pp. 147-153, ISSN 0300-5526

Huang, S., Sirikhachornkit, A., Su, X., Faris, J., Gill, B., Haselkorn, R. & Gornicki, P. (2002). Genes Encoding Plastid Acetyl-CoA Carboxylase and 3-Phosphoglycerate Kinase of the Triticum/Aegilops Complex and the Evolutionary History of Polyploid Wheat. *Proceedings of the National Academy of Sciences*, Vol.99, No.12, (June 2002), pp. 8133- 8138, ISSN 0027-8424

Hull, R., Cherry, W. & Johnson, I. (1956). The Adaptation and Maintenance of Mammalian Cells to Continuous Growth in Tissue Culture. *The Anatomical Record: Advances in Integrative Anatomy and Evolutionary Biology,* Vol.124, (1956), pp. 490–499, ISSN 0003-276X,

Hull, R., Cherry, W. & Tritch, O. (1962). Growth Characteristics of Monkey Kidney Cell Strains LLC-MK1, LLC-MK2, and LLC-MK2 (NCTC-3196) and Their Utility in Virus Research. *The Journal of Experimental Medicine*, Vol.115, (May 1962), pp. 903- 918, ISSN 0022-1007

Hussein, S., Batada, N., Vuoristo, S., Ching, R., Autio, R., Närvä, E., Ng, S., Sourour, M., Hämäläinen, R., Olsson, C., Lundin, K., Mikkola, M., Trokovic, R., Peitz, M., Brüstle, O., Bazett-Jones, D., Alitalo, K., Lahesmaa, R., Nagy, A. & Otonkoski, T. (2011). Copy Number Variation and Selection during Reprogramming to Pluripotency. *Nature*, Vol.471, No.7336, (March 2011), pp. 58-62, ISSN 0028-0836

Ingram, V. (1956). A Specific Chemical Difference between the Globins of Normal Human and Sickle-cell Anaemia Haemoglobin. *Nature*, Vol.178, No.4537, (October 1956), pp. 792-794, ISSN 0028-0836

Ingram, V. (1957). Gene Mutations in Human Haemoglobin: The Chemical Difference between Normal and Sickle Cell Haemoglobin. *Nature*, Vol.180, No.4581, (August 1957), pp. 326-328, ISSN 0028-0836

International Human Genome Sequencing Consortium. (2001). Initial Sequencing and Analysis of the Human Genome. *Nature*, Vol.409, No.6822, (February 2001), pp. 860–921, ISSN 0028-0836

International Human Genome Sequencing Consortium. (2004). Finishing the Euchromatic Sequence of the Human Genome. *Nature*, Vol.431, No.7011, (October 2004), pp. 931– 945, ISSN 0028-0836

International Silkworm Genome Consortium. (2008). The Genome of a Lepidopteran Model Insect, the Silkworm Bombyx mori. *Insect Biochemistry and Molecular Biology*, Vol.38, No.12, (December 2008), pp. 1036-1045, ISSN 0965-1748

Jackson, S. & Bartek, J. (2009). The DNA-damage Response in Human Biology and Disease. *Nature*, Vol.461, No.7267, (October 2009), pp. 1071-1078, ISSN 0028-0836

Jacobsen, J., Bawden, C., Rudiger, S., McLaughlan, C., Reid, S., Waldvogel, H., MacDonald, M., Gusella, J., Walker, S., Kelly, J., Webb, G., Faull, R., Rees, M. & Snell, R. (2010). An Ovine Transgenic Huntington's Disease Model. *Human Molecular Genetics*, Vol.19, No.10, (May 2010), pp. 1873-1882, ISSN 0964-6906

Jayandharan, G., Aslanidi, G., Martino, A., Jahn, S., Perrin, G., Herzog, R. & Srivastava, A. (2011). Activation of the NF-(kappa)B Pathway by Adeno-associated Virus (AAV) Vectors and Its Implications in Immune Response and Gene Therapy. *Proceedings of the National Academy of Sciences*, Vol.108, No.9, (March 2011), pp. 3743-3748, ISSN 0027-8424

Jiang, H., Patel, P., Kohlmaier, A., Grenley, M., McEwen, D. & Edgar, B. (2009). Cytokine/Jak/Stat Signaling Mediates Regeneration and Homeostasis in the Drosophila Midgut. *Cell*, Vol.137, No.7, (June 2009), pp. 1343-1355, ISSN 0092-8674

Jin, Y., Mazza, C., Christie, J., Giliani, S., Fiorini, M., Mella, P., Gandellini, F., Stewart, D., Zhu, Q., Nelson, D., Notarangelo, L. & Ochs, H. (2004). Mutations of the Wiskott- Aldrich Syndrome Protein (WASP): Hotspots, Eeffect on Transcription, and Translation and Phenotype/Genotype Correlation. *Blood, Journal of the American Society of Hematology*, Vol.104, No.13, (December 2004), pp. 4010-4019, ISSN 0006- 4971

Joenje, H., Lo ten Foe, J., Oostra, A., van Berkel, C., Rooimans, M., Schroeder-Kurth, T., Wegner, R., Gille, J., Buchwald, M. & Arwert, F. (1995). Classification of Fanconi Anemia Patients by Complementation Analysis: Evidence for a Fifth Genetic Subtype. *Blood, Journal of the American Society of Hematology*, Vol.86, No.6, (September 1995), pp. 2156-2160, ISSN 0006-4971

Joenje, H. & Patel, K. (2001). The Emerging Genetic and Molecular Basis of Fanconi Anaemia. *Nature Reviews Genetics*, Vol.2, No.6, (June 2001), pp. 446-457, ISSN 1471- 0056

Johansson, P., Sveger, T., Ahlfors, K., Ekstrand, J. & Svensson, L. (1996). Reovirus Type 1 Associated with Meningitis. *Scandinavian Journal of Infectious Diseases*, Vol. 28, No.2, (1996), pp. 117-120, ISSN 0036-5548

Johnson, R. & Jasin, M. (2001). Double-strand-break-induced Homologous Recombination in Mammalian Cells. *Biochemical Society Transactions*, Vol.29, No.part 2, (May 2001), pp. 196-201, ISSN 0300-5127

Jones, L., Davies, C., Green, B. & Nuttall, P. (1987). Reassortment of Thogoto Virus (A Tick- borne Influenza-like Virus) in a Vertebrate Host. *Journal of General Virology*, Vol.68, No.part 5, (May 1987), pp. 1299-1306, ISSN 0022-1317

Juliano, R. & Ling, V. (1976). A Surface Glycoprotein Modulating Drug Permeability in Chinese Hamster Ovary Cell Mutants. *Biochimica et Biophysica Acta*, Vol.455, No.1, (November 1976), pp. 152–162, ISSN 0005-2736.

Kadyk, L. & Hartwell, L. (1992). Sister Chromatids Are Preferred over Homologs as Substrates for Recombinational Repair in *Saccharomyces cerevisiae*. *Genetics*, Vol.132, No.2, (October 1992), pp. 387-402, ISSN 0016-6731

Kang, E., Choi, U., Theobald, N., Linton, G., Long Priel, D., Kuhns, D. & Malech, H. (2010). Retrovirus Gene Therapy for X-linked Chronic Granulomatous Disease Can Achieve Stable Long-term Correction of Oxidase Activity in Peripheral Blood Neutrophils. *Blood, Journal of the American Society of Hematology*, Vol.115, No.4, (January 2010), pp. 783-791, ISSN 0006-4971

Karasin, A., Schutten, M., Cooper, L., Smith, C., Subbarao, K., Anderson, G., Carman, S. & Olsen, C. (2000). Genetic Characterization of H3N2 Influenza Viruses Isolated from Pigs in North America, 1977-1999: Evidence for Wholly Human and Reassortant Virus Genotypes. *Virus Research*, Vol.68, No.1, (June 2000), pp. 71-85, ISSN 0168- 1702

Kerem, B., Rommens, J., Buchanan, J., Markiewicz, D., Cox, T., Chakravarti, A., Buchwald, M. & Tsui, L. (1989). Identification of the Cystic Fibrosis Gene: Genetic Analysis. *Science*, Vol.245, No.4922, (September 1989), pp. 1073-1080, ISSN 0036-8075

Kim, Y., Lach, F., Desetty, R., Hanenberg, H., Auerbach, A. & Smogorzewska, A. (2011). Mutations of the SLX4 Gene in Fanconi Anemia. *Nature Genetics*, Vol.43, No.2, (February 2011), pp. 142-146, ISSN 1061-4036

Kobayashi, H., Dorai, T., Holland, J. & Ohnuma, T. (1994). Reversal of Drug Sensitivity in Multidrug-resistant Tumor Cells by an MDR1 (PGY1) Ribozyme. *Cancer Research*, Vol.54, No.5, (March 1994), PP. 1271-1275, ISSN 0008-5472

Kobayashi, H., Takemura, Y., Holland, J. & Ohnuma, T. (1998). Vincristine Saturation of Cellular Binding Sites and Its Cytotoxic Activity in Human Lymphoblastic Leukemia Cells: Mechanism of Inoculum Effect. *Biochemical Pharmacology*, Vol.55, No.8, (April 1998), pp. 1229-1234, ISSN 0006-2952

Koenig, M., Hoffman, E., Bertelson, C., Monaco, A., Feener, C. & Kunkel, L. (1987). Complete Cloning of the Duchenne Muscular Dystrophy (DMD) cDNA and Preliminary Genomic Organization of the DMD Gene in Normal and Affected Individuals. *Cell*, Vol.50, No.3, (July 1987), pp. 509-517, ISSN 0092-8674

Koenig, M., Monaco, A. & Kunkel, L. (1988). The Complete Sequence of Dystrophin Predicts a Rod-shaped Cytoskeletal Protein. *Cell*, Vol.53, No.2, (April 1988), pp. 219-228, ISSN 0092-8674

Koller, B. & Smithies, O. (1992). Altering Genes in Animals by Gene Targeting. *Annual Review of Immunology*, Vol.10, (1992), pp. 705-730, ISSN 0732-0582

Krause, D., Theise, N., Collector, M., Henegariu, O., Hwang, S., Gardner, R., Neutzel, S. & Sharkis, S. (2001). Multi-organ, Multi-lineage Engraftment by a Single Bone Marrow-derived Stem Cell. *Cell*, Vol.105, No.3, (May 2001), pp. 369-377, ISSN 0092- 8674

Kuehn, M., Bradley, A., Robertson, E. & Evans, M. (1987). A Potential Animal Model for Lesch-Nyhan Syndrome through Introduction of HPRT Mutations into Mice. *Nature*, Vol.326, No.6110, (March 1987), pp. 295-298, ISSN 0028-0836

Kunkel, L. (2004). William Allan Award Address. Cloning of the DMD Gene. *The American Journal of Human Genetics*. Vol.76, No.2, (February 2005), pp. 205-214, ISSN 0002- 9297

Lam, A. & Dean, A. (2010). Progress and Prospects: Nuclear Import of Nonviral Vectors. *Gene Therapy*, Vol.17, No.4, (April 2010), pp. 439-447, ISSN 0969-7128

Laughlin, M., Barker, J., Bambach, B., Koc, O., Rizzieri, D., Wagner, J., Gerson, S., Lazarus, H., Cairo, M., Stevens, C., Rubinstein, P. & Kurtzberg, J. (2001). Hematopoietic Engraftment and Survival in Adult Recipients of Umbilical-cord Blood from Unrelated Donors. *The New England Journal of Medicine*, Vol.344, No.24, (June 2001), pp. 1815-1822, ISSN 0028-4793

Laughlin, M., Eapen, M., Rubinstein, P., Wagner, J., Zhang, M., Champlin, R., Stevens, C., Barker, J., Gale, R., Lazarus, H., Marks, D., van Rood, J., Scaradavou, A. & Horowitz, M. (2004). Outcomes after Transplantation of Cord Blood or Bone Marrow from Unrelated Donors in Adults with Leukemia. *The New England Journal of Medicine*, Vol.351, No.22, (November 2004), pp. 2265-2275, ISSN 0028-4793

Laurent, L., Ulitsky, I., Slavin, I., Tran, H., Schork, A., Morey, R., Lynch, C., Harness, J., Lee, S., Barrero, M., Ku, S., Martynova, M., Semechkin, R., Galat, V., Gottesfeld, J., Izpisua Belmonte, J., Murry, C., Keirstead, H., Park, H., Schmidt, U., Laslett, A., Muller, F., Nievergelt, C., Shamir, R. & Loring, J. (2011). Dynamic Changes in the Copy Number of Pluripotency and Cell Proliferation Genes in Human ESCs and iPSCs during Reprogramming and Time in Culture. *Cell Stem Cell*, Vol.8, No.1, (January 2011), pp. 106-118, ISSN 1934-5909

Lederberg, J. & Tatum, E. (1946). Gene Recombination in *Escherichia coli. Nature*, Vol.158, No.4016, (October 1946), pp. 558, ISSN 0028-0836

Lederberg, J. (1947). Gene Recombination and Linked Segregations in Escherichia coli. *Genetics*, Vol.32, No.5, (September 1947), pp. 505-525, ISSN 0016-6731

Lederberg, J. (1987a). Gene Recombination and Linked Segregations in Escherichia coli. *Genetics*, Vol.117, No.1, (September 1987), pp. 1-4, ISSN 0016-6731

Lederberg, J. (1987b). Genetic Recombination in Bacteria: A Discovery Account. *Annual Review of Genetics*, Vol.21, (1987), pp. 23-46, ISSN 0066-4197

Lee, T., Matthews, D. & Blair, G. (2005). Novel Molecular Approaches to Cystic Fibrosis Gene Therapy. *Biochemical Journal*, Vol.387, No.part 1, (April 2005), pp. 1-15, ISSN 0264-6021

Levitus, M., Rooimans, M., Steltenpool, J., Cool, N., Oostra, A., Mathew, C., Hoatlin, M., Waisfisz, Q., Arwert, F., de Winter, J. & Joenje, H. (2004). Heterogeneity in Fanconi Anemia: Evidence for 2 New Genetic Subtypes. *Blood, Journal of the American Society of Hematology*, Vol.103, No.7, (April 2004), pp. 2498-2503, ISSN 0006-4971

Levitus, M., Joenje, H. & de Winter J. (2006). The Fanconi Anemia Pathway of Genomic Maintenance. *Cellular Oncology*, Vol.28, No.1-2, (April 2006), pp. 3-29, ISSN 1570- 5870

Levy, S., Sutton, G., Ng, P., Feuk, L., Halpern, A., Walenz, B., Axelrod, N., Huang, J., Kirkness, E., Denisov, G., Lin, Y., MacDonald, J., Pang, A., Shago, M., Stockwell, T., Tsiamouri, A., Bafna, V., Bansal, V., Kravitz, S., Busam, D., Beeson, K., Mclntosh, T., Remington, K., Abril, J., Gill, J., Borman, J., Rogers, Y-H., Frazier, M., Scherer, S., Strausberg, R. & Venter, J. (2007). The Diploid Genome Sequence of an Individual Human. *PLoS Biology*, Vol.5, No.10, (October 2007), pp. e254. doi:10.1371/journal.pbio.0050254, ISSN 1544-9173

Li., Q., Xie, X., Zhi, Q., Ma, L., Abudourexiti, A., Liu, Y., Dangzheng, W., Zhang, Y., Hashan, ., Wang, C., Hamatai, S., Liang, G., He, Y., Zhao, Z., Chen, B., Xu, P. & Zuo, J. (1992). First Isolation of 8 Strains of New Orbivirus (BANNA) from Patients with Innominate Fever in Xinjiang. *Endemic Disease Bulletin*, Vol.7, No.1, (February 1992), pp. 77-81, ISSN 1000-3711

Li, X. & Heyer, W. (2008). Homologous Recombination in DNA Repair and DNA Damage Tolerance. *Cell Research*, Vol.18, No.1, (January 2008), pp. 99-113, ISSN 1001-0602

Li, Z., Düllmann, J., Schiedlmeier, B., Schmidt, M., von Kalle, C., Meyer, J., Forster, M., Stocking, C., Wahlers, A., Frank, O., Ostertag, W., Kühlcke, K., Eckert, H., Fehse, B.

& Baum, C. (2002). Murine Leukemia Induced by Retroviral Gene Marking. *Science*, Vol.296, No.5567, (April 2002), pp. 497, ISSN 0036-8075

Liang, X., Huang, Z., Wang, Y. & Xiao, P. (1985). Studies on the Interference of a Non- pathogenic Virus with the Japanese B Encephalitis Virus. *Acta Academiae Medicinae Sinicae*, Vol.7, No.4, (August 1985), pp. 290-293, ISSN 1000-503X

Liao, J., Wu, Z., Wang, Y., Cheng, L., Cui, C., Gao, Y., Chen, T., Rao, L., Chen, S., Jia, N., Dai, H., Xin, S., Kang, J., Pei, G. & Xiao, L. (2008). Enhanced Efficiency of Generating Induced Pluripotent Stem (iPS) Cells from Human Somatic Cells by a Combination of Six Transcription Factors. *Cell Research* Vol.18, No.5, (May 2008), pp. 600-603, ISSN 1001-0602

Lister, R., Pelizzola, M., Kida, Y., Hawkins, R., Nery, J., Hon, G., Antosiewicz-Bourget, J., O'Malley, R., Castanon, R., Klugman, S., Downes, M., Yu, R., Stewart, R., Ren, B., Thomson, J., Evans, R. & Ecker, J. (2011). Hotspots of Aberrant Epigenomic Reprogramming in Human Induced Pluripotent Stem Cells. *Nature*, Vol.471, No.7336, (March 2011), pp. 68-73, ISSN 0028-0836

Lo Ten Foe, J., Kwee, M., Rooimans, M., Oostra, A., Veerman, A., van Weel, M., Pauli, R., Shahidi, N., Dokal, I., Roberts, I., Altay, C., Gluckman, E., Gibson, R., Mathew, C., Arwert, F. & Joenje, H. (1997). Somatic Mosaicism in Fanconi Anemia: Molecular Basis and Clinical Significance. *European Journal of Human Genetics*, Vol.5, No.3, (May-June 1997), pp. 137-148, ISSN 1018-4813

Lobitz, S. & Velleuer, E. (2006). Guido Fanconi (1892-1979): A Jack of All Trades. *Nature Reviews Cancer*, Vol.6, No.11, (November 2006), pp. 893-898, ISSN 1474-175X

Locati, M. & Murphy, P. (1999). Chemokines and Chemokine Receptors: Biology and Clinical Relevance in Inflammation and AIDS. *Annual Review of Medicine*, Vol.50, (1999), pp. 425-440, ISSN 0066-4219

Lodish, H., Berk, A., Kaiser, C., Krieger, M., Scott, M., Bretscher, A., Ploegh, H. & Matsudaira, P. (2008). *Molecular Cell Biology*, W.H. Freeman and Company, ISBN-10 0-7167-7601-4, New York, USA

Lorenz, A. & Whitby, M. (2006). Crossover Promotion and Prevention. *Biochemical Society Transactions,* Vol.34, No.part 4, (August 2006), pp. 537-541, ISSN 0300-5127

Ludtke, J., Zhang, G., Sebestyen, M. & Wolff, J. (1999). A Nuclear Localization Signal Can Enhance both the Nuclear Transport and Expression of 1 kb DNA. *Journal of Cell Science*, Vol.112, No.part 12, (June 1999), pp. 2033-2041, ISSN 0021-9533

Lundstrom, K. (2001). Alphavirus Vectors for Gene Therapy Applications. *Current Gene Therapy*, Vol.1, No.1, (May 2001), pp. 19-29, ISSN 1566-5232

Lundstrom, K. (2005). Biology and Application of Alphaviruses in Gene Therapy. *Gene Therapy*, Vol.12, No.supplement 1, (October 2005), pp. S92-S97, ISSN 0969-7128

Luster, A. (1998). Chemokines-Chemotactic Cytokines that Mediate Inflammation. *The New England Journal of Medicine*, Vol.338, No.7, (February 1998), pp. 436-445, ISSN 0028- 4793

Ma, W., Lager, K., Lekcharoensuk, P., Ulery, E., Janke, B., Solórzano, A., Webby, R., García- Sastre, A. & Richt, J. (2010). Viral Reassortment and Transmission after Co-infection of Pigs with Classical H1N1 and Triple-reassortant H3N2 Swine Influenza Viruses. *Journal of General Virology*, Vol.91, No.part 9, (September 2010), pp. 2314-2321, ISSN 0022-1317

MacDonald, M., Ambrose, C., Duyao, M., Myers, R., Lin, C., Srinidhi, L., Barnes, G., Taylor, S., James, M., Groot, N., MacFarlane, H., Jenkins, B., Anderson, M., Wexler, N. & Gusella, J. (1993). A Novel Gene Containing a Trinucleotide Repeat That Is Expanded and Unstable on Huntington's Disease Chromosomes. *Cell*, Vol.72, No.6, (March 1993), pp. 971-983, ISSN 0092-8674

Mankad, A., Taniguchi, T., Cox, B., Akkari, Y., Rathbun, R., Lucas, L., Bagby, G., Olson, S., D'Andrea, A., Grompe, M. (2006). Natural Gene Therapy in Monozygotic Twins with Fanconi Anemia. *Blood, Journal of the American Society of Hematology*, Vol.107, No.8, (April 2006), pp. 3084-3090, ISSN 0006-4971

Martinez-Agosto, J., Mikkola, H., Hartenstein, V. & Banerjee, U. (2007). The Hematopoietic Stem Cell and Its Niche: A Comparative View. *Genes & Development*, Vol.21, No.23, (December 2007), pp. 3044-3060, ISSN 0890-9369

Mastaglio, S., Stanghellini, M., Bordignon, C., Bondanza, A., Ciceri, F. & Bonini, C. (2010). Progress and Prospects: Graft-versus-host Disease. *Gene Therapy*, Vol.17, No.11, (November 2010), pp. 1309–1317, ISSN 0969-7128

Matthijnssens, J., Taraporewala, Z., Yang, H., Rao, S., Yuan, L., Cao, D., Hoshino, Y., Mertens, P., Carner, G., McNeal, M., Sestak, K., Van Ranst, M. & Patton, J. (2010). Simian Rotaviruses Possess Divergent Gene Constellations that Originated from Interspecies Transmission and Reassortment. *Journal of Virology*, Vol. 84, No.4, (February 2010), pp. 2013-2026, ISSN 0022-538X

Maunula, L. & Von Bonsdorff, C. (2002). Frequent Reassortments May Explain the Genetic Heterogeneity of Rotaviruses: Analysis of Finnish Rotavirus Strains. *Journal of Virology*, Vol. 76, No.23, (December 2002), pp. 11793-11800, ISSN 0022-538X

McClain, M., Spendlove, R. & Lennette, E. (1967). Infectivity Assay of Reoviruses: Comparison with Immunofluorescent Cell Count and Plaque Methods. *The Journal of Immunology*, Vol.98, No.6, (June 1967), pp. 1301-1308, ISSN 0022-1767

McCulloch, E., Siminovitch, L. & Till J. (1964). Spleen-colony Formation in Anemic Mice of Genotype WW. *Science*, Vol.144, No.1620, (May 1964), pp. 844-846, ISSN 0036-8075

Ménoret, S., Remy, S., Usal, C., Tesson, L. & Anegon, I. (2010). Generation of Transgenic Rats by Microinjection of Short DNA Fragments. *Methods in Molecular Biology*, Vol.597, (2010), pp. 81-92, ISSN 1064-3745

Mir, L. (2009). Nucleic Aacids Electrotransfer-based Gene Therapy (Electrogenetherapy): Past, Current, and Future. *Molecular Biotechnology*, Vol.43, No.2, (October 2009), pp. 167-176, ISSN 1073-6085

Modlich, U., Kustikova, O., Schmidt, M., Rudolph, C., Meyer, J., Li, Z., Kamino, K., von Neuhoff, N., Schlegelberger, B., Kuehlcke, K., Bunting, K., Schmidt, S., Deichmann, A., von Kalle, C., Fehse, B. & Baum C. (2005). Leukemias following Retroviral Transfer of Multidrug Resistance 1 (MDR1) Are Driven by Combinatorial Insertional Mutagenesis. *Blood, Journal of the American Society of Hematology,* Vol.105, No.11, (June 2005), pp. 4235-4246, ISSN 0006-4971

Mohd Jaafar, F., Attoui, H., Mertens, P., de Micco, P. & de Lamballerie, X. (2005). Structural Organization of an Encephalitic Human Isolate of Banna Virus (Genus Seadornavirus, Family Reoviridae). *Journal of General Virology*, Vol.86, No.part 4, (April 2005), pp. 1147-1157, ISSN 0022-1317

Monaco, A., Neve, R., Colletti-Feener, C., Bertelson, C., Kurnit, D. & Kunkel, L. (1986). Isolation of Candidate cDNAs for Portions of the Duchenne Muscular Dystrophy Gene. *Nature*, Vol.323, No.6089, (October 1986), pp. 646-650, ISSN 0028-0836

Mortellaro, A., Hernandez, R., Guerrini, M., Carlucci, F., Tabucchi, A., Ponzoni, M., Sanvito, F., Doglioni, C., Di Serio, C., Biasco, L., Follenzi, A., Naldini, L., Bordignon, C., Roncarolo, M. & Aiuti, A. (2006). Ex vivo Gene Therapy with Lentiviral Vectors Rescues Adenosine Deaminase (ADA)-deficient Mice and Corrects Their Immune and Metabolic Defects. *Blood, Journal of the American Society of Hematology*, Vol.108, No.9, (November 2006), pp. 2979-2988, ISSN 0006-4971

Moscow, J. & Cowan, K. (1988). Multidrug Resistance. *Journal of the National Cancer Institute,* Vol.80, No.1, (March 1988), pp. 14-20, ISSN 0027-8874

Müller, L. & Williams, D. (2009). Finding the Needle in the Hay Stack: Hematopoietic Stem Cells in Fanconi Anemia. *Mutation Research - Fundamental and Molecular Mechanisms of Mutagenesis*, Vol.668, No.1-2, (July 2009), pp. 141-149, ISSN 0027-5107

Muntoni, F., Torelli, S. & Ferlini, A. (2003). Dystrophin and Mutations: One Gene, Several Proteins, Multiple Phenotypes. *The Lancet Neurology*, Vol.2, No.12, (December 2003), pp. 731-740, ISSN 1474-4422

Murdoch, C. & Finn, A. (2000). Chemokine Receptors and Their Role in Inflammation and Infectious Diseases. *Blood, Journal of the American Society of Hematology,* Vol.95, No.10, (May 2000), pp. 3032-3043, ISSN 0006-4971

Nabeshima, T., Thi Nga, P., Guillermo, P., Parquet Mdel, C., Yu, F., Thanh Thuy, N., Minh Trang, B., Tran Hien, N., Sinh Nam, V., Inoue, S., Hasebe, F. & Morita, K. (2008). Isolation and Molecular Characterization of Banna Virus from Mosquitoes, Vietnam. *Emerging Infectious Diseases*, Vol.14, No.8, (August 2008), pp. 1276-1279, ISSN: 1080-6059

Nayak, S. & Herzog, R. (2010). Progress and Prospects: Immune Responses to Viral Vectors. *Gene Therapy*, Vol.17, No.3, (March 2010), pp. 295-304, Erratum in: *Gene Therapy*, Vol.17, No.2, (February 2010), pp. 294, ISSN 0969-7128

Nelson, S., Crosbie, R., Miceli, M. & Spencer, M. (2009). Emerging Genetic Therapies to Treat Duchenne Muscular Dystrophy. *Current Opinion in Neurology*, Vol.22, No.5, (October 2009), pp. 532-538, ISSN 1350-7540

Neumann, E., Schaefer-Ridder, M., Wang, Y. & Hofschneider, P. (1982). Gene Transfer into Mouse Lyoma Cells by Electroporation in High Electric fields. *The EMBO Journal*, Vol.1, No.7, (1982), pp. 841-845, ISSN 0261-4189

Nibert, M., Furlong, D. & Fields, B. (1991). Mechanisms of Viral Pathogenesis. Distinct Forms of Reoviruses and Their Roles during Replication in Cells and Host. *The Journal of Clinical Investigation*, Vol.88, No.3, (September 1991), pp. 727-734, ISSN 0021-9738

Niidome, T. & Huang L. (2002). Gene Therapy Progress and Prospects: Nonviral Vectors. *Gene Therapy*, Vol.9, No.24, (December 2002), pp. 1647-1652, ISSN 0969-7128

Nishio, H., Takeshima, Y., Narita, N., Yanagawa, H., Suzuki, Y., Ishikawa, Y., Ishikawa, Y., Minami, R., Nakamura, H. & Matsuo, M. (1994). Identification of a Novel First Exon in the Human Dystrophin Gene and of a New Promoter Located More than 500 kb Upstream of the Nearest Known Promoter. *Journal of Clinical Investigation*, Vol.94, No.3, (September 1994), pp. 1037-1042, ISSN 0021-9738

Norkin, L. (1995). Virus Receptors: Implications for Pathogenesis and the Design of Antiviral Agents. *Clinical Microbiology Reviews*, Vol.8, No.2, (April 1995), pp. 293-315, ISSN 0893-8512

Notini, A., Craig, J. & White, S. (2008). Copy Number Variation and Mosaicism. *Cytogenetic and Genome Research*, Vol.123, No.1-4, (2008), pp. 270-277, ISSN 1424-8581

Nuttall, P. & Moss, S. (1989). Genetic Reassortment Indicates a New Grouping for Tick- borne Orbiviruses. *Virology,* Vol.171, No.1, (July 1989), pp. 156-161, ISSN 0042-6822

O'Brien, S., Menotti-Raymond, M., Murphy, W., Nash, W., Wienberg, J., Stanyon, R., Copeland, N., Jenkins, N., Womack, J. & Marshall Graves, J. (1999). The Promise of Comparative Genomics in Mammals. *Science*, Vol.286, No.5439, (October 1999), pp. 458-462, 479-481, ISSN 0036-8075

Octaviani, C., Li, C., Noda, T. & Kawaoka, Y. (2011). Reassortment between Seasonal and Swine-origin H1N1 Influenza Viruses Generates Viruses with Enhanced Growth Capability in Cell Culture. *Virus Research,* Vol.156, No.1-2, (March 2011), pp. 147- 150, ISSN 0168-1702

Osawa, M., Hanada, K., Hamada, H. & Nakauchi, H. (1996). Long-term Lymphohematopoietic Reconstitution by a Single CD34-low/Negative Hematopoietic Stem Cell. *Science,* Vol.273, No.5272, (July 1996), pp. 242-245, ISSN 0036-8075

Park, H., Kim, J., Ju, Y., Gokcumen, O., Mills, R., Kim, S., Lee, S., Suh, D., Hong, D., Kang, H., Yoo, Y., Shin, J., Kim, H., Yavartanoo, M., Chang, Y., Ha, J., Chong, W., Hwang, G., Darvishi, K., Kim, H., Yang, S., Yang, K., Kim, H., Hurles, M., Scherer, S., Carter, N., Tyler-Smith, C., Lee, C. & Seo, J. (2010). Discovery of Common Asian Copy Number Variants Using Integrated High-resolution Array CGH and Massively Parallel DNA Sequencing. *Nature Genetics,* Vol.42, No.5, (May 2010), pp. 400-405, ISSN 1061-4036

Park, I., Zhao, R., West, J., Yabuuchi, A., Huo, H., Ince, T., Lerou, P., Lensch, M. & Daley, G. (2008). Reprogramming of Human Somatic Cells to Pluripotency with Defined Factors. *Nature,* Vol.451, No.7175, (January 2008), pp. 141-146, ISSN 0028-0836

Parmar, K., D'Andrea, A. & Niedernhofer, L. (2009). Mouse Models of Fanconi Anemia. *Mutation Research - Fundamental and Molecular Mechanisms of Mutagenesis,* Vol.668, No.1-2, (July 2009), pp. 133-140, ISSN 0027-5107

Pastan, I. & Gottesman, M. (1987). Multidrug Resistance in Human Cancer. *The New England Journal of Medicine,* Vol.316, No.22, (May 1987), pp. 1388-1393, ISSN 0028-4793

Pawliuk, R., Westerman, K., Fabry, M., Payen, E., Tighe, R., Bouhassira, E., Acharya, S., Ellis, J., London, I., Eaves, C., Humphries, R., Beuzard, Y., Nagel, R. & Leboulch, P. (2001). Correction of Sickle Cell Disease in Transgenic Mouse Models by Gene Therapy. *Science,* Vol.294, No.5550, (December 2001), pp. 2368-2371, ISSN 0036-8075

Pennisi E. (1998). Training Viruses to Attack Cancers. *Science,* Vol.282, No.5392, (November 1998), pp. 1244-1246, ISSN 0036-8075

Peters, C. (2005). Marburg and Ebola--Arming Ourselves against the Deadly Filoviruses. *The New England Journal of Medicine,* Vol.352, No.25, (June 2005), pp. 2571-2573, ISSN 0028-4793

Piotrowski, A., Bruder, C., Andersson, R., Diaz de Ståhl, T., Menzel, U., Sandgren, J., Poplawski, A., von Tell, D., Crasto, C., Bogdan, A., Bartoszewski, R., Bebok, Z., Krzyzanowski, M., Jankowski, Z., Partridge, E., Komorowski, J. & Dumanski, J. (2008). Somatic Mosaicism for Copy Number Variation in Differentiated Human Tissues. *Human Mutation,* Vol.29, No.9, (September 2008), pp. 1118-1124, ISSN 1059-7794

Pitisuttithum, P., Tansuphasawadikul, S., Simpson, A., Howe, P. & White, N. (2001). A Prospective Study of AIDS-associated Cryptococcal Meningitis in Thailand Treated with High-dose Amphotericin B. *Journal of Infection,* Vol.43, No.4, (November 2001), pp. 226-233, ISSN 0163-4453

Qasim, W., Gaspar, H. & Thrasher, A. (2009). Progress and Prospects: Gene Therapy for Inherited Immunodeficiencies. *Gene Therapy,* Vol.16, No.11, (November 2009), pp. 1285–1291, ISSN 0969-7128

Rambaut, A., Pybus, O., Nelson, M., Viboud, C., Taubenberger, J. & Holmes, E. (2008). The Genomic and Epidemiological Dynamics of Human Influenza A Virus. *Nature,* Vol.453, No.7195, (May 2008), pp. 615-619, ISSN 0028-0836

Ramesh, N., Antón, I., Hartwig, J. & Geha, R. (1997). WIP, a Protein Associated with Wiskott-Aldrich Syndrome Protein, Induces Actin Polymerization and Redistribution in Lymphoid Cells. *Proceedings of the National Academy of Sciences,* Vol.94, No.26, (December 1997), pp. 14671-14676, ISSN 0027-8424

Raper, S., Chirmule, N., Lee, F., Wivel, N., Bagg, A., Gao, G., Wilson, J. & Batshaw M. (2003). Fatal Systemic Inflammatory Response Syndrome in a Ornithine Transcarbamylase Deficient Patient following Adenoviral Gene Transfer. *Molecular Genetics and Metabolism,* Vol.80, No.1-2, (September-October 2003), pp.148-158, ISSN 1096-7192

Reid, S., Schindler, D., Hanenberg, H., Barker, K., Hanks, S., Kalb, R., Neveling, K., Kelly, P., Seal, S., Freund, M., Wurm, M., Batish, S., Lach, F., Yetgin, S., Neitzel, H., Ariffin, H., Tischkowitz, M., Mathew, C., Auerbach, A. & Rahman, N. (2007). Biallelic Mutations in PALB2 Cause Fanconi Anemia Subtype FA-N and Predispose to Childhood Cancer. *Nature Genetics,* Vol.39, No.2, (February 2007), pp. 162-164, ISSN 1061-4036

Ridinger, D., Spendlove, R., Barnett, B., George, D. & Roth, J. (1982). Evaluation of Cell Lines and Immunofluorescence and Plaque Assay Procedures for Quantifying Reoviruses in Sewage. *Applied and Environmental Microbiology,* Vol.43, No.4, (April 1982), pp. 740-746, ISSN 0099-2240

Riordan, J., Deuchars, K., Kartner, N., Alon, N., Trent, J. & Ling, V. (1985). Amplification of P-glycoprotein Genes in Multidrug-resistant Mammalian Cell Lines. *Nature*, Vol.316, No.6031, (August 1985). pp. 817–819, ISSN 0028-0836

Riordan, J., Rommens, J., Kerem, B-S., Alon, N., Rozmahel, R., Grzelczak, Z., Zielenski, J., Lok, S., Plavsic, N., Chou, J-L., Drumm, M., Iannuzzi, M., Collins, F. & Tsui, L-C. (1989). Identification of the Cystic Fibrosis Gene: Cloning and Characterization of Complementary DNA. *Science*, Vol.245, No.4922 (September 1989), pp. 1066-1073, Erratum in: *Science*, Vol.245, No.4925, (September 1989), pp. 1437, ISSN 0036-8075

Roberts R. (2001). Dystrophins and Dystrobrevins. *Genome Biology*, Vol.2, No.4, (April 2001), pp. reviews3006.1–reviews3006.7, ISSN 1465-6906

Robertson, E., Bradley, A., Kuehn, M. & Evans, M. (1986). Germ-line Transmission of Genes Introduced into Cultured Pluripotential Cells by Retroviral Vector. *Nature*, Vol.323, No.6087, (October 1986), pp. 445-448, ISSN 0028-0836

Rocha, V., Labopin, M., Sanz, G., Arcese, W., Schwerdtfeger, R., Bosi, A., Jacobsen, N., Ruutu, T., de Lima, M., Finke, J., Frassoni, F., Gluckman, E., Acute Leukemia Working Party of European Blood and Marrow Transplant Group & Eurocord- Netcord Registry. (2004). Transplants of Umbilical-cord Blood or Bone Marrow from Unrelated Donors in Adults with Acute Leukemia. *The New England Journal of Medicine*, Vol.351, No.22, (November 2004), pp. 2276-2285, ISSN 0028-4793

Rogaev, E., Sherrington, R., Rogaeva, E., Levesque, G., Ikeda, M., Liang, Y., Chi, H., Lin, C., Holman, K., Tsuda, T., Mar, L., Sorbi, S., Nacmias, B., Piacentini, S., Amaducci, L., Chumakov, I., Cohen, D., Lannfelt, L., Fraser, P., Rommens, J. & St George-Hyslop, P. (1995). Familial Alzheimer's Disease in Kindreds with Missense Mutations in a Gene on Chromosome 1 Related to the Alzheimer's Disease Type 3 Gene. *Nature*, Vol.376, No.6543, (August 1995), pp. 775-778, ISSN 1061-4036

Rommens, J., Iannuzzi, M., Kerem, B-S., Drumm, M., Melmer, G., Dean, M., Rozmahel, R., Cole, J., Kennedy, D., Hidaka, N., Zsiga, M., Buchwald, M., Riordan, J., Tsui, L-C. & Collins, F. (1989). Identification of the Cystic Fibrosis Gene: Chromosome Walking and Jumping. *Science*, Vol.245, No.4922, (September 1989), pp. 1059-1065, ISSN 0036-8075

Rouquet, P., Froment, J., Bermejo, M., Kilbourn, A., Karesh, W., Reed, P., Kumulungui, B., Yaba, P., Délicat, A., Rollin, P. & Leroy, E. (2005). Wild Animal Mortality Monitoring and Human Ebola Outbreaks, Gabon and Republic of Congo, 2001-2003. *Emerging Infectious Diseases*, Vol.11, No.2, (February 2005), pp. 283-290, ISSN 1080-6059

Rowntree, R. & Harris, A. (2003). The Phenotypic Consequences of CFTR Mutations. *Annals of Human Genetics*, Vol.67, No.part 5, (September 2003), pp. 471-485, ISSN 0003-4800

Rozee, K. & Easterbrook, K. (1970). Application of Freeze-etching Method to the Study of Reovirus-infected LLC-MK2 Cells. *Applied Microbiology*, Vol.19, No.6, (June 1970), pp. 997-1000, ISSN 0099-2240

Sakharkar, M., Chow, V. & Kangueane, P. (2004). Distributions of Exons and Introns in the Human Genome. *In Silico Biology*, Vol.4, No.4, (June 2004), pp. 387-393, ISSN 1434-3207

Schiff, L., Nibert, M & Tyler, K. (2007). Orthoreoviruses and Their Replication, In: *Fields Virology, 5th Edition, Volume One*, David M. Knipe, Peter M. Howley, Diane E. Griffin, Robert A. Lamb, Malcolm A. Martin, Bernard Roizman, pp.2388-2471, Lippincott Williams & Wilkins, ISBN-10 0781760607, Philadelphia, USA

Seggewiss, R., Pittaluga, S., Adler, R., Guenaga, F., Ferguson, C., Pilz, I., Ryu, B., Sorrentino, B., Young, W 3rd., Donahue, R., von Kalle, C., Nienhuis, A. & Dunbar, C. (2006). Acute Myeloid Leukemia Is Associated with Retroviral Gene Transfer to Hematopoietic Progenitor Cells in a Rhesus Macaque. *Blood, Journal of the American Society of Hematology*, Vol.107, No.10, (May 2006), pp. 3865-3867, ISSN 0006-4971

Selb, B. & Weber, B. (1994). A Study of Human Reovirus IgG and IgA Antibodies by ELISA and Western Blot. *Journal of Virological Methods*, Vol.47, No.1-2, (April 1994), pp. 15- 25, ISSN 0166-0934

Sherrington, R., Rogaev, E., Liang, Y., Rogaeva, E., Levesque, G., Ikeda, M., Chi, H., Lin, C., Li, G., Holman, K., Tsuda, T., Mar, L., Foncin, J., Bruni, A., Montesi, M., Sorbi, S., Rainero, I., Pinessi, L., Nee, L., Chumakov, I., Pollen, D., Brookes, A., Sanseau, P., Polinsky, R., Wasco, W., Da Silva, H., Haines, J., Perkicak-Vance, M., Tanzi, R., Roses, A., Fraser, P., Rommens, J., St George-Hyslop, P.(1995). Cloning of a Gene Bearing Missense Mutations in Early-onset Familial Alzheimer's Disease. *Nature*, Vol.375, No.6534, (June 1995), pp. 754-760, ISSN 1061-4036

Siminovitch, L., McCulloch, E. & Till, J. (1963). The Distribution of Colony-forming Cells among Spleen Colonies. *Journal of Cellular and Comparative Physiology*, Vol.62, No.3, (December 1963), pp. 327-336, ISSN 0095-9898

Smith, D., Lapedes, A., de Jong, J., Bestebroer, T., Rimmelzwaan, G., Osterhaus, A. & Fouchier, R. (2004). Mapping the Antigenic and Genetic Evolution of Influenza Virus. *Science*, Vol.305, No.5682, (July 2004), pp. 371-376, ISSN 0036-8075

Smogorzewska, A., Matsuoka, S., Vinciguerra, P., McDonald, E. 3rd., Hurov, K., Luo, J., Ballif, B., Gygi, S., Hofmann, K., D'Andrea, A. & Elledge, S. (2007). Identification of the FANCI Protein, a Monoubiquitinated FANCD2 Paralog Required for DNA Repair. *Cell*, Vol.129, No.2, (April 2007), pp. 289-301, ISSN 0092-8674

Somiari, S., Glasspool-Malone, J., Drabick, J., Gilbert, R., Heller, R., Jaroszeski, M. & Malone, R. (2000). Theory and in vivo Application of Electroporative Gene Delivery. *Molecular Therapy*, Vol.2, No.3, (September 2000), pp. 178-187, ISSN 1525-0016

Song, L., Chen, B. & Zhao Z. (1995). Isolation and Identification of New Members of Coltivirus from Mosquitoes Collected in China. *Chinese Journal of Experimental and Clinical Virology*, Vol.9, No.1, (March 1995), pp. 7-10, ISSN 1003-9279

Song, L. & Chen, B. (1995). Arboviral Encephalitides (I). *Chinese Journal of Vector Biology and Control*, Vol.6, No.6, (December 1995), pp. 477-480, ISSN 1003-4692

Song, L. & Chen, B. (1996). Arboviral Encephalitides (II). *Chinese Journal of Vector Biology and Control*, Vol.7, No.1, (February 1996), pp. 66-69, ISSN 1003-4692

Song, L., Ohnuma, T. & Holland, J. (1999a). Effect of Trehalose on Cell Kills Effects of Anticancer Agents or Hyperthermia against Human Tumor Cell Lines. *Mount Sinai Journal of Medicine*, Vol.66, No.5-6, (October/November 1999), pp. 388, ISSN 0027- 2507

Song, L., Gelman, I., Holland, J. & Ohnuma, T. (1999b). Study of Reovirus-induced Cytopathic Effect (CPE) and Ras Protein in Human Tumor Cell Lines. *Mount Sinai Journal of Medicine*, Vol.66, No.5-6, (October/November 1999), pp. 388, ISSN 0027- 2507

Song, L., Gelman, I., Holland, J. & Ohnuma, T. (2000). Study of Reovirus-induced Cytopathic Effect (CPE), Ras, Doublestranded RNA-activated Protein Kinase (PKR) and eIF2α in Human Tumor Cell Lines. *Proceedings of the American Association for Cancer Research*, Vol.41, (2000), pp. 350, ISSN 0569-2261

Song, L., Mandecki, W. & Goldman, E. (2003). Expression of Non-open Reading Frames Isolated from Phage Display due to Translation Reinitiation. *Journal of the Federation of American Societies for Experimental Biology*, Vol.17, No.12, (September 2003), pp. 1674-1681, ISSN 0892-6638

Song, L. (2009). A Possible Approach for Stem Cell Gene Therapy of Fanconi Anemia.*Current Gene Therapy*, Vol.9, No.1, (February 2009), pp. 26-32, ISSN 1566-5232

Song, L., Ohnuma, T., Gelman, I. & Holland, J. (2009). Reovirus Infection of Cancer Cells Is Not due to Activated Ras Pathway. *Cancer Gene Therapy*, Vol.16, No.4, (April 2009), pp. 382, ISSN 0929-1903

Song, L. (2010). Comment on the AACR News of "Reovirus May Be a Novel Approach to Prostate Cancer Treatment, March 9, 2010", In: *AACR News*, March 9, 2010,

06.06.2011, Available from: http://aacrnews.wordpress.com/2010/03/09/reovirus-may-be-a-novel-approach- to-prostate-cancer-treatment/#comments

Song, L-H., Zhou, Y., He, J., Zhu, H., Huang, R., Mao, P. & Duan, Q. (2008). Comparative Sequence Analyses of a New Mammalian Reovirus Genome and the Mammalian Reovirus S1 Genes from Six New Serotype 2 Human Isolates. *Virus Genes*, Vol.37, No.3, (December 2008), pp. 392-399, ISSN 0920-8569

Spangrude, G., Heimfeld, S. & Weissman, I. (1988). Purification and Characterization of Mouse Hematopoietic Stem Cells. *Science*, Vol.241, No.4861, (July 1988), pp. 58-62, Erratum in: *Science*, Vol.244, No.4908, (June 1989), pp. 1030, ISSN 0036-8075

Stanford, K., Etrie, W. & Likely, G. (1948). The Growth in vitro of Single Isolated Tissue Cells. *The Journal of the National Cancer Institute*, Vol.9, No.3, (December 1948), pp. 229–246, ISSN 0027-8874

Stegemann, S. & Bock, R. (2009). Exchange of Genetic Material between Cells in Plant Tissue Grafts. *Science*, Vol.324, No.5927, (May 2009), pp. 649-651, ISSN 0036-8075

Stoepker, C., Hain, K., Schuster, B., Hilhorst-Hofstee, Y., Rooimans, M., Steltenpool, J., Oostra, A., Eirich, K., Korthof, E., Nieuwint, A., Jaspers, N., Bettecken, T., Joenje, H., Schindler, D., Rouse, J. & de Winter, J. (2011). SLX4, a Coordinator of Structure- specific Endonucleases, Is Mutated in a New Fanconi Anemia Subtype. *Nature Genetics*, Vol.43, No.2, (February 2011), pp. 142-146, ISSN: 1061 4036

Sun, Y., Qin, K., Wang, J., Pu, J., Tang, Q., Hu, Y., Bi, Y., Zhao, X., Yang, H., Shu, Y. & Liu, J. (2011). High Genetic Compatibility and Increased Pathogenicity of Reassortants Derived from Avian H9N2 and Pandemic H1N1/2009 Influenza Viruses. *Proceedings of the National Academy of Sciences*, Vol.108, No.10, (March 2011), pp. 4164-4169, ISSN 0027-8424

Sutlu, T. & Alici, E. (2009). Natural Killer Cell-based Immunotherapy in Cancer: Current Insights and Future Prospects. *Journal of Internal Medicine*, Vol.266, No.2, (August 2009), pp. 154-181, ISSN 0954-6820

Swann, J. & Smyth, M. (2007). Immune Surveillance of Tumors. *The Journal of Clinical Investigation,* Vol.117, No.5, (May 2007), pp. 1137-1146, ISSN 0021-9738

Tai, J., Williams, J., Edwards, K., Wright, P., Crowe, J. Jr. & Dermody, T. (2005). Prevalence of Reovirus-specific Antibodies in Young Children in Nashville, Tennessee. *The Journal of Infectious Diseases*, Vol.191, No.8, (April 2005), pp. 1221-1224, ISSN 1537-6613

Takahashi, K., Tanabe, K., Ohnuki, M., Narita, M., Ichisaka, T., Tomoda, K. & Yamanaka, S. (2007). Induction of Pluripotent Stem Cells from Adult Human Fibroblasts by Defined Factors. *Cell*, Vol.131, No.5, (November 2007), pp. 861-872, ISSN 0092-8674

Taniguchi, T. & D'Andrea, A. (2006). Molecular Pathogenesis of Fanconi Anemia: Recent Progress. *Blood, Journal of the American Society of Hematology,* Vol.107, No.11, (June 2006), pp. 4223-4233, ISSN 0006-4971

Tao, S., Cai, Z., Yang, D., Wang, H., Liu, Q., Fan, Y., Fan, X. & Chen, B. (1999). New Subtype of Coltivirus Isolated from Mosquitoes in the Northeast Part of China. *Chinese Journal of Experimental and Clinical Virology*, Vol.13, No.3, (Septmember 1999), pp. 228-230, ISSN 1003-9279 Tatum, E. & Lederberg J. (1947). Gene Recombination in the Bacterium *Escherichia coli*. *Journal of Bacteriology*, Vol.53, No.6, (June 1947), pp. 673-684, ISSN 0021-9193

Terry, L., Shows, E. & Wente, S. (2007). Crossing the Nuclear Envelope: Hierarchical Regulation of Nucleocytoplasmic Transport. *Science.* Vol.318, No.5855, (November 2007), pp. 1412-1416, ISSN 0036-8075

Terzi, D. & Zachariou, V. (2008). Adeno-associated Virus-mediated Gene Delivery Approaches for the Treatment of CNS Disorders. *Biotechnology Journal*, Vol.3, No.12, (December 2008), pp. 1555-1563, ISSN 1860-7314

Tesson, L., Cozzi, J., Ménoret, S., Rémy, S., Usal, C., Fraichard, A. & Anegon, I. (2005). Transgenic Modifications of the Rat Genome. *Transgenic Research*, Vol.14, No.5, (October 2005), pp. 531-546, ISSN 0962-8819

The 1000 Genomes Project Consortium. (2010). A Map of Human Genome Variation from Population-scale Sequencing. *Nature*, Vol.467, No.7319, (October 2010), pp. 1061- 1073, ISSN 0028-0836

The 1000 Genomes Consortium. (2011). Mapping Copy Number Variation by Population- scale Genome. *Nature*, Vol.470, No. 7332, (February 2011), pp. 59-65, ISSN 0028-0836

Theerthagiri, G., Eisenhardt, N., Schwarz, H. & Antonin, W. (2010). The Nucleoporin Nup188 Controls Passage of Membrane Proteins across the Nuclear Pore Complex. *The Journal of Cell Biology*, Vol.189, No.7, (June 2010), pp. 1129-1142, ISSN 0021-9525

Thomson, J., Itskovitz-Eldor, J., Shapiro, S., Waknitz, M., Swiergiel, J., Marshall, V. & Jones, J. (1998). Embryonic Stem Cell Lines Derived from Human Blastocysts. *Science*, Vol.282, No.5391, (November 1998), pp. 1145-1147, Erratum in: *Science*, Vol.282, No.5395, (December 1998), pp. 1827, ISSN 0036-8075

Till, J., McCulloch, E. & Siminovitch, L. (1964). A Stochastic Model of Stem Cell Proliferation, Based on the Growth of Spleen Colony-forming Cells. *Proceedings of the National Academy of Sciences*, Vol.51, No.1, (January 1964), pp. 29–36, ISSN 0027-8424

Tischkowitz, M. & Hodgson, S. (2003). Fanconi Anaemia. *Journal of Medical Genetics*, Vol.40, No.1, (January 2003), pp. 1-10, ISSN 0022-2593

Tischkowitz, M., Easton, D., Ball, J., Hodgson, S. & Mathew, C. (2008). Cancer Incidence in Relatives of British Fanconi Anaemia Patients. *BMC Cancer*, Vol.8, (September 2008), pp. 257, doi: 10.1186/1471-2407-8-257, ISSN 1471-2407

Tran, E. & Wente, S. (2006). Dynamic Nuclear Pore Complexes: Life on the Edge. *Cell*, Vol.125, No.6, (June 2006), pp. 1041-1053, ISSN 0092-8674

Tran, V., Zhang, Z., Yagi, M., Nishiyama, A., Habara, Y., Takeshima, Y. & Matsuo, M. (2005). A Novel Cryptic Exon Identified in the 3' Region of Intron 2 of the Human Dystrophin Gene. *Journal of Human Genetics*, Vol.50, No.8, (2005), pp. 425-433, ISSN 1434-5161

Tycko, B. (1994). Genomic Imprinting: Mechanism and Role in Human Pathology. *The American Journal of Pathology*, Vol.144, No.3, (March 1994), pp. 431-443, ISSN 0002-9440

Tyler, K., Barton, E., Ibach, M., Robinson, C., Campbell, J., O'Donnell, S., Valyi-Nagy, T., Clarke, P., Wetzel, J. & Dermody, T. (2004). Isolation and Molecular Characterization of a Novel Type 3 Reovirus from a Child with Meningitis. *The Journal of Infectious Diseases*, Vol.189, No.9, (May 2004), pp. 1664-1675, ISSN 1537-6613

Vajta, G. & Gjerris, M. (2006). Science and Technology of Farm Animal Cloning: State of the Art. *Animal Reproduction Science*, Vol.92, No.3-4, (May 2006), pp. 211-230, ISSN 0378-4320

van den Wollenberg, D., van den Hengel, S., Dautzenberg, I., Cramer, S., Kranenburg, O. & Hoeben, R. (2008). A Strategy for Genetic Modification of the Spike-encoding Segment of Human Reovirus T3D for Reovirus Targeting. *Gene Therapy*, Vol.15, No.24, (December 2008), pp. 1567-1578, ISSN 0969-7128

van Houdt, W., Smakman, N., van den Wollenberg, D., Emmink, B., Veenendaal, L., van Diest, P., Hoeben, R., Borel Rinkes, I. & Kranenburg, O. (2008). Transient Infection of Freshly Isolated Human Colorectal Tumor Cells by Reovirus T3D Intermediate Subviral Particles. *Cancer Gene Therapy*, Vol.15, No.5, (May 2008), pp. 284-292, ISSN 0929-1903

Veeriah, S., Taylor, B., Meng, S., Fang, F., Yilmaz, E., Vivanco, I., Janakiraman, M., Schultz, N., Hanrahan, A., Pao, W., Ladanyi, M., Sander, C., Heguy, A., Holland, E., Paty, P., Mischel, P., Liau, L., Cloughesy, T., Mellinghoff, I., Solit, D. & Chan, T. (2010). Somatic Mutations of the Parkinson's Disease-associated Gene PARK2 in Glioblastoma and Other Human Malignancies. *Nature Genetics*, Vol.42, No.1, (January 2010), pp. 77-82, ISSN 1061-4036

Venter, J., Adams M., Myers, E., Li, P., Mural, R., Sutton, G., Smith, H., Yandell, M., Evans, C., Holt, R., Gocayne, J., Amanatides, P., Ballew, R., Huson, D., Wortman, J., Zhang, Q., Kodira, C., Zheng, X., Chen, L., Skupski, M., Subramanian, G., Thomas, P., Zhang, J., Gabor Miklos, G., Nelson, C., Broder, S., Clark, A., Nadeau, J., McKusick, V., Zinder, N., Levine, A., Roberts, R., Simon, M., Slayman, C., Hunkapiller, M., Bolanos, R., Delcher, A., Dew, I., Fasulo, D., Flanigan, M., Florea, L., Halpern, A., Hannenhalli, S., Kravitz, S., Levy, S., Mobarry, C., Reinert, K., Remington, K., Abu- Threideh, J., Beasley, E., Biddick, K., Bonazzi, V., Brandon, R., Cargill, M., Chandramouliswaran, I., Charlab, R., Chaturvedi, K., Deng, Z., Di Francesco, V., Dunn, P., Eilbeck, K., Evangelista, C., Gabrielian, A., Gan, W., Ge, W., Gong, F., Gu, 1. ., Guan, P., Heiman, T., Higgins, M., Ji, R., Ke, Z., Ketchum, K., Lai, Z., Lei, Y., Li, Z., Li, J., Liang, Y., Lin, X., Lu, F., Merkulov, G., Milshina, N., Moore, H., Naik, A., Narayan, V., Neelam, B., Nusskern, D., Rusch, D., Salzberg, S., Shao, W., Shue, B., Sun, J., Wang, Z., Wang, A., Wang, X., Wang, J., Wei, M., Wides, R., Xiao, C., Yan, C., Yao, A., Ye, J., Zhan, M., Zhang, W., Zhang, H., Zhao, Q., Zheng, L., Zhong, F., Zhong, W., Zhu, S., Zhao, S., Gilbert, D., Baumhueter, S., Spier, G., Carter, C., Cravchik, A., Woodage, T., Ali, F., An, H., Awe, A., Baldwin, D., Baden, H., Barnstead, M., Barrow, I., Beeson, K., Busam, D., Carver, A., Center, A., Cheng, M., Curry, L., Danaher, S., Davenport, L., Desilets, R., Dietz, S., Dodson, K., Doup, L., Ferriera, S., Garg, N., Gluecksmann, A., Hart, B., Haynes, J., Haynes, C., Heiner, C., Hladun, S., Hostin, D., Houck, J., Howland, T., Ibegwam, C., Johnson, J., Kalush, F., Kline, L., Koduru, S., Love, A., Mann, F., May, D., McCawley, S., McIntosh, T., McMullen, I., Moy, M., Moy, L., Murphy, B., Nelson, K., Pfannkoch, C., Pratts, E., Puri, V., Qureshi, H., Reardon, M., Rodriguez, R., Rogers, Y., Romblad, D., Ruhfel, B., Scott, R., Sitter, C., Smallwood, M., Stewart, E., Strong, R., Suh, E., Thomas, R., Tint, N., Tse, S., Vech, C., Wang, G., Wetter, J., Williams, S., Williams M., Windsor, S., Winn-Deen, E., Wolfe, K., Zaveri, J., Zaveri, K., Abril, J., Guigó, R., Campbell, M., Sjolander, K., Karlak, B., Kejariwal, A., Mi, H., Lazareva, B., Hatton, T., Narechania, A., Diemer, K., Muruganujan, A., Guo, N., Sato, S., Bafna, V., Istrail, S., Lippert, R., Schwartz, R., Walenz, B., Yooseph, S., Allen, D., Basu, A., Baxendale, J., Blick, L., Caminha, M., Carnes-Stine, J., Caulk, P., Chiang, Y., Coyne, M., Dahlke, C., Mays, A., Dombroski, M., Donnelly, M., Ely, D., Esparham, S., Fosler, C., Gire, H., Glanowski, S., Glasser, K., Glodek, A., Gorokhov, M., Graham, K., Gropman, B., Harris, M., Heil, J., Henderson, S., Hoover, J., Jennings, D., Jordan, C., Jordan, J., Kasha, J., Kagan, L., Kraft, C., Levitsky, A., Lewis, M., Liu, X., Lopez, J., Ma, D., Majoros, W., McDaniel, J., Murphy, S., Newman, M., Nguyen, T., Nguyen, N., Nodell, M., Pan, S., Peck, J., Peterson, M., Rowe, W., Sanders, R., Scott, J., Simpson, M., Smith, T., Sprague, A., Stockwell, T., Turner, R., Venter, E., Wang, M., Wen, M., Wu, D., Wu, M., Xia, A., Zandieh, A. & Zhu, X. (2001) The Sequence of the Human Genome. *Science*, Vol.291, No.5507, (February 2001), pp. 1304–1351, ISSN 0036-8075

Vincent, A., Lager, K., Ma, W., Lekcharoensuk, P., Gramer, M., Loiacono, C. & Richt, J. (2006). Evaluation of Hemagglutinin Subtype 1 Swine Influenza Viruses from the United States. *Veterinary Microbiology*, Vol.118, No.3-4, (December 2006), pp. 212- 222, ISSN 0378-1135

Virgin, S. (2007). Pathogenesis of Viral Infection. In: *Fields Virology, 5th Edition, Volume One,* David M. Knipe, Peter M. Howley, Diane E. Griffin, Robert A. Lamb, Malcolm A. Martin, Bernard Roizman, pp. 414-497, Lippincott Williams & Wilkins, ISBN-10: 0781760607, Philadelphia, USA

Wada, T., Schurman, S., Otsu, M., Garabedian, E., Ochs, H., Nelson, D. & Candotti, F. (2001). Somatic Mosaicism in Wiskott--Aldrich Syndrome Suggests in vivo Reversion by a DNA Slippage Mechanism. *Proceedings of the National Academy of Sciences*, Vol.98, No.15, (July 2001), pp. 8697-8702, ISSN 0027-8424

Wada, T., Konno, A., Schurman, S., Garabedian, E., Anderson, S., Kirby, M., Nelson, D. & Candotti, F. (2003). Second-site Mutation in the Wiskott-Aldrich Syndrome (WAS) Protein Gene Causes Somatic Mosaicism in Two WAS Siblings. *The Journal of Clinical Investigation*, Vol. 111, No.9, (May 2003), pp. 1389-1397, ISSN 0021-9738

Waisfisz, Q., Morgan, N., Savino, M., de Winter, J., van Berkel, C., Hoatlin, M., Ianzano, L., Gibson, R., Arwert, F., Savoia, A., Mathew, C., Pronk, J. & Joenje, H. (1999). Spontaneous Functional Correction of Homozygous Fanconi Anaemia Alleles Reveals Novel Mechanistic Basis for Reverse Mosaicism. *Nature Genetics,* Vol.22, No.4, (August 1999), pp. 379–383, ISSN 1061-4036

Wang, G., Bunnell, B., Painter, R., Quiniones, B., Tom, S., Lanson, N. Jr., Spees, J., Bertucci, D., Peister, A., Weiss, D., Valentine, V., Prockop, D. & Kolls, J. (2005). Adult Stem Cells from Bone Marrow Stroma Differentiate into Airway Epithelial Cells: Potential Therapy for Cystic Fibrosis. *Proceedings of the National Academy of Sciences*, Vol.102, No.1, (January 2005), pp. 186-191, ISSN 0027-8424

Wang, J., Wang, W., Li, R., Li, Y., Tian, G., Goodman, L., Fan, W., Zhang, J., Li, J., Zhang, J., Guo, Y., Feng, B., Li, H., Lu, Y., Fang, X., Liang, H., Du, Z., Li, D., Zhao, Y., Hu, Y., Yang, Z., Zheng, H., Hellmann, I., Inouye, M., Pool, J., Yi, X., Zhao, J., Duan, J., Zhou, Y., Qin, J., Ma, L., Li, G., Yang, Z., Zhang, G., Yang, B., Yu, C., Liang, F., Li, W., Li, S., Li, D., Ni, P., Ruan, J., Li, Q., Zhu, H., Liu, D., Lu, Z., Li, N., Guo, G., Zhang, J., Ye, J., Fang, L., Hao, Q., Chen, Q., Liang, Y., Su, Y., San, A., Ping, C., Yang, S., Chen, F., Li, L., Zhou, K., Zheng, H., Ren, Y., Yang, L., Gao, Y., Yang, G., Li, Z., Feng, X., Kristiansen, K., Wong, G., Nielsen, R., Durbin, R., Bolund, L., Zhang, X., Li, S., Yang, H. & Wang, J. (2008). The Diploid Genome Sequence of an Asian Individual. *Nature*, Vol.456, No.7218, (November 2008), pp. 60-65, ISSN 0028-0836

Wang, J., Chen P-J., Wang G. & Keller, L. (2010). Chromosome Size Differences May Affect Meiosis and Genome Size. *Science*, Vol.329, No.5989, (July 2010), pp. 293, ISSN 0036- 8075

Webby, R., Swenson, S., Krauss, S., Gerrish, P., Goyal, S. & Webster, R. (2000). Evolution of Swine H3N2 Influenza Viruses in the United States. *Journal of Virology*, Vol.74, No.18, (September 2000), pp. 8243-8251, ISSN 0022-538X

Weil, C. (2002). Finding the Crosswalks on DNA. *Proceedings of the National Academy of Sciences,* Vol.99, No.9, (April 2002), pp. 5763-5765, ISSN 0027-8424

Weissman, I. & Shizuru, J. (2008). The Origins of the Identification and Isolation of Hematopoietic Stem Cells, and Their Capability to Induce Donor-specific Transplantation Tolerance and Treat Autoimmune Diseases. *Blood, Journal of the American Society of Hematology*, Vol.112, No.1, (November 2008), pp. 3543-3553, ISSN 0006-4971

Whitby, M. (2005). Making Crossovers during Meiosis. *Biochemical Society Transactions*, Vol.33, No.part 6, (December 2005), pp. 1451-1455, ISSN 0300-5127

Whitney, M., Thayer, M., Reifsteck, C., Olson, S., Smith, L., Jakobs, P., Leach, R., Naylor, S., Joenje, H. & Grompe, M. (1995). Microcell Mediated Chromosome Transfer Maps the Fanconi Anaemia Group D Gene to Chromosome 3p. *Nature Genetics*, Vol.11, No.3, (November 1995), pp. 341-343, ISSN 1061-4036

Willadsen, S. (1986). Nuclear Transplantation in Sheep Embryos. *Nature*, Vol.320, No.6057, (March 1986), pp. 63-65, ISSN 0028-0836

Willers, H., Dahm-Daphi, J. & Powell, S. (2004). Repair of Radiation Damage to DNA. *British Journal of Cancer*, Vol.90, No.7, (April 2004), pp. 1297-1301, ISSN 0007-0920

Wilmut, I., Schnieke, A., McWhir, J., Kind, A. & Campbell, K. (1997). Viable Offspring Derived from Fetal and Adult Mammalian Cells. *Nature*, Vol.385, No.6619, (February 1997), pp. 810–813, ISSN 0028-0836

Wilson, A., Murphy, M., Oskarsson, T., Kaloulis, K., Bettess, M., Oser, G., Pasche, A., Knabenhans, C., Macdonald, H. & Trumpp, A. (2004). c-Myc Controls the Balance between Hematopoietic Stem Cell Self-renewal and Differentiation. *Genes & Development*, Vol.18, No.22, (November 2004), pp. 2747-2763, ISSN 0890-9369

Wu, L., Sun, C., Ryan, T., Pawlik, K., Ren, J. & Townes, T. (2006). Correction of Sickle Cell Disease by Homologous Recombination in Embryonic Stem Cells. *Blood, Journal of the American Society of Hematology*, Vol.108, No.4, (August 2006), pp. 1183-1188, ISSN 0006-4971

Xia, Q., Zhou, Z., Lu, C., Cheng, D., Dai, F., Li, B., Zhao, P., Zha, X., Cheng, T., Chai, C., Pan, G., Xu, J., Liu, C., Lin, Y., Qian, J., Hou, Y., Wu, Z., Li, G., Pan, M., Li, C., Shen, Y., Lan, X., Yuan, L., Li, T., Xu, H., Yang, G., Wan, Y., Zhu, Y., Yu, M., Shen, W., Wu, D., Xiang, Z., Yu, J., Wang, J., Li, R., Shi, J., Li, H., Li, G., Su, J., Wang, X., Li, G., Zhang, Z., Wu, Q., Li, J., Zhang, Q., Wei, N., Xu, J., Sun, H., Dong, L., Liu, D., Zhao, S., Zhao, X., Meng, Q., Lan, F., Huang, X., Li, Y., Fang, L., Li, C., Li, D., Sun, Y., Zhang, Z., Yang, Z., Huang, Y., Xi, Y., Qi, Q., He, D., Huang, H., Zhang, X., Wang, Z., Li, W., Cao, Y., Yu, Y., Yu, H., Li, J., Ye, J., Chen, H., Zhou, Y., Liu, B., Wang, J., Ye, J., Ji, H., Li, S., Ni, P., Zhang, J., Zhang, Y., Zheng, H., Mao, B., Wang, W., Ye, C., Li, S., Wang, J., Wong, G. & Yang, H. (2004). A Draft Sequence for the Genome of the Domesticated Silkworm (Bombyx mori). *Science*, Vol.306, No.5703, (December 2004), pp. 1937-1940, ISSN 0036-8075

Xu, L., Tao, S., Cao, Y., Wang, H., Yang, D., He, Y., Liu, Q. & Chen, B. (2003). Genotyping of the Chinese Isolates of Coltivirus. *Chinese Journal of Experimental and Clinical Virology*, Vol.17, No.4, (December 2003), pp. 346-350, ISSN 1003-9279

Xu, P., Wang, Y., Zuo, J., Lin, J. & Xu, P-M. (1990). New Arboviruses Isolated from Patients with Unknown Fever and Encephalitis in Yunnan Province. *Chinese Journal of Virology*, Vol.6, No.1, (1990), pp. 27-33, ISSN 1000-8721

Xu, X., Smith, C., Mungall, B., Lindstrom, S., Hall, H., Subbarao, K., Cox, N. & Klimov, A. (2002). Intercontinental Circulation of Human Influenza A (H1N2) Reassortant Viruses during the 2001-2002 Influenza Season. *The Journal of Infectious Diseases*, Vol.186, No.10, (November 2002), pp. 1490-1493, ISSN 0022-1899

Xu, X., Lindstrom, S., Shaw, M., Smith, C., Hall, H., Mungall, B., Subbarao, K., Cox, N. & Klimov, A. (2004). Reassortment and Evolution of Current Human Influenza A and B Viruses. *Virus Research*, Vol.103, No.1-2, (July 2004), pp. 55-60, ISSN 0168-1702

Yamaizumi, M., Horwich, A. & Ruddle, F. (1983). Expression and Stabilization of Microinjected Plasmids Containing the Herpes Simplex Virus Thymidine Kinase Gene and Polyoma Virus DNA in Mouse Cells. *Molecular and Cellular Biology*, Vol.3, No.4, (April 1983), pp. 511-522, ISSN 1098-5549

Yang, S., Cheng, P., Banta, H., Piotrowska-Nitsche, K., Yang, J., Cheng, E., Snyder, B., Larkin, K., Liu, J., Orkin, J., Fang, Z., Smith, Y., Bachevalier, J., Zola, S., Li, S., Li, X. & Chan, A. (2008). Towards a Transgenic Model of Huntington's Disease in a Non- human Primate. *Nature*, Vol.453, No.7197, (June 2008), pp. 921-924, ISSN 0028-0836

Yim, S., Kim, T., Hu, H., Kim, J., Kim, B., Lee, J., Han, B., Shin, S., Jung, S. & Chung, Y. (2010). Copy Number Variations in East-Asian Population and Their Evolutionary and Functional Implications. *Human Molecular Genetics*, Vol.19, No.6, (March 2010), pp. 1001-1008, ISSN 0964-6906

Yin, W., Barkess, G., Fang, X., Xiang, P., Cao, H., Stamatoyannopoulos, G. & Li, Q. (2007). Histone Acetylation at the Human beta-globin Locus Changes with Developmental Age. *Blood, Journal of the American Society of Hematology*, Vo.110, No.12, (December 2007), pp. 4101-4107, ISSN 0006-4971

You, Z., Wang, Y., Zhao, Z-G., He, Y., Xu, P., Zhao, Z., Cheng, Y., Cheng, W. & Zhang, H. (1990). Isolation and Identification of Two Strains of Orbivirus in Hainan Province. *Chinese Journal of Virology*, Vol.6, No.3, (1990), pp. 272-276, ISSN 1000-8721

Young, W., Merz, T., Ferguson-Smith, M. & Johnston, A. (1960). Chromosome Number of the Chimpanzee, Pan troglodytes. *Science*, Vol.131, No.3414, (June 1960), pp. 1672- 1673, ISSN 0036-8075

Youssoufian, H. & Pyeritz, R. (2002). Mechanisms and Consequences of Somatic Mosaicism in Humans. *Nature Reviews Genetics*, Vol.3, No.10, (October 2002), pp. 748-758, ISSN 1471-0056

Yu, J., Vodyanik, M., Smuga-Otto, K., Antosiewicz-Bourget, J., Frane, J., Tian, S., Nie, J., Jonsdottir, G., Ruotti, V., Stewart, R., Slukvin, I. & Thomson, J. (2007). Induced Pluripotent Stem Cell Lines Derived from Human Somatic Cells. *Science*, Vol.318, No.5858, (December 2007), pp. 1917-1920, ISSN 0036-8075

Yu, X-J., Liang, M-F., Zhang, S-Y., Liu, Y., Li, J-D., Sun, Y-L., Zhang, L., Zhang, Q-F., Popov, V., Li, C., Qu, J., Li, Q., Zhang, Y-P., Hai, R., Wu, W., Wang, Q., Zhan, F-X., Wang, X-J., Kan, B., Wang, S-W., Wan, K-L., Jing, H-Q., Lu, J-X., Yin, W-W., Zhou, H., Guan, X-H., Liu, J-F., Bi, Z-Q., Liu, G-H., Ren, J., Wang, H., Zhao, Z., Song, J-D., He, J-R., Wan, T., Zhang, J-S., Fu, X-P., Sun, L-N., Dong, X-P., Feng, Z-J., Yang, W-Z., Hong, T., Zhang, Y., Walker, D., Wang, Y. & Li, D-X. (2011). Fever with Thrombocytopenia Associated with a Novel Bunyavirus in China. *The New England Journal of Medicine,* Vol.364, No.16, (April 2011), pp. 1523-1532, ISSN 0028-4793

Zhang, Z., Habara, Y., Nishiyama, A., Oyazato, Y., Yagi, M., Takeshima, Y. & Matsuo, M. (2007). Identification of Seven Novel Cryptic Exons Embedded in the Dystrophin Gene and Characterization of 14 Cryptic Dystrophin Exons. *Journal of Human Genetics,* Vol.52, No.7, (July 2007), pp. 607-617, ISSN 1434-5161

Zhao, T., Zhang, Z., Rong, Z. & Xu, Y. (2011). Immunogenicity of Induced Pluripotent Stem Cells. *Nature*, Vol.474, No.7350, (May 2011), pp. 212-215, ISSN 0028-0836

Zhou, N., Senne, D., Landgraf, J., Swenson, S., Erickson, G., Rossow, K., Liu, L., Yoon, K., Krauss, S. & Webster, R. (1999). Genetic Reassortment of Avian, Swine, and Human Influenza A Viruses in American Pigs. *Journal of Virology*, Vol.73, No.10, (October 1999), pp. 8851-8856, ISSN 0022-538X

Zhu, Q., Watanabe, C., Liu, T., Hollenbaugh, D., Blaese, R., Kanner, S., Aruffo, A., Ochs, H. (1997). Wiskott-Aldrich Syndrome/X-linked Thrombocytopenia: WASP Gene Mutations, Protein Expression, and Phenotype. *Blood, Journal of the American Society of Hematology*, Vol.90, No.7, (October 1997), pp. 2680-2689, ISSN 0006-4971

Zielenski, J. (2000). Genotype and Phenotype in Cystic Fibrosis. *Respiration*, Vol.67, No.2, (2000), pp. 117-133, ISSN 0025-7931

7 Chapter

ALTERNATIVE METHODS FOR TRANSFERRING GENES WITHOUT USING VIRUSES IN THE TREATMENT OF LYSOSOMAL STORAGE DISORDERS

1. INTRODUCTION: AN OVERVIEW

The lysosomal storage disorders (LSDs) are a collection of nearly 50 hereditary diseases. These diseases are characterized by mutations and reduced activity of lysosomal enzymes or, less commonly, non-lysosomal proteins that play a role in protein maturation or the formation of lysosomes (Meikle et al, 2004). The majority of LSDs follow an autosomal recessive inheritance pattern, while there are a few exceptions such as Hunter syndrome (which is X-linked recessive), Danon disease (which is X-linked dominant), and Fabry disease (which is X-linked and affects a significant number of heterozygous females).

The lysosome is responsible for storing unique material that is either undegraded or partially degraded, typically the substrate of the malfunctioning enzyme. The LSDs are classified into different categories (table 1) based on the type of substrate. These categories include mucopolysaccharidoses (characterized by the accumulation of mucopolysaccharides, also known as glycosaminoglycans), lipidoses (accumulation of lipids), glycogenoses (accumulation of glycogen), and oligosaccharidoses (accumulation of small sugar chains). However, there are several clinical commonalities identified both within and within each group, notwithstanding this classification. Typically, these diseases affect multiple systems in the body. Common clinical characteristics of many lysosomal storage disorders (LSDs) include enlarged organs, impaired central nervous system function, and coarse hair and facial features. The majority of LSDs exhibit a progressive nature, leading to severe morbidity and higher mortality rates. However, it is important to note that there are notable variances across various diseases and even among individuals with the same disease (Walkley 2009). Lysosomal enzymes are found throughout the body, however the storage of substrates typically occurs in cells, tissues, and organs that have a greater rate of substrate breakdown.

Bellettato and Scarpa (2010) have proposed that the main genetic abnormality and the accumulation of substances are the initial causes of a series of interconnected processes that result in various symptoms of the disease. In this specific situation, the accumulation of secondary substrate, disruptions in Calcium balance, and disturbances in lipid transportation would all play a role in the development of the disease. The lysosome's involvement in vesicle trafficking leads to various other effects, such as antigen presentation, innate immunity, and signal transduction. These effects contribute to the inflammatory and autoimmune disruptions reported in lysosomal storage disorders (Parkinson-Lawrence et al., 2010). Furthermore, other fundamental processes such as unfolded protein response, reticulum stress, oxidative stress, and autophagy blockage are also involved in the development of the disease (Vitner et al., 2010).

The occurrence and frequency of these diseases varied across various countries and areas. For instance, the prevalence of GM1 Gangliosidosis is generally estimated to be 1 in every 100,000 to 1 in every 200,000 individuals. However, in certain countries like Malta (1 in every 3,700) and the South of Brazil (1 in every

13,317), the prevalence is significantly greater (Baiotto et al, 2011). According to extensive population research, the overall occurrence of the LSDs ranges from 1 in 5,000 to 1 in 7,700 (Fuller et al, 2006).

Until a few decades ago, the treatment choices for LSDs were limited to supportive measures. Currently, there are targeted therapies accessible for a select group of LSDs, but a few of them are still in the experimental stage or have restricted efficacy. Table 1 presents treatment choices that encompass both approved therapies and those currently being evaluated in clinical trials, including the possibility of compassionate use. In this scenario, support measures and palliative care are not regarded as viable treatment choices.

Table 1. List of lysosomal storage diseases with their respective OMIM accession number, gene and enzyme deficiency and treatment options (including experimental ones).

Disease	OMIN	Gene	Enzyme	Available Treatments
Aspartylglucosaminuria	208400	*AGA*	N-aspartyl-betaglucosaminidase	HSCT
Canavan disease	271900	*ASPA*	Aspartocylase	None
Cystinosis	219800	*CTNS*	Cystinosin	Cysteamine (drug)
Danon disease	300257	*LAMP2*	Lysosomal-associated membrane protein 2	None
Fabry disease	301500	*GLA*	A-galactosidase A	ERT
Farber disease	228000	*ASAH1*	Ceramidase	HSCT
Fucosidosis	230000	*FUCA1*	α-L-fucosidase	HSCT
Galactosialidosis	256540	*CTSA*	Cathepsin A	None
Gaucher disease	230800	*GBA*	acid β-glucosidase	ERT, GT, HSCT, PCT, SSI
GM1 gangliosidosis	230600	*GLB1*	β-Galactosidase	HSCT
Krabbe disease	245200	*GALC*	galactocerebrosidase	HSCT
Lysosomal Acid Lipase Deficiency	278000	*LIPA*	Lysosomal acid lipase	HSCT
α- mannosidosis	248500	*MAN2B1*	α -D –mannosidase	HSCT
β- mannosidosis	248510	*MANBA*	β -D-mannosidase	None
Metachromatic Leucodystrophy	250100	*ARSA*	Arylsulphatase-A	HSCT
Metachromatic Leucodystrophy	249900	*ARSA*	Saposin-B	HSCT
Mucolipidosis type I	256550	*NEU1*	Sialidase	None
Mucolipidosis types II/III	252500	*GNPTAB*	N-acetylglucosamine-1-Phosphotransferase	None
Mucolipidosis type IIIC	252605	*GNPTG*	N-acetylglucosamine-1-phosphotransferase γ-subunit	None
Mucolipidosis type IV	252650	*MCOLN1*	Mucolipin 1	None
Mucopolysaccharidosis type I	607014	*IDUA*	α-L-iduronidase	ERT, HSCT

Disease	OMIN	Gene	Enzyme	Available Treatments
Mucopolysaccharidosis type II	309900	*IDS*	Iduronate sulfatase	ERT, GT, HSCT
Mucopolysaccharidosis type IIIA	252900	*SGSH*	Heparan-N-sulfatase	None
Mucopolysaccharidosis type IIIB	252920	*NAGLU*	α- Nacetylglucosaminidase	None
Mucopolysaccharidosis	252930	*HGSNAT*	AcetylCoa-glucosamine-N-	None
type IIIC Mucopolysaccharidosis			acetyltransferase	None
type IIID Mucopolysaccharidosis	253000	*GALNS*	N-acetylgalactosamine-6-Sulphatase	ERT
Mucopolysaccharidosis type IVB	253010	*GLB1*	β-Galactosidase	None
Mucopolysaccharidosis type VI	253200	*ARSB*	N-acetylgalactosamine-4-Sulphatase	ERT
Mucopolysaccharidosis type VII	253220	*GUSB*	β-Glucuronidase	None
Mucopolysaccharidosis type IX	601492	*HYAL1*	Hyaluronidase	None
Multiple sulphatase Deficiency	272200	*SUMF1*	Formylglicine-generatingenzyme	None
Neuronal ceroid lipofuscinosis 1	256730	*PPT1*	Palmitoyl protein thioesterase-1	HCST
Neuronal ceroid lipofuscinosis 2	204500	*TPP1*	Tripeptydil-peptidase I	GT
Neuronal ceroid lipofuscinosis 3	204200	*CLN3*	CLN3 protein	None
Neuronal ceroid lipofuscinosis 5	256731	*CLN5*	CLN5 protein	None
Neuronal ceroid lipofuscinosis 6	601780	*CLN6*	CLN6 protein	None
Neuronal ceroid lipofuscinosis 8	600143	*CLN8*	CLN8 protein	None
Niemann-Pick disease A/B	257200	*SMPD1*	Acid sphingomyelinidase	None
Niemann-Pick disease C1	257220	*NPC1*	NPC1 protein	SSI
Niemann-Pick disease C2	607625	*NPC2*	NPC2 protein	SSI
Pompe disease	232300	*GAA*	Alpha-glucosidase	ERT, GT
Prosaposin deficiency	176801	*PSAP*	Prosaposin	None
Pycnodysostosis	265800	*CTSK*	Cathepsin K	Hormone therapy

Disease	OMIN	Gene	Enzyme	Available Treatments
Sandhoff disease	268800	*HEXB*	Hexosaminidase B	PCT
Schindler disease	609241	*NAGA*	Alpha-Nacetylgalactosaminidase	None
Sialic acid storage disease	269920	*SLC17A5*	Sialin UDP-N	None
Sialuria	269921	*GNE*	Acetylglucosamine-2-epimerase	None
Tay-Sachs disease	272800	*HEXA*	Hexosaminidase A	PCT, SSI
Abbreviations: GT – Gene Therapy; HSCT – Hematopoetic Stem Cell Transplantation; PCT – Pharmacological Chaperone Therapy; OMIM – Online Mendelian Inheritance in Man; SSI – Specific Substrate Inhibition; ERT – Enzyme Replacement Therapy				

Hematopoietic Stem Cell Transplantation (HSCT) and Enzyme Replacement Therapy (ERT) are the two most commonly utilized therapy approaches. Hematopoietic stem cell transplantation (HSCT) has demonstrated efficacy in treating certain disorders, particularly when administered promptly to prevent permanent damage. However, there are still constraints that hinder the effectiveness of this treatment, including the requirement for early identification, the challenges of finding a suitable donor within a short timeframe, and the significant rates of illness and death associated with the procedure (Malatack et al., 2003). Thus, despite the significant progress made in this therapy over the past three decades (Boelens et al 2010), it is not being utilized and is instead being replaced by ERT whenever it is accessible. Enzyme replacement therapy has been authorized for an increasing variety of lysosomal storage disorders, particularly those that do not affect the central nervous system. Research has demonstrated that it effectively decreases certain internal symptoms including hepatosplenomegaly and enhances respiratory function (Sifuentes et al, 2007). However, it remains a significant problem to target organs that are hard to access, such as the brain and bones. Novel methods of enzyme delivery have been experimented with in order to reach the central nervous system (CNS), such as intrathecal enzyme replacement therapy (ERT) (Munoz-Rojas et al, 2008; Munoz-Rojas et al, 2010).

Additional therapeutic methods currently in clinical practice or being tested include Specific Substrate Inhibition (SSI) and Pharmacological Chaperone Therapy (PCT). SSI seeks to minimize storage by inhibiting substrate production. PCT employs tiny compounds that have the ability to stabilize the mutated enzyme and facilitate its evasion of proteasomal degradation, thereby partially restoring the enzyme's activity. The existence of limitations in all these treatments (as shown in table 2) justifies the need for the development of gene therapy methods for these disorders.

Table 2. Pros and limitations of different therapeutic approaches for lysosomal storage diseases.

Approach	Brief Description	Pros	Limitations
Intravenous enzyme replacement therapy	Intravenous injection of a recombinant version of the missing enzyme	Ameliorates several visceral symptoms	High cost Does not correct difficult-to-reach sites Need of repeated injections
Intrathecal enzyme replacememnt therapy	Intrathecal injection of a recombinant version of the missing enzyme	Reaches the CNS	High Cost Need of repeated injections
Hematopoietic stem cell transplantation	Non-autologous transplantation of CD 34+ cells	Able to correct visceral and CNS symptoms if performed early	Efficacy uncertain Limited efficacy Relatively high morbidity and mortality rates

Approach	Brief Description	Pros	Limitations
Specific substrate Inhibition	Use of drugs that can inhibit the synthesis of the undegraded material	Orally administered Reaches the CNS	Limited efficacy High cost
Pharmacological chaperone therapy	Drugs that can stabilize the mutated protein, allowing some enzyme activity	Orally administered Reaches the CNS	Works only in patients with specific mutations
Stop-codon readthrough Therapy	Use of molecules that can suppress stop-codon mutations	Orally administered Reaches the CNS Low cost	Works only in stopcodon mutations Clinical trials yet to be Performed
Gene Therapy	Administration of a normal copy of the mutated gene	Potentially effective with single injection	Safety concerns Efficacy uncertain
Abbreviations: CNS – Central Nervous System			

2. RATIONALE FOR GENE THERAPY IN LSDS

The rationale for gene therapy and other enzyme-based approaches for treatment of LSDs was first introduced almost five decades ago by Christian de Duve, and can be summarized in the following sentence from his original work "In our pathogenic speculations and in our therapeutic attempts, it may be well to keep in mind that any substance which is taken up intracellularly in an endocytic process is likely to end up within lysosomes." (de Duve, 1964). His work and other pioneer studies showing cross-correction between fibroblasts from patients with Mucopolysaccharidosis type I and type II (deficient in alpha-L- iduronidase and iduronate-sulphatase, respectively) established that the enzyme produced in one cell could be uptaken by a deficient cell, thus restoring its phenotype (Fratantoni et al, 1968). Later studies identified that this uptake was a saturable, receptor-mediated process, and the main actor of this process was the mannose-6-phosphate (M6P) receptor localized in the plasmatic membrane. The post-translational modification of addition of the M6P to the protein was discovered to be a signal not only to endocytosis but also for targeting nascent hydrolases to lysosomes (Fisher et al, 1980; Sly et al., 1981).

These pivotal discoveries in the field of endocytosis and targeting of lysosomal enzymes provided the basis for treatments like HSCT and ERT. In the same way, LSDs may be considered good targets for gene therapy, despite their multisystemic involvement. The correction of a few cells could lead to the enzyme being secreted into the circulation and uptaken by the deficient cells, resulting in widespread correction of the biochemical defect (figure 1).

Since long term gene expression is desirable, most clinical and preclinical trials used viral based vectors. Initial studies on fibroblasts showed promising data using retroviruses (Anson et al, 1992). However, when tests in animal models started, it became clear that some organs as the brain would not be easily corrected, as the enzyme could not cross the blood- brain-barrier (Elinwood et al, 2004). This is a major hurdle as most LSD shows some degree of CNS involvement. Nevertheless, CNS targeted approaches could be envisaged to overcome this obstacle. For instance, Worgall et al. (2008) showed a slowing of progression of Neuronal Ceroid Lipofuscinosis 2 in ten children treated with serotype 2 adeno- associated virus expressing *CLN2* cDNA. Another clinical trial, for Pompe disease, also used adeno-associated-based vector, but in this case serotype 1 (NCT00976352- www.clinicaltrials.gov). Other two trials performed in the late 1990s used retroviral vectors for Gaucher (Dunbar et al., 1998) and Mucopolysaccharidosis type II (NCT00004454 - www.clinicaltrials.gov).

Safety issues related to immune response of the adenoviral vectors (Wilson, 2009) and the possibility of insertional mutagenesis of the retroviral vectors (Hacein-Bey-Abina S et al., 2008) led researchers to develop a series of studies in parallel using non-viral approaches to treat lysosomal storage disorders.

3. NON-VIRAL APPROACHES

Non-viral vectors have important safety advantages over viral approaches, including their reduced pathogenicity and capacity for insertional mutagenesis, as well as their low cost and ease of production (Fraga et al., 2008). The application of non-viral vectors to humans has, however, been held back by the poor efficiency of their delivery to cells and the transient expression of the transgenes. As new strategies are being developed for the application of non-viral vectors as nucleic acid delivery systems (Resina et al., 2009; Wasungu and Hoekstra, 2006), progress is being made in the application of this kind of therapy to LSDs.

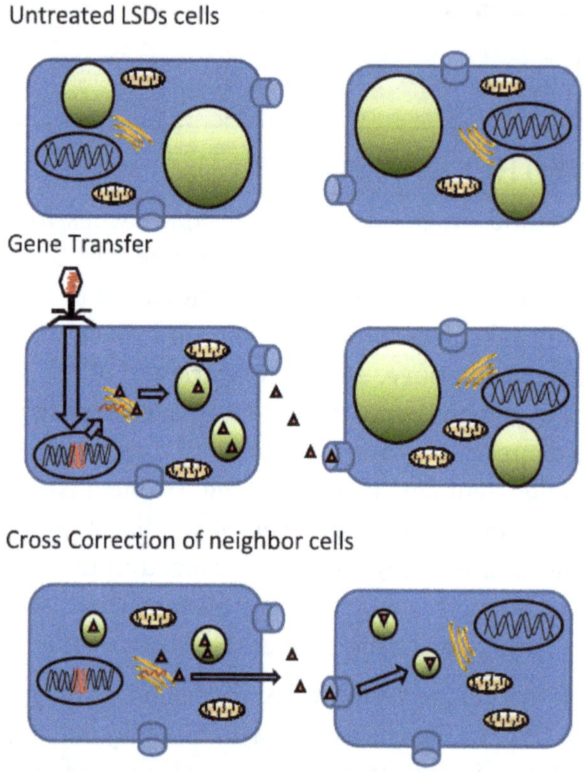

Fig. 1. Rationale for Gene Therapy in Lysosomal Storage Disorders: Cross-correction. The upper part of the figure shows two cells from a patient with LSD, with large lysosomes (green) due to accumulation of undegraded material. The gene transfer of a normal copy of the missing enzyme (red structure in the nucleus), allows the enzyme (red triangles) to be produced, and degrade the material accumulated in the lysosomes. Part of the enzyme is secreted from the recombinant cell and is captured by neighbouring cells via the mannose-6- phosphate receptors (light blue structures located on the cell membrane), reaching the lysosomes and being able to correct their phenotype, in a process called cross-correction. Note that the cell on the right was never transduced, but is able to capture the enzyme from the circulation or from the neighbouring cell.

3.1 Naked plasmid transfection

The direct injection of a plasmid containing the gene of interest and the regulatory mechanisms to ensure its expression is the simplest form of gene therapy. It has some advantages over the virus mediated gene transfer systems. First, DNA preparation is simple and can be performed at relatively low cost. In addition, the safety concerns are much lower and large amounts of DNA can be transferred. However, the major limitation of this method is that it requires a local administration and the level of transgene expression is relatively low and restricted to the injection site (Glover et al., 2005).

Hydrodynamic injection (Liu et al, 1999) is an experimental method capable to achieve efficient gene transfer and high level of transgene expression by systemic administration. In this procedure, a large volume of saline containing plasmid DNA is injected in a short period of time. The large volume and high injection rate forces the DNA solution into the liver, probably due to the permeability of liver fenestrae. A small hepatotoxicity, probably due to the large volume of saline, is observed and resolves within a few days. Even though this procedure is widely used in mice by tail vein injection, its feasibility has been demonstrated in larger animal models using a balloon occlusion catheter-based method to mimic hydrodynamic injection (Brunetti-Pierri et al, 2007; Kamimura et al, 2009).

This approach has been used in a proof-of-concept study to show the importance of coexpression of the formylglycine-generating enzyme for synthesis and secretion of functional Arylsulfatase A in a mouse model of Metachromatic Leukodystrophy (Takakusaki et al., 2005). This enzyme is a posttranslational modifying enzyme essential for activating multiple forms of sulfatases including Arylsulfatase A and therefore limits the amount of functional enzyme that can be secreted from transduced cells.

It has also been used in Mucopolysaccharidosis (MPS) type I (Camassola et al, 2005) and type VII (Richard et al, 2009). In the MPS I animals, storage content was reduced and enzyme activity was elevated in the liver and spleen. For the MPS VII model, a beneficial effect on the pathology was also observed, as liver-directed gene transfer led to the correction of secondary enzymatic elevations and to the reduction of GAGs storage in peripheral tissues and brain, as well as to histological correction in many tissues.

3.2 Liposomes and nanotechnology

Liposomes are lipid particles that resemble the cell membrane. Liposome-based gene delivery was first reported by Felgner in 1987, and is still one of the major techniques for non viral gene delivery into cells (Niidome & Huang, 2002). Lipoplexes are formed by the interaction of anionic nucleic acids binding to the surface of cationic lipids, forming multilamellar lipid–nucleic acid complexes where the negatively charged nucleic acid remains trapped inside the lipid bilayer (Dass, 2004). Since its discovery, different lipid formulations have been tested and modified. For example, stealth liposomes sterically stabilized with methoxypoly(ethylene glycol)distearoylphosphatidylethanolamine conjugates (PEG-DSPE) have long circulation half-lives following intravenous injection (Moreira et al., 2001). In addition, linear polycations such as linear, branched and dendritic vectors based on poly(ethylenimine) (PEI), poly(L-lysine) (PLL) and a range of poly(ethylene glycol) (PEG) were also developed (Hunter, 2006). Although linear PEI shows greater *in vivo* efficiency because of a dynamic structure change of the complex under high salt concentrations as found in blood (Niidome & Huang, 2002), it also possesses greater toxicity (Morille et al., 2008).

Cationic nanoemulsions have been more recently considered as potential systems for nucleic acid delivery. The interest in these systems is justified by the fact that they are biocompatible and able to form complexes with DNA protecting it from enzymatic degradation (Nam et al., 2009). Other practical advantages include ease of production and the potential for repeated administration (Al-Dorsari and Gao, 2009). We have investigated the influence of phospholipids on the properties of cationic nanoemulsions/pDNA complexes. Complexes containing the phospholipids DSPC (1,2-distearoyl-sn-glycero-3-phosphocholine) and DSPE (1,2- distearoyl-sn-glycero-3- phosphoethanolamine) were less toxic in comparison with the formulations obtained with lecithin, DOPC (1,2-dioleoyl-sn-glycero-3-phosphocholine) and DOPE

(1,2-dioleoyl-sn glycero-3-phosphoethanolamine). In addition, higher amounts of reporter DNA were detected for the formulation obtained with the DSPC phospholipid (Fraga et al., in press).

These cationic macromolecules can readily condense DNA or RNA into stable nanostructures for use in gene delivery (Hunter, 2006) but also for other nanotechnology-based approaches useful for the treatment of LSDs (Muro, 2010). Nanomaterials, as a result of their small size and their large surface area offer great promise for neuro-therapeutics (Ragnail et al., 2011) and thus may be a valid option for a large number of LSDs that affect the CNS.

Liposome-mediated gene transfer for LSD has been performed *in vitro* using patient's fibroblasts as target cells. Estruch et al. (2001) delivered therapeutic genes by integrin- mediated uptake into fibroblasts from patients with Fucosidosis and Fabry disease. The vectors consisted of a complex of lipofectin and a peptide containing an integrin-targeting domain and a poly-lysine domain to which plasmid DNA was bound. Transfected cells produced the corresponding enzyme at levels which were 10-40% of the total activity in cultures of normal fibroblasts. Although 95-98% of this activity was secreted, it did not appear to affect the viability of the cells. Our group used Lipofectamine to transduce fibroblasts from GM1 Gangliosidosis patients with the beta-galactosidase gene. Treated cells showed 33 to 100- fold increases in enzyme activity compared to untreated fibroblasts. However, after seven days enzyme activity was back to uncorrected values (Balestrin et al., 2008). When Geneticin was added to the medium (figure 2), stable expression at therapeutic levels was observed (mean 300 nmoles/h/mg prot) for 30 days, although at values lower than the normal range (mean 1,300 nmoles/h/mg prot).

In vivo, PEG-coated liposomes have been modified with monoclonal antibodies in order to reach the CNS. A liposome is coated with peptidomimetic monoclonal antibodies that undergo receptor-mediated transcytosis across the blood-brain barrier on the endogenous peptide receptor transporters (Pardridge, 2007). These Trojan horses (figure 3) may use the insulin or transferrin receptor, and since the MAb binding site is different from the binding site of the endogenous ligand, there is no interference of endogenous ligand transport (Skarlatos et al, 1995).

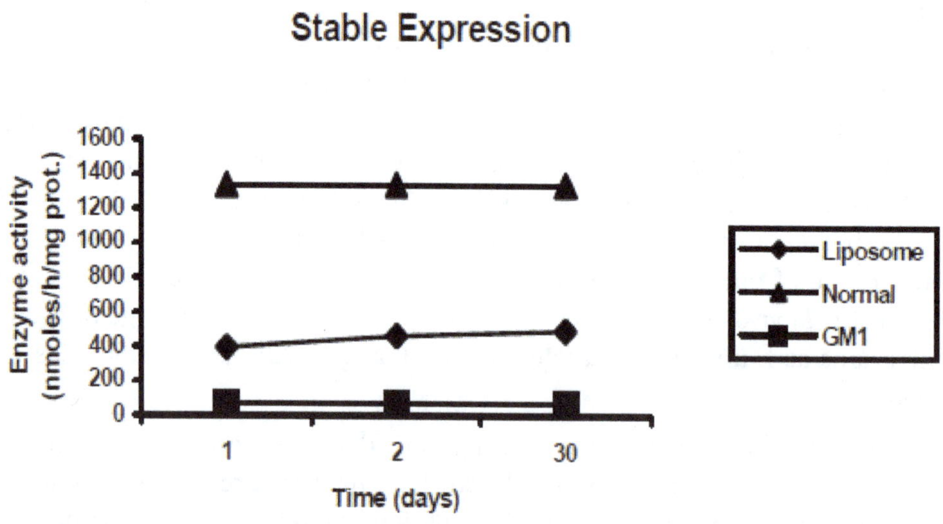

Fig. 2. Stable expression of β-Galactosidase in fibroblasts from G1 Gangliosidosis patients after *in vitro* liposome-based gene transfer. Mean values of liposome-treated cells: 300 nmoles/h/mg prot.; mean values of normal fibroblasts: 1,300 nmoles/h/mg prot.; mean values of affected fibroblasts: 68 nmoles/h/mg prot. (Balestrin et al., 2008).

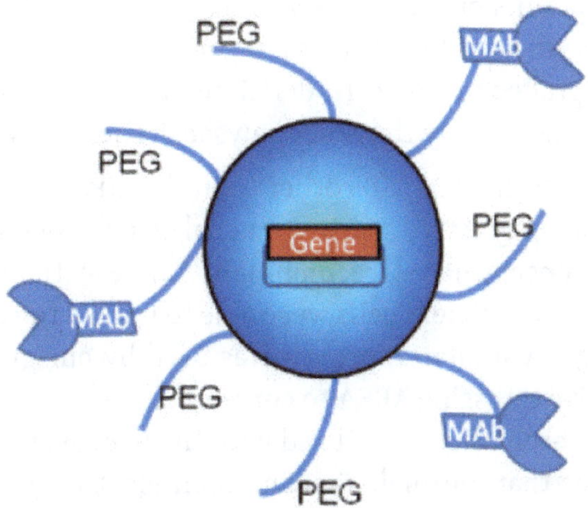

Fig. 3. Schematic view of a Trojan horse liposome. A stealth PEGylated liposome is complexed with monoclonal antibodies (MAb) that undergo receptor-mediated transcytosis across the blood-brain barrier.

This approach has been used to deliver a non-viral plasmid DNA to the brain across the blood-brain-barrier after intravenous administration of liposomes coated with monoclonal antibody to the mouse transferrin receptor in a mouse model of Mucopolysaccharidosis type VII (Zhang et al, 2008). The enzyme activity was increased greater than ten-fold in brain, liver, spleen, lung, and kidney, but not in heart. A similar strategy has been used by Osborn et al (2008) for Mucopolysaccharidosis type I, although they used a plasmid bearing a fusion gene consisting of Transf errin (Tf) and α-L-iduronidase. The fusion product consisted of an enzymatically active protein that was transported into the CNS by TfR-mediated endocytosis. Short-term treatment resulted in a decrease in GAGs in the cerebellum of Mucopolysaccharidosis type I mice.

1.3 Cell microencapsulation

Cell microencapsulation is an approach in which cells are trapped in a semipermeable membrane, allowing the exchange of metabolites and nutrients between them and the external environment. The membrane prevents the access of the immune system to the cells, without the need for continued immunosuppression of the host (Uludag et al., 2000). Furthermore, this technique allows the localized and controlled release, and long term duration of therapeutic products derived from the microencapsulated cells (Orive et al., 2002). Microencapsulation has become an important system for cellular preservation (Mayer et al., 2010) and a potential strategy for the controlled delivery of therapeutic products (Orive et al., 2003). Alginate has been the most important encapsulation polymer due to its abundance, easy gelling properties and biocompatibility. Agarose, chitosan, and hyaluronic acid are other polymers used for microencapsulation (Orive et al., 2003).

Cell microencapsulation presents the potential to deliver the therapeutic product of interest directly to the Central Nervous System (CNS). This has been achieved by different groups for brain tumors (Kuijlen et al 2007) and neurodegenerative diseases (Spuch et al 2010). A phase I clinical trial was conducted in Huntington's patients without signs of toxicity (Bloch et al 2004). This approach delivers the gene of interest in the spinal fluid, similar to the intrathecal enzyme replacement therapy. Thus, cell encapsulation can be suitable for the treatment of LSD, once the deficient enzyme could be released for long term directly in the CNS (Matte et al., 2011).

In order to obtain larger amounts of secreted enzyme, the encapsulated cells should be genetically modified to over-express the enzyme of interest. This enzyme would then be released to the extracapsular space (Bressel et al 2008) and uptaken by adjacent deficient cells (figure 4). This strategy has been used experimentally for different LSD, especially the Mucopolysaccharidosis (MPS).

Three *in vitro* studies were performed in LSDs other than the MPS, one in Fabry disease and the other two in Metachromatic Leukodystrophy (MLD). Naganawa et al (2002) co-cultured fibroblasts from patients with Fabry disease with microencapsulated recombinant Chinese Hamster Ovary cells (CHO) over-expressing alpha-galactosidase. The deficient cells were able to uptake the enzyme decreasing their levels of globotriaosylceramide storage. A similar approach was used by our group to test the ability of Baby Hamster Kidney (BHK) cells over-expressing ARSA to correct the deficiency of this enzyme in human skin fibroblasts from MLD patients. Fibroblasts co-cultured with the encapsulated cells for four weeks showed levels of enzyme activity higher than normal. Transmission electron microscopy showed evidence of normalization of the lysosomal ultra structure, suggesting that the secreted enzyme was able to degrade the substrate (Lagranha et al., 2008). Consiglio et al (2007) collected the conditioned media of C2C12 cells over-expressing ARSA encapsulated in polyether-sulfone polymer and used it to treat oligodendrocytes from MLD mice. The deficient cells internalized the enzyme and it was normally sorted to the lysosomal compartment, reaching 80% of physiological levels and restoring sulfatide metabolism.

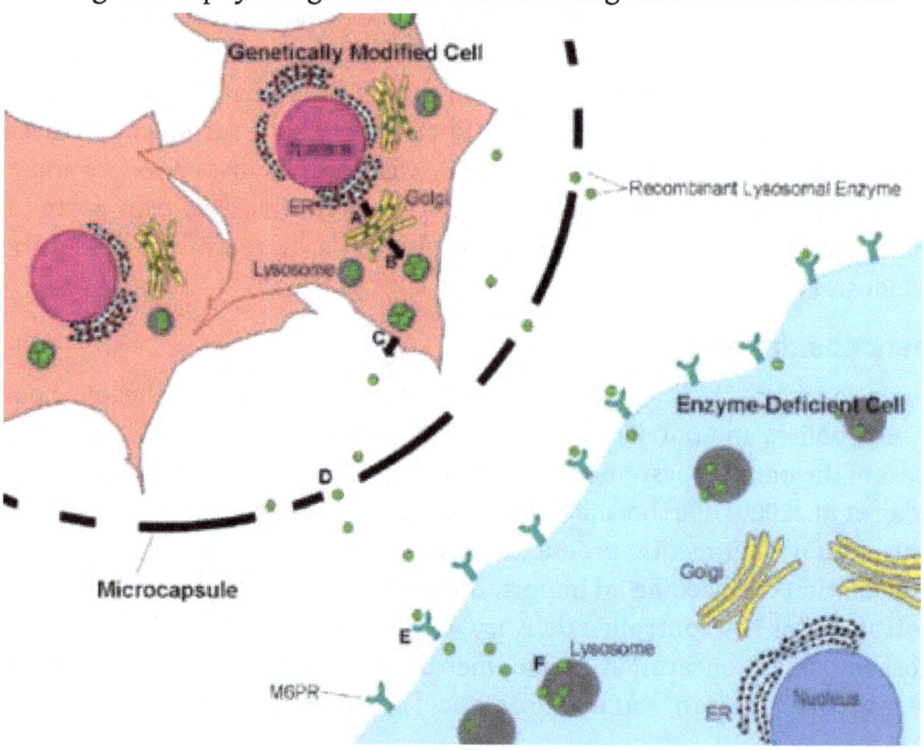

Fig. 4. Traffic of lysosomal enzymes throughout the encapsulated cells. The nascent lysosomal enzymes are glycosylated in the endoplasmic reticulum (ER) of the genetically modified cells. (A) The enzymes are phosphorylated at the residue of mannose-6 in the Golgi apparatus. (B) Most enzymes are transported to mature lysosomes. (C) Some, however, are secreted to the extracellular environment and (D) to outside of the microcapsules. (E) Phosphorylated enzymes bind to mannose-6-phosphate receptors (M6PR) of the enzyme-deficient cells (F) where they are endocyted and subsequently targeted to the lysosomes (Matte et al., 2011).

Both *in vitro* and *in vivo* studies have been performed in the MPS types I, II and VII. For MPS II two studies were performed. The first was a proof of principle in which Hunter primary fibroblasts were co-cultured with alginate microcapsules containing C2C12 cell clones over- expressing IDS. After 5 days of co-culture this strategy was able to increase IDS activity inside the deficient fibroblasts to levels similar to normal (Tomanin et al 2002). The second study was a pre-clinical experiment in which APA (alginate-poli-L-lysine-alginate) microcapsules containing 1.5 x 106 allogeneic C2C12 myoblasts over-expressing IDS were implanted in the peritoneum of the MPS II mouse model. An increase in IDS activity in plasma was observed, along with a reduction on urinary GAG between the fourth and the sixth week of treatment. After 8 weeks, a reduction of 30% in the amount of GAG accumulated in the liver and 38% in the kidney were shown (Friso et al 2005).

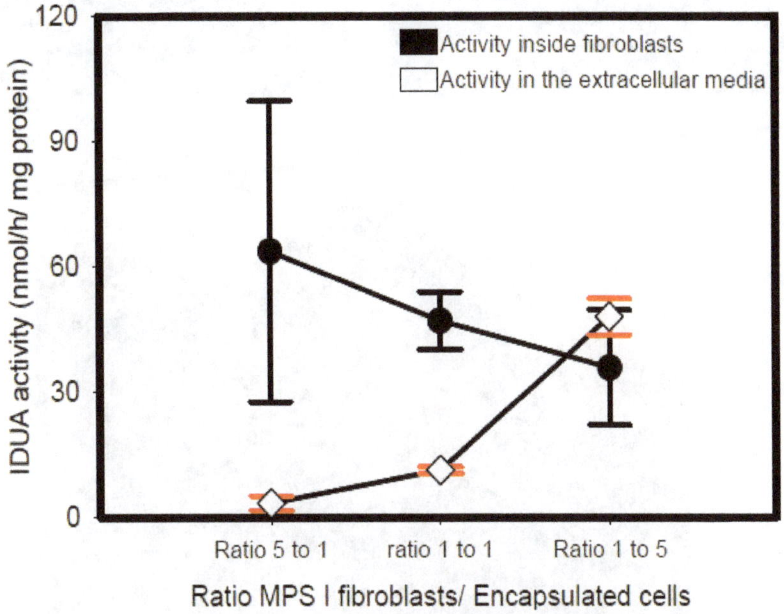

Fig. 5. IDUA enzyme activity in MPS I fibroblasts co-cultured with recombinant BHK cells overexpressing IDUA at different ratios.

To evaluate the usefulness of this technique to treat MPS VII, Ross et al (2000a) injected APA encapsulated non-autologous cells overexpressing Gusb in the peritoneum of MPS VII mice. The results showed the presence of Gusb in the plasma 24 hours after implantation, reaching 66% of physiological levels by 2 weeks post implantation. Activity of Gusb was also detected in liver and spleen for the duration of the 8-week experiment. Accumulation of GAG was significantly reduced in liver and spleen sections and urinary GAG content reached normal levels. In another study, enzyme released by APA encapsulated 2A50 fibroblasts implanted directly into the lateral ventricles of the brain of MPS VII mice was delivered throughout most of the CNS, reversing the histological pathology (Ross et al., 2000b). *In vitro* studies were performed by Nakama et al (2006) who encapsulated immortalized recombinant human amniotic epithelial cells with MPS VII human and mouse fibroblasts and high GUSB activity was detected in the medium. Addition of mannose-6-phosphate led to decreased enzyme activity, suggesting that enzyme uptake was mediated by mannose-6-phosphate receptor.

Our group has shown that the correction of MPS I fibroblasts by recombinant encapsulated BHK cells is also mediated by mannose-6-phosphate receptor. The effect of the ratio fibroblasts:encapsulated cells was also analysed (figure 5). The amount of enzyme uptaken by the fibroblasts is essentially the same under the different ratios (5:1; 1:1; 1:5) although the enzyme activity in the medium increases, as more enzyme is released.

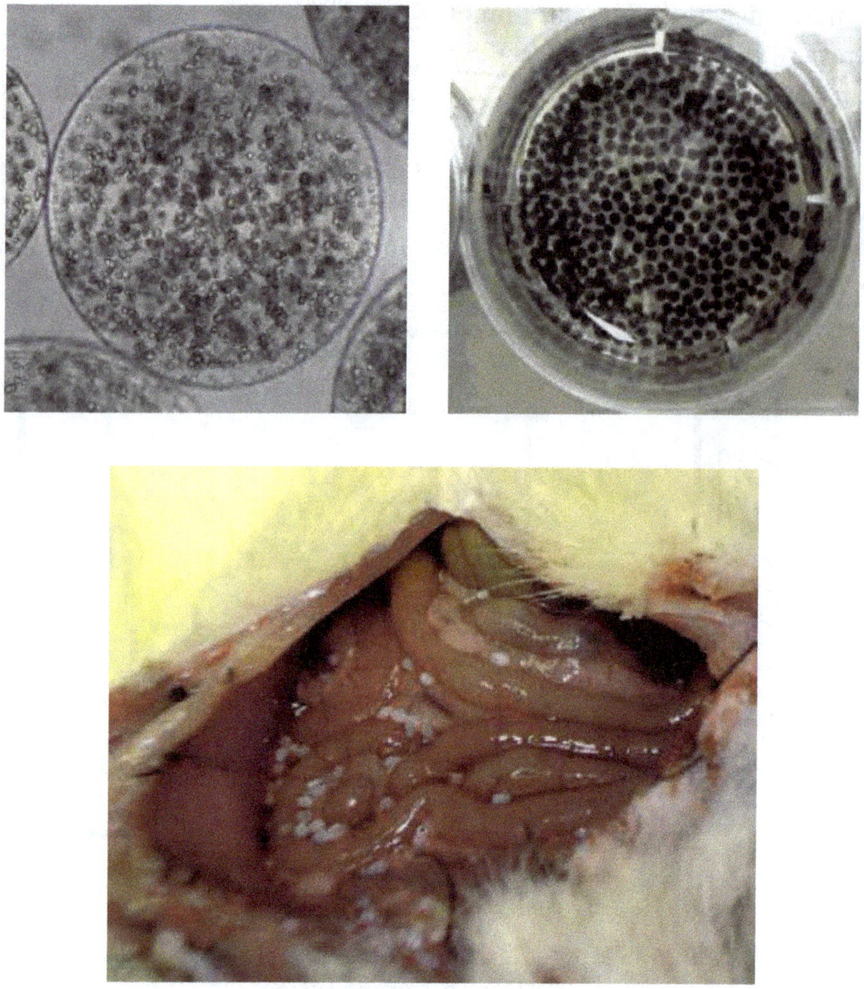

Fig. 6. Microcapsules used for the treatment of MPS I mice. Left upper panel: photomicrography of alginate beads containing recombinant BHK that overexpress IDUA (small round dots). Right upper panel: Macroscopic aspect of the microcapsules (black dots) in a 96-well plate. Capsules were ink stained to help visualization. Lower panel: Aspect of the capsules in the peritoneum ten days after implantation. Note that some capsules are attached to the intestine.

1.4 Transposon-based systems

For human gene therapy we can enumerate some important aspects of transposon systems (special emphasis in this chapter will be given to the *sleeping beauty* transposon) that make them appealing as a vector: (1) the integrated gene has stable expression, providing long- lasting expression of a therapeutic gene, which, as already mentioned, is essential for lysosomal storage diseases; (2) the transposase directs the integration of single copies of a DNA sequence into chromatin and (3) the system is binary (the transposon is not autonomous or able to transpose on its own) (Hacket et al, 2005).

The most studied transposon system, which has been used in pre-clinical studies for treatment of lysosomal disorders, is the *Sleeping beauty (SB)* system. Sleeping beauty transposon is a type of mobile element that belongs to the *Tc1/Mariner* class and that is able to transpose via movement of a DNA element in a simple cut-and-paste manner. For that, a precise piece of DNA is excised from one DNA molecule and moved to another site in the same or in a different DNA molecule (Plasterk, 1993). This reaction is catalyzed by the protein transposase, which can be supplied *in trans* by another plasmid for gene therapy purposes.

The SB transposon system consists of two components: (i) a transposon, made up of the gene of interest flanked by inverted repeats (IRs), and (ii) a source of transposase (figure 7). For Sleeping Beauty-mediated transposition, the transposase can recognize the ends of the IRs, excises the gene of interest from the delivered plasmid DNA, and then inserts it into another DNA site. Based on studies in about 2,000 integration events in either mouse or human genomes, transposons seem to integrate into random sites, including exons, introns and intergenic sequences (Carlson et al, 2003; Horie et al, 2003; Hacket et al, 2005). This is a potential problem, since it may lead to an event of insertional mutagenesis. A complete list of insertion positions of the SB tranposon can be found at the Mouse Transposon Insertion database at http://mouse.ccgb.umn.edu/transposon/ (Roberg- Perez et al, 2003).

The use of SB transposons as gene therapy approach in LSD is still recent. The first published work was conducted by Aronovich et al (2007) who studied the effects of intravenous hydrodynamic injection of the SB transposon into mice with Mucopolysaccharidosis types I or VII. Without immunomodulation, initial enzyme activities in plasma reached levels higher than 100-fold that of wild-type (WT). However, both GUSB (MPS VII) and IDUA (MPS I) levels fell to background within 4 weeks post- injection.

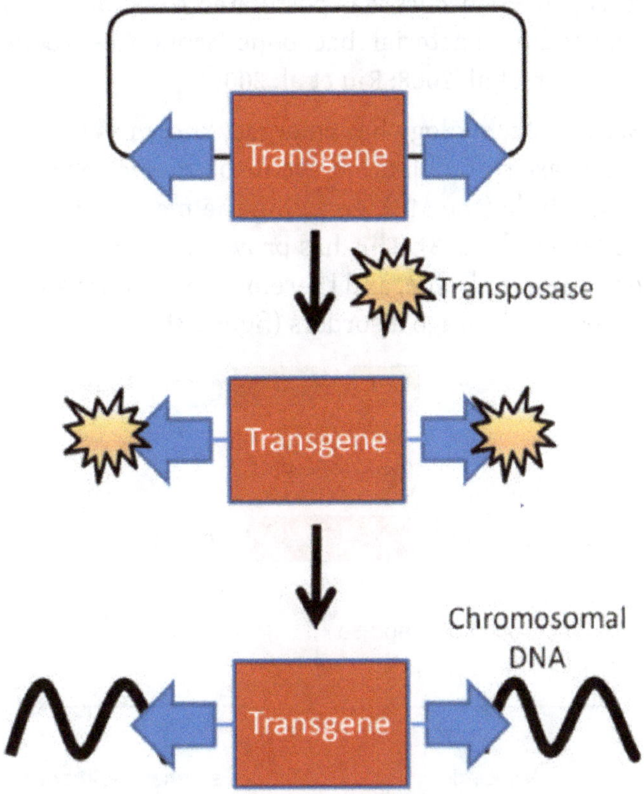

Fig. 7. SB transposon system. This simplified version of the SB transposon system shows the cut-and-paste system used by the SB transposase to insert the DNA into the host genome. The gene of interest is flanked by two inverted repeats regions (IRs, arrows) which will be recognized by the transposase (usually given in *trans* in a second plasmid) and allow the transgene to be inserted into the host genome. This way only the transgene will have prolonged expression, as transposase gene expression is transient. The SB transposon is a non-viral method of gene delivery that allows integration of the transgene in the host cell.

A second group of animals was performed with immunomodulation only in MPS I mice. Plasma IDUA persisted for over 3 months at up to 100-fold WT activity in one-third of the mice, which was sufficient to reverse lysosomal pathology in the liver and, partially, in distant organs. Histological and immunohistochemical examination of liver sections in IDUA transposon-treated WT mice revealed inflammation 10 days post-injection consisting predominantly of mononuclear cells, which can be seen as a potential side-effect.

A posterior study by the same group (Aronovich et al, 2009) was performed in another MPS I strain, which is immunodeficient (NOD/SCID mice). Using the same approach from the previous experiment, they were able to show a persistent elevation (100-fold normal) in the plasma IDUA levels for 18 weeks. Also, IDUA activity was present in all organs analyzed, including the brain. The SB transposon system proved efficacious in correcting several clinical manifestations of MPS I mice, including bone abnormalities, hepatomegaly, and accumulation of foamy macrophages in bone marrow and synovium. In 2008 the first human clinical trial using the SB transposon was approved in the USA for treatment of cancer (Williams, 2008), however although promising no clinical trials have been conducted in LSDs so far.

1.5 Minicircle gene therapy

Sustained *in vivo* transgene expression from plasmids can be difficult to achieve due to gene silencing. The mechanism by which this process occurs was postulated to be due to the deposition of repressive heterochromatin on the noncoding bacterial backbone sequences required for plasmid bacterial preparation and propagation (Chen et al, 2008; Riu et al, 2005).

Based on those findings, a new technology has emerged, known as the minicircle (MC) gene therapy. This system uses a phiC31 integrase recombination event to remove the bacterial backbone elements of the plasmid resulting in a DNA circle (the MC), encoding the mammalian expression cassette of choice and a small attR footprint (Chen et al, 2003). This has proven to be resistant to gene silencing *in vivo*, is maintained as an extrachromosomal episome, and therefore represents an interesting platform for gene replacement strategies for lysosomal storage disorders (figure 8).

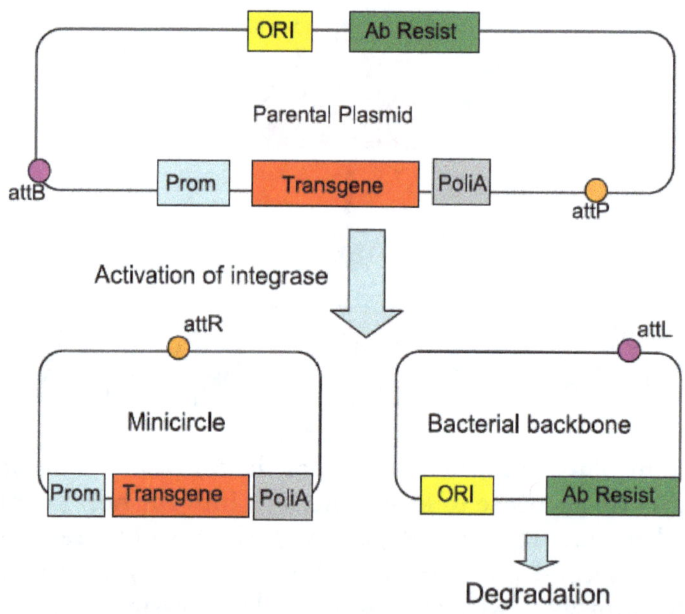

Fig. 8. Production of a minicircle plasmid. This simplified version of the process shows the parental plasmid containing both the gene of interest and the bacterial components, including origin of replication (ORI), genes that confer resistance to antibiotic (Ab resist) and sites that allow attachment of the integrase (att B and P). After activation of the integrase, a cis-recombination event occurs, separating the gene of interest and its regulatory elements from the bacterial backbone, which then is degraded.

This technology was recently used for treatment of MPS I mice in a proof-of-concept study (Osborn et al, 2011). In this study, the researchers performed a hydrodynamic injection of a minicircle plasmid containing the IDUA gene combined with immunomodulation, achieving stable expression of the transgene, increased IDUA tissue levels and reduction in GAG storage. As a recent technology, this is the only study performed on LSDs so far, but results are encouraging and should be tested on other diseases soon.

4. CONCISE SUMMARY

Currently, the majority of lysosomal storage disorders lack a viable treatment. Furthermore, the existing treatments lack the ability to rectify all the symptoms associated with these complex disorders affecting multiple systems of the body. Although gene therapy methods are somewhat limited compared to other fields such as cancer, they have demonstrated significant potential in treating a wide range of disorders. Both proof of concept trials and therapeutically useful findings have been achieved in certain circumstances.

In recent years, advancements in vector design and techniques, such as nanotechnology, transposons, and minicircle approaches, have been gradually addressing the limitations of non-viral gene transfer. These limitations include transient expression of the transgene and low transfection efficiency, as discussed in this chapter. Novel strategies to facilitate the DNA's evasion of endosomal degradation and traversal of the nuclear membrane are currently being developed, although they were not the focus of this chapter. However, these enhancements are facilitating the progression of non-viral gene therapy towards clinical trials in LSDs, which are anticipated to occur in the future. Non-viral vectors are a more secure option compared to viral particles, making them an attractive alternative for treating lysosomal storage disorders and other monogenic diseases. However, there is still a significant distance to be covered before reaching clinical implementation. Nevertheless, progress is being made consistently by the scientific community.

REFERENCES

Al-Dosari, M.S., Gao X. (2009). Nonviral gene delivery: principle, limitations, and recent progress. *AAPS J*, Vol 11, No 4, pp 671-681.

Anson, D.S., Bielicki, J., Hopwood, J.J. (1992). Correction of mucopolysaccharidosis type I fibroblasts by retroviral-mediated transfer of the human alpha-L-iduronidase gene. *Hum Gene Ther*, Vol 3, No 4, pp 371-379.

Aronovich, E.L., Bell, J.B., Belur, L.R., Gunther, R., Koniar, B., Erickson, D.C., Schachern, P.A., Matise, I., McIvor, R.S., Whitley, C.B., Hackett, P.B. (2007) Prolonged expression of a lysosomal enzyme in mouse liver after Sleeping Beauty transposon-mediated gene delivery: implications for non-viral gene therapy of mucopolysaccharidoses. *J Gene Med*, Vol 9, No 5, pp 403-415.

Aronovich EL, Bell JB, Khan SA, Belur LR, Gunther R, Koniar B, Schachern PA, Parker JB, Carlson CS, Whitley CB, McIvor RS, Gupta P, Hackett PB. (2009) Systemic correction of storage disease in MPS I NOD/SCID mice using the sleeping beauty transposon system. *Mol Ther*.Vol 17, No 7, pp 1136-1144.

Baiotto, C., Sperb, F., Matte ,U., Silva, C.D., Sano, R., Coelho, J.C., Giugliani, R. (2011) Population analysis of the GLB1 gene in south Brazil. *Gen Mol Biol*, Vol 34, No 1, pp 45-48.

Baldo, G., Quoos Mayer, F., Burin, M., Carrillo-Farga, J., Matte, U., Giugliani, R. (2011) Recombinant Encapsulated Cells Overexpressing Alpha-L-Iduronidase Correct Enzyme Deficiency in Human Mucopolysaccharidosis Type I Cells. *Cells Tissues Organs*, epub.

Balestrin, R.C., Baldo, G., Vieira, M.B., Sano, R., Coelho, J.C., Giugliani, R., Matte, U. (2008) Transient high-level expression of beta-galactosidase after transfection of fibroblasts from GM1 gangliosidosis patients with plasmid DNA. *Braz J Med Biol Res*, Vol 41, No 4, PP 283-288.

Barsoum, S.C., Milgram, W., Mackay, W., Coblentz, C., Delaney, K.H., Kwiecien, J.M., Kruth, S.A., Chang P.L. (2003) Delivery of recombinant gene product to canine brain with the use of microencapsulation. *J Lab Clin Med*, Vol 142, No 6, pp 399-413.

Bellettato, C.M., Scarpa, M. (2010) Pathophysiology of neuropathic lysosomal storage disorders. *J Inherit Metab Dis*, Vol 33, No4, PP 347-362.

Bloch, J., Bachoud-Lévi, A.C., Déglon, N., Lefaucheur, J.P., Winkel, L., Palfi, S., Nguyen, J.P., Bourdet, C., Gaura, V., Remy, P., Brugières, P., Boisse, M.F., Baudic, S., Cesaro, P., Hantraye, P., Aebischer, P., Peschanski, M. (2004). Neuroprotective gene therapy for Huntington's disease, using polymer-encapsulated cells engineered to secrete human ciliary neurotrophic factor: results of a phase I study. *Hum Gene Ther*, Vol 15, No 10, pp 968-975.

Bressel, T., Paz, A.H., Baldo, G., Lima, E.O.C., Matte, U., Saraiva-Pereira, M. (2008) An effective device for generating alginate microcapsules. *Genet. Mol. Biol*, Vol 31, No 1, pp 136-140.

Brunetti-Pierri, N., Stapleton, G.E., Palmer, D.J., Zuo, Y., Mane, V.P., Finegold, M.J., Beaudet, A.L., Leland, M.M., Mullins, C.E., Ng, P. (2007) Pseudo-hydrodynamic delivery of helper-dependent adenoviral vectors into non-human primates for liver-directed gene therapy. *Mol Ther*, Vol 15, No4, pp 732- 740.

Camassola, M., Braga, L.M., Delgado-Cañedo, A., Dalberto, T.P., Matte, U., Burin, M., Giugliani, R., Nardi, N.B. (2005) Nonviral in vivo gene transfer in the mucopolysaccharidosis I murine model. *J Inherit Metab Dis*, Vol 28, No 6, pp 1035-1043.

Carlson, C.M., Dupuy, A.J., Fritz, S., Roberg-Perez, K.J., Fletcher, C.F., Largaespada, D.A. (2003) Transposon mutagenesis of the mouse germline. *Genetics* Vol 165, No 1 pp 243–256.

Chen, Z.Y., He, C.Y., Ehrhardt, A. and Kay, M.A. (2003) Minicircle DNA vectors devoid of bacterial DNA result in persistent and high-level transgene expression in vivo. *Mol Ther*, Vol 8, No 3, pp 495–500.

Chen, Z.Y., Riu, E., He, C.Y., Xu, H. and Kay, M.A. (2008) Silencing of episomal transgene expression in liver by plasmid bacterial backbone DNA is independent of CpG methylation. *Mol Ther*, Vol 16, No 3, pp 548–556.

Consiglio, A., Martino, S., Dolcetta, D. (2007) Metabolic correction in oligodendrocytes derived from metachromatic leukodystrophy mouse model by using encapsulated recombinant myoblasts. *J Neurol Sci*, Vol 255, No 1-2, pp 7-16.

Dass, C.R.(2004) Lipoplex-mediated delivery of nucleic acids: factors affecting in vivo transfection. *J Mol Med*, Vol 82, No 9, pp 579–591.

de Duve, C. (1964) From cytases to lysosomes. *Fed Proc*, Vol 23, pp 1045–1049.

Dunbar, C.E., Kohn, D.B., Schiffmann, R., Barton, N.W., Nolta, J.A., Esplin, J.A., Pensiero, M.,Long, Z., Lockey, C., Emmons, R.V., Csik, S., Leitman, S., Krebs, C.B., Carter, C., Brady, R.O., Karlsson, S. (1998) Retroviral transfer of the glucocerebrosidase gene into CD34+ cells from patients with Gaucher disease: in vivo detection of transduced cells without myeloablation. *Hum Gene Ther*, Vol 9, No 17, pp 2629-2640.

Ellinwood, N.M., Vite, C.H., Haskins, M.E. (2004) Gene therapy for lysosomal storage diseases: the lessons and promise of animal models. *J Gene Med*. Vol 6, No 5, pp 481-506.

Estruch, E.J., Hart, S.L., Kinnon, C., Winchester, B.G. (2001) Non-viral, integrin-mediated gene transfer into fibroblasts from patients with lysosomal storage diseases. *J Gene Med*, Vol 3, No 5, pp 488-497.

Fischer, H.D., Gonzalez-Noriega, A., Sly, W.S., Morre, D.J. (1980) Phosphomannosyl-enzyme receptors in rat liver. Subcellular distribution and role in intracellular transport of lysosomal enzymes. J Biol Chem. Vol 255, No 20, pp 9608–15.

Fraga, M., Laux, M., Zandoná, B., Santos, G.R., Giuberti, C.S., Oliveira, M.C., Matte, U., Teixeira, H.F. (2008) Optimization of stearylamine-based nanoemulsions obtained by spontaneous emulsification process as nucleic acids delivery systems. *J Drug Del Sci Tech* Vol 18, No 6, pp 398-403.

Fraga, M., Bruxel, F., Lagranha, V., Teixeira, H.F., Matte, U. Influence of phospholipid composition on DNA/cationic emulsions complexes: Physicochemical properties, cytotoxicity and transfection on Hep G2 cells. *International Journal of Nanomedicine, in press.*

Fratantoni, J.C., Hall, C.W., Neufeld, E.F. (1968) Hurler and Hunter syndromes: mutual correction of the defect in cultured fibroblasts. *Science*, Vol 162, No 853, pp 570–572.

Friso, A., Tomanin, R., Alba, S. (2005) Reduction of GAG storage in MPS II mouse model following implantation of encapsulated recombinant myoblasts. *J Gene Med*, Vol 7, No 11, pp 1482-1491.

Fuller, M., Meikle, P.J., Hopwood, J.J. (2006) Epidemiology of lysosomal storage diseases: an overview. In *Fabry Disease: Perspectives from 5 Years of FOS.* Mehta A, Beck M, Sunder- Plassmann G, editors. Oxford: Oxford PharmaGenesis.

Glover D, Lipps H, Jans D (2005) Towards safe, non-viral therapeutic gene expression in humans. *Nature Reviews Genetics,* Vol 6, No 4, 299-310.

Hacein-Bey-Abina, S., Garrigue, A., Wang, G.P., Soulier, J., Lim, A., Morillon, E., Clappier, E., Caccavelli, L., Delabesse, E., Beldjord, K., Asnafi, V., MacIntyre, E., Dal Cortivo, L., Radford, I., Brousse, N., Sigaux, F., Moshous, D., Hauer, J., Borkhardt, A., Belohradsky, B.H., Wintergerst, U., Velez, M.C., Leiva, L., Sorensen, R., Wulffraat, N., Blanche, S., Bushman, F.D., Fischer, A., Cavazzana-Calvo, M. (2008) Insertional oncogenesis in 4 patients after retrovirus-mediated gene therapy of SCID-X1. *J Clin Invest*, Vol 118, No 9, pp 3132-3142.

Hackett, P.B., Ekker, S.C., Largaespada, D.A., McIvor, R.S. (2005) Sleeping beauty transposon- mediated gene therapy for prolonged expression. *Adv Genet*, Vol 54, pp 189-232.

Horie, K., Yusa, K., Yae, K., Odajima, J., Fischer, S. E., Keng, V. W., Hayakawa, T., Mizuno, S., Kondoh, G., Ijiri, T., Matsuda, Y., Plasterk, R. H., and Takeda, J. (2003) Characterization of Sleeping Beauty transposition and its application to genetic screening in mice. *Mol. Cell. Biol*, Vol 23, No 24, pp 9189–9207.

Hunter, C. (2006) Molecular hurdles in polyfectin design and mechanistic background to polycation induced cytotoxicity. *Advanced Drug Delivery Reviews,* Vol 58, pp 1523–1531.

Kamimura, K., Zhang, G., Liu, D. (2009) Image-guided, intravascular hydrodynamic gene delivery to skeletal muscle in pigs. *Mol Ther*, Vol 18, No 1, pp 93-100.

Kuijlen, J.M., de Haan, B.J., Helfrich, W., de Boer, J.F., Samplonius, D., Mooij, J.J., de Vos, P. (2006) The efficacy of alginate encapsulated CHO-K1 single chain-TRAIL producer cells in the treatment of brain tumors. *J Neurooncol* Vol 78, No 1, pp 31-39.

Lagranha, V.L., Baldo, G., de Carvalho, T.G. Burin M, Saraiva-Pereira, M.L., Matte, U., Giugliani, R. (2008) In vitro correction of ARSA deficiency in human skin fibroblasts from metachromatic leukodystrophy patients after treatment with microencapsulated recombinant cells. *Metab Brain Dis* Vol 23, No 4, pp 469-84.

Liu, F., Song, Y., Liu, D. (1999) Hydrodynamics-based transfection in animals by systemic administration of plasmid DNA. *Gene Ther*, Vol 6, No 7, pp 1258-1266.

Malatack, J.J., Consolini, D.M., Bayever, E. (2003) The status of hematopoietic stem cell transplantation in lysosomal storage disease. *Pediatr Neurol*, Vol 29, No 5, pp 391-403.

Matte U, de Carvalho T, Mayer F, Lagranha V, Giugliani R. (2011) Cell Microencapsulation: a Potential Tool for the Treatment of Neuronopathic Lysosomal Storage Diseases. *Journal of Inherited Metabolic Diseases,* epub.

Mayer, F.Q., Baldo, G., de Carvalho, T.G., Lagranha, V.L., Giugliani, R., Matte, U. (2010) Effects of cryopreservation and hypothermic storage on cell viability and enzyme activity in

recombinant encapsulated cells overexpressing alpha-L-iduronidase. *Artif Organs*, Vol 34, No 5, pp 434-439.

Meikle, P.J., Fietz, M.J., Hopwood, J.J. (2004) Diagnosis of lysosomal storage disorders: current techniques and future directions. *Expert Rev Mol Diagn.* Vol 4, No 5, pp 677–691.

Moreira, J., Gaspar, R., Allen T. (2001) Targeting Stealth liposomes in a murine model of human small cell lung cancer. *Biochimica et Biophysica Acta,* Vol 1515, pp 167-176.

Morille, M., Passirani, C., Vonarbourg, A., Clavreul, A., Benoit, J.P. (2008) Progress in developing cationic vectors for non-viral systemic gene therapy against cancer. *Biomaterials,* Vol, 29, No 24-25, pp 3477-3496.

Muñoz-Rojas, M.V., Horovitz, D.D., Jardim, L.B., Raymundo, M., Llerena, J.C. Jr, de Magalhães, T.S., Vieira, T.A., Costa, R., Kakkis, E., Giugliani, R. (2010) Intrathecal administration of recombinant human N-acetylgalactosamine 4-sulfatase to a MPS VI patient with pachymeningitis cervicalis. *Mol Genet Metab,* Vol 99, No 4, pp 346-350.

Munoz-Rojas, M.V., Vieira, T., Costa, R., Fagondes, S., John, A., Jardim, L.B., Vedolin, L.M., Raymundo, M., Dickson, P.I., Kakkis, E., Giugliani, R. (2008) Intrathecal enzyme replacement therapy in a patient with mucopolysaccharidosis type I and symptomatic spinal cord compression. *Am J Med Genet A*, Vol 146A, Vol 19, pp 2538-2544.

Muro, S. (2010) New biotechnological and nanomedicine strategies for treatment of lysosomal storage disorders. *Wiley Interdiscip Rev Nanomed Nanobiotechnol*, Vol 2, No 2, pp 189- 204.

Naganawa, Y., Ohsugi, K., Kase, R., Date, I., Sakuraba, H., Sakuragawa, N. (2002) In vitro study of encapsulation therapy for Fabry disease using genetically engineered CHO cell line. *Cell Transplant*, Vol 11, No 4, pp 325-329.

Nakama, H, Ohsugi, K, Otsuki, T., Date, I., Kosuga, M., Okuyama, T., Sakuragawa, N. (2006) Encapsulation cell therapy for mucopolysaccharidosis type VII using genetically engineered immortalized human amniotic epithelial cells. *Tohoku J Exp Med*, Vol 209, No 1, pp 23-32.

Nam, H.Y., Park, J.H., Kim, K., Kwon, I.C., Jeong, S.Y. (2009) Lipid-based emulsion system as non-viral gene carriers. *Arch Pharm Res* Vol 32, No 5, pp 639-646.

Niidome, T. and Huang, L. (2002) Gene Therapy Progress and Prospects: Nonviral vectors. *Gene Therapy* Vol 9, No 24, pp 1647–1652.

Orive, G., Ponce, S., Hernandez, R.M., Gascon, A.R., Igartua, M., Pedraz, J.L. (2002) Biocompatibility of microcapsules for cell imobilization alaborated with different type of alginates. *Biomaterials* Vol 23, No 18, pp 3825-3831.

Orive, G., Gascon, R.A., Hernandez, R.M., Igartua, M., Pedraz, J.L. (2003) Cell microencapsulation technology for biomedical purposes: novel insights and challenges. *Trends in Pharmacological Sciences*, Vol 24, No 5, pp 207-210.

Osborn, M.J., McElmurry, R.T., Peacock, B., Tolar, J., Blazar, B.R. (2008) Targeting of the CNS in MPS-IH using a nonviral transferrin-alpha-L-iduronidase fusion gene product. *Mol Ther*, Vol 16, No 8, pp 1459-1466.

Osborn, M.J., McElmurry, R.T., Lees, C.J., Defeo, A.P., Chen, Z.Y., Kay, M.A., Naldini, L., Freeman, G., Tolar, J., Blazar, B.R. (2011) Minicircle DNA-based Gene Therapy Coupled With Immune Modulation Permits Long-term Expression of α-L-Iduronidase in Mice With Mucopolysaccharidosis Type I. *Mol Ther*, 2011 Vol 19, No 3, pp 450-460.

Pardridge, W.M. (2007) Blood-brain barrier delivery of protein and non-viral gene therapeutics with molecular Trojan horses. *J Control Release*, Vol 122, No 3, pp 345-348.

Parkinson-Lawrence, E.J., Shandala, T., Prodoehl, M., Plew, R., Borlace, G.N., Brooks, D.A. (2010) Lysosomal storage disease: revealing lysosomal function and physiology. *Physiology (Bethesda)*, Vol 25, No 2, pp 102-115.

Plasterk, R.H. (1993) Molecular mechanisms of transposition and its control. *Cell*, Vol 74, No 5, pp 781–786.

Ragnaill, M.N., Brown, M., Ye, D., Bramini, M., Callanan, S., Lynch, I., Dawson, K.A. (2011) Internal benchmarking of a human blood-brain barrier cell model for screening of nanoparticle uptake and transcytosis. *Eur J Pharm Biopharm*. Vol 77, No 3, pp 360-367.

Resina, S., Prevot, P., Thierry, A.R. (2009) Physico-chemical characteristics of lipoplexes influence cell uptake mechanisms and transfection efficacy. *PLoS One* Vol 4, No 6, e6058.

Richard, M., Arfi, A., Seguin, J., Gandolphe, C., Scherman, D. (2009) Widespread biochemical correction of murine mucopolysaccharidosis type VII pathology by liver hydrodynamic plasmid delivery. *Gene Ther*, Vol 16, No 6, pp 746-756.

Riu, E., Grimm, D., Huang, Z. and Kay, M.A. (2005). Increased maintenance and persistence of transgenes by excision of expression cassettes from plasmid sequences in vivo. *Hum Gene Ther* Vol 16, No 5, pp 558–570.

Roberg-Perez, K., Carlson, C.M., Largaespada, D.A. (2003) MTID: a database of Sleeping Beauty transposon insertions in mice. *Nucleic Acids Res*, Vol 31, No 1, pp 78-81.

Ross, C.J., Bastedo, L., Maier, S.A., Sands, M.S., Chang, P.L. (2000a) Treatment of a lysosomal storage disease, mucopolysaccharidosis VII, with microencapsulated recombinant cells. *Hum Gene Ther*, Vol 11, No 15, pp 2117-2127.

Ross, C.J., Ralph, M., Chang, P.L. (2000b) Somatic gene therapy for a neurodegenerative disease using microencapsulated recombinant cells. *Exp Neurol*, Vol 166, No 2, pp 276-286.

Sifuentes, M., Doroshow, R., Hoft, R., Mason, G., Walot, I., Diament, M., Okazaki, S., Huff, K., Cox, G.F., Swiedler, S.J., Kakkis, E.D. (2007) A follow-up study of MPS I patients treated with laronidase enzyme replacement therapy for 6 years. *Mol Genet Metab*, Vol 90, No 2, pp 171-180.

Skarlatos, S., Yoshikawa, T., Pardridge, W.M. (1995) Transport of [125I]transferrin through the rat blood-brain barrier. *Brain Res*, Vol 683, No 2, pp 164-171.

Sly, W.S., Fischer, H.D., Gonzalez-Noriega, A., Grubb, J.H., Natowicz, M. (1981) Role of the 6- phosphomannosyl-enzyme receptor in intracellular transport and adsorptive pinocytosis of lysosomal enzymes. *Methods Cell Biol,* Vol 23, pp 191-214.

Spuch. C., Antequera, D., Portero, A., Orive, G., Hernández, R.M., Molina, J.A., Bermejo- Pareja, F., Pedraz, J.L., Carro, E. (2010) The effect of encapsulated VEGF-secreting cells on brain amyloid load and behavioral impairment in a mouse model of Alzheimer's disease. *Biomaterials*, Vol 31, No21, pp 5608-5618.

Takakusaki, Y., Hisayasu, S., Hirai, Y., Shimada, T. (2005) Coexpression of formylglycine- generating enzyme is essential for synthesis and secretion of functional arylsulfatase A in a mouse model of metachromatic leukodystrophy. *Hum Gene Ther*, Vol 16, No 8, pp 929-936.

Tomanin, R., Friso, A., Alba, S., Piller Puicher, E., Mennuni, C., La Monica, N., Hortelano, G., Zacchello, F., Scarpa M. (2002) Non-viral transfer approaches for the gene therapy of mucopolysaccharidosis type II (Hunter syndrome). *Acta Paediatr Suppl*, Vol 91, No 439, pp 100-104.

Uludag, H., Vos, P., Tresco, P. (2000) Technology of mammalian cell encapsulation. *Adv Drug Delivery Reviews*, Vol 42, No 1-2, pp 29-64.

Vitner, E.B., Platt, F.M., Futerman, A.H. (2010) Common and uncommon pathogenic cascades in lysosomal storage diseases. *J Biol Chem*, Vol 285, No 27, pp 20423-20427.

Walkley, S.U. (2009) Pathogenic cascades in lysosomal disease-Why so complex? *J Inherit Metab Dis*, Vol 32, No 2, pp 181-189.

Wasungu, L., Hoekstra, D. (2006) Cationic lipids, lipoplexes and intracellular delivery of genes. *Journal of Controlled Release*, Vol. 116, No 2, pp 255-264.

Williams, D.A. (2008). Sleeping beauty vector system moves toward human trials in the United States. *Mol Ther*, Vol 16, No 9, pp 1515-1516.

Wilson JM. (2009) Lessons learned from the gene therapy trial for ornithine transcarbamylase deficiency. *Mol Genet Metab*,Vol 96, No 4, pp 151-157.

Worgall, S., Sondhi, D., Hackett, N.R., Kosofsky, B., Kekatpure, M.V., Neyzi, N., Dyke, J.P., Ballon, D., Heier, L., Greenwald, B.M., Christos, P., Mazumdar, M., Souweidane, M.M., Kaplitt, M.G., Crystal, R.G. (2008) Treatment of late infantile neuronal ceroid lipofuscinosis by CNS administration of a serotype 2 adeno-associated virus expressing CLN_2 cDNA. *Hum Gene Ther*, Vol 19, No 5, pp 463-474.

Zhang, Y., Wang, Y., Boado, R.J., Pardridge, W.M. (2007) Lysosomal enzyme replacement of the brain with intravenous non-viral gene transfer. *Pharm Res,* Vol 25, No 2, pp 400- 406.

8 Chapter

ELECTROGENE TRANSFER IS A METHOD USED FOR DNA VACCINATION

1. INTRODUCTION: AN OVERVIEW

Vaccination: Traditional and new generation vaccines

Vaccination has long been recognized as a crucial approach to preventing the spread of infectious diseases in both humans and animals. The establishment of basic principles for vaccine development occurred when Edward Jenner inoculated James Phipps with a bovine poxvirus in 1796 to provide protection against the closely related human pathogen smallpox virus. Almost a century later, Pasteur developed a live attenuated vaccine against rabies (Fraser and Rappuoli 2005). A vaccine is typically a substance containing weakened or dead microorganisms, or purified components of a pathogen, such as proteins, toxins, or polysaccharides, that is given to stimulate the immune system and provide protection against a particular disease.

There are two types of immunization that have inherent differences: prophylactic vaccination triggers a response against an antigen that the immune system has not encountered before, resulting in the creation of long-term memory cells and protective effectiveness. On the other hand, therapeutic vaccination stimulates the immune system in response to a persistently present antigen, leading to the elimination of an existing infection.

Vaccines manufactured using standard methods can prevent certain infectious diseases. These approaches involve growing the microorganism in a controlled laboratory environment and then using biochemical, microbiological, and serological techniques to separate individual components. Every antigen is manufactured in a pure state either by extracting it directly from the bacteria or by utilizing DNA recombinant technology. Subsequently, it is evaluated to determine its capacity to stimulate an immune response (Serruto and Rappuoli 2006).

The utilization of traditional methods formed the foundation of vaccinology and resulted in significant accomplishments, including the complete elimination of smallpox and the near eradication of diseases such as diphtheria, tetanus, poliomyelitis, pertussis, measles, mumps, rubella, and invasive Haemophilus influenzae B. These achievements have substantially improved the quality of life and increased life expectancy (Andre 2003). However, they have significant drawbacks, including being time-consuming and impractical in certain situations due to the challenge of cultivating certain microorganisms in a controlled environment. Additionally, even when attenuated, they can still lead to harmful or undesired immune responses (Purcell et al. 2007). Furthermore, it is important to note that the antigens produced during infection may not be replicated in laboratory settings. Additionally, the proteins that are most plentiful and easily purified may not necessarily be effective as protective antigens. Moreover, it is worth mentioning that only a limited number of molecules can be isolated and tested simultaneously.

Over the past ten years, there has been a significant transformation in the methodology and progress of vaccine design and development. These advancements encompass novel delivery systems designed to enhance the safety and immunogenicity of conventional vaccinations, innovative approaches to find protective antigens, and the development of enhanced adjuvants.

Given that new diseases are bound to arise as a result of evolutionary processes such as mutation, gene exchange, interspecies transfer, or human exposure to new settings, it is imperative to have more dependable methods in place to immediately address these potential dangers (Plotkin 2005). Advanced technologies including genomics, proteomics, functional genomics, and synthetic chemistry enable the rational identification of antigens, the manufacture of complex glycans, and the creation of tailored carrier proteins. This results in the identification, creation, and testing of novel vaccines, which can be utilized not only for combating infectious diseases but also for treating autoimmune disorders, allergies, chronic inflammatory diseases, and cancer. Several vaccine modalities are now being investigated, including subunit vaccines, synthetic peptide vaccines, ex-vivo loaded dendritic cells (DCs), and genetic immunization. These modalities will be described here.

1.1 Subunit vaccines

Subunit vaccines have surpassed conventional attenuated or killed vaccinations in some aspects, such as safety and production. The primary methods employed for the production of these vaccines involve the utilization of bacteria, yeast, insect, or mammalian cells. Nevertheless, the manufacture of recombinant vaccine proteins in these expression systems often incurs high costs, necessitating the use of large-scale fermenters and rigorous purification methods. Globally, there are only a limited number of facilities equipped to manufacture kilos of a particular protein for use as an immunogen. The establishment, validation, and ultimate certification of new production facilities require several years and significant investments in both financial and human resources.

Furthermore, certain antigens necessitate post-translational changes that are unattainable through all expression techniques. Over the past ten years, researchers have created non-fermentative alternatives that rely on living organisms to address these issues and offer affordable technology for vaccine manufacturing. Insects and plants have been modified to produce subunit vaccines, which offers distinct advantages over traditional fermentative systems. These advantages include faster development time, easier scaling-up of production, and improved cost-efficiency (Brun et al., 2011). Although recombinant technology has seen advancements, the production of these vaccines continues to be challenging due to their intrinsic toxicity for the bacterial/viral expression system. Examples of such vaccines include Human Papilloma Virus type 16-E2 and wild-type p53 (van der Burg et al. 2006).

1.2 Synthetic peptide vaccines

The identification of specific epitopes within protective proteins enables the development of peptide vaccines as an alternative method to employing the entire protein as a vaccine. Peptide epitopes that have been chosen are the smallest portion of a protein antigen that can trigger an immunological response. They are used to specifically focus immune responses towards the activation of T-cell immunity. An optimal peptide vaccination should incorporate epitopes that are recognized by both B and T cells, while also considering the MHC restriction of the T-cell response. Occasionally, there can be significant overlap between B and T-cell epitopes within the sequence of an antigen. However, in other circumstances, these epitopes may be located in distinct sections of the antigen or even in different antigens from the targeted pathogen. The ease of manufacturing clinical-grade peptides enables quick modifications in the design of peptide vaccines, facilitating the rapid implementation of new immunological ideas. This is particularly advantageous for the development of vaccines targeting rapidly mutating viruses like influenza (Brun et al., 2011). Although this strategy has potential benefits, the progress in creating effective peptide vaccines has been primarily hindered by challenges related to stability, low immunogenicity of basic peptides, and the MHC polymorphism of the host species (Tam 1996).

1.3 *Ex-vivo* loaded dendritic cells

Vaccines based on Antigen Presenting Cells (APCs) are being extensively investigated in the field of vaccine research. This method involves extracting dendritic cells (DC) from the patient, exposing them to antigens or introducing genes that encode these antigens, and then reintroducing the modified DC back into the patient. This vaccination technique has the capacity to enhance the presentation of antigens through the major histocompatibility complex class I pathway and subsequently stimulate the proliferation of cytotoxic T lymphocytes that specifically target the tumor. DC vaccines have shown a remarkable capacity to induce measurable immune responses in translational experiments including melanoma patients. Nevertheless, these responses frequently do not succeed in eliciting significant clinical outcomes. Given the challenges in determining the specific roles of DCs and effector T cells in these scenarios, a comprehensive examination of the in vivo interactions between different immune cell types may be necessary to achieve a full comprehension of the involvement of DCs (Palucka et al. 2007).

1.4 Genetic vaccination

In recent times, novel vaccination techniques, such as those utilizing gene transfer, have surfaced. Genetic vaccination is derived from gene therapy. The primary purpose of genetic vaccination is to introduce a gene into the host that codes for the antigen of the disease target. This is done with the intention of stimulating a targeted immune response. On the other hand, the objective of gene therapy is to facilitate the creation of a protein that is either absent or faulty in the host. So far, the bulk of clinical trials for gene therapy have focused on cancer (66.5%), cardiovascular illnesses (9.1%), and infectious diseases (6.5%). A total of 85 gene therapy trials have been conducted for infectious disorders, with the majority focusing on human immunodeficiency virus infection, tetanus, cytomegalovirus, and adenovirus infections (Chiarella et al., 2008a).

Present methods of gene transfer in mammals involve encapsulating the DNA into carriers for gene delivery. The optimal vehicles for gene delivery should be both safe and capable of protecting DNA in both extracellular and intracellular environments. They should also efficiently transport DNA to the proper cellular compartments, ensuring high and durable expression levels.

Currently, viral vectors surpass non-viral systems in terms of efficiency, achieving high levels of effectiveness, estimated at approximately 90%, for both gene delivery and expression. Nevertheless, the clinical use of gene therapy has been restricted due to concerns regarding immunogenicity, inflammatory reactions, high scale-up costs, and the risk of integrating into the host genome. This has led to a decrease in the use of recombinant virus-based vectors for DNA delivery in preclinical and clinical protocols compared to the past. For instance, in the year 2000, approximately 75% of clinical protocols involving gene therapy utilized these vectors (Chiarella et al., 2008a). In recent years, several clinical trials have highlighted various disadvantages associated with the use of viral vectors as antigen delivery systems. These drawbacks include toxicity, recombination, pre-existing host immunity, increased immunogenicity compared to the target antigen, and limited DNA carrying capacity (Harrington et al., 2002; Ramirez et al., 2000). Recent advancements in our understanding of the biology of the immune system have prompted the exploration of non-viral approaches, such as naked DNA vaccination, as a safer and promising method for introducing foreign antigens into the host and stimulating an immunological response.

Currently, non-viral technologies, particularly those utilizing plasmid DNA delivery, have gained popularity in both academic research facilities and medical environments.

2. DNA VACCINATION

2.1 DNA vaccines: An emerging field

In 1990, Wolff and collaborators found that bacterial plasmid DNA encoding a reporter gene could result in *in vivo* expression of the encoded protein after simple intramuscular injection without the need for more complex vectors (Wolff et al. 1990). Following Wolff's findings a new era of vaccination started.

Naked-DNA vaccines are for definition vectors based on bacterial plasmids engineered to express the disease-specific antigen using promoter elements active in mammalian cells, without the addition of surrounding chemicals or a viral coat. The main advantages of naked DNA vaccines are safety and production in large amount, as well as stability at different temperatures and, more important, flexibility in design, since multiple antigenic targets or multiple independent cytokines or co-stimulatory sequences can be incorporated into a single DNA vector. They are also likely to be attractive from a health economics perspective: they are relatively easy to manufacture in large quantities in contrast to the complicated processes requested for attenuated virus vaccines, and do not require any special transportation or storage conditions that could hinder their widespread distribution as it happens for live pathogens-based vaccines which need to be distributed and stored in cold conditions. The genes encoding the antigens can be chemically constructed without deriving them from live virulent organisms so avoiding for operators and patients the risks of exposure to dangerous pathogens. They are commonly delivered by a simple intramuscular injection. In mammals the skeletal muscle represents approximately the 30% of the body mass, and muscle fibres are ideal targets for DNA transfection. These are stable and large syncytial cells containing several nuclei that can actively take part in immune reactions. For the easy accessibility of the skeletal muscle and good vasculature, the delivery of DNA vaccine into this organ is highly preferable (Wiendl et al. 2005). Immunisation with DNA induces all three arms of adaptive immunity (antibodies, helper T cells, CTLs), and even innate immune responses can be easily and rapidly made while maintaining fidelity to the immunological aspects necessary for a pathogen, yet excluding other undesirable proteins or immune responses (Liu 2011).

Table 1. Advantages and disadvantages of plasmid DNA vaccines.

Characteristic	Advantage/disadvantage of plasmid DNA vaccines
Antigen	In vivo antigen synthesis with native conformation
Antigen presentation	MHC-I, MHC-II, cross-priming
Immune Response	Humoural and cytotoxic
Manufacture	Easy and fast
Stability	Stable at various temperature (RT)
Risk	Does not induce the disease related to the encoded antigen
Applicability	Prophylaxis and therapy of disease
Indication of use	Infectious disease, allergy, cancer, autoimmune disease
Safety	Low risk of recombination and inflammation
Immunogenicity	Weak

In table 1 the advantages and disadvantages of DNA vaccines are listed.

On the other hand, a major disadvantage of plasmid DNA vaccines is their poor immunogenicity when administered as an unformulated intramuscular injection. Large quantities of DNA are required to induce only modest immunogenicity and many efforts have focussed on the development of new technologies aimed at increasing the DNA vaccine potency (Chiarella et al. 2008a). That said, better strategies are needed in designing more effective vectors and combined protocols so as to induce a strong immune response to weakly immunogenic antigens. These strategies comprise new insights in studying the mechanism of action and induction of the immune response in a host injected with a DNA- based vaccine. Recently, most relevant patented strategies have been developed to enhance the plasmid DNA vaccine immunogenicity taking into account DNA plasmid construction, epitope and antigen choice, selection and use of new adjuvants and different delivery methods.

2.2 Mechanism of action and induction of the immune response

The crucial event responsible for the initiation of an immune response against a foreign antigen is recognition by specialized cells namely the antigen presenting cells (APCs), uptake and presentation of the antigen to naïve lymphocytes and induction of effector T helper (Th), cytotoxic (CTL) and B lymphocytes.

In this context the mechanism of action of DNA vaccines looks very simple. Once the DNA vaccine is delivered into the skeletal muscle, the plasmid DNA is taken up by the resident DCs and by the muscle fibres. While transfected muscle cells behave as permanent antigen reservoir as well as target of immune effector cells (Payette et al. 2001), resident DCs have the property to leave the muscle tissue and move to the closest draining lymph nodes in order to process and present the antigen to T lymphocytes (**Fig. 1**). DCs are specialized in capturing extracellular antigens by receptor-mediated endocytosis and pinocytosis mechanisms and following antigen uptake they undergo a complex multi-step maturation process. DC maturation depends also on the microbial and pathogens-derived signals which increase their capacity to migrate towards the draining lymph node. While DCs move to the lymphoid organs, they interact with various chemokines which contribute further to their maturation process (Palucka et al. 2010). Once in the lymph nodes, DCs shift from an antigen-capturing cell to a T sensitizing cell, being capable to present antigen in association with the class I and class II MHC molecules to CTLs and Th lymphocytes. Interaction between the DC and the T lymphocyte induces formation of the immunological synapse (IS) *via* complex MHC- antigen- T cell receptor (TCR) resulting in the clonal expansion of the T lymphocyte and differentiation in T memory cell. Professional DCs can also capture antigens released in the interstitial space by skeletal muscle fibres or in form of apoptotic bodies activating the cross-presentation pathway (**Fig. 1**) (Russo et al. 2000).

This route allows presentation of extracellular/exogenous antigens through the MHC-I restriction pathways (Kurts et al. 2010). Therefore, extracellular antigens which normally induce a humoural immune response can also access to the MHC-I compartment through endoplasmic reticulum, leading to simultaneous stimulation of the CTL immune response. Antigen synthesized by DC or skeletal muscle cell can also be released in the extracellular environment and activate directly the B lymphocytes through antigen-antibody interaction (**Fig.1**).

Considering the mechanism described above, plasmid DNA vaccines are able to stimulate all the principal effector cells of the adaptive immune system but due to the presence of CpG islands intrinsic to the DNA structure they can also can mimic some aspects of live infection, activating important signals of the innate immune system (Matzinger 2007). Theoretically genetic immunisation by plasmid DNA vaccines seems to confer the same broad immunological advantages as immunisation with live, attenuated vaccines does, without the accompanying safety concerns associated with live infection, such as reversion to the virulent form and/or incomplete inactivation of live vaccines. However there are some obstacles that make naked DNA vaccination less potent than traditional vaccines.

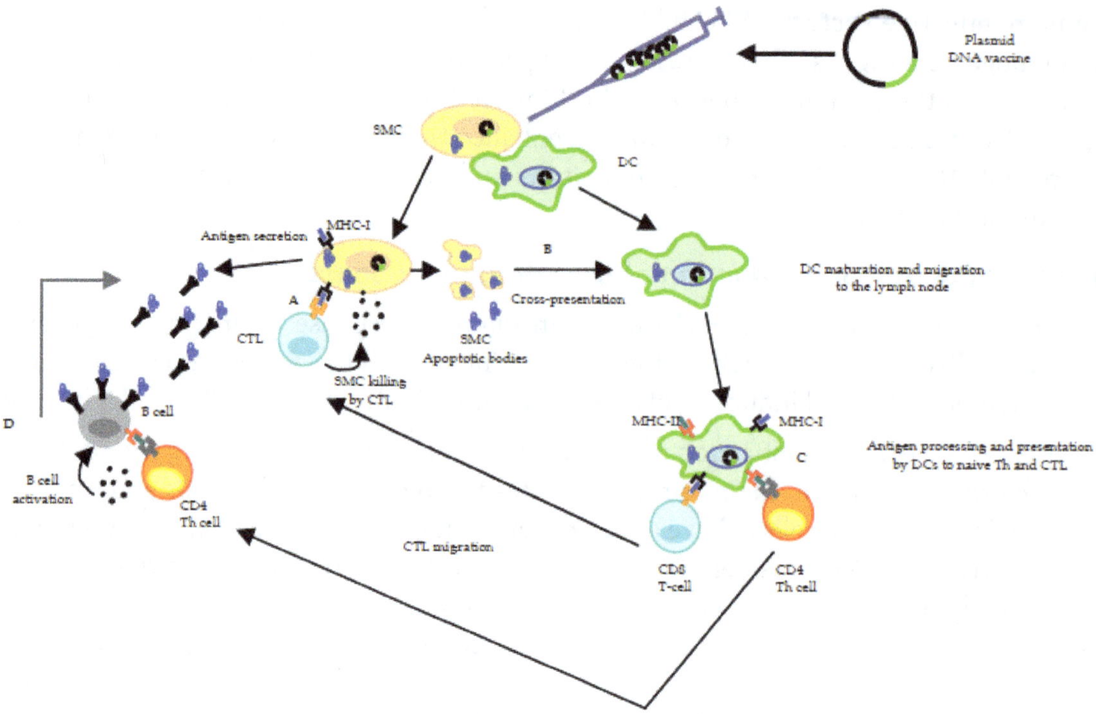

Fig. 1. Mechanism of action of plasmid DNA vaccination.

The antigen sequence is cloned into a bacterial plasmid vector specific for vaccination. The DNA vaccine is administered by intramuscular injection. After plasmid uptake by muscle cells, the gene coding for the antigen is translated into a protein. Transfected skeletal muscle cells can present the antigen to CTL through MHC-I molecules (**A**) as well as through DCs cross presentation (**B**). Plasmid is also uptaken by resident DCs which synthesize the antigen and present it in association with MHC-I molecules and MHC-II molecules to CD8+ T and CD4+ T-helper lymphocytes (**C**). Furthermore, antigen is released in the extracellular environment inducing the production of antibodies by B lymphocytes whose activation is mediated by the CD4+ T-helper cells (**D**). SMC, skeletal muscle cell; DC, dendritic cell; CTL, cytotoxic lymphocyte.CD8 T cell, CD8 T lymphocyte; CD4 Th cell, CD4 T helper lymphocyte; B cell, B lymphocyte.

2.3 DNA plasmid construction

DNA plasmid vaccines offer several advantages when compared to viral vectors or live attenuated vaccines. First of all they are easy to construct and to manipulate which is an important characteristic required to vaccines against pandemic diseases. They are also very stable at room temperature, do not require particular storage conditions which makes them ideal candidates for long-term delivery in other countries. The antigen can be chemically synthesized and cloned directly into the plasmid vaccine simplifying the operations of amplification by molecular techniques and avoiding to work with potentially dangerous live antigen source. The DNA constructs can be easily made to encode for modified proteins deprived of regions that might be dangerous or toxic to the recipient cell or that might suppress the immune response. DNA is a highly flexible molecule, the basic construct can be manipulated in several ways by genetic engineering in order to increase antigen expression, immunogenicity and uptake by recipient cells. All the modifications can affect both the vector backbone and the gene sequence incorporated into the plasmid, which can include adjuvant-like sequences with stimulating activity on the immune system (Abdulhaqq and Weiner 2008). By using this approach, enhanced antigen specific immune responses were observed, suggesting that this could be a general method for targeting antigen to selected cell types. Different strategies can be used for enhancing the plasmid DNA vaccine potency. A first strategy allows to improve the vector construct *i.e.* by working on the plasmid backbone design and construction; a second strategy allows to improve the codon usage in order to maximize the antigen synthesis.

An expression vector for genetic vaccination generally consists of the following elements: 1) a promoter/enhancer sequence; 2) the gene of interest encoding the target antigen; 3) a polyadenylation/transcriptional terminator sequence; 4) a resistance gene for plasmid selection and an origin of replication (Ori) in order to allow production of high numbers plasmid copies.

The capacity of a plasmid DNA vector to drive gene expression is obtained by an optimal combination of all these requisites, and by the possibility of introducing various modifications into the plasmid backbone. The promoters mostly used for gene expression vectors are the cytomegalovirus immediately-early promoter (CMV) the simian virus SV40 early promoter (SV40) and the Rous sarcoma virus promoter (RSV). The CMV promoter is the most popular, as it drives gene expression in a wide range of cells and tissues (Lundquist et al. 1999). Plasmid DNA vectors can also contain tissue specific, synthetic and controllable promoters, whose sequences are designed for a specific use (Papadakis et al. 2004). When the expression of a gene is desired in certain tissues, promoters that control expression in a cell- tissue-specific manner are used. For instance the alpha skeletal muscle actin promoter is specifically used for selective expression in skeletal muscle cells whereas endothelial cell-specific promoters are used to drive expression in the tumour-associated endothelial cells (Dong and Nor 2009). The flexibility of DNA vector permits investigators to exploit the concept of gene expression optimisation by creating synthetic promoters, such as enhancer/promoters composed of numerous combinations of various regulatory sequences (Edelman et al. 2000; Li et al. 1999). One of these regulatory synthetic sequence, the hybrid CMV-Ub promoter, was found to have higher expression than the natural muscle promoters (Yew et al. 2001). In order to display gene expression kinetics, naturally regulated systems were developed by incorporating sequence elements that respond to the local environment of the given cell or tissue or that are regulated by small molecule drugs (Yew 2005). Thus transgene expression can be regulated by modulating expression of these transcriptional factors or by altering their activity through drug administration. However, regulation of *in vivo* transgene expression by such approaches is unreliable, mainly due to the low levels of control associated with the complexity of these systems. Appropriate choice of regulatory elements and vector backbone can lead the gene expression kinetics from a few days to several months. Nuclear localization of plasmid DNA is another prerequisite for the effective antigen expression. To this purpose our laboratory developed a series of plasmids with a functional nuclear translocation sequence (NTS) (Ciafre et al. 1998). All the characteristics of plasmid DNA vaccines mentioned above are summarized in **Table 1**.

The codon usage is an important issue to consider in the DNA vaccine manufacture. Plasmid DNA are totally dependent on the host cell machinery for protein transcription and translation. Since codon usage of bacterial organisms is different from the codon usage of mammalian, it is mandatory to optimize codon usage in DNA vaccine to allow maximum antigen expression (Bojak et al. 2002). Various codon usage approaches are now commonly exploited in both non-human primate studies and clinical trials. This strategy has been successfully used to optimize the sequence of mycobacterial antigens (e.g., Ag85B) improving protein expression and thereby enhancing the immunogenicity of DNA vaccines against M. tuberculosis (Ko et al. 2005).

1.4 Epitope and antigen selection

The first requisite for a DNA vaccine to induce effective immune response is related to the choice of the target antigen. Sometimes the immunogenic determinant of a certain pathogen is unknown hence selection of antigen sequences has to be included as first step in the design of an epitope-based vaccine. The goal is to identify relevant T cell epitopes, able to bind to MHC class I and II molecules that are both effective and sufficient in vaccine protection against pathogen challenge or, in the case of cancer vaccines, T cell epitopes of malignant antigens that are not ignored by the immune system. In particular for tumours, highly immunogenic antigen determinants remain to be identified for most cancers types.

The "direct immunological approach" which consists of deriving tumour cell lines from malignant biopsies, isolating the cancer antigens and expanding the human CTLs specific for that given antigen is now substituted by the "reverse immunological approach" (Sette and Rappuoli 2010). With this new method a candidate cancer antigen expressed on a tumour is selected by *in silico* studies. First tumour antigens are identified by exploiting immune assays based on the availability of specific polyclonal and monoclonal antibodies. Otherwise the serological analysis of recombinantly expressed clones technology (SEREX) is also supportive to identify novel tumour antigens where the blood serum of patients affected by neoplastic diseases is screened against tumour antigen cDNA expression libraries (Jager et al. 2004). Techniques employed in the molecular biology field are also helpful. The analysis of the human transcriptome based on DNA and RNA microarrays allows identification of cancer antigens in a high-throughput system. As following step, putative antigenic determinants are predicted with the aid of bioinformatics. This makes it possible to identify a variety of epitopes within an antigen sequence and to choose the best candidates for the binding to specific MHC molecules. This system works well for epitope discovery, and predictions of the MHC class I pathway is being further improved by integration with prediction tools for proteasomal cleavage and Transporter associated with Antigen Processing (TAP) binding (Larsen et al. 2007) . Furthermore, native epitopes that do not fit perfectly into the MHC groove can be modified at specific sites to increase their affinity to the MHC molecule of interest, leading to the generation of what are called "heteroclitic" epitopes (Dyall et al. 1998).

The antigen determinant can be modified either by increasing the affinity of the binding to the MHC molecules or by augumenting the binding ability of the peptide–MHC complex for the TCR. The first approach is the most widely used (Dudek NL et al. 2010). This strategy is commonly exploited in the design of cancer vaccines, allowing to convert a subdominant into a dominant epitope by making it more competitive in the binding to specific MHC alleles, and thereby enhancing the potency of the vaccine. The primary and/or secondary anchor residues of an epitope can be replaced with specific aminoacids that provide much of the specificity of binding to the MHC molecule. Epitope "enhancement" is possible for both classes of human MHC resulting in priming of the CD8+ cytotoxic T-lymphocytes (CTL) which can recognise the target epitope on tumour or infected cells and in activation of CD4+ T helper cells whose role is crucial for promoting humoural and cytolytic responses, regardless the CTL epitope enhancement. The selection of biologically relevant epitopes within an antigen sequence is performed with different bioinformatic softwares.

Several databases of MHC binding peptides now exist and a number of programs performing such predictions are available on the web. We experienced the use of SYFPEITHI, BIMAS and PROPRED I-II programs (www.syfpeithi.de; www- bimas.cit.nih.gov; www.imtech.res.in/raghava/propred)(Parker et al. 1994; Rammensee et al. 1999; Singh and Raghava 2001). They are based on different algorithms that provide estimation of the binding of a certain peptide sequence to a wide spectrum of human MHC molecules. Unfortunately, a major drawback of these programs is their intrinsic feature of being 'predictive' in estimating the binding affinity between the MHC molecule and the antigenic epitope, with approximately 70% reliability. Another limit is they cannot calculate the binding of the MHC-peptide complex to the T cell receptor, which is a crucial point to verify the immunogenicity of the tumour antigenic determinant.

Once immunogenic peptides are predicted *in silico* they need to be verified experimentally. This goal is achieved by performing *in vitro* assays to confirm their stability and binding capacity. After that, human CTL isolated from patients are assayed *in vitro* to verify their capability to recognize specifically the selected epitope on the tumour cells. This result represents the validation of the tumour antigens. Once the antigenic epitope has been validated, it can be taken in consideration for DNA vaccines manufacture and for initiating vaccination trials.

2.5 Adjuvants

It is widely accepted immunogenicity of DNA vaccines is generally weak in comparison to that of traditional vaccines. Although plasmid vaccines are capable to induce a complete immune response involving activation of CTLs, Th and B lymphocytes and a certain activation of the innate compartment of the immune system, these vaccines show low potency and efficacy when administered as unformulated injection. DNA vaccines like all the subunit vaccines, which are made of purified or recombinant antigens, require additional components to help stimulating a "comprehensive" immune reaction. Such "help" is provided by substances and components termed adjuvants.

Adjuvants of current use, either in man or in animals, have for the most part been developed empirically, without a clear understanding of their cellular and molecular mechanisms of action. Indeed adjuvants were historically defined as "the dirty little secret of the immunologist" (Janeway 1989).

Table 2. Adjuvants used in vaccination with naked DNA.

Class of adjuvants	Adjuvant name	Nature of the adjuvant
Genetic adjuvants	Interleukin-2	Cytokine
	Interleukin-12	Cytokine
	Granulocyte Monocyte- Colony Stimulating Factor	Cytokine
	T-helper epitopes of toxins	Peptides
Adjuvants targeting Pattern Recognition Receptors	Monophosphoryl Lipid A AS02 AS01 QS-21 CpG-DNA	Lipid derivative Oil in water emulsion Liposomal formulation Saponin Oligodeoxynucleotides
Aluminium-based compounds	Aluminium Phosphate Aluminium Hydroxide	Mineral salt Mineral salt

The main function of an adjuvant is to create a depot of antigen at the injection site, resulting in a gradual release of small quantities of antigen over a long period of time. The adjuvant also serves as a vehicle for delivering the antigen to the lymph organs, where antigenic epitopes can be presented to T cells by professional APCs. However, recent advances in the immunological research suggest that most, if not all, adjuvants enhance T and B cell responses by engaging components of the innate immune system, rather than by exerting direct effects on the lymphocytes. In DNA vaccinations, adjuvants are used to achieve qualitative alteration of the immune response. Adjuvants confer to DNA or to subunit vaccines the ability to promote an immune response which might not occur in their absence. Here we describe certain classes of adjuvants most widely used in DNA vaccinations. A summary of the adjuvants described bellow is presented in **Table 2.**

2.5.1 Genetic adjuvants

Genetic adjuvants are molecules such as cytokines, chemokines and co-stimulatory factors that may be cloned into the DNA plasmid vaccine and expressed *in vivo*. These adjuvants can be encoded on the same vector expressing the antigen or inserted into a separate vector and co-injected with the vaccine. This method provides adjuvant activity at the site of antigen production, with lasting effect from transfected cells. Among various factors, cytokines are highly preferred as genetic adjuvants because they act on cells involved in the host defense and can be used to modulate immune responses. Co-delivery of cytokines in DNA vaccine formulations has been used extensively for a wide range of infectious diseases such as malaria, leishmania, schistosoma to enhance the T cell mediated responses (Ivory and Chadee 2004). One of the earliest cytokines to be incorporated into a DNA vaccine was IL-2, a well known T-cell growth factor included in several immunotherapy protocols. Addition of this cytokine to a plasmid vaccination vector resulted in enhanced antibody responses in low responder mice against malaria (Good et al. 1988) and increased production of antibodies directed against the complementary determining hypervariable region 3 of the Ig heavy chain in human B-cell lymphoma (Rinaldi et al. 2001). IL-2 contributed to increase the efficacy of a DNA vaccine against a simian immunodeficiency virus when it was fused with the immunoglobulin Fc fragment, resulting in augmentation of the cytokine half-life (Barouch et al. 2000). However, the use of IL-2 is now being limited by the emerging evidence that this cytokine can play a major role in maintaining self-tolerance and in supporting survival of CD25+ CD4+ regulatory T cells (T-regs) (Bayer et al. 2005; Setoguchi et al. 2005).

IL-12 is another cytokine used in DNA vaccination. It acts on T and NK cells by inducing the generation of CTLs through T-helper 1 cell activation and IFN-γ production. The beneficial effect of IL-12 in pre-clinical experimental tumour models suggested the possibility of using IL-12 as an anti-tumour agent in clinical trials. Despite some toxicity associated with certain doses of IL-12 when administered as a drug in patients affected by melanoma and colon cancer, some clinical responses were observed; this indicates that IL-12 can be used in clinical protocols of cancer therapy where a toxic effect of the cytokine could be acceptable (Atkins et al. 1997; Gollob et al. 2003). Granulocyte/macrophage colony-stimulatory factor (GM-CSF) is probably the most attractive adjuvant for DNA vaccines for its ability to recruit antigen-presenting cells to the site where antigen synthesis occurs as well as for its capacity to stimulate DC maturation. Plasmid DNA vaccines were constructed fusing GM-CSF to the S antigen of Hepatitis B Virus (HBV) to vaccinate HBV-transgenic mice. This fusion construct worked well in conferring protection from the HBV to both normal and transgenic mice (Qing et al. 2010). In another study the utility of GM-CSF as a DNA vaccine adjuvant for glycoprotein B (gB) of pseudorabies virus (PrV) was evaluated in the vaccination of a murine model. Mouse co-inoculation with a vector expressing GM-CSF enhanced the protective immunity against PrV infection. This immunity was caused by the induction of increased humoural and cellular immunity in response to PrV antigen (Yoon et al. 2006). A DNA vaccine encoding the GM-CSF gene and a DNA vaccine encoding the H1N1 influenza (A/New Caledonia/20/99) HA antigen were co-administered by particle-mediated epidermal delivery in Rhesus Macaques. After three immunizations the DNA vaccines were shown to significantly enhance both the systemic and mucosal immunogenicity of the HA influenza vaccine (Loudon et al. 2010).

Among genetic adjuvants the pathogen-derived immune-enhancing proteins are noteworthy for their ability to stimulate the immune system when they are fused with target antigens. Modified bacterial toxins, such as anthrax, diphteria and pertussis toxins, are being used in vaccination as effective carriers to deliver foreign epitopes which stimulate protective CTL responses in mammalian cells (Ballard et al. 1996; Carbonetti et al. 1999). However, the ability of modified toxins to activate the host immune system does not reside only in the delivery effect exerted on the fused antigen (Stevenson et al. 2004).

The tetanus toxin Fragment C (FrC) is one of the widely used genetic adjuvant as a fusion partner for foreign antigens. This protein was found to increase the immunogenicity of the Schistosoma mansoni glutathione S-transferase antigen when administered as genetic fusion in a live Aro-attenuated vaccine strain of Salmonella (Khan et al. 1994b) and similar results were obtained when a vaccine construct consisting of a portion of P28 glutathione S- transferase was administering intravenously as C-terminal fusion to tetanus toxin FrC in a live Aro-attenuated vaccine strain of Salmonella (Khan et al. 1994a). In cancer vaccination, a domain of the tetanus toxin FrC fused to a single antigenic determinant was demonstrated able to induce an anti-tumoural CTL mediated response in vaccinated mice (Rice et al. 2002). Likewise, in vaccination against B cell lymphoma, DNA vaccines containing the idiotypic determinants of the Ig variable region provided protective immunity against the tumour when expressed as single-chain variable fragment (sc-Fv) fused to tetanus toxin FrC (King et al. 1998). Reproducible data from several published papers, show that the high immunogenicity of the tetanus toxin FrC depends on two main attributes; 1) a conformational sequence-dependent effect ; 2) the presence of promiscuous T-helper epitopes within the protein (Umland et al. 1997). Our group has analysed the sequence of the tetanus toxin FrC, and has identified numerous T-helper epitopes in the protein domain: 1) the universal p30 T-helper epitope (FNNFTVSFWLRVPKVSASHLE aa 947-967), a strong promiscuous immunogenic T-helper epitope consisting of at least three distinct overlapping helper peptides, each of which is presented in association with multiple HLA class II alleles [42]; 2) the p21TT helper epitope (IREDNNITLKLDRCNN aa 1064-1079); 3) the p23TT epitope (VSIDKFRIFCKALNPK aa 1084-1099), 4) the pGINGKA epitope (PGINGKAIHLVNNESSE aa 916-932) [47]; 5) the p32TT epitope (LKFIIKRYTPNNEID aa 1173-1188); 6) the pGQI epitope (GQIGNDPNRDIL aa 1273-1284 (Chiarella et al. 2007). Since CD4+ T-helper cells support both cell and humoural immunity, it seems that the antigen fusion to promiscuous T-helper peptides contributes to the activation of these lymphocytes.

The result is enhancement of the immune response mediated by T-helper cells and this might explain the strong potency of the tetanus toxin FrC domain as vaccine adjuvant. Specific T-cell epitopes of FrC that are universally immunogenic, have also been widely exploited in peptide vaccination as they have been demonstrated to enhance the humoural immune response. In particular, T-helper epitopes were successfully used as vaccine carriers to induce humoural response against polysaccharide antigens when used in form of string-of-beads (Baraldo et al. 2005). This approach is based on the concept that response to the subset of antigens and epitopes, and not to the whole organism, can be sufficient for host protection. Furthermore, the availability of bioinformatic tools and softwares for prediction of the antigen binding to the human MHC molecules helps in the design of DNA vaccines. In multi-epitope vaccination, more than one CTL epitope belonging to a certain antigen of a specific disease can be linked to a series of promiscuous MHC-II binding T-helper epitope, to generate a string-of-beads vaccine, with or without intervening spacers. In several reports, vaccination with DNA constructs consisting of a single promiscuous T-helper epitope fused to antigen determinants has proved to be effective in stimulating a strong immune response when weakly immunogenic CTL epitopes are either co-injected or chemically linked to the T-helper sequence (Tymciu et al. 2004). Synthetic universal pan HLA-DR-binding T helper epitopes such as the PADRE were conceived and they were successfully used in making DNA vaccines against infectious diseases. PADRE is a synthetic universal peptide that binds to the more common HLA-

DR molecules of the human population. Its efficacy in increasing immunogenicity of CTL and B epitopes was demonstrated to be higher (Alexander et al. 1994). The sequence of PADRE has been deduced from the core sequence of the ovalbumin master T-helper peptide (aa 323-339) and adapted for binding to the more representative human MHC class II molecules (del Guercio et al. 1997).

2.5.2 Adjuvants targeting pattern recognition receptors

Traditional vaccines based on live attenuated pathogens and inactivated whole pathogens have been extremely successful in preventing many common infectious diseases. The potent immunogenicity of such vaccines depends on the presence of "endogenous adjuvants" which are simply molecular portions of the pathogenic agent. *i.e.* now defined with the name of Pathogen Associated Molecular Pattern (PAMPs). These are lipopolysaccharide (LPS), CpG-containing oligonucleotides (CpG), and peptidoglycans. LPS is a Gram-negative membrane molecule consisting of a hydrophilic polysaccharide and a lipophilic phospholipid (lipid A). The lipophilic portion of LPS is such a potent stimulus on the pro-inflammatory cytokine production that it can lead to septic shock (Heine et al. 2001) whereas monophosphoryl lipid A (MPL) is a lipid A derivative included in many adjuvant formulations with good adjuvanticity and lower toxicity (Ismaili et al. 2002). MPL was the first TLR ligand approved for human use in the hepatitis B vaccine, Fendrix (GlaxoSmithKline Biologicals, Rixensart, Belgium) (Baldridge and Crane 1999). The adjuvanticity of these adjuvant molecules depends on their ability to bind and activate the TLR4. As a result, many TLR agonists or lipid A mimetics displaying TLR4-dependent immunostimulating functions, have been synthesized and proposed as new adjuvants (Johnson et al. 1999). AS02 is an oil-in-water emulsion containing MPL and QS-21, a saponin-derived immunostimulator, induces strong antibody and Th1 responses. AS02 is being evaluated in clinical trials in vaccines against malaria, human papillomavirus (HPV), HBV, tuberculosis, and HIV (Vandepapeliere et al. 2007). Smilarly to AS02, AS01 is a liposomal formulation containing MPL that induces potent humoural and cell-mediated responses, including cytotoxic T lymphocyte responses, and is being evaluated in clinical trials of vaccine against malaria.

In bacterial DNA there is a high frequency of unmethylated CpG dinucleotide sequences in comparison to the human genome which makes perception of plasmids as "foreign" elements by the human host. This is the only characteristic that confers a certain level of immunogenicity to plasmid DNA vaccines since the CpG unmethylated sequence is recognized by the Toll Like Receptor 9 preferentially expressed on the host APCs (Liu and Ulmer 2005). CpG-DNA is used as immunostimulatory potentiator in both pre-clinical peptide and DNA vaccination trials leading to activation of innate immunity and cytokine-dependent promotion of the Th-1 response. The CpG motifs, present in bacterial DNA, consist of an unmethylated CpG dinucleotide flanked by two 5' purines and two 3' pyrimidines. During an infection, the release of unmethylated CpG-DNA from bacteria serves as a danger signal stimulating the immune system of the host (Krieg 2002).

Bacterial DNA and synthetic unmethylated CpG oligodeoxynucleotides trigger an immunostimulatory cascade that culminates in maturation, differentiation and proliferation of several immune cells creating a pro-inflammatory and Th1-biased environment. The immunostimulatory activity of this CpG-DNA is species-specific. As a result, sequences specific for the host are designed and optimized for selective TLR9 binding. In particular, *in vivo* CpG-DNA half-life has been improved by replacing phosphodiester CpG-DNA with a nuclease-resistant phosphothioate oligodeoxynucleotide although it may induce immune reaction leading to an anti-DNA immune response (Ciafre et al. 1995). Addition of the immunostimulatory CpG-DNA to peptide vaccines, tilts the balance towards a Th-1 immune response, which is often accompanied by a significant increase in IgG2a production in comparison to IgG1. The general effect of CpG-DNA addition is augmentation of the antibody serum titer against antigens and production of Th-1 cytokines such as

IFN-γ (Klinman et al. 1999). By contrast, in DNA plasmid vaccination, the immunomodulating effect of bacterial CpG oligodeoxynucleotides on T-helper cell balance, shows great variability. Such variability often depends on the route of immunisation.

2.5.3 Aluminium-based compounds

Aluminium salt adjuvants (aluminium hydroxide, aluminium phosphate) are generally used in combination with protein antigens as they form a precipitated or adsorbed vaccine. An efficient, safe and well tolerated adjuvant in humans is aluminium hydroxide (Alum). It has been approved for clinical use. Although traditionally thought to function primarily by forming a long-lasting depot for antigen and by promoting their uptake by APCs, it is now clear that innate immune stimulation plays a primary role in the adjuvant activity of alum (Lambrecht et al. 2009). Aluminium hydroxide is used primarily to enhance antibody production and does not utilize TLR for its function in vivo (Gavin et al. 2006).This adjuvant induces a Th-2 biased immune response in mice whereas in humans it stimulates also a Th-1 type immunity. *In vitro* studies demonstrated that Alum can activate the inflammosome pathway to produce IL1-β (Li et al. 2007).

Recently, aluminium hydroxide has been also shown to work well in DNA vaccination in pre-clinical models (Kenney and Edelman 2003). However, the enhancing effect of aluminium hydroxide on the immune response elicited by DNA vaccines, is not related to the levels of antigen expression. Rather, it seems to affect antigen after *in vivo* expression, suggesting the adjuvanticity of this substance is strictly related to the antigen delivery mechanism (Ulmer et al. 1999).

2.6 Delivery methods

Different methods for enhancing naked DNA vaccine delivery into host cells have been studied. The best investigated strategies are based on chemical or physical devices aimed at facilitating the DNA entry into recipient cells.

2.6.1 Microparticles

Microparticle-based methods operate a DNA condensation and complexation in particles (O'Hagan et al. 2004). The encapsulation of plasmid DNA into micro- or nanospheres can provide protection from the environment prior to delivery and aid in targeting to a specific cell type for efficient delivery.

The major advantage of particulate delivery is that synthetic microparticles have excellent potential for targeting cells of the immune system stimulating antigen uptake. It has been demonstrated that particles of about 1-10 μm in diameter are preferred for their size that is readily phagocytosed by dendritic cells and other antigen-presenting cells. They are readily internalised by phagocytic cells of the immune system, leading to an enhanced antigen presentation to the immune effector cells. Furthermore, microparticulates appear to improve delivery of DNA to APCs by facilitating trafficking to the local lymphoid tissue via the afferent lymph and antigen uptake by dendritic cells (Denis-Mize et al. 2000; Denis-Mize et al. 2003; Dupuis et al. 2000). Moreover, antigen and adjuvant molecules can be delivered to the same cell at the same time being entrapped together in biodegradable microparticles such as poly-lactide-co-glycolide (PLG) or chitosan, or complexed with non-ionic block copolymers or polycations such as polyethyleneimine. Microparticulate adjuvants are currently tested in some clinical trials against human immunodeficiency virus (HIV), hepatitis B virus (HBV) and influenza (Fuller et al. 2006). DNA entrapment or encapsulation into biodegradable microspheres for DNA vaccine delivery has been illustrated in patent WO0203961 (Johnson 2003).

2.6.2 Cationic lipids/liposomes

Cationic Solid Lipid Nanoparticles (SLNs) have been recently proposed as alternative carriers for DNA delivery, due to many technological advantages such as large-scale production from substances generally recognized as safe, good storage stability and possibility of steam sterilization and lyophilisation. Cationic lipids are amphiphilic molecules composed of one or two fatty acid side chains (acyl) or alkyl, a linker and a hydrophilic amino group. The hydrophobic part can be cholesterol-derived moieties. In aqueous media, cationic lipids are assembled into a bilayer vesicular-like structure (liposomes). Liposomes/DNA complex is usually termed a lipoplex (Bolhassani et al. 2011). The future success of cationic SLNs for administration of genetic material will depend on their ability to efficiently cross the physiological barriers, selectively targeting a specific cell type in vivo and expressing therapeutic genes (Bondi and Craparo 2010).

2.6.3 Biolistic particle delivery

In order to address accelerating micro-projectiles into intact cells or tissues, is generally used a biolistic apparatus described in patent US6004287 (Loomis 1999). Application of this strategy to DNA vaccines resulted in the invention of a new DNA delivery technology that made it possible to move naked DNA plasmid into target cells on an accelerated particle carrier. This specific delivery system is based on the use of the gene gun device that, under pressurized helium, is capable of delivering plasmid DNA-coated gold beads to the epidermal layer of skin as described in patent US6436709 (Lin 2002). Because the DNA carrier is introduced directly into the skin cells, delivery of plasmid DNA vaccines using this strategy reduces the amount of DNA needed to induce immune responses. Robust immunogenicity has been shown in many different preclinical models and in clinical trials predominantly for infectious diseases (Fuller et al. 2006). In contrast to intramuscular or intradermal injection by needle, the gene gun delivery system releases plasmid DNA directly into the cells of the epidermis (Yang et al. 1990).

Intradermal injection is becoming increasingly popular, as the dense network of antigen-presenting cells in the skin, absent in muscle, provides a favourable environment for induction of antigen uptake. This network of Langerhans cells (LCs) can help in the priming of both cellular and humoural immune responses. Importantly, direct transfection of Langerhans cells is carried out with very small doses of plasmid DNA (i.e. 1-10 μg), suggesting that minimum amounts of vector are required to induce the immune response. The advantage of using low doses of plasmid DNA is particularly attractive for prophylactic vaccines against infectious diseases, where a simple and rapid delivery is the main pre-requisite. Gene gun delivery has recently been used with success in a trial against the influenza virus, inducing sero-protective levels of antibody and it has been used in trials against HBV and HIV infections (Fuller et al. 2006). A further implementation of the biolistic delivery was obtained also by creating improved injection device suitable for application in human tissues. Patent US6730663 describes a flexible multi-needle injector device with a wide surface area as well as a modified injector device to be used for injection through an endoscopic device. Such a method leads to a deep injection of DNA within tissues (Hennighausen 2004).

2.6.4 Electropermeabilization

The use of electric pulses as a safe tool to deliver therapeutic molecules to tissues and organs has been rapidly developed over the last decade. This technology leads to a transient increase in the permeability of cell membranes when exposed to electric field pulses. This process is commonly known as electropermeabilization or electroporation (EP) (Chiarella et al 2010; Favard et al. 2007; Mir et al. 1999). The simultaneous publications in 1998 by Aihara and Miyazaki and Harrison and co-workers, demonstrated EP as being a more efficient method for gene transfer into muscle than the simple i.m. injection of DNA (Aihara and Miyazaki 1998; Harrison et al. 1998). The strategy is not only promising for enhancing the gene delivery of therapeutic proteins and drugs. Infectious disease, cancer gene therapy and chemotherapy are other fields of application, making electrochemogenetherapy relevant in a variety of research branches and promising in the gene therapy field (Mir 2008; Wells 2004). The advantage of DNA electrotransfer is dual. On the one hand, a high number of muscle cells are transfected with the DNA vaccine; on the other hand the damaged muscle cells release danger signals that favour antigen presenting cell recruitment, thus enhancing the immune response (Chiarella et al. 2008b). For this reason we consider the electropermeabilization such an important and very promising tool in the future of DNA vaccination therapy that it deserves a dedicated section of this book chapter.

3. ELECTROGENE TRANSFER

3.1 Mechanisms and application of gene transfer by electric fields

Naked DNA vaccination emerged as a promising approach for introducing foreign antigens into the host to induce protective immunity. The delivery of DNA vaccine into skeletal muscle is highly preferable as this organ is not only a passive site but can actively take part in immune reactions. However, one important limitation of intramuscular (i.m.) genetic vaccines is their weak performance in large animals as regards the low DNA transfection efficiency of the tissue (Escoffre et al. 2010). For this reason, novel and safe delivery systems have been developed to further improve the vaccine efficiency and immunogenicity. In several reports, electroporation-mediated DNA delivery was described as an effective tool in eliciting immune response in small and large animal models (Babiuk et al. 2002; Otten et al. 2004; Peng et al. 2007), with numerous studies proving that this technique is effective in the stimulation of humoural and cellular immunity (Dupuis et al. 2000; Tollefsen et al. 2002; Widera et al. 2000). Interest in the application of EP to DNA vaccination protocols is greatly growing in these last years for several considerations. It has been demonstrated that EP allows an augmented uptake of DNA in tissue cells especially if used in combination with hyaluronidase (McMahon et al. 2001). A higher DNA uptake *in vivo* is possible thanks to the enhancement of cell membrane permeabilization and electrophoretic movement of DNA molecules into the target cells. Moreover, if EP is applied in muscle cells, these work as a platform for antigen production within the skeletal muscle (Shirota et al. 2007). A combination of both these events facilitates target cell transfection, this resulting in a higher synthesis of the gene of interest and in an intensification of the immune response to the encoded protein.

Many studies have reported the beneficial effect of EP on the activated response by the immune system against the transferred antigen in several animal models (Tsang et al. 2007). Respect to a simple administration of DNA vaccines through i.m. injection, EP is responsible for a significant increase in antibody titre (Buchan et al. 2005), antigen-specific T-cell frequency and induction of several T-cell effector functions (Bachy et al. 2001; Capone et al. 2006). In particular, a study performed on DNA vaccination mediated by EP demonstrated that the concomitant injection of plasmid DNA and EP is crucial for the adjuvant effect exerted by EP, which is responsible for eliciting antigen-specific IgG2a antibody production and Th-1 biased immune responses (Gronevik et al. 2005). Another study demonstrated that a single i.m. DNA vaccination in combination with EP enhanced significantly the onset and the duration of the primary antibody response affecting immune memory (Tsang et al. 2007). The use of EP for delivering a DNA vaccine encoding anthrax toxin protective agent has been demonstrated able in a rapid induction of antibodies against the antigen in 2 weeks following a single immunization in several experimental animals (Luxembourg et al. 2008). Despite these evidences, details on possible mechanisms responsible for the positive effect of EP on the immune response to DNA vaccines were not completely characterised (Escoffre et al. 2009).

Recently EP has been reported as crucial event through which, inducing transient morphological changes and a local moderate damage in the treated muscle, is possible to generate an early production of endogenous cytokines responsible for signalling danger at the local level. The activation of a danger pro-inflammatory pathway and the recruitment of inflammatory cells result in T lymphocyte migration, indicating electropermeabilization *per se* is able to recruit and trigger cells involved in antigen presentation (Chiarella et al. 2008b). Due to these immunostimulating effects, EP is now recognised as a good adjuvant (Chiarella et al. 2007), helpful in DNA vaccination for increasing the potency and safety of this therapeutic approach due to its property to induce a higher DNA uptake and its ability to stimulate both humoural and cellular immunity. At present, numerous findings are clarifying EP mechanism (Golzio et al. 2010), also showing the easy applicability of EP to large animals. In this view, many studies are concentrated to find the most appropriate and tolerable parameters that will make EP suitable for humans (Tjelle et al. 2008; Tjelle et al. 2006). To this purpose, various electroporating devices have been developed for animal and human use (**Fig. 2**).

Because these initial results seem promising, several clinical trials based on DNA vaccination assisted by electropermeabilization are under investigation and the efficacy and tolerability of EP will be deeply studied in the near future both in preclinical and clinical gene therapy protocols.

3.2 Application of electrogene transfer in DNA vaccination protocols

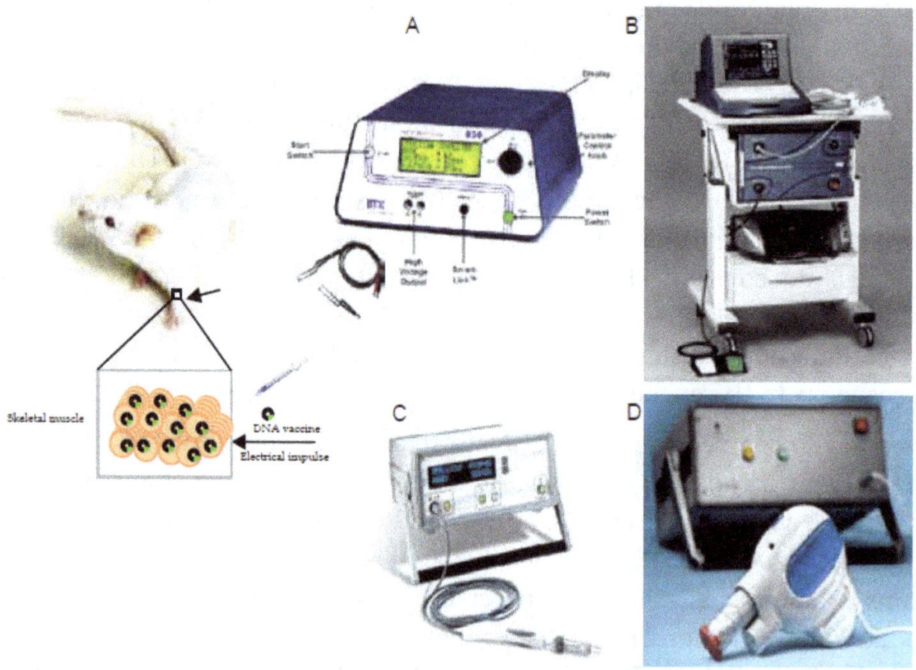

Fig. 2. Administration of DNA vaccine by electropermeabilization of the muscle

The use of EP technology for DNA vaccination in the clinical settings takes advantage from the development of new protocols, which are effective in administrating the vaccine with the minimum discomfort and maximum tolerability for the patient.

The type of EP device, intensity of electrical stimulations, number of electrical pulses and administrations, and choice of the target organ, are important parameters taken into consideration for the design of clinical protocols.

ecause the need to develop non-invasive or minimally invasive genetic vaccination methods has become an important issue, several studies have been focussed on the needle-free injection of DNA vaccines. A recent patent, issued in 2007, reports the combination of needle-free injection and electroporation, demonstrating that this non- invasive strategy is sufficient to introduce the DNA vaccine in a form suitable for electrotransfer into a region of the host tissue. This needle-free injection may be used in combination with suitable non invasive electrode configurations (Hofmann 2007). Safe EP protocols for human disease therapies are under investigation and many preclinical studies are described in numerous works investigating the potential of EP in both infectious and tumour diseases.

The DNA vaccine is injected into the skeletal muscle of the mouse limb and penetration into the target tissue is achieved by electropermeabilization of the muscle after DNA is injected with a syringe. The electrical impulse is applied by electrodes in contact with the skeletal muscle. Different electroporator devices are shown in the figure. A) BTX ECM 830 Harvard Apparatus; B) IGEA Cliniporator; C) Inovio MedPulser® EPT; D) Ichor TriGridTM.

HIV vaccine administration has been conducted in several animal models. Electrically- mediated delivery technology has been applied to DNA vaccines against HIV virus, and substantially higher immune responses have been achieved in mice and rabbits following vaccination with DNA encoding HIV genes. Vaccines were administered with constant electric current or constant electric voltage, causing up to 20-fold higher immune responses in comparison to the application of DNA vaccines alone (Selby 2000). In another study in mice, *in vivo* EP amplified cellular and humoural immune responses to a HIV type 1 Env DNA vaccine, enabled a 10- fold reduction in vaccine dose, and resulted in increased recruitment of inflammatory cells (Liu et al. 2008). Another study on the development of plasmid DNA vaccine able in eliciting robust cell-mediated immune response to multiple HIV type 1 (HIV-1)-derived antigens has been conducted in Rhesus macaques. Vaccination in combination with *in vivo* EP led to a more rapid onset and enhanced vaccine-specific immune responses (Luckay et al. 2007).

Also the hepatitis C virus disease is object of investigation. It was demonstrated that gene electrotransfer of a novel candidate DNA vaccine encoding an optimised version of the non structural region of HCV (from NS3 to NS5B) induces substantially more potent, broad, and long-lasting CD4+ and CD8+ cellular immunity than a simple naked DNA injection in mice and in Rhesus macaques. As already discussed, the T-cell responses elicited by the DNA- based electroporation strategy can be useful in prophylactic vaccine approaches against HCV and this work supports this hypothesis (Capone et al. 2006). Because the administration of a plasmid cocktail, encoding antigen and adjuvants in combination with EP, is proposed as an efficient genetic immunisation strategy, the same group designed a protocol in which Hepatitis C virus (HCV) E2 and cytokine encoding plasmids have been co-injected in the mouse quadriceps with or without EP. The vaccination outcome has been evaluated by analysis of antigen-specific cellular- mediated or antibody-mediated immunity. The co-injection of cytokine and HCV E2-encoding plasmids followed by EP, strongly enhanced T- or B-cell responses to various levels, depending on the particular combination used (Arcuri et al. 2008).

Respect to cancer, strong cellular immune responses can be induced in both mice and non- human primates, following the administration with EP of a novel HPV18 DNA vaccine encoding an E6/E7 fusion consensus protein (Yan et al. 2008).

Improvement in the efficacy of a cancer vaccine administered by electroporation, could increase its chances for clinical success. A demonstration of the inhibition of tumour growth has been reported by Curcio and collaborators. They demonstrated that a vaccination protocol using a plasmid encoding the extracellular and transmembrane domains of the Neu oncogene delivered by electroporation, prevents longterm tumour formation in cancer- prone transgenic mice (Curcio et al. 2008).

Since electropermeabilization is considered a promising delivery system for plasmid DNA vaccination, several clinical trials are now experimenting EP as a medical technology in human patients affected by infectious as well as cancer diseases (Bodles-Brakhop et al. 2009).

A summary of the clinical trials performed by DNA vaccination and EP are shown in the following tables.

Table 3. Clinical trials in infectious diseases.

Clinical trial	Condition	Pathogen	Intervention	Phase
NCT01169077	Malaria	Plasmodium falciparum	EP-1300	I
NCT00685412	Human Papillomavirus infection	HPV (E6-E7)	VGX-3100 CELLECTRA	I
NCT00563173	Chronic Hepatitis C Virus Infection	HCV	CHRONVAC-C®	I/IIa
NCT00545987	AIDS (prophylactic)	HIV	ADVAX TriGrid™	I
NCT01082692	AIDS (therapeutic)	HIV-1	PENNVAX-B CELLECTRA	I

Table 4. Clinical trials in cancer diseases.

Clinical trial	Condition	Antigen	Intervention	Phase
NCT01064375	Colorectal Cancer	CEA	tetwtCEA DNA wt CEA with tetanus toxoid Th epitope Derma Vax (electroporation device)	I/II
NCT00471133	Intraocular Melanoma	Tyrosinase	TriGrid	I
NCT01138410	Melanoma	Antibody (SCIB-1)	EP device	I/II
NCT00859729	Prostate cancer	PSA	pVAXrcPSAv53l DERMA VAX™	I/II
GTAC No 89	Prostate cancer	PSMA	EP device (Tjelle 2006)	I/II (Closed 1.4.2008)
NCT00753415	Colon cancer Breast Cancer Melanoma	V934/V935 hTERT	V934/V935 DNA	I

The results seem to be promising and applicable to a large cohort of diseases in the next future. Therefore electroporation would appear to be the more efficient technology for local injection of plasmid DNA vaccine into the tissue (Kato and Nakamua 1965; Wells 2010).

4. CONCISE SUMMARY

Over the past ten years, there has been a significant transformation in the methodology and progress of vaccine design and development. Although there have been significant achievements in preventing infections, there is currently no availability of vaccinations against intracellular pathogens that require cell-mediated immunity. This is concerning because infectious diseases like tuberculosis and malaria continue to pose a major threat in developing countries. In recent years, the research conducted by Wolff and colleagues has demonstrated the potential benefits of immunization using naked plasmid DNA that encodes antigens. This strategy, known as DNA vaccination, shows promise as a therapeutic method for combating infectious diseases and cancer. Undoubtedly, enhancing the effectiveness of vaccines has been a primary objective in the advancement of DNA vaccination techniques. Electropermeabilization has demonstrated the ability to enhance both the quantity of transfected cells and the amount of plasmids that enter each cell. As a result, electropermeabilization is now considered a potential method for delivering plasmid DNA vaccines. The combination of intramuscular DNA vaccination with electropermeabilization has been found to effectively activate both the humoral and cellular immune response in the host, while also increasing the expression of the encoded antigen. Multiple studies have demonstrated that EP exhibits adjuvant-like characteristics when used in conjunction with plasmid DNA administration. This method is currently being utilized in both preclinical animal protocols and human investigations. Investigations are currently underway to evaluate the pain and stress caused by the treatment, which suggests that this strategy is applicable and promising. Due to the absence of any adverse effects associated with the delivery approach, we highly endorse efforts to enhance the effectiveness of DNA vaccines provided using electropermeabilization in clinical procedures. This novel methodology has the potential to significantly enhance the likelihood of achieving positive outcomes in clinical trials involving human subjects.

REFERENCES

Abdulhaqq, S. A., and Weiner, D. B. (2008). "DNA vaccines: developing new strategies to enhance immune responses." *Immunol Res*, 42(1-3), 219-32.

Aihara, H., and Miyazaki, J. (1998). "Gene transfer into muscle by electroporation in vivo."*Nat Biotechnol*, 16(9), 867-70.

Alexander, J., Sidney, J., Southwood, S., Ruppert, J., Oseroff, C., Maewal, A., Snoke, K., Serra, H. M., Kubo, R. T., Sette, A., and et al. (1994). "Development of high potency universal DR-restricted helper epitopes by modification of high affinity DR- blocking peptides." *Immunity*, 1(9), 751-61.

Andre, F. E. (2003). "Vaccinology: past achievements, present roadblocks and future promises." *Vaccine*, 21(7-8), 593-5.

Arcuri, M., Cappelletti, M., Zampaglione, I., Aurisicchio, L., Nicosia, A., Ciliberto, G., and Fattori, E. (2008). "Synergistic effect of gene-electro transfer and adjuvant cytokines in increasing the potency of hepatitis C virus genetic vaccination." *J Gene Med*, 10(9), 1048-54.

Atkins, M. B., Robertson, M. J., Gordon, M., Lotze, M. T., DeCoste, M., DuBois, J. S., Ritz, J., Sandler, A. B., Edington, H. D., Garzone, P. D., Mier, J. W., Canning, C. M., Battiato, L., Tahara, H., and Sherman, M. L. (1997). "Phase I evaluation of intravenous recombinant human interleukin 12 in patients with advanced malignancies." *Clin Cancer Res*, 3(3), 409-17.

Babiuk, S., Baca-Estrada, M. E., Foldvari, M., Storms, M., Rabussay, D., Widera, G., and Babiuk, L. A. (2002). "Electroporation improves the efficacy of DNA vaccines in large animals." *Vaccine*, 20(27-28), 3399-3408.

Bachy, M., Boudet, F., Bureau, M., Girerd-Chambaz, Y., Wils, P., Scherman, D., and Meric, C. (2001). "Electric pulses increase the immunogenicity of an influenza DNA vaccine injected intramuscularly in the mouse." *Vaccine*, 19(13-14), 1688-93.

Baldridge, J. R., and Crane, R. T. (1999). "Monophosphoryl lipid A (MPL) formulations for the next generation of vaccines." *Methods*, 19(1), 103-7.

Ballard, J. D., Collier, R. J., and Starnbach, M. N. (1996). "Anthrax toxin-mediated delivery of a cytotoxic T-cell epitope in vivo." *Proc Natl Acad Sci U S A*, 93(22), 12531-4.

Baraldo, K., Mori, E., Bartoloni, A., Norelli, F., Grandi, G., Rappuoli, R., Finco, O., and Del Giudice, G. (2005). "Combined conjugate vaccines: enhanced immunogenicity with the N19 polyepitope as a carrier protein." *Infect Immun*, 73(9), 5835-41.

Barouch, D. H., Craiu, A., Kuroda, M. J., Schmitz, J. E., Zheng, X. X., Santra, S., Frost, J. D., Krivulka, G. R., Lifton, M. A., Crabbs, C. L., Heidecker, G., Perry, H. C., Davies, M. E., Xie, H., Nickerson, C. E., Steenbeke, T. D., Lord, C. I., Montefiori, D. C., Strom, T. B., Shiver, J. W., Lewis, M. G., and Letvin, N. L. (2000). "Augmentation of immune responses to HIV-1 and simian immunodeficiency virus DNA vaccines by IL-2/Ig plasmid administration in rhesus monkeys." *Proc Natl Acad Sci U S A*, 97(8), 4192-7.

Bayer, A. L., Yu, A., Adeegbe, D., and Malek, T. R. (2005). "Essential role for interleukin-2 for CD4(+)CD25(+) T regulatory cell development during the neonatal period." *J Exp Med*, 201(5), 769-77.

Bodles-Brakhop, A. M., Heller, R., and Draghia-Akli, R. (2009). "Electroporation for the delivery of DNA-based vaccines and immunotherapeutics: current clinical developments." *Mol Ther*, 17(4), 585-92.

Bojak, A., Wild, J., Deml, L., and Wagner, R. (2002). "Impact of codon usage modification on T cell immunogenicity and longevity of HIV-1 gag-specific DNA vaccines." *Intervirology*, 45(4-6), 275-86.

Bolhassani, A., Safaiyan, S., and Rafati, S. (2011). "Improvement of different vaccine delivery systems for cancer therapy." *Mol Cancer*, 10, 3.

Bondi, M. L., and Craparo, E. F. (2010). "Solid lipid nanoparticles for applications in gene therapy: a review of the state of the art." *Expert Opin Drug Deliv*, 7(1), 7-18.

Brun, A., Barcena, J., Blanco, E., Borrego, B., Dory, D., Escribano, J. M., Le Gall-Recule, G., Ortego, J., and Dixon, L. K. (2011). "Current strategies for subunit and genetic viral veterinary vaccine development." *Virus Res*, 157(1), 1-12.

Buchan, S., Gronevik, E., Mathiesen, I., King, C. A., Stevenson, F. K., and Rice, J. (2005). "Electroporation as a "prime/boost" strategy for naked DNA vaccination against a tumor antigen." *J Immunol*, 174(10), 6292-8.

Capone, S., Zampaglione, I., Vitelli, A., Pezzanera, M., Kierstead, L., Burns, J., Ruggeri, L., Arcuri, M., Cappelletti, M., Meola, A., Ercole, B. B., Tafi, R., Santini, C., Luzzago, A., Fu, T. M., Colloca, S., Ciliberto, G., Cortese, R., Nicosia, A., Fattori, E., and Folgori, A. (2006). "Modulation of the immune response induced by gene electrotransfer of a hepatitis C virus DNA vaccine in nonhuman primates." *J Immunol*, 177(10), 7462- 71.

Carbonetti, N. H., Irish, T. J., Chen, C. H., O'Connell, C. B., Hadley, G. A., McNamara, U., Tuskan, R. G., and Lewis, G. K. (1999). "Intracellular delivery of a cytolytic T- lymphocyte epitope peptide by pertussis toxin to major histocompatibility complex class I without involvement of the cytosolic class I antigen processing pathway." *Infect Immun*, 67(2), 602-7.

Chiarella, P., Fazio, V.M., and Signori, E. (2010). "Application of electroporation in DNA vaccination protocols." *Curr Gene Ther*,10 (4), 281-6.

Chiarella, P., Massi, E., De Robertis, M., Fazio, V. M., and Signori, E. (2008a). "Strategies for effective naked-DNA vaccination against infectious diseases." *Recent Pat Antiinfect Drug Discov*, 3(2), 93-101.

Chiarella, P., Massi, E., De Robertis, M., Sibilio, A., Parrella, P., Fazio, V. M., and Signori, E. (2008b). "Electroporation of skeletal muscle induces danger signal release and antigen-presenting cell recruitment independently of DNA vaccine administration." *Expert Opin Biol Ther*, 8(11), 1645-57.

Chiarella, P., Massi, E., De Robertis, M., Signori, E., and Fazio, V. M. (2007). "Adjuvants in vaccines and for immunisation: current trends." *Expert Opin Biol Ther*, 7(10), 1551- 62.

Chu, C., Parker, R., Highlands, L., and Morales J. (2010). "Phase I of Human Papillomavirus (HPV) DNA Plasmid (VGX-3100) + Electroporation for CIN 2 or 3." *Date of access: 04/04/2011. Available from: Clinical Trials.gov.*

Ciafre, S. A., Rinaldi, M., Gasparini, P., Seripa, D., Bisceglia, L., Zelante, L., Farace, M. G., and Fazio, V. M. (1995). "Stability and functional effectiveness of phosphorothioate modified duplex DNA and synthetic 'mini-genes'." *Nucleic Acids Res*, 23(20), 4134- 42.

Ciafre, S. A., Rinaldi, M., Vespignani, I., Parrella, P., Seripa, D., Signori, E., Ria, F., Farace, M. G., and Fazio, V. M. (1998). "A plasmid family containing two different expression cassettes suitable for immunomodulation and genetic immunization." *Plasmid*, 40(1), 84-9.

Curcio, C., Khan, A. S., Amici, A., Spadaro, M., Quaglino, E., Cavallo, F., Forni, G., and Draghia-Akli, R. (2008). "DNA immunization using constant-current electroporation affords long-term protection from autochthonous mammary carcinomas in cancer-prone transgenic mice." *Cancer Gene Ther*, 15(2), 108-14.

del Guercio, M. F., Alexander, J., Kubo, R. T., Arrhenius, T., Maewal, A., Appella, E., Hoffman, S. L., Jones, T., Valmori, D., Sakaguchi, K., Grey, H. M., and Sette, A. (1997). "Potent immunogenic short linear peptide constructs composed of B cell epitopes and Pan DR T helper epitopes (PADRE) for antibody responses in vivo." *Vaccine*, 15(4), 441-8.

Denis-Mize, K. S., Dupuis, M., MacKichan, M. L., Singh, M., Doe, B., O'Hagan, D., Ulmer, J. B., Donnelly, J. J., McDonald, D. M., and Ott, G. (2000). "Plasmid DNA adsorbed onto cationic microparticles mediates target gene expression and antigen presentation by dendritic cells." *Gene Ther*, 7(24), 2105-12.

Denis-Mize, K. S., Dupuis, M., Singh, M., Woo, C., Ugozzoli, M., O'Hagan, D. T., Donnelly, J. J., 3rd, Ott, G., and McDonald, D. M. (2003). "Mechanisms of increased immunogenicity for DNA-based vaccines adsorbed onto cationic microparticles." *Cell Immunol*, 225(1), 12-20.

Dong, Z., and Nor, J. E. (2009). "Transcriptional targeting of tumor endothelial cells for gene therapy." *Adv Drug Deliv Rev*, 61(7-8), 542-53.

Dudek, N.L., Perlmutter, P., Aguilar, M.I., Croft, N.P., and Purcell, A.W. (2010). "Epitope discovery and their use in peptide based vaccines." Curr Pharm Des, 16(28), 3149-57.

Dupuis, M., Denis-Mize, K., Woo, C., Goldbeck, C., Selby, M. J., Chen, M., Otten, G. R., Ulmer, J. B., Donnelly, J. J., Ott, G., and McDonald, D. M. (2000). "Distribution of DNA vaccines determines their immunogenicity after intramuscular injection in mice." *J Immunol*, 165(5), 2850-8.

Dyall, R., Bowne, W. B., Weber, L. W., LeMaoult, J., Szabo, P., Moroi, Y., Piskun, G., Lewis, J. J., Houghton, A. N., and Nikolic-Zugic, J. (1998). "Heteroclitic immunization induces tumor immunity." *J Exp Med*, 188(9), 1553-61.

Edelman, G. M., Meech, R., Owens, G. C., and Jones, F. S. (2000). "Synthetic promoter elements obtained by nucleotide sequence variation and selection for activity." *Proc Natl Acad Sci U S A*, 97(7), 3038-43.

Ertl, P. F. (2007). *Patent No. US2007248679.*

Escoffre, J. M., Portet, T., Wasungu, L., Teissie, J., Dean, D., and Rols, M. P. (2009). "What is (still not) known of the mechanism by which electroporation mediates gene transfer and expression in cells and tissues." *Mol Biotechnol*, 41(3), 286-95.

Escoffre, J. M., Teissie, J., and Rols, M. P. (2010). "Gene transfer: how can the biological barriers be overcome?" *J Membr Biol*, 236(1), 61-74.

Favard, C., Dean, D. S., and Rols, M. P. (2007). "Electrotransfer as a non viral method of gene delivery." *Curr Gene Ther*, 7(1), 67-77.

Fraser, C. M., and Rappuoli, R. (2005). "Application of microbial genomic science to advanced therapeutics." *Annu Rev Med*, 56, 459-74.

Fuller, D. H., Loudon, P., and Schmaljohn, C. (2006). "Preclinical and clinical progress of particle-mediated DNA vaccines for infectious diseases." *Methods*, 40(1), 86-97.

Gavin, A. L., Hoebe, K., Duong, B., Ota, T., Martin, C., Beutler, B., and Nemazee, D. (2006). "Adjuvant-enhanced antibody responses in the absence of toll-like receptor signaling." *Science*, 314(5807), 1936-8.

Gollob, J. A., Veenstra, K. G., Parker, R. A., Mier, J. W., McDermott, D. F., Clancy, D., Tutin, L., Koon, H., and Atkins, M. B. (2003). "Phase I trial of concurrent twice-weekly recombinant human interleukin-12 plus low-dose IL-2 in patients with melanoma or renal cell carcinoma." *J Clin Oncol*, 21(13), 2564-73.

Golzio, M., Escoffre, J. M., Portet, T., Mauroy, C., Teissie, J., Dean, D. S., and Rols, M. P. (2010) "Observations of the mechanisms of electromediated DNA uptake--from vesicles to tissues." *Curr Gene Ther*, 10(4), 256-66.

Good, M. F., Pombo, D., Lunde, M. N., Maloy, W. L., Halenbeck, R., Koths, K., Miller, L. H., and Berzofsky, J. A. (1988). "Recombinant human IL-2 overcomes genetic nonresponsiveness to malaria sporozoite peptides. Correlation of effect with biologic activity of IL-2." *J Immunol*, 141(3), 972-7.

Gothelf, A., Mir, L. M., and Gehl, J. (2003). "Electrochemotherapy: results of cancer treatment using enhanced delivery of bleomycin by electroporation." *Cancer Treat Rev*, 29(5), 371-87.

Gronevik, E., Mathiesen, I., and Lomo, T. (2005). "Early events of electroporation-mediated intramuscular DNA vaccination potentiate Th1-directed immune responses." *J Gene Med*, 7(9), 1246-54.

Harrington, L. E., Most Rv, R., Whitton, J. L., and Ahmed, R. (2002). "Recombinant vaccinia virus-induced T-cell immunity: quantitation of the response to the virus vector and the foreign epitope." *J Virol*, 76(7), 3329-37.

Harrison, R. L., Byrne, B. J., and Tung, L. (1998). "Electroporation-mediated gene transfer in cardiac tissue." *FEBS Lett*, 435(1), 1-5.

Heine, H., Rietschel, E. T., and Ulmer, A. J. (2001). "The biology of endotoxin." *Mol Biotechnol*, 19(3), 279-96.

Hennighausen, L., and Furth, P.A. (2004). "Targeting gene expression to living tissue using jet injection." *Patent No. US 6730663*.

Hofmann, G. A., Rabussay, D.P., and Zhang L. (2007). "Intradermal delivery of active agents by needle-free injection and electroporation." *Patent No. US 7171264*.

Ismaili, J., Rennesson, J., Aksoy, E., Vekemans, J., Vincart, B., Amraoui, Z., Van Laethem, F., Goldman, M., and Dubois, P. M. (2002). "Monophosphoryl lipid A activates both human dendritic cells and T cells." *J Immunol*, 168(2), 926-32.

Ivory, C., and Chadee, K. (2004). "DNA vaccines: designing strategies against parasitic infections." *Genet Vaccines Ther*, 2(1), 17.

Jager, D., Taverna, C., Zippelius, A., and Knuth, A. (2004). "Identification of tumor antigens as potential target antigens for immunotherapy by serological expression cloning." *Cancer Immunol Immunother*, 53(3), 144-7.

Janeway, C. A., Jr. (1989). "Approaching the asymptote? Evolution and revolution in immunology." *Cold Spring Harb Symp Quant Biol*, 54 Pt 1, 1-13.

Johnson, D. A., Keegan, D. S., Sowell, C. G., Livesay, M. T., Johnson, C. L., Taubner, L. M., Harris, A., Myers, K. R., Thompson, J. D., Gustafson, G. L., Rhodes, M. J., Ulrich, J. T., Ward, J. R., Yorgensen, Y. M., Cantrell, J. L., and Brookshire, V. G. (1999). "3-O- Desacyl monophosphoryl lipid A derivatives: synthesis and immunostimulant activities." *J Med Chem*, 42(22), 4640-9.

Johnson, M. E., Mossman, S., Evans, L., and Cecil T. (2003). "Microspheres and adjuvants for DNA vaccine delivery." *Patent No. WO 0203961*.

Kenney, R. T., and Edelman, R. (2003). "Survey of human-use adjuvants." *Expert Rev Vaccines*, 2(2), 167-88.

Khan, C. M., Villarreal-Ramos, B., Pierce, R. J., Demarco de Hormaeche, R., McNeill, H., Ali, T., Chatfield, S., Capron, A., Dougan, G., and Hormaeche, C. E. (1994a). "Construction, expression, and immunogenicity of multiple tandem copies of the Schistosoma mansoni peptide 115-131 of the P28 glutathione S-transferase expressed as C-terminal fusions to tetanus toxin fragment C in a live aro-attenuated vaccine strain of Salmonella." *J Immunol*, 153(12), 5634-42.

Khan, C. M., Villarreal-Ramos, B., Pierce, R. J., Riveau, G., Demarco de Hormaeche, R., McNeill, H., Ali, T., Fairweather, N., Chatfield, S., Capron, A., and et al. (1994b). "Construction, expression, and immunogenicity of the Schistosoma mansoni P28 glutathione S-transferase as a genetic fusion to tetanus toxin fragment C in a live Aro attenuated vaccine strain of Salmonella." *Proc Natl Acad Sci U S A*, 91(23), 11261-5.

King, C. A., Spellerberg, M. B., Zhu, D., Rice, J., Sahota, S. S., Thompsett, A. R., Hamblin, T. J., Radl, J., and Stevenson, F. K. (1998). "DNA vaccines with single-chain Fv fused to fragment C of tetanus toxin induce protective immunity against lymphoma and myeloma." *Nat Med*, 4(11), 1281-6.

Klinman, D. M., Barnhart, K. M., and Conover, J. (1999). "CpG motifs as immune adjuvants." *Vaccine*, 17(1), 19-25.

Ko, H. J., Ko, S. Y., Kim, Y. J., Lee, E. G., Cho, S. N., and Kang, C. Y. (2005). "Optimization of codon usage enhances the immunogenicity of a DNA vaccine encoding mycobacterial antigen Ag85B." *Infect Immun*, 73(9), 5666-74.

Krieg, A. M. (2002). "CpG motifs in bacterial DNA and their immune effects." *Annu Rev Immunol*, 20, 709-60.

Kurts, C., Robinson, B. W., and Knolle, P. A. (2010). "Cross-priming in health and disease." *Nat Rev Immunol*, 10(6), 403-14.

Lambrecht, B. N., Kool, M., Willart, M. A., and Hammad, H. (2009). "Mechanism of action of clinically approved adjuvants." *Curr Opin Immunol*, 21(1), 23-9.

Larsen, M. V., Lundegaard, C., Lamberth, K., Buus, S., Lund, O., and Nielsen, M. (2007). "Large-scale validation of methods for cytotoxic T-lymphocyte epitope prediction." *BMC Bioinformatics*, 8, 424.

Li, H., Nookala, S., and Re, F. (2007). "Aluminum hydroxide adjuvants activate caspase-1 and induce IL-1beta and IL-18 release." *J Immunol*, 178(8), 5271-6.

Li, X., Eastman, E. M., Schwartz, R. J., and Draghia-Akli, R. (1999). "Synthetic muscle promoters: activities exceeding naturally occurring regulatory sequences." *Nat Biotechnol*, 17(3), 241-5.

Lin, H. J., Huang, K.J., Yang, H.T., and Chen W.C. (2002). "Low-pressure accelerated particle gene gun." *Patent No US 6436709.*

Liu, J., Kjeken, R., Mathiesen, I., and Barouch, D. H. (2008). "Recruitment of antigen- presenting cells to the site of inoculation and augmentation of human immunodeficiency virus type 1 DNA vaccine immunogenicity by *in vivo* electroporation." *J Virol*, 82(11), 5643-9.

Liu, M. A. (2011). "DNA vaccines: an historical perspective and view to the future." *Immunol Rev*, 239(1), 62-84.

Liu, M. A., and Ulmer, J. B. (2005). "Human clinical trials of plasmid DNA vaccines." *Adv Genet*, 55, 25-40.

Loomis, D. J., Celeste, J., and Sanford, J. (1999). "Biolistic apparatus for delivering substances into cells and tissues." *Patent of USA, No US 6004287.*

Loudon, P. T., Yager, E. J., Lynch, D. T., Narendran, A., Stagnar, C., Franchini, A. M., Fuller, J. T., White, P. A., Nyuandi, J., Wiley, C. A., Murphey-Corb, M., and Fuller, D. H. (2010). "GM-CSF increases mucosal and systemic immunogenicity of an H1N1 influenza DNA vaccine administered into the epidermis of non-human primates." *PLoS One*, 5(6), e11021.

Luckay, A., Sidhu, M. K., Kjeken, R., Megati, S., Chong, S. Y., Roopchand, V., Garcia-Hand, D., Abdullah, R., Braun, R., Montefiori, D. C., Rosati, M., Felber, B. K., Pavlakis, G. N., Mathiesen, I., Israel, Z. R., Eldridge, J. H., and Egan, M. A. (2007). "Effect of plasmid DNA vaccine design and in vivo electroporation on the resulting vaccine- specific immune responses in rhesus macaques." *J Virol*, 81(10), 5257-69.

Lundquist, C. A., Meier, J. L., and Stinski, M. F. (1999). "A strong negative transcriptional regulatory region between the human cytomegalovirus UL127 gene and the major immediate-early enhancer." *J Virol*, 73(11), 9039-52.

Luxembourg, A., Hannaman, D., Nolan, E., Ellefsen, B., Nakamura, G., Chau, L., Tellez, O., Little, S., and Bernard, R. (2008). "Potentiation of an anthrax DNA vaccine with electroporation." *Vaccine*, 26(40), 5216-22.

Matzinger, P. (2007). "Friendly and dangerous signals: is the tissue in control?" *Nat Immunol*, 8(1), 11-3.

McMahon, J. M., Signori, E., Wells, K. E., Fazio, V. M., and Wells, D. J. (2001). "Optimisation of electrotransfer of plasmid into skeletal muscle by pretreatment with hyaluronidase -- increased expression with reduced muscle damage." *Gene Ther*, 8(16), 1264-70.

Miller, A. M., Ozenci, V., Kiessling, R., and Pisa, P. (2005). "Immune monitoring in a phase 1 trial of a PSA DNA vaccine in patients with hormone-refractory prostate cancer." *J Immunother*, 28(4), 389-95.

Mir, L. M. (2008). "Application of electroporation gene therapy: past, current, and future." *Methods Mol Biol*, 423, 3-17.

Mir, L. M., Bureau, M. F., Gehl, J., Rangara, R., Rouy, D., Caillaud, J. M., Delaere, P., Branellec, D., Schwartz, B., and Scherman, D. (1999). "High-efficiency gene transfer into skeletal muscle mediated by electric pulses." *Proc Natl Acad Sci U S A*, 96(8), 4262-7.

O'Hagan, D. T., Singh, M., and Ulmer, J. B. (2004). "Microparticles for the delivery of DNA vaccines." *Immunol Rev*, 199, 191-200.

Otten, G., Schaefer, M., Doe, B., Liu, H., Srivastava, I., zur Megede, J., O'Hagan, D., Donnelly, J., Widera, G., Rabussay, D., Lewis, M. G., Barnett, S., and Ulmer, J. B. (2004). "Enhancement of DNA vaccine potency in rhesus macaques by electroporation." *Vaccine*, 22(19), 2489-93.

Palucka, A. K., Ueno, H., Fay, J. W., and Banchereau, J. (2007). "Taming cancer by inducing immunity via dendritic cells." *Immunol Rev*, 220, 129-50.

Palucka, K., Banchereau, J., and Mellman, I. (2010). "Designing vaccines based on biology of human dendritic cell subsets." *Immunity*, 33(4), 464-78.

Papadakis, E. D., Nicklin, S. A., Baker, A. H., and White, S. J. (2004). "Promoters and control elements: designing expression cassettes for gene therapy." *Curr Gene Ther*, 4(1), 89- 113.

Parker, K. C., Bednarek, M. A., and Coligan, J. E. (1994). "Scheme for ranking potential HLA- A2 binding peptides based on independent binding of individual peptide side- chains." *J Immunol*, 152(1), 163-75.

Pavlenko, M., Roos, A. K., Lundqvist, A., Palmborg, A., Miller, A. M., Ozenci, V., Bergman, B., Egevad, L., Hellstrom, M., Kiessling, R., Masucci, G., Wersall, P., Nilsson, S., and Pisa, P. (2004). "A phase I trial of DNA vaccination with a plasmid expressing prostate-specific antigen in patients with hormone-refractory prostate cancer." *Br J Cancer*, 91(4), 688-94.

Payette, P. J., Weeratna, R. D., McCluskie, M. J., and Davis, H. L. (2001). "Immune-mediated destruction of transfected myocytes following DNA vaccination occurs via multiple mechanisms." *Gene Ther*, 8(18), 1395-400.

Peng, B., Zhao, Y., Xu, L., and Xu, Y. (2007). "Electric pulses applied prior to intramuscular DNA vaccination greatly improve the vaccine immunogenicity." *Vaccine*, 25(11), 2064-73.

Plotkin, S. A. (2005). "Vaccines: past, present and future." *Nat Med*, 11(4 Suppl), S5-11.

Purcell, A. W., McCluskey, J., and Rossjohn, J. (2007). "More than one reason to rethink the use of peptides in vaccine design." *Nat Rev Drug Discov*, 6(5), 404-14.

Qing, Y., Chen, M., Zhao, J., Hu, H., Xu, H., Ling, N., Peng, M., and Ren, H. (2010). "Construction of an HBV DNA vaccine by fusion of the GM-CSF gene to the HBV-S gene and examination of its immune effects in normal and HBV-transgenic mice." *Vaccine*, 28(26), 4301-7.

Ramirez, J. C., Gherardi, M. M., Rodriguez, D., and Esteban, M. (2000). "Attenuated modified vaccinia virus Ankara can be used as an immunizing agent under conditions of preexisting immunity to the vector." *J Virol*, 74(16), 7651-5.

Rammensee, H., Bachmann, J., Emmerich, N. P., Bachor, O. A., and Stevanovic, S. (1999). "SYFPEITHI: database for MHC ligands and peptide motifs." *Immunogenetics*, 50(3- 4), 213-9.

Rice, J., Buchan, S., and Stevenson, F. K. (2002). "Critical components of a DNA fusion vaccine able to induce protective cytotoxic T cells against a single epitope of a tumor antigen." *J Immunol*, 169(7), 3908-13.

Rinaldi, M., Ria, F., Parrella, P., Signori, E., Serra, A., Ciafre, S. A., Vespignani, I., Lazzari, M., Farace, M. G., Saglio, G., and Fazio, V. M. (2001). "Antibodies elicited by naked DNA vaccination against the complementary-determining region 3 hypervariable region of immunoglobulin heavy chain idiotypic determinants of B-lymphoproliferative disorders specifically react with patients' tumor cells." *Cancer Res*, 61(4), 1555-62.

Roos, A. K., King, A., and Pisa, P. (2008). "DNA vaccination for prostate cancer." *Methods Mol Biol*, 423, 463-72.

Russo, V., Tanzarella, S., Dalerba, P., Rigatti, D., Rovere, P., Villa, A., Bordignon, C., and Traversari, C. (2000). "Dendritic cells acquire the MAGE-3 human tumor antigen from apoptotic cells and induce a class I-restricted T cell response." *Proc Natl Acad Sci U S A*, 97(5), 2185-90.

Selby, M., Walsh, R., Pertile, T., Goldbeck, C., Ulmer, J., and Liu MA. (2000). "Electrically- mediated enhancement of DNA vaccine immunity and efficacy *in vivo*." *Patent No. WO/2000/045823*.

Seo, S. H., Jin, H. T., Park, S. H., Youn, J. I., and Sung, Y. C. (2009). "Optimal induction of HPV DNA vaccine-induced CD8+ T cell responses and therapeutic antitumor effect by antigen engineering and electroporation." *Vaccine*, 27(42), 5906-12.

Serruto, D., and Rappuoli, R. (2006). "Post-genomic vaccine development." *FEBS Lett*, 580(12), 2985-92.

Setoguchi, R., Hori, S., Takahashi, T., and Sakaguchi, S. (2005). "Homeostatic maintenance of natural Foxp3(+) CD25(+) CD4(+) regulatory T cells by interleukin (IL)-2 and induction of autoimmune disease by IL-2 neutralization." *J Exp Med*, 201(5), 723-35.

Sette, A., and Rappuoli, R. (2010). "Reverse vaccinology: developing vaccines in the era of genomics." *Immunity*, 33(4), 530-41.

Shirota, H., Petrenko, L., Hong, C., and Klinman, D. M. (2007). "Potential of transfected muscle cells to contribute to DNA vaccine immunogenicity." *J Immunol*, 179(1), 329- 36.

Singh, H., and Raghava, G. P. (2001). "ProPred: prediction of HLA-DR binding sites."

Bioinformatics, 17(12), 1236-7.

Stevenson, F. K., Rice, J., Ottensmeier, C. H., Thirdborough, S. M., and Zhu, D. (2004). "DNA fusion gene vaccines against cancer: from the laboratory to the clinic." *Immunol Rev*, 199, 156-80.

Tam, J. P. (1996). "Recent advances in multiple antigen peptides." *J Immunol Methods*, 196(1), 17-32.

Tjelle, T. E., Rabussay, D., Ottensmeier, C., Mathiesen, I., and Kjeken, R. (2008). "Taking electroporation-based delivery of DNA vaccination into humans: a generic clinical protocol." *Methods Mol Biol*, 423, 497-507.

Tjelle, T. E., Salte, R., Mathiesen, I., and Kjeken, R. (2006). "A novel electroporation device for gene delivery in large animals and humans." *Vaccine*, 24(21), 4667-70.

Tollefsen, S., Tjelle, T., Schneider, J., Harboe, M., Wiker, H., Hewinson, G., Huygen, K., and Mathiesen, I. (2002). "Improved cellular and humoral immune responses against Mycobacterium tuberculosis antigens after intramuscular DNA immunisation combined with muscle electroporation." *Vaccine*, 20(27-28), 3370-8.

Tsang, C., Babiuk, S., van Drunen Littel-van den Hurk, S., Babiuk, L. A., and Griebel, P. (2007). "A single DNA immunization in combination with electroporation prolongs the primary immune response and maintains immune memory for six months." *Vaccine*, 25(30), 5485-94.

Tymciu, S., Durieux-Alexandrenne, C., Wijkhuisen, A., Creminon, C., Frobert, Y., Grassi, J., Couraud, J. Y., and Boquet, D. (2004). "Enhancement of antibody responses in DNA vaccination using a vector encoding a universal T-helper cell epitope." *DNA Cell Biol*, 23(6), 395-402.

Ulmer, J. B., DeWitt, C. M., Chastain, M., Friedman, A., Donnelly, J. J., McClements, W. L., Caulfield, M. J., Bohannon, K. E., Volkin, D. B., and Evans, R. K. (1999). "Enhancement of DNA vaccine potency using conventional aluminum adjuvants." *Vaccine*, 18(1-2), 18-28.

Umland, T. C., Wingert, L. M., Swaminathan, S., Furey, W. F., Schmidt, J. J., and Sax, M. (1997). "Structure of the receptor binding fragment HC of tetanus neurotoxin." *Nat Struct Biol*, 4(10), 788-92.

van der Burg, S. H., Bijker, M. S., Welters, M. J., Offringa, R., and Melief, C. J. (2006). "Improved peptide vaccine strategies, creating synthetic artificial infections to maximize immune efficacy." *Adv Drug Deliv Rev*, 58(8), 916-30.

Vandepapeliere, P., Lau, G. K., Leroux-Roels, G., Horsmans, Y., Gane, E., Tawandee, T., Merican, M. I., Win, K. M., Trepo, C., Cooksley, G., Wettendorff, M., and Ferrari, C. (2007). "Therapeutic vaccination of chronic hepatitis B patients with virus suppression by antiviral therapy: a randomized, controlled study of co- administration of HBsAg/AS02 candidate vaccine and lamivudine." *Vaccine*, 25(51), 8585-97.

Wells, D. J. (2004). "Gene therapy progress and prospects: electroporation and other physical methods." *Gene Ther*, 11(18), 1363-9.

Wells, D. J. (2010). "Electroporation and ultrasound enhanced non-viral gene delivery in vitro and in vivo." *Cell Biol Toxicol*, 26(1), 21-8.

Widera, G., Austin, M., Rabussay, D., Goldbeck, C., Barnett, S. W., Chen, M., Leung, L., Otten, G. R., Thudium, K., Selby, M. J., and Ulmer, J. B. (2000). "Increased DNA vaccine delivery and immunogenicity by electroporation *in vivo*." *J Immunol*, 164(9), 4635-40.

Wiendl, H., Hohlfeld, R., and Kieseier, B. C. (2005). "Immunobiology of muscle: advances in understanding an immunological microenvironment." *Trends Immunol*, 26(7), 373- 80.

Wolff, J. A., Malone, R. W., Williams, P., Chong, W., Acsadi, G., Jani, A., and Felgner, P. L. (1990). "Direct gene transfer into mouse muscle in vivo." *Science*, 247(4949 Pt 1), 1465-8.

Yan, J., Harris, K., Khan, A. S., Draghia-Akli, R., Sewell, D., and Weiner, D. B. (2008). "Cellular immunity induced by a novel HPV18 DNA vaccine encoding an E6/E7 fusion consensus protein in mice and rhesus macaques." *Vaccine*, 26(40), 5210-5.

Yang, N. S., Burkholder, J., Roberts, B., Martinell, B., and McCabe, D. (1990). "In vivo and in vitro gene transfer to mammalian somatic cells by particle bombardment." *Proc Natl Acad Sci U S A*, 87(24), 9568-72.

Yew, N. S. (2005). "Controlling the kinetics of transgene expression by plasmid design." *Adv Drug Deliv Rev*, 57(5), 769-80.

Yew, N. S., Przybylska, M., Ziegler, R. J., Liu, D., and Cheng, S. H. (2001). "High and sustained transgene expression in vivo from plasmid vectors containing a hybrid ubiquitin promoter." *Mol Ther*, 4(1), 75-82.

Yoon, H. A., Aleyas, A. G., George, J. A., Park, S. O., Han, Y. W., Lee, J. H., Cho, J. G., and Eo, S. K. (2006). "Cytokine GM-CSF genetic adjuvant facilitates prophylactic DNA vaccine against pseudorabies virus through enhanced immune responses." *Microbiol Immunol*, 50(2), 83-92.

Chapter 9 USING ULTRASOUND AND MICROBUBBLES FOR GENE DELIVERY

1. INTRODUCTION: AN OVERVIEW

Gene therapy and siRNA treatment have the ability to effectively treat a diverse range of ailments. The stability of genetic material is typically not maintained and is commonly hydrolyzed after being administered intravascularly. Delivering genes can be challenging since complete genetic information often needs to be transferred into cells to have a therapeutic effect. In the context of gene therapy, it is typically necessary for the gene construct to reach the intranuclear sub-compartment of the cell in order to produce the intended biological outcome. Viruses have developed the ability to transport DNA and RNA into cells. However, gene therapy via viral vectors has been linked to effects that are inherent in biological systems (Marshall, Muruve, Thomas). Non-viral technologies have the ability to transmit genetic materials without the negative biological effects associated with viral systems. Generally, non-viral gene delivery technologies have proven to be less efficient than viral-based systems, resulting in lower levels of gene expression (Litzinger). Microbubbles, such as acoustically active carriers, when used in combination with ultrasound (as demonstrated by Unger and Zhou), have the potential to be highly effective in targeted gene therapy and delivery of genetic elements like siRNA (as shown by Zhigang).

2. OVERVIEW OF GENE DELIVERY USING MICROBUBBLES AND ULTRASOUND

Figure 1 illustrates the fundamental structure of a microbubble. Microbubbles are made up of gas and have a shell made of stabilizing materials such as lipids, albumin, or biocompatible polymers. Biomedical applications use particles that vary in size, ranging from a few microns to several hundred nanometers in diameter. The initial biomedical use involved employing ultrasonic contrast agents for echocardiography. The FDA has approved two contrast agents, Definity® and Optison®, for use in echocardiography. These agents are made up of phospholipid-coated perfluoropropane microbubbles and albumin-coated perfluoropropane microbubbles, respectively. They are sold by the company Lantheus in Billerica, MA.

Microbubbles are utilized as ultrasound contrast agents because to their exceptional ability to efficiently scatter sound waves back. This is mostly down to the significant impedance mismatch between liquid and gas, making microbubbles great acoustic reflectors. They possess exceptional sound energy reflection properties, making them exceptional contrast agents for biological ultrasound imaging. In addition, the implementation of specialized ultrasonic pulse sequences such as 2nd harmonic imaging and phase inversion harmonic imaging has effectively enhanced ultrasound imaging by reducing substantial levels of noise caused by tissue reflection.

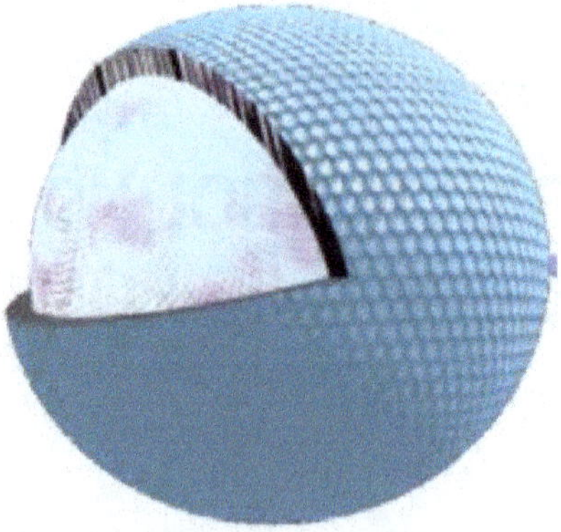

Fig. 1. Depicts a microbubble coated with a film of stabilizing material.

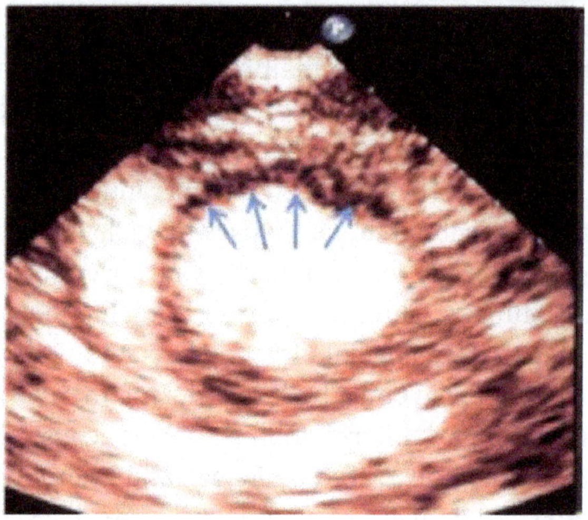

Fig. 2. Shows a contrast-enhanced echocardiogram image of a porcine heart after occlusion of the left anterior descending coronary artery (model of acute myocardial infarction). The area of decreased perfusion in the left ventricular wall is clearly seen on the post contrast images but was not detectable without contrast.

When ultrasound encounters the acoustic interface of a microbubble not only may the ultrasound be scattered, but also, because of their size and the fortuitous insonation frequencies used clinically, the microbubbles also can oscillate (stable cavitation). Depending upon the acoustic pressure, oscillating microbubbles may rupture (cavitate - described further in section 4, inertial cavitation). On the basis of cavitation it was discovered that microbubbles had therapeutic applications for gene and drug delivery and treatment of vascular thrombosis. Ultrasound effects on cell membranes may be two-fold; 1) ultrasound itself can enhance membrane permeability, thereby allowing more diffusion from the extracellular milieu (sonoporation); or, 2) can be used in conjunction with microbubbles to enhance localized delivery via microbubble bursting and subsequent radiation force induced particle penetration through the membrane surface. Cavitation can be used to increase cell permeability and local delivery of materials such as DNA for gene therapy.

A number of preclinical studies have been performed for gene (described further below) and drug delivery, but phospholipid-coated microbubbles have entered human clinical trials for treatment of vascular thrombosis. In these studies microbubbles have been administered intravenously and been shown to permeate a thrombus. Ultrasound energy has then been applied to the site of the clot to cavitate the microbubbles and dissolve the thrombus.

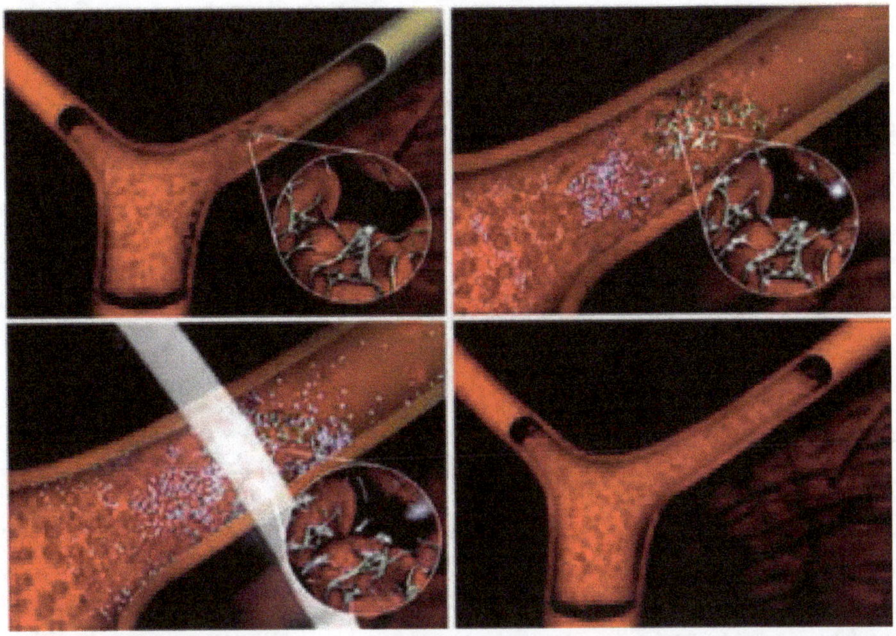

Fig. 3. Depicts and occlusive thrombus in an arterial blood vessel (upper left). Microbubbles are infused IV and permeate the clot (upper right). This can be seen on ultrasound imaging. High energy ultrasound is focused on the site of thrombus (lower left) and the microbubbles cavitate, dissolving the clot and restoring blood flow.

3. DETAILS OF MICROBUBBLES FOR GENE DELIVERY

Microbubbles are mainly composed of fluorinated gases. Air and nitrogen are relatively water-soluble and hence will diffuse into the blood and the bubbles will then rapidly shrink and eventually collapse from Laplace pressures. This is not to say that air and nitrogen cannot be used to make microbubbles for biomedical ultrasound applications, but that stabilizing materials will need to be more robust to preserve the microbubbles.

Table 1. Potential Compounds for Making Gaseous Cores of Microbubbles

Compound	Molecular Weight	Aqueous Solubility (Ostwald's Coefficient)	Boiling Point °C
Nitrogen	28	18071	-196
Oxygen	32	4865	-183
Sulfur Hexafluoride	146	5950	-64
Perfluoropropane	188	583	-36.7
Perfluorobutane	238	<500	-1.7
Perfluoropentane	288	>24 and <500	29
Perfluorohexane	338	24	56.6

As opposed to oxygen and nitrogen being relatively soluble in water, perfluorocarbons are virtually insoluble in the aqueous milieu. In fact perfluorocarbons are amphiphobic. The higher the molecular weight of the fluorinated compound the less water soluble and we would predict that microbubbles prepared from that gas should be correspondingly more stable in the blood stream (all else equal). Note that perfluoropentane has a boiling point of 29oC and that perfluorohexane volatilizes at 56.6oC and therefore will be a liquid at biological temperature (presumably due to van der Waal's attractions).

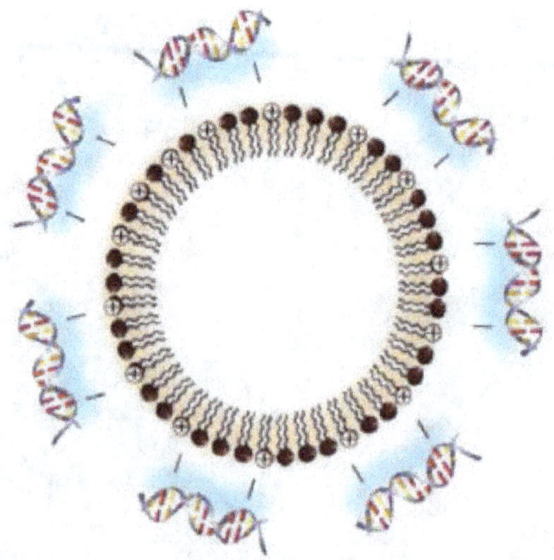

Fig. 4. Depicts a microbubble for gene delivery. The gaseous core is coated by a monolayer of phospholipid containing cationic lipid imparting a net positive charge to the microbubble. DNA, as a polyanion, is adsorbed electrostatically to the exterior surface of the microbubble.

With respect to the microbubble membrane, the above design demonstrates how DNA is adsorbed to the surface via electrostatic adhesion with cationic lipids inserted in the membrane. However, note also that there is no steric protection of the DNA, making it susceptible to biodegradation/hydrolysis. As shown below in Figure 5, microbubbles can be designed to incorporate PEG'ylated lipid (e.g. 8-10 mole percent PEG'ylated lipid) and we have still found that cationic PEG'ylated microbubbles will still adsorb useful payloads of DNA.

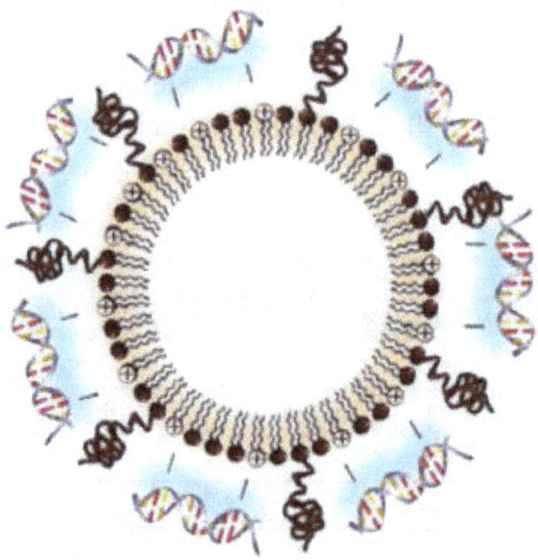

Fig. 5. Depicts a PEG'ylated cationic microbubble binding DNA.

In vivo studies with plasmid DNA using the construct depicted in Figure 5 have shown high levels of gene expression in the zones of insonation.

We predict, however, that targeted constucts that would bind to cellular targets should be more effective for gene delivery. Since microbubbles are micron-sized structures, they are not expected to extravasate from the intravscular space. However, for the purposes of delivering genes to regions in close proximity to targeted tissues, they can be targeted to eptitopes expressed on endothelial cells. Because they can also be engulfed by phagocytic cells, such as immune cells, and targeted as intracellular passengers, the nanoparticles fused to microbubble surfaces as described below (Figure 8) can also be decorated with cell surface identifiers/targets, uptake enhancers, and even intracellular targets in order to provide additional selectivity. We have prepared targeted microbubbles to a variety of different targets. Depicted below is the design for a targeted microbubble to the integrin αvβIII, Expressed on the endothelial surface in angiogenesis.

The microbubble construct depicted above combines several features of nanotechnology. The parent microbubble is PEG'ylated to impart stealth properties in order to prevent reticuloendothelial system elimination and includes targeting ligands directed to epitopes expressed on endothelial cells. It could be monitored as an ultrasound contrast agent and then activated with higher energy ultrasound using low MI ultrasound for imaging to monitor delivery to the target site and high MI ultrasound energy for cavitation. Upon activation the nanoparticles containing DNA would be expected to extravasate from the vasculature into the interstitial tissues.

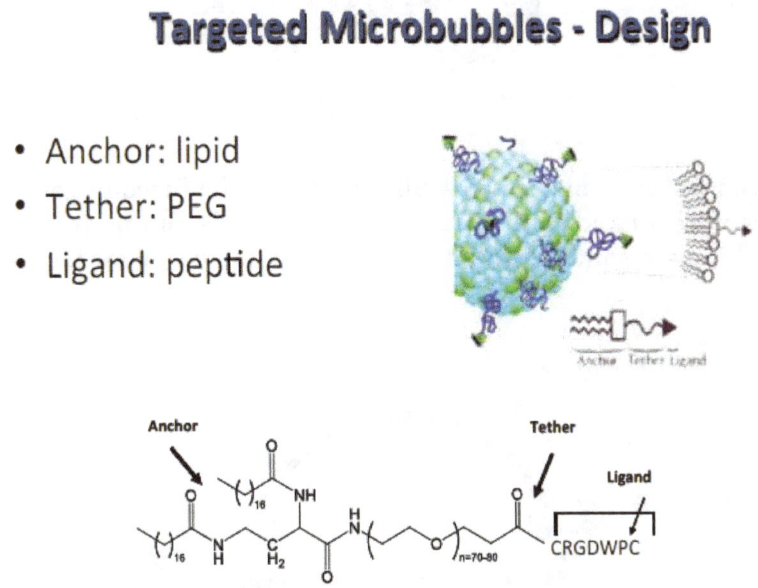

Fig. 6. Depicts a targeted microbubble. The cyclo-CRGDC analog is attached to a lipid anchor via a PEG spacer to form a bioconjugate. The bioconjugate comprises about from about 0.1 to 5 mole percent of the lipid coating the microbubble. There is additional PEG'ylated lipid from about 5-10 mole percent in the lipid coating the microbubble.

The nanoparticles could be constructed to contain targeting ligands and/or cell penetration-enhancing agents to facilitate uptake by target cells. A challenge is to design linkers to attach the nanoparticles to the surface of the parent microbubbles. One approach would be to biotinylate the nanoparticles and to have avidin tethers attached to the parent microbubbles taking advantage of the avidin/biotin interaction. This approach could be utilized in proof-of-principle studies but would not be clinically translatable. A maleimide labeled spacer might be affixed to the microbubble and the nanoparticles might be thiolated to covalently bind the nanoparticle to the surface of the microbubbles in a potentially biocompatible manner. Another approach would be to use electrostatic interaction between the nanoparticle and the surface of the microbubbles but this must be optimized to ensure that the DNA is bound until it reaches the target site. Note also that if the bubble has excess cationic charge this may adversely affect biodistribution. Note also that it is more difficult to bind low molecular weight genetic material such as si- RNA using merely electrostatic interaction.

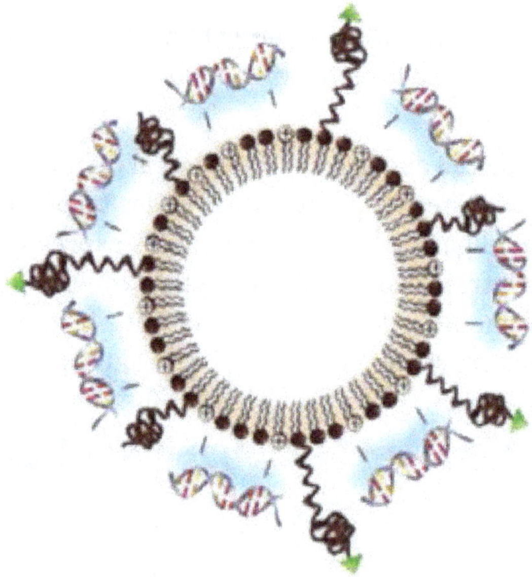

Fig. 7. Depicts a targeted cationic microbubble binding DNA. The bioconjugates are designed to bind to endothelial epitopes upregulated in disease. The microbubble can be followed and monitored by ultrasound imaging and activated with ultrasound energy for local delivery.

Fig. 8. Depicts a unique microbubble construct for gene delivery. The parent microbubble is PEG'ylated but the DNA (or siRNA) is condensed into nanoparticles that are bound to the surface of the microbubble. The nanoparticles have targeting ligands to bind to cell specific epitopes. Note also that the construct could be modified to comprise endothelial targeting moieties on the microbubble so that the microbubble could bind to endothelial epitopes.

4. ULTRASOUND – PARAMETERS AND BIOEFFECTS

Ultrasound is a commonly used modality for biomedical imaging, only exceeded by X- rays in overall worldwide use. For medical imaging the power or intensity of ultrasound that can be used is limited by regulatory guidelines. The intensity of the ultrasound can be described by the mechanical index (MI) which is related to the peak negative pressure of the ultrasound wave divided by the square root of the center frequency of the ultrasound. The FDA-approved mechanical index for most body imaging ultrasound (e.g. cardiac and abdominal) is limited to an MI < 1.9. For neurovascular imaging it is limited to MI < 1.0 and for ophthalmic ultrasound to MI < 0.8.

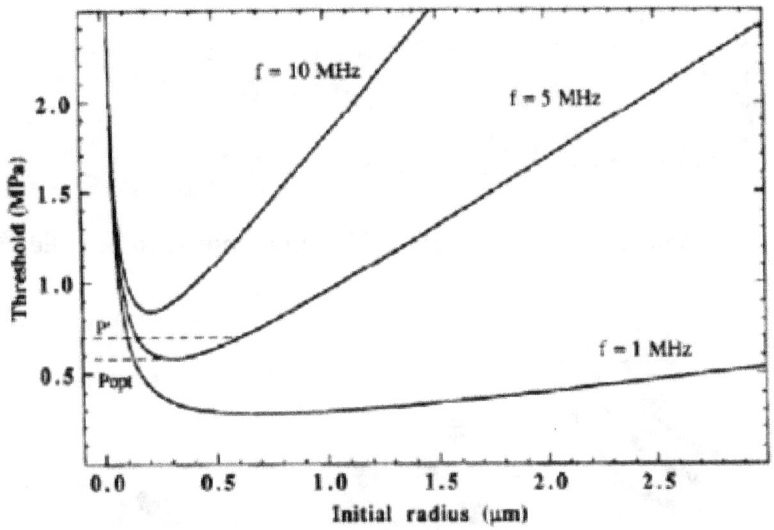

Fig. 9. Microbubbles Lower Thresh-hold of Ultrasound Energy for Cavitation (Apfel).

Cavitation is a phenomenon in which ultrasound exposure at the resonance frequency of the microbubble will induce expansion and collapse of microbubbles (Apfel, Hallow). This can occur spontaneously at high acoustic pressure in the absence of exogenous microbubbles. Sufficiently high acoustic power is sufficient to create a microbubble nidus in situ and cause expansion and collapse, i.e. cavitation (Marmottant). Cavitation causes acoustic streaming and local shock waves that may radiate on the order of microns or larger depending upon the acoustic intensity and other factors (Mehier-Humbert). Cavitation can be used to increase cell permeability (i.e. sonoporation)(Deng), open the blood brain barrier or destroy tissues (i.e. sonoablation)(Conger, Feril). The acoustic pressure can be controlled to create the desired effects (Forsberg). Within the ranges of allowable acoustic pressures, in the absence of exogenous microbubbles, biomedical ultrasound imaging does not generally cause violent cavitation and ultrasound imaging is generally quite safe, particularly compared to other technologies such as X-ray imaging that uses ionizing radiation.

As shown below in Figure 9, microbubbles lower the threshold of ultrasound energy necessary for cavitation to occur. The effect is frequency dependent with a greater proportional effect at 1 MHz than at 10 MHz. Note that at 1 MHz, cavitation occurs at MI < 0.5 within the allowable ultrasound power limits for biomedical imaging. Note that microbubbles in the size range of 0.5 to 2.0 microns are quite effective in lowering the thresh-hold of energy for cavitation. Stable microbubbles for gene delivery can certainly be made in this size range. If necessary, for therapeutic applications, higher levels of ultrasound energy can be employed. Higher levels of ultrasound energy are used therapeutically for hyperthermia and focused ultrasound surgery (Brujan).

Figure 10, below, shows images from ultra-high speed videomicroscopy of a single bubble in response to a single high MI pulse of ultrasound energy. The microbubble expands, collapses, and fragments. The daughter fragments then undergo one additional cycle of expansion, collapse and disappear. In this process genetic materials might be released from a gene- carrying microbubble. The cavitation process will also create acoustic jets and streaming which might be used for delivering the genetic material to the target tissue or cells.

Figure 11 depicts a gene carrying microbubble in response to cavitation. The stabilizing wall material of the microbubble fragments and the genetic material is ejected with the cavitation ballistically, thereby extravasating from the vasculature to the target tissue.

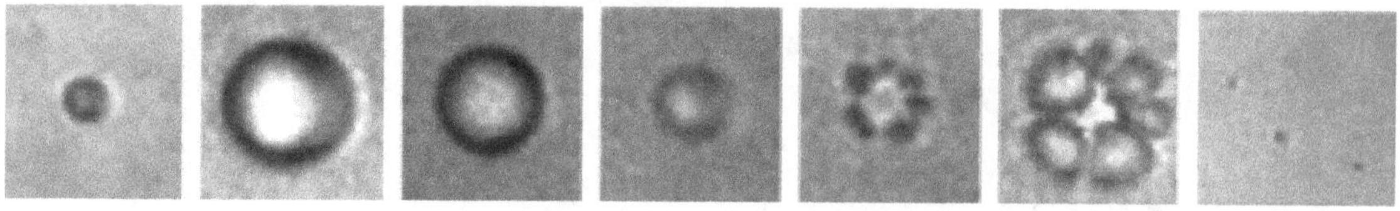

Fig. 10. Videomicroscopy images of a single bubble in response to a single ultrasonic wave.

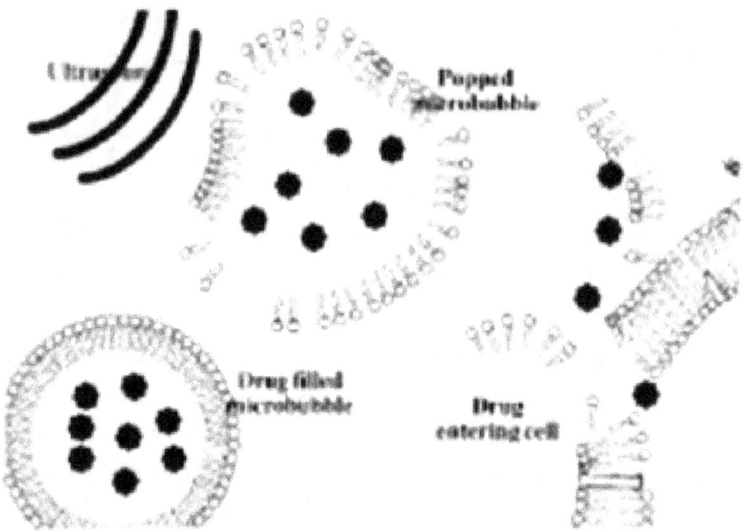

Fig. 11. A drug carrying-microbubble cavitating in response to ultrasound.

Another major mechanism in ultrasound that may be useful for drug and gene delivery is the mechanical force of microbubbles. Ultrasound exerts a radiation (pushing) force that can move microparticles and improve cellular delivery of biomaterials. Figure 12 below shows avidin-coated microbubbles flowing in blood over a biotinylated plate. The microbubbles flow and do not appear to bind but after application of relatively low MI ultrasound the microbubbles are pushed by the acoustic waves and bind to the surface of the plate. The same process can be used to improve cellular uptake of biomaterials such as DNA. The (pushing) radiation force of ultrasound occurs at lower energies than necessary for cavitation and can be used to increase efficacy from gene delivery.

An acoustic transducer that generally uses a piezoelectric material to convert electrical energy into ultrasound waves creates ultrasound waves. The transducer design varies depending upon the biomedical application. Figure 13 (below) shows the design of a three dimensional transducer for imaging and treating the heart. This transducer can be used to image the heart and visualize microbubble as they enter the myocardial circulation. Ultrasound energy can be applied to the heart to cavitate the microbubbles or for the radiation force to improve myocardial delivery. The transducer is currently being used in pre-clinical studies for microbubble enhanced sonothrombolysis to treat myocardial infarction and planned for use in clinical studies. The same transducer design could potentially be used in clinical studies for microbubble-mediated gene delivery to treat the heart.

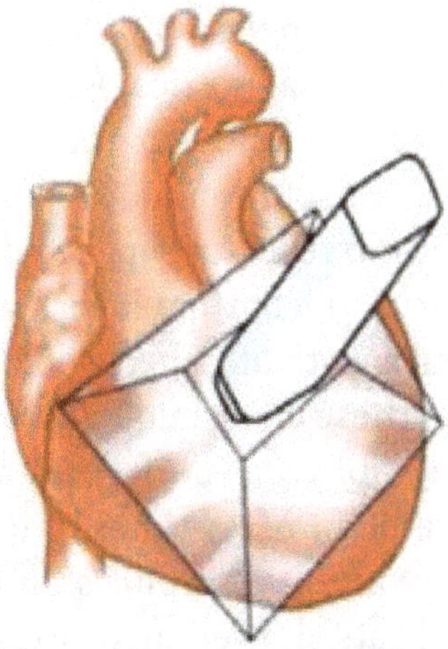

Fig. 12. Drawing of three-dimensional transducer for imaging and treating the heart.

Despite its ease of use, unwanted bioeffects can be experienced using ultrasound at high acoustic outputs. High levels of ultrasound energy with cavitation may cause cell damage, cell death and apoptosis (Miller). Ultrasound may also heat tissues and cause coagulative necrosis at high temperatures (Ter Haar). We emphasize, however, that ultrasound power levels within the limits for diagnostic ultrasound are generally safe. Ultrasound power levels can be optimized for gene therapy to maximize gene expression while minimizing unwanted bioeffects (Chen, Rahim).

Despite its ease of use, unwanted bioeffects can be experienced using ultrasound at high acoustic outputs. High levels of ultrasound energy with cavitation may cause cell damage, cell death and apoptosis (Miller). Ultrasound may also heat tissues and cause coagulative necrosis at high temperatures (Ter Haar). We emphasize, however, that ultrasound power levels within the limits for diagnostic ultrasound are generally safe. Ultrasound power levels can be optimized for gene therapy to maximize gene expression while minimizing unwanted bioeffects (Chen, Rahim).

5. STUDIES USING MICROBUBBLES AND ULTRASOUND FOR GENE DELIVERY

In this section we summarize studies that have been performed using microbubbles and ultrasound for gene delivery (Klibanov, Liu, Newman). Over the period of more that a decade a number of different studies have been performed, in vitro and in vivo. In vitro studies showed that ultrasound increased the efficacy of transfection with cationic liposomes in cell culture studies to express reporter genes. Saphenous vein grafts have been transfected with ultrasound and MBs (Kodama) to enhance graft survival after implantation (Akowuah). In vivo studies have been performed with reporter genes showing increased expression of the reporter gene after IV administration of microbubbles either binding or in association with the reporter genes. The zones of highest expression have been in the regions of tissues of insonation (except that cationic microbubbles have also been accumulated by liver, lungs, spleen and phagocytic organs). Most of the *in vivo* studies Have been performed with reporter genes. A few studies have performed with therapeutic genes. A wide variety of different tissues (Hauff) and cells have shown enhanced transfection with ultrasound including neuronal cells (Fischer) and skeletal muscle (Liang).

A number of groups have studied ultrasound and microbubbles to transfect tumors (Michel, Anwer). With collaborators we performed a study assessing tumor regression with transfection of the IL-12 gene with ultrasound and cationic liposomes (Anwer). In this study the liposomes were lyophilized and may have contained nitrogen gas. The tumors were insonated with 1 MHz ultrasound. Increased expression of IL-2 was observed in the insonated tumors and statistically significant tumor regression.

In addition to phospholipid coated microbubbles, MBs can be stabilizes with denatured serum albumin (e.g. Optison®, GE Healthcare Medical Diagnostics, Princeton, NJ). Albumin binds a variety of molecules and also appears to bind DNA. In one study the plasmid of AdCMV-b- Gal was attached to the microbubbles (Shohet). In these studies the solution of AdCMV-b-Gal was added to the microbubble suspension and mixed for 2 hours at 4°C. The mixture was separated into 2 distinct layers. The upper layer consisted of microbubbles with attached virus; the bottom layer, which contained unattached virus, was discarded. The concentration of microbubbles with attached AdCMV-β-Gal was 1.2x109 bubbles/mL; the mean diameter was 3.5 um. The viral titer of these microbubbles was determined. Rats were anesthetized, and MBs were administered IV. Echocardiography was performed at 1.3 MHz with a mechanical index of 1.5. Images were ECG-triggered to deliver a burst of 3 frames of ultrasound every 4 to 6 cardiac cycles. The hearts of all 6 rats in the experimental group showed blue staining with 5- bromo-4-chloro-3-indolyl-β-D-galactopyranoside. None of the control rats showed myocardial staining, which confirmed that the destruction of the microbubbles containing the virus was responsible for the observed β-galactosidase expression in the rat myocardium.

Initiated in 2001, researchers began using gas-filled, albumin-shelled MBs (Optison) and non-viral carriers to increase nascent serum HDL cholesterol in mice and Sprague-Dawley rats. Initial feasibility studies were performed using reporter genes, 5-bromo-4-chloro-3- indolyl-β-D-galactopyranoside and GFP. Subsequent studies now use ApoA-I DNA plasmids in combination with Optison. In a series of rat studies performed with partial support from an NIH SBIR award (44HL095238-01), the results revealed that an average peak response of HDL-C elevation occurred within 24 hours of treatment and 88% of the responses occurred within 48 hours. The control animal group included ultrasound only, apoA-I DNA only, and microbubble only rats. These data demonstrated a rapid incorporation of the plasmid into the cell and efficient ApoA-I protein production that was later incorporated into serum HDL-C. Summarized data revealed an average increase of 16.8% on day one post treatment. Due to wide normal variation in individual rat HDL cholesterol values, all data is plotted as a percentage from an individual animal's baseline. When examining the overall response to treatment in mg/dL, the average peak response to treatment was 14.0% (p-value <0.0001) as seen in Figure 1 (below). This response was observed uniformly across different animals and ranged from 40mg/dL HDL-C baseline to 100mg/dL HDL-C baseline.

Ongoing studies include additional design of experiments for the following: (1) optimization of sonoporation utility for raising HDL cholesterol, (2) efficient energy delivery algorithms, (3) modes and methods of delivery, and (4) mechanistic analyses of the intra- cellular plasmid location.

Ongoing studies have been performed with lipid coated MBs binding plasmid DNA with genes to treat diabetes. In vivo studies in rats have shown long-term normalization of blood glucose levels in diabetic rats. Studies are planned to test this system in primates. Encouraging work has been performed in models of hypercholesterolemia showing the potential to perform gene therapy for H1Alpha to improve the capacity of the liver to produce HDL as treatment for atherosclerosis. Promising work continues to progress in the use of MBs binding genes to treat hemophilia. Cationic MBs binding DNA (similar to Figure 7) have been tested with ultrasound and compared to DefinityR MBs which do not bind DNA. Expression of Factor IX transgenes is more robust with the cationic MBs binding DNA than with MBs not binding DNA. In both the HI1Alpha/HDL and hemophilia treatment applications the target organ is the liver. The goal in both of these programs and in many others is to generate sustained transgene expression in the liver (Guo). Catheter mediated administration may prove attractive to maximize efficiency of delivery to hepatocytes and catheters can be deployed for delivery to other organs such as the kidneys, heart and blood vessels.

Ultrasound can be targeted to any organ for which it is possible to create an acoustic window. Most tissues are readily accessible to ultrasound. Ultrasound can be applied across the intact skin or, if need be, via specialized probes. While ultrasound is blocked by air/tissue or air/fluid interfaces, even in the lung, it is possible to create acoustic windows to the bronchi with bronchoscopes and water filled balloons. Ultrasound can be targeted precisely to tissues to treat volumes of tissues ranging to more than 500 ml to less than a ml. Most studies have shown relatively low toxicity due to ultrasound and MBs. It also is possible to repeat treatment, potentially indefinitely.

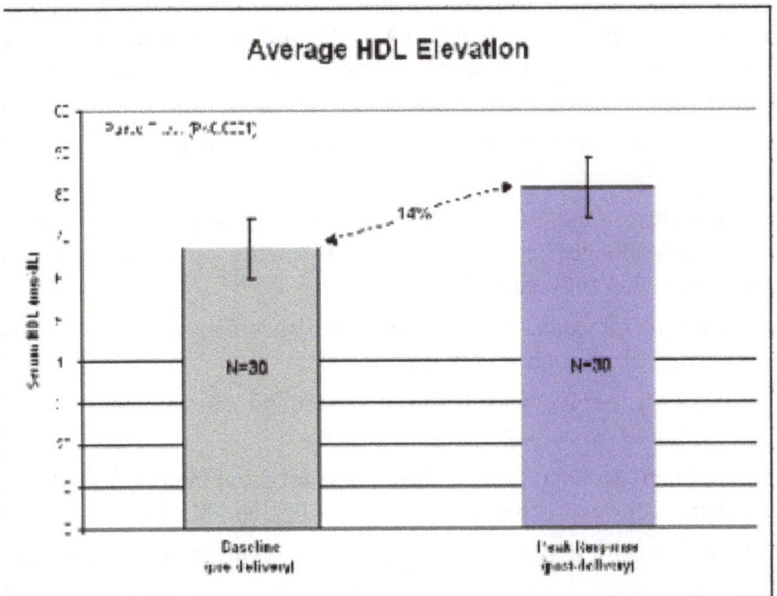

Fig. 13. Baseline to peak response measurement in 30 treated rats recorded in raw data values of mg/dL. A statically significant 14% increase was measured (p-value= 0.0001).

Although ultrasound and gene delivery investigations using MBs were conducted over a decade ago, none of the ultrasound gene delivery projects with MBs have progressed to clinical trials as of the time this chapter was prepared. The field is now in its early stages of development. However, the fact that this technology is well-tolerated, easy to use, and allows for high levels of expression in the specific tissue it targets suggests that it should be seriously considered for clinical use.

6. CONCISE SUMMARY

Ultrasound-assisted gene delivery via acoustically active carriers, like as microbubbles, is a promising area of research that offers the promise for achieving high levels of gene expression in specific tissues, while avoiding the negative biological effects associated with viral-based carriers. The field is still in its nascent stage, as no clinical investigations have been conducted at the time of preparing this publication. The clinical advancement of this auspicious domain will necessitate the discovery of biological targets in medically underserved areas and the establishment of interdisciplinary collaborations among material scientists, biologists, and biopharmaceutical companies.

REFERENCES

Akowuah EF, Gray C, Lawrie A, Sheridan PJ, Su CH, Bettinger T et al. Ultrasound-mediated delivery of TIMP-3 plasmid DNA into saphenous vein leads to increased lumen size in a porcine interposition graft model. Gene Therapy 2005; 12: 1154–1157.

Anwer K, Kao G, Proctor B, Anscombe I, Florack V, Earls R, Wilson E, McCreery T, Unger E, Rolland A, Sullivan SM. Ultrasound enhancement of cationic lipid-mediated gene transfer to primary tumors following systemic administration. Gene Ther. 2000;7 (21):1833-9.

Apfel, RE, Holland, CK. Microbubbles Lower Thresh-hold of Ultrasound Energy for Cavitation. Ultrasound Med Biol. 1991;17(2):179-85.

Brujan EA, Ikeda T, Matsumoto Y. Jet formation and shock wave emission during collapse of ultrasound-induced cavitation bubbles and their role in the therapeutic applications of high-intensity focused ultrasound. Phys Med Biol 2005; 50: 4797–4809.

Chen S, Shohet RV, Bekeredjian R, Frenkel P, Grayburn PA. Optimization of ultrasound parameters for cardiac gene delivery of adenoviral or plasmid deoxyribonucleic acid by ultrasound targeted microbubble destruction. J Am Coll Cardiol 2003; 42: 301–308.

Conger AD, Ziskin MC, Wittels H. Ultrasonic effects on mammalian multicellular tumor spheroids. .I Clin Ultrasound 1981;9:167-174.

Deng CX, Sieling F, Pan H, Cui J. Ultrasound-induced cell membrane porosity. Ultrasound Med Biol 2004; 30: 519–526. 15 Zarnitsyn VG, Prausnitz MR. Physical parameters influencing optimization of ultrasound-mediated DNA transfection. Ultrasound Med Biol 2004; 30: 527–538.

Feril LB, Kondo T, Zhao QL, Ogawa R, Tachibana K, Kudo N et al. Enhancement of ultrasound-induced apoptosis and cell lysis by echo-contrast agents. Ultrasound Med Biol 2003; 29: 331–337.

Forsberg F, Merton DA, Goldberg BB. In vivo destruction of ultrasound contrast microbubbles is independent of the mechanical index. J Ultrasound Med 2006; 25: 143–144

Guo DP, Li XY, Sun P, Tang YB, Chen XY, Chen Q et al. Ultrasound-targeted microbubble destruction improves the low density lipoprotein receptor gene expression in HepG2 cells. Biochem Biophys Res Commun 2006; 343: 470–474.

FischerAJ,StankeJJ,OmarG,AskwithCC,BurryRW.Ultrasound- mediated gene transfer into neuronal cells. J Biotechnol 2006; 122: 393–411.

Hallow DM, Mahajan AD, McCutchen TE, Prausnitz MR. Measurement and correlation of acoustic cavitation with cellular bioeffects. Ultrasound Med Biol 2006; 32: 1111–1122.

Hauff P, Seemann S, Reszka R, Schultze-Mosgau M, Reinhardt M, Buzasi T et al. Evaluation of gas-filled microparticles and sonoporation as gene delivery system: feasibility study in rodent tumor models. Radiology 2005; 236: 572–578.

Klibanov AL. Microbubble contrast agents: targeted ultrasound imaging and ultrasound- assisted drug-delivery applications. Invest Radiol 2006; 41: 354–362.

Kodama T, Tan PH, Offiah I, Partridge T, Cook T, George AJ et al. Delivery of oligodeoxynucleotides into human saphenous veins and the adjunct effect of ultrasound and microbubbles. Ultrasound Med Biol 2005; 31: 1683–1691.

Liang HD, Lu QL, Xue SA, Halliwell M, Kodama T, Cosgrove DO et al. Optimisation of ultrasound-mediated gene transfer (sonoporation) in skeletal muscle cells. Ultrasound Med Biol 2004; 30: 1523–1529.

Liu Y, Miyoshi H, Nakamura M. Encapsulated ultrasound microbubbles: therapeutic application in drug/gene delivery. J Control Release 2006; 114: 89–99.

Litzinger DC, Brown JM, WalaI, Kaufman SA, et al. Fate of cationic liposomes and their complex with oligonucleotive in vivo. Biochimica et Biophysica Acta (BBA) - Biomembranes. 1996;1281(2):2139-149.

Marmottant P, Hilgenfeldt S. Controlled vesicle deformation and lysis by single oscillating bubbles. Nature 2003; 423: 153–156. 12 Sundaram J, Mellein BR, Mitragotri S. An experimental and theoretical analysis of ultrasound-induced permeabilization of cell membranes. Biophys J 2003; 84: 3087–3101. 13 Brujan EA. The role of cavitation microjets in the therapeutic applications of ultrasound. Ultrasound Med Biol 2004; 30:381–387.

Marshall E. Gene Therapy Death Prompts Review of Adenovirus Vector. Science.1999;286(5448):2244-2245.

Mehier-Humbert S, Bettinger T, Yan F, Guy RH. Ultrasound- mediated gene delivery: kinetics of plasmid internalization and gene expression. J Control Release 2005; 104: 203–211.

Michel MS, Erben P, Trojan L, Schaaf A, Kiknavelidze K, Knoll T et al. Acoustic energy: a new transfection method for cancer of the prostate, cancer of the bladder and benign kidney cells. Anticancer Res 2004; 24: 2303–2308.

Miller MW, Miller DL, Brayman AB. A review of *in vitro* bioeffects of inertial ultrasonic cavitation from a mechanistic perspective. Ultrasound in Med. & Biol.. Vol. 22. No. 9, pp. 1131-l 154,

Muruve DA, Barnes MJ, Stillman IE, Libermann TA. Adenoviral gene therapy leads to rapid induction of multiple chemokines and acute neutrophil dependent hepatic injury in vivo. Hum Gene Ther.1999;(10): 965–976.

Newman CMH, Bettinger T. Gene therapy progress and prospects: Ultrasound for gene transfer Gene Therapy. 2007;14 ; 465–475.

Rahim AA, Taylor SL, Bush NL, TerHaar GR, Bamber JC, Porter CD. Spatial and acoustic pressure dependence of microbubble-mediated gene delivery targeted using focused ultrasound. J Gene Med 2006; 8: 1347–1357.

Shohet RV, Chen S, Zhou Y-T, Wang Z, Meidell RS, Unger RH, Grayburn PA. Echocardiographic Destruction of Albumin Microbubbles Directs Gene Delivery to the Myocardium. Circulation 2000;101:2554-2556.

TerHaar G. Therapeutic applications of ultrasound . Prog Biophys Mol Biol 2007; 93: 111–129.

Thomas CE, EhrhardtA, Kay MA. progress and problems withthe use of viral vectors for gene therapy. 2003;4: www.nature.com/reviews/genetics p346-358.

Unger, E C, Mccreery Tp, Sweitzer Rh. Ultrasound Enhances Gene Expression of Liposomal Transfection Investigative Radiology.1997;32:723-727.

Zhigang W, Zhiyu L, Haitao R, Hong R, Qunxia Z, Ailong H et al. Ultrasound-mediated microbubble destruction enhances VEGF gene delivery to the infarcted myocardium in rats. Clin Imaging 2004; 28: 395–398.

Zhou QH, Miller DL, Carlisle RC, Seymour LW, Oupicky D. Ultrasound-enhanced transfection activity of HPMA-stabilized DNA polyplexes with prolonged plasma circulation. J Control Release 2005; 106: 416–427.

10 Chapter

GENE DELIVERY USING ULTRASOUND

1. INTRODUCTION: AN OVERVIEW

Human gene therapy shows significant potential in treating both inherited genetic abnormalities and many disease conditions, including cancer, viral infections, stroke, and myocardial infarctions. The achievement can be attained through the introduction of a functional gene into specific cells that have genetic deficiencies or mutations. Alternatively, it can be accomplished by delivering a therapeutic substance, such as agents that target cancer-causing oncogenes, growth factor genes, antisense oligonucleotides (ODN), or small interfering RNA (siRNA), to rectify the disease condition. This can be done using either viral or nonviral vectors. Viral gene therapy has demonstrated success in numerous animal illness models (Snyder 1999) and has advanced to the stage of clinical trials (Hacein-Bey-Abina et al. 2002; Kay et al. 2000). Nevertheless, there are still notable challenges to overcome, such as immunological reactions (Manno et al., 2006) or the development of tumors (Hacein-Bey-Abina et al., 2003). Utilizing a method that does not include viruses would offer a more secure and reliable procedure. The capacity of therapeutic ultrasound (US) to induce minimally invasive nonviral gene transfer has been acknowledged for a considerable time. A mounting body of research suggests that substantial improvement in transgene expression can be attained by employing high frequency acoustic radiation. US, in addition to its widely recognized function of offering affordable and immediate imaging capacity, has also been utilized for therapeutic purposes for a significant period of time (Herzog et al. 1999).

The predominant therapeutic usage entails the application of low acoustic intensities with the purpose of generating heat in deep tissues, such as in the field of sports medicine. HIFU (high intensity focused ultrasound) is situated at the opposite extreme of the acoustic intensity spectrum. It has the capability to either ablate tissues (Fischer et al., 2010) or liquefy them (Hall et al., 2009). Intermediate levels of ultrasonic energy have been utilized in various systems, along with externally introduced microbubbles [MBs], to exploit the acoustically-induced behavior of the MBs in order to produce specific biological effects. The latter typically entails modifying the permeability of natural barriers to allow the passage of substances that are normally unable to cross (such as medications or large molecules). Various gene therapies have been tried through direct injection of gene vectors into muscle tissue or organ tissue; these vectors quickly enter the interstitial space and then have to cross the plasma membrane of the specific cells they are targeting. In the United States, contrast agents are often delivered intravascularly. When a gene vector is present, the initial obstacle met is the vascular endothelium. Subsequently, there are other vascular anatomical characteristics (such as the basement membrane, smooth muscle layer, etc.) followed by the outer cell membrane of the cells that one aims to target. Ultimately, the DNA must be transported across the nuclear membrane in order to get access to the nucleus and for optimal gene expression.

This review will mostly discuss the application of ultrasound targeted microbubble destruction (UTMD) for delivering foreign DNA, medicines, or photo reactive nanoparticles into specific host tissues and cells. Numerous publications have explored these applications. The research on this subject is expanding rapidly due to the introduction of microbubble-based US contrast agents in the 1980s. It was soon discovered that acoustic cavitation might have both positive and negative impacts on biological systems.

The utilization of acoustically-activated microbubbles for therapeutic purposes continues to be a captivating subject. Undoubtedly, exploiting the dynamic properties of acoustic microbubbles is considered a very desirable achievement in the field of acoustically-guided gene (or medication) delivery, as mentioned by Lindner in 2009. Nevertheless, it is crucial to approach claims of significant accomplishments with UTMD with a healthy dose of skepticism, as evidenced by studies such as Villanueva (2009). Replication studies are scarce, and conflicting results are not uncommon. It is important to be skeptical, particularly when evaluating US studies that lack transparency. An increasingly clear understanding of gene therapy facilitated by the United States is developing. Using UTMD techniques, the problem ultimately stems from the ability of ultrasound (US) to induce dynamic behaviors in microbubbles (MBs). Endogenous microbubbles (MBs) are not present in the majority of bodily tissues (Carstensen et al., 2000; Gross et al., 1985). MBs that are delivered externally are typically inactive without exposure to sound waves, but can be induced to vibrate gently or exhibit intense but highly localized movement when subjected to acoustic fields. Targeting can be accomplished by giving microbubbles (MBs) and gene vectors together and subjecting the specific tissue to ultrasound (US). This will ensure that bubble activation only occurs at the targeted region. Bioeffects occur when microbubbles (MBs) are pushed at pressure amplitudes that are capable of inducing nonlinear oscillations in the bubbles.

In this discussion, we will cover the topics comprehensively, with the aim of providing the reader with a wide comprehension of the approaches, applications, apparent mechanisms, and some insights into the accomplishments that have been made. The majority of our discussion is centered around in vivo studies, as the use of ultrasound (US) to assist gene therapy in living organisms presents a more difficult problem compared to using US to promote cell permeabilization in laboratory settings or even in preserved tissues outside of the body (see; for example, Kodama et al., 2005). We have made an effort to avoid using technical language and have not included mathematical explanations of the subjects covered here, as these may be found in other sources. Several basic abbreviations are employed: US (ultrasound); MB (microbubble); Pa or Pr (acoustic pressure amplitude or rarefaction pressure, respectively); and pDNA (plasmid DNA). Additionally, it is imperative to discuss US exposure measures. The parameter of interest in certain applications of therapeutic ultrasound, such as tissue heating, is the acoustic intensity measured in units of W/cm2. The presence or absence of exogenous MBs and the peak acoustic intensity, namely the peak acoustic pressures, primarily influence the occurrence and kind of acoustic cavitation in vivo. These values are often measured in units of megaPascals (MPa) or kiloPascals (kPa), with 1 MPa being equivalent to 10 atmospheres (atm). The acoustic intensity increases proportionally to the square of the pressure amplitude. The level of auditory reporting in the gene therapy literature facilitated by the US differs significantly. Acoustic exposures will be described mostly in terms of the pressure amplitude (Pa) or the peak rarefactional acoustic pressure (Pr), which may be inferred by us at times. However, our intention here is not to be strictly quantitative.

2. ULTRASOUND-TARGETED MICROBUBBLE DESTRUCTION: PRINCIPLES AND TECHNIQUES IN PHYSICS AND TECHNOLOGY

There are many potential applications for UTMD gene delivery; a large body of work has been conducted using both *in* Utilize vitro and in vivo model systems to investigate the mechanisms and efficacy of UTMD in producing therapeutic effects. The primary objective of most studies is to utilize technology for the purpose of achieving minimally-invasive treatment of disease or condition. However, in vitro UTMD transfection techniques are currently being employed for cell-based therapeutics, as stated by Otani et al. (2009). Section 5 will cover the detailed uses of specific organs and clinical situations. It is important to mention that in studies involving UTMD-based gene delivery, a significant amount of research has been dedicated to using model or surrogate systems. For example, the focus has been on delivering reporter genes rather than therapeutic ones. Nevertheless, certain model systems have shown therapeutic gene transfer effects, namely in terms of reducing tumor bulk.

Therapeutic ultrasound has the capacity to improve minimally invasive gene therapies. UTMD approaches has numerous desirable qualities, especially for gene therapies that utilize naked DNA vectors. The advantages of this treatment are as follows: (1) all components of the treatment have low toxicity, (2) the vectors used have low immunogenicity, (3) the procedure is minimally invasive (for example, the vector and gas bodies can be administered intravascularly, and for organs that can be accessed sonographically, therapeutic ultrasound can be applied through the skin), (4) the treatment can be repeated multiple times, (5) specific organs can be targeted with high precision, and (6) the technique can be applied to a wide range of organs due to its sonographic accessibility. Nevertheless, the issue of low efficiency persists.

2.1 What is ultrasound targeted microbubble destruction?

In this article, we aim to present an overview of the common types of microbubbles (MBs) used in gene therapies facilitated by the United States, the various ways in which these MBs can be destroyed or activated, and the mechanisms involved in microvascular damage, extravasation, and the uptake of gene vectors induced by ultrasound-targeted microbubble destruction (UTMD). For a more detailed analysis of the physics, refer to the publication of Wu and Nyborg in 2008.

Generally, the targeting component of UTMD therapies relies on the observation that microbubbles (MBs) in tissues only respond dynamically when the area is subjected to ultrasound (US). The inclusion of ligands in the MB shell can increase the accumulation of the MBs and any cargo they transport in a specific area. However, it is the targeted acoustic exposure that triggers the activation of the MBs. As previously stated, it is possible to enhance the retention of MB shells in specific areas of interest by including site-specific ligands. The field in question has garnered significant interest, as noted by Ferrara et al. in 2009. Acoustic manipulation can be used to direct bubbles towards a specific area of interest. An acoustic field that is spreading out applies a radiation force on microbubbles (MBs), as demonstrated by Sarvazyan et al. in 2010. This force has the ability to drive the MBs in the same direction as the wave is moving. According to Emmer et al. (2007), when the MB experiences radial oscillations, the bubble moves forward abruptly, decelerating as it expands and accelerating as it contracts. The utilization of radiation force enables the concentration of microbubbles (MBs) along a particular wall of an intracavitary space or blood vessel, as demonstrated by Horie et al. in 2010.

The destruction of microbubbles (MB) in ultrasound-targeted microbubble destruction (UTMD) is triggered by acoustic exposure. This process can lead to the release of lipid or aqueous-phase drugs, as demonstrated by Smith et al. in 2010. Additionally, it can cause minor damage to microvessels, enabling substances that are typically impermeable, such as drugs, plasmids, cells, or fragments of microbubbles, to exit the blood vessels and enter the interstitial space. MB activation is also linked to cellular permeabilization.

2.2 Ultrasound contrast agents and other stabilized gas bodies

Microbubble US contrast agent evolution from early, agitated saline or sugar solutions to protein-shelled agents containing air to lipid- or polymer-shelled agents using relatively insoluble gases was rapid. The evolution of these agents continues (Qin *et al.* 2009), as do new applications for them (Cosgrove & Harvey 2009). US contrast agents are typically micron-sized gas bodies which are stabilized against diffusion by a shell (Overvelde *et al.* 2010; Sarkar *et al.* 2009); most contrast agent MBs have mean diameters of ~2 μm. New agents currently under development for imaging or therapy are smaller still, and are considered to be nanobubbles (Krupka *et al.* 2009). Multi-layered structures in which a gas body stabilized by a lipid monolayer is contained in an aqueous compartment bounded by a lipid bilayer (echogenic liposomes) are also under development. *When used for imaging, US contrast agents are unique amongst contrast agents in that they respond dynamically to the signal used to interrogate tissues for their presence.* It is this same property that is exploited for UTMD therapies or therapy models.

1.3 Microbubble destruction: Shell disruption and shell/gas body fragmentation

There are three principal mechanisms by which US or time can destroy a shelled MB (Chomas *et al.* 2001). In increasing order of 'violence' to the MBs, these are: (1) static diffusion, in which gas dissolves into the surrounding host fluid. This may be rapid if the bubble is 'free' (without stabilizing shell) or very slow if a stabilizing shell is present; (2) acoustically-forced shell disruption, which leads to accelerated diffusive loss of gas relative to unperturbed shelled bubbles; and (3) shell fragmentation and rapid loss of gas.

A micron-sized free air bubble can be expected to dissolve in water in ~30 ms (Sarkar *et al.* 2009), which is why relatively insoluble gases such as perfluorocarbons are now used in modern contrast agents, and a stabilizing shell is employed. The principal mechanism by which bubble shell materials stabilize MBs against diffusion is by reducing the surface tension at the bubble surface. Shell properties also affect the dynamic responses and stability of MBs at Pa below the shell fragmentation threshold (Emmer *et al.* 2007; Ferrara *et al.* 2009).

Acoustically-forced MB dissolution can occur when the Pa is sufficient to drive MB oscillations which stretch and compress the shell sufficiently to produce relatively small defects in the shell, which may re-seal (Huang 2008). Very high physical stresses develop in the shells as the MB oscillates, even when the driving pressures are only a few hundred kPa (Stride & Saffari 2003). Small shell defects lead to more rapid dissolution of the gas body than would otherwise occur, but complete gas dissolution may require many acoustic cycles (Smith *et al.* 2010). For examples, Optison MBs exposed to 3.5 MHz US of Pa > 0.15 MPa undergo accelerated loss of gas, indicating shell compromise (Porter *et al.* 2006). Shell disruption Pa thresholds for albumin-stabilized Optison and surfactant stabilized Sonazoid are similar; at 1.1 MHz, these are 0.13 MPa or 0.15 MPa, respectively.

At 3.5 MHz, these thresholds are somewhat higher (Optison: 0.48 MPa; Sonazoid: 0.58 MPa) (Chen *et al.* 2003). Others have reported a somewhat lower threshold (0.15 MPa) for Optison shell disruption at 3.5 MHz (Porter *et al.* 2006). Likewise, (Borden *et al.* 2005) studied the behavior of lipid monolayer-encapsulated MBs at 2.25 MHz using single-cycle pulses, and found that the Pr threshold for slow MB dissolution to be in the range of 0.4 – 0.6 MPa, and the Pr threshold for MB fragmentation to be 0.8 MPa. A Pa of around 0.3 MPa was also found to be the threshold for soft-shelled contrast agent MB disruption (de Jong *et al.* 2009). Echogenic liposomes also appear to have two pressure thresholds; one associated with compromise of the shell of the interior gas body, and a higher one associated with disruption of the outer lipid bilayer (Smith *et al.* 2010). Rapid fragmentation occurs on a time scale of microseconds when Pa is sufficient to buckle and completely rupture the shell (Marmottant *et al.* 2005), creating free gas bodies which may then dissolve, grow by rectified diffusion, coalesce into larger bubbles, or dissolve, depending stochastically on exposure conditions.

2.4 Microbubble dynamics: An overview of how gas bodies respond to ultrasound

At very low Pa, MB volume oscillations are related linearly to Pa. At modest Pas of a few hundred kPa (or less), MB volume oscillations become non-linearly related to Pa, with bubble expansion being relatively slow and bubble collapse much faster, being governed by the inertia of the in-rushing surrounding fluid to a greater extent than by the compressive phase of the applied pressure field; hence the phenomenon of bubbles undergoing acoustically-driven expansions followed by rapid, inertially-dominated collapse is termed 'inertial cavitation'. This typically occurs when a MB has been driven by the rarefaction phase of the acoustic wave to a diameter roughly 2 – 3 times the initial diameter (Chomas *et al.* 2001). At very modest Pas, inertial cavitation can be stable and sustained for an almost indefinite number of acoustic cycles. For example, (Church & Carstensen 2001) found that surfactant-coated Sonazoid MBs could undergo repetitive inertial collapses and rebounds when driven by 2.5 MHz US at acoustic pressures greater than about 0.3 – 0.4 MPa; for this agent, irreversible post-collapse fragmentation occurred at a Pa of ~ 1.5 MPa. Others have reported that a 3 µm bubble exposed to 2.25 MHz US can be expected to undergo expansion, inertial collapse, and fragmentation at a Pa just over 0.3 MPa (Chomas *et al.* 2001). The albumin-shelled, first-generation US contrast agent Albunex began to emit acoustic signatures characteristic of nonlinear oscillations at acoustic pressures as low as 0.005 – 0.010 MPa (Krishna & Newhouse 1997). At even modest Pa (*e.g.*, 0.6 MPa at 1 MHz), some contrast agent bubbles can expand to 10 times or larger than their equilibrium radius; under such conditions, inertial collapse is expected. Fragmentation of the inertially-collapsing bubble occurs at some instant near the time of minimum bubble radius {Postema & Schmitz 2007). Inertial bubble collapse is so rapid and the gas so compressed that the maximum temperature inside the bubble can reach more than 5000 K (Apfel & Holland 1991). Light emissions (Matula 2003) and reactive free radicals may be produced (Okada *et al.* 2009). MB rebound and fragmentation can re-emit acoustic energy at many times the excitation pressures. These emissions also contain higher frequency spectral components which can produce local heating. In the present context, however, it appears that the most important determinant of UTMD gene delivery methods is that *bubble expansion and collapse can produce local tissue distortions or damage which are presumed to be the principal mechanism by which microvascular bioeffects become manifest, and target cell permeabilization occurs.*

Cavitation bubbles can interact with nearby boundaries; these interactions can produce high velocity fluid jets and induce bubble translation, with the directions of fluid jets and translation depending on the boundary's properties. For rigid boundaries, bubbles closer to the boundary than about twice the fully-expanded radius of the bubble (Kodama & Tomita 2000) may collapse asymmetrically, with a high velocity water jet 'punching through' the bubble and impinging on the rigid boundary. Near rigid, planar boundaries, an oscillating MB translates toward the boundary and if collapse jetting occurs, the jet is directed toward the rigid surface (Plesset & Chapman 1971) with sufficient water hammer pressure to damage the surface (Blake & Gibson 1987). *In vivo*, however, this condition would seem rare.

MBs driven to oscillate near pressure release boundaries (an air-water interface, for example), collapse jets which occur are directed away from the boundary, as is the direction of bubble translation (Chahine 1977; Robinson *et al.* 2001).

Oscillating MBs collapsing near a 'soft', planar boundary (*e.g.*, gels or large vessel walls) also undergo translation and jetting, but the directions of jetting and translation can be either toward or away from the boundary, being determined by bubble size, bubble distance from the boundary, and the mechanical properties of the boundary (see. *e.g.*, (Kodama & Tomita 2000; Shima *et al.* 1989). Deformation of the elastic boundary stores energy which is 'returned' as the bubble collapses; this creates a hydrodynamic pressure gradient which in turn produces fluid flow away from the elastic boundary. If the pressure gradient is large enough, bubble collapse jets directed away from the boundary, and bubble translation away from the boundary, can result (Blake & Gibson 1987). Liquid jets were observed in an early experimental study of acoustically forced MB behavior in a 200 μm diameter cellulose tube (Postema *et al.* 2004). In vessel-simulating gel tunnels, at 1.7 MHz, the Pa at which Optison MBs began to emit broadband noise characteristic of inertial collapse and rebound was weakly dependent on tunnel diameter (~0.8 MPa in 90 μm diameter tunnels *vs.* ~0.6 MPa in 800 μm tunnels) (Sassaroli & Hynynen 2007). Single-MB dynamics observations in actual microvessels were acquired using a rat cecum model; MBs translated toward the vessel walls and a toroidal bubble morphology consistent with the formation of a microjet was observed (Caskey *et al.* 2007).

Where bioeffects attributable to inertial cavitation occur, cavitation jets directed toward cells or tissues have long been assumed to be causative. Vessel 'stretching' during bubble expansion has also been proposed as a mechanism of vessel damage (Miao *et al.* 2008). Very high speed imaging of MBs driven to growth and inertial collapse in microvessels of rat mesentery (Chen *et al.* 2010; Chen *et al.* 2011) indicate that (1) MB collapse jets form frequently in the intravascular environment, (2) jets are typically directed *away* from the nearest blood vessel wall, (3) blood vessel expansion in response to bubble growth is minimal, and (4) tissues 'follow' the collapsing bubbles such that inward vessel distortions are much greater than outward distortions, with inward vessel wall motion having speeds of 5 – 10 m/s and the events occurring on microsecond time scales. This is perhaps one mechanism by which vessel permeabilization occurs. In any case, permeabilization of biological transport barriers is associated with some cell killing, both *in vitro* (Brayman *et al.* 1999) and *in vivo* (Ferrara 2008; Miller *et al.* 2011; Price *et al.* 1998; Skyba *et al.* 1998)

Finally, it is worth mentioning that MB-excitation techniques for cellular permeabilization need not use US to achieve MB activation or cell permeabilization; targeted laser illumination can also be used effectively. For a recent research paper on this topic, using a liposome MB contrast agent, see (Zhou *et al.* 2010).

2.5 Effect of different shell compositions or agents on gene delivery

It is difficult to compare the efficacy of different contrast agents in US-mediated gene delivery between different studies, because MB type, concentration as injected, rate of injection, total dose, animal model, US exposure source, exposure conditions, and even the way the MBs are handled vary widely and can influence outcome. MB concentration is a determinant of UTMD-mediated pDNA expression in mouse liver (Miao *et al.* 2005; Shen *et al.* 2008). Different agents have different 'native' concentrations, and agents with similar gas content but dissimilar shells can have different efficacies in enhancing gene delivery under otherwise comparable conditions (*cf.* effects of Optison and PESDA MBs; (Pislaru *et al.* 2003)). Relatively large differences between contrast agent types in UTMD gene delivery have been observed, but the reasons are not always clear. With equalized MB concentrations and either Optison, SonoVue or Sonazoid contrast agents, UTMD-mediated transfer of phosphorodiamidate morpholino oligomer to a dystrophin-deficient murine heart model induced dystrophin-positive cells in the hearts when harvested 7 days after treatment.

Optison and Sonazoid were equivalent in effectiveness, producing about four times greater expression than did SonoVue (Alter *et al.* 2009).

2.6 Sensitivity of stabilized gas bodies to technique

Bubbles are buoyant, fragile, can coalesce into larger bubbles, or 'disappear' as gas is lost to diffusion. In many *in vivo* studies of UTMD, the MBs and gene vectors are infused slowly; this prolongs the time window in which acoustic treatments can occur, but aggravates the problem of time-dependent changes in MB distribution and delivery (Kaya *et al.* 2009). It is worth noting that MBs can be destroyed by static pressure or tension; an over-pressure of only ~0.3 MPa can destroy them (Stringham *et al.* 2009). MBs can also be destroyed by drawing up or injecting an MB suspension *too rapidly* when using a small gauge needle (Talu *et al.* 2008); thus one can unwittingly alter experimental outcome by 'over-enthusiasm'.

3. BREACHING THE PHYSICAL BARRIERS TO GENE DELIVERY USING ULTRASOUND

In order to achieve efficient gene expression following ultrasound-mediated gene delivery, multiple barriers need to be overcome to allow pDNA to enter into the nucleus of target cells including penetrating vascular and cellular membranes as well as trafficking through different intracellular compartments.

US contrast agents are intended to be intravascular agents. Whether the gene vector is administered as a simple mixture with the gas body suspension, or is in some way linked to the gas bodies, they are also intravascular agents in vascular UTMD methods. The first barrier encountered is the vascular endothelium. The next are other vascular anatomical features (*e.g.*, the basement membrane, smooth muscle layer, *etc.*) and then the outer cell membrane of the cells one hopes to target. Some intracellular membranous compartments must also be traversed. However, MBs have been used with intramuscular or intraparenchymal injections of vectors, and some successes reported. In UTMD-based gene transfer methods using vascular approaches, to mediate gene therapy *via* acoustic excitation, gas bodies must first exert their influence from within the vascular lumen.

Most of the available evidence from *in vivo* studies indicate that vessel permeabilization effects occur principally in the microcirculation; larger vessels are too robust to be penetrated by cavitation events, even if their vascular endothelium can be effectively destroyed by intraluminal inertial cavitation (Hwang *et al.* 2005). Extravasation of dyes, nanoparticles or macromolecules through microvessels is almost always accompanied by extravasation of red cells (see below); since these have diameters on the order of 6 μm, breaches in the endothelial wall can be quite large. However, there is also some evidence that more subtle effects, such as partial opening of the tight junctions between endothelial cells, can also contribute. Assuming that the gene vector escapes the intravascular compartment and enters the interstitium, it must then enter the surrounding cells; thus the plasma membrane is the second major barrier encountered by the vectors. Lethal effects of cavitation occurring in the cardiac microcirculation can extend outward into the myocardium (Miller *et al.* 2011); there is good reason to expect that sub lethal poration of cells located within a few cell diameters of the intravascular cavitation event(s) also occurs.

3.1 Extravasation of dyes, nanoparticles and cells

Evans blue [EB] is an azo dye which binds serum albumin with high affinity, and is normally unable to pass through the endothelium. Extravasation of EB through reversible or irreversible capillary modification has been observed in small animal (rat) hearts exposed to low-MHz frequency US with the use of various contrast agent MBs, accompanied by premature ventricular contractions (Li *et al.* 2004). The effect was sensitive to Pa (apparent threshold somewhat less than 1.6 MPa at 1.5 MHz) and to the concentration of the injected MBs (Miller et al. 2004). The effect appears to be of mechanical origin (Miller *et al.* 2004) With injection of dilute Definity suspensions, extravasation of EB from canine hearts, and cell killing of cardiomycetes within the ultrasonically-interrogated area also occurs under diagnostically realistic exposure conditions (Miller *et al.* 2006).

Erythrocyte extravasation has been observed in rat kidney glomerular capillaries exposed to diagnostic US with contrast agent MBs in the circulation (Miller et al. 2010a, Miller et al., 2010b); such damage was associated with tissue inflammation at 1 week and fibrosis at 4 weeks (Miller *et al.* 2009). A threshold Pa of ~0.8 MPa was indicated (Miller *et al.* 2007). Glomerular capillary hemorrhage with contrast-enhanced diagnostic US also occurs in large animals (pigs) exposed to 1.5 MHz, 1.7 MPa US (Miller *et al.* 2010). Petechial hemorrhage on intestinal blood vessels in an *in vivo* murine model was observed with the contrast agent Albunex, with apparent Pa thresholds of 0.4 MPa at 0.4 MHz, 0.8 MPa at 1.1 MHz, or 2.3 MPa at 2.3 MHz; the thresholds were lower with the agent Levovist than with Albunex (Miller & Gies 1998a, Miller & Gies 1998b). At 2.25 MHz, a threshold Pa of between 0.85 and 1.0 MPa was indicated for vessel damage in exteriorized rat cremaster muscle containing Definity MBs (Samuel *et al.* 2009). Similarly, intravital observation of red cell extravasation from rat cremaster muscle capillaries containing MB contrast agent was observed more than a decade ago (Price *et al.* 1998; Skyba *et al.* 1998). Microscopic observations of US and MB-induced red cell extravasation have been reported in glass catfish (Maruvada & Hynynen 2004). In exteriorized rat spinotrapezius muscle, 1 MHz US at a Pa of 0.75 MPa was shown to result in extravasation of 100 nm diameter microspheres co-injected with contrast agent MBs (Song *et al.* 2002). In rats, *in vivo* exposure of the kidneys following injection of MBs resulted in red cell extravasation from the glomerular capillaries into Bowman's space with an apparent threshold Pa of ~ 0.73 MPa (at ~ 1 MHz), with nearly 40% of histological sections taken from the focal plane showing extravasation at a Pa of 1.8 MPa (Miller *et al.* 2007). Intravital microscopy and concurrent cavitation detection was used to study the relationship between bubble dynamics and extravasation of red cells from rat cremaster muscle using 2.25 MHz US, and Definity contrast agent infused *via* the tail vein.

Vascular damage and acoustic emissions from the MBs were correlated. The greatest amount of red cell extravasation and the greatest cumulated bubble acoustic emissions occurred at 10 Hz pulse repetition frequency **[PRF]**, indicating that the time for tissue refill with bubbles following each pulse was ~100 ms. In experiments in which pulses were applied at 100 Hz PRF and the Pa varied from 0 – 2.0 MPa, there was no vascular damage at Pr ≤ 0.85 MPa, but unambiguous damage occurred at Pr ≥ 1.0 MPa (Samuel *et al.* 2009). Finally, there is evidence that low amplitude US induced MB oscillations (1 MHz, Pr = 0.l MPa, SonoVue bubbles) can increase the permeability of primary endothelial monolayer cultures *in vitro*. Insonation produced immediate influx of Ca^{2+} ions into the cells, indicating poration of the endothelial plasma membranes. The effect was essentially abolished by application of catalase, strongly suggesting a role for extracellularly-produced H2O2 associated with nonlinear bubble oscillations. Moreover, histochemical staining for a protein associated with gap junctions showed an approximately 50% increase immediately after insonation, but returned to control levels within 30 minutes of insonation (Juffermans *et al.* 2009).

3.2 Transient poration of the cell membrane

Two mechanisms by which US and MBs facilitate poration of cell membranes are prominent; these are: (1) cavitational; *e.g.*, the opening of transient holes in membranes in consequence to local shear forces exerted on membranes by fluid flow ('micro streaming') around oscillating bubbles, local shock waves (which produce large pressure gradients across a cell), or cavitation microjets, or (2) endocytosis (Doinikov & Bouakaz 2010; Walton & Shohet 2009). Other mechanisms have been proposed, of course. However, it appears that shear stresses associated with bubble activation is probably the principal mechanism. A third mechanism by which normally 'tight' physiological barriers may be permeabilized is by sonochemical stimulation.

Transient pores of a few hundred nanometers in diameter and lifetimes of several seconds can be formed in cell membranes by acoustically-driven, single MB oscillations (Deng *et al.* 2004; Han *et al.* 2007;

Zhou *et al.* 2008; Zhou *et al.* 2009). Voltage clamp studies of *Xenopus* oocytes exposed to 1 MHz US showed that without MBs, there was no change in current at Pr as high as 1.2 MPa. With Optison MBs, opening and resealing of individual pores was observable even at Pr as low as 0.2 MPa. The transmembrane current was carried by influx of Ca2+ ions. The transmembrane current was greater at 0.4 MPa than at 0.2 MPa; in both cases, pore lifetime was ~2 s. More generally, the effect was Pa dependent, increasing slowly with increasing Pa over the range of 0.3 – 0.55 MPa, and then inflecting sharply upward at higher Pa (Deng *et al.* 2004).

Pore opening showed a high level of temporal correlation with inertial cavitation noise (Zhou *et al.* 2008). Pore size was estimated as 220 ± 80 (mean ± SD) nanometers, and refined estimates of pore resealing times indicated closure in 3 – 5 s (Zhou *et al.* 2009). *In vitro* studies of cells in contact with SonoVue MBs excited by 1 MHz US at Pa of 0.05 – 2.50 MPa have shown Ca2+ influx associated with poration (Juffermans *et al.* 2008). Sonoporation to naked DNA *in vitro* is correlated with inertial cavitation activity (Qiu *et al.* 2010), and depends in part on the fluidity of the membrane at the time of UTMD treatment (Zarnitsyn & Prausnitz 2004), with higher reporter gene transfection rates at 37°C than at 21°C. This is consistent with the need for porated membranes to reseal rapidly in order to maintain viability. Still others have shown that 1 MHz US at Pa as low as 0.1 MPa can permeabilize cells *in vitro* to pDNA (Rahim *et al.* 2006). In cell suspensions containing Definity MBs, the Pa thresholds for cell permeabilization to 70 kDa FITC-dextran, propidium iodide (indicating cell death) and MB destruction were 75 kPa at 0.5 MHz, 200 kPa at 2.5 MHz, and 600 kPa at 5 MHz (Karshafian *et al.* 2009). However, the idea that cellular poration results from inertial cavitation has been challenged on the basis of observations of cavitation noise and permeabilization of *in vitro* cell monolayers to 500 kDa FITC dextran; the supporting data indicate that permeabilization precedes the occurrence of broadband noise associated with inertial collapse and rebound cavitation. The authors conclude that the forces associated with microstreaming around activated bubbles is the principal mechanism of sonoporation (Forbes *et al.* 2008). More study is clearly needed.

US and MB-induced cell lysis and reversible poration in other cell types *in vitro* are strongly correlated with inertial cavitation (Chen *et al.* 2003a; Chen *et al.* 2003b; Lai *et al.* 2006). Cell surface antigens may be stripped off the surface of viable cells during such events (Brayman *et al.* 1999). While *in vitro* UTMD-based transfection with naked DNA is often associated with high cell mortality, this is not always the case (Wang *et al.* 2009).

The use of the collapse jets of MBs generated by laser pulses to selectively and directionally sonoporate individual cells has reached a high level of sophistication, producing pore sizes of ~ 200 nm (Sankin *et al.* 2010), which may someday prove useful for cell therapies. This is very similar to pore size determined by sonoporation of *Xenopus* oocytes using 1.07 MHz US of 0.3 MPa Pa with Definity MBs; *viz.*, a mean pore size of 220 ± 80 nm diameter (Zhou Y *et al.* 2009). Lipoplexes, which are normally taken up by cells *via* endocytosis, are taken up poorly when 'PEGylated' (coated with polyethylene glycol). However, when PEGylated pDNA/lipolexes were attached to acoustically activated MBs, large increases in PEGylated lipoplex transfection were achieved in an *in vitro* model system, relative to ether 'free' PEGylated lipolexes, or lipolexes + MB simple (un-linked) mixtures (Lentacker *et al.* 2009). The investigators also used a number of endocytosis inhibitors; their results indicate that endocytosis was not the primary mechanism involved. They postulate that lipoplex-loaded MBs collapsing near cell monolayers in culture plates released the lipolexes, resulting in high local lipoplex concentrations, that some of the released lipolexes became entrained in MB collapse jets and were 'injected' into the cells with the fluid jet (see also (Miller 2000)). However, Meijering and colleagues, using fluorescently labeled dextrans of ~4 – 500 kDa, present data which indicate that the principal mechanism by which UTMD treatment (1 MHz US, 0.2 MPa Pr, SonoVue

MBs) facilitates macromolecule delivery across the plasma membrane is *via* induction of endocytosis. Sonoporation was also observed, but did not seem to be the mechanism for macromolecular uptake (Meijering *et al.* 2009). It is clear that the our understanding of the mechanism(s) of UTMD-enhanced uptake of macromolecules remains incomplete.

3.3 Entry of pDNA into the nucleus

Following delivery of pDNA across the plasma membrane of cells into cytoplasm, pDNA may travel through multiple cellular compartments and finally enter the nucleus *via* diffusion (Liang *et al.* 2004) or other assisted mechanism to produce efficient transgene expression. UTMD may facilitate overcoming some of these barriers most likely *via* cavitation bioeffects. It was reported that long-term exposure to therapeutic ultrasound (1 MHz, 2 W/cm2, 30% duty cycle for 30 mins) can overcome the rate-limiting step of driving DNA into the cell nucleus (Duvshani-Eshet & Machluf 2005). One thousand fold higher gene expression levels of luciferase was achieved with minimal loss in cell viability (<20%) in three different cell types (BHK, LNCaP, and BCE). These data implied that therapeutic US is the main driving force delivering pDNA not only to the cell cytoplasm but also to the nucleus. However, in this case, no MB was used. The same group (Duvshani-Eshet *et al.* 2006) showed that adding Optison further increased transfection levels. Confocal and atomic force microscopy studies indicated that long-term therapeutic US application localizes the pDNA in cell and nucleus regardless of Optison addition. In addition, the use Optison did not affect the kinetics of protein expression, indicating Optison did not affect DNA trafficking to the nucleus. They hypothesized that US application by itself plays a major role in delivering DNA to the nucleus.

An interesting recent report used Doxorubicin (DOX) as a molecular nanotheranostic agent to study UTMD-mediated intracellular delivery and nuclear trafficking (Mohan & Rapoport 2010). DOX is a popular research tool due to its inherent fluorescence and was encapsulated in poly(ethylene glycol)-co-polycaprolactone (PEG-PCL) micelles or PEG-PCL stabilized perfluorocarbon nanodroplets in this study. US triggered DOX trafficking into cell nuclei; the trafficking was further enhanced in the presence of phase-transition nanodroplets which become gas MBs upon US exposure of sufficient Pr. This was believed to be due to cavitation induced transient permeabilization of both plasma and nuclear membranes, thus allowing DOX penetration into the cell nuclei. Whether MBs and/or nanodroplet emulsions can significantly augment DNA delivery to the nucleus needs further investigation.

Once pDNA enters the nucleus, some DNA can be condensed by histones and form persistent nucleosome-like structures. Persistent gene expression from these stable episomal pDNA genomes requires the introduction of specific cis-acting elements in the gene transfer constructs. We have demonstrated that incorporation of locus control region and intron elements into specific gene transfer constructs can achieve persistent expression of therapeutic-levels of coagulation proteins in the liver following nonviral gene therapy (Miao *et al.* 2001; Miao *et al.* 2003; Ye *et al.* 2003). The locus control region (Yant *et al.* 2003) can contribute to the open chromatin structure of the pDNA genome and avoid silencing of the transgene expression cassette (Miao *et al.* 2005). Furthermore, addition of an intron element significantly enhanced the transcription efficiency of stable mRNA (Miao *et al.* 2005). Thus, gene transfer of episomal pDNA into slow-dividing or terminally differentiated cells facilitated by UTMD has high potential to achieve a therapeutic effect to treat specific diseases. For gene transfer into dividing cells where integration of target genes are required, incorporation of other systems such as sleeping beauty transposons (Aronovich *et al.* 2011; Yant *et al.* 2007) or PhiC31 integrase (Keravala *et al.* 2011) with UTMD-mediated nonviral gene transfer methodology can lead to long-term transgene expression.

4. THE VECTORS: PACKAGING DESIRED GENES FOR ULTRASOUND-MEDIATED DELIVERY

4.1 Naked DNA: Plasmids

Naked plasmid DNA provides many advantages as a nonviral gene transfer vector, including: (1) ease of preparation, (2) cost-effectiveness, (3) minimum toxicity, and (4) it is least immunogenic of the vectors. Indeed, the immunogenic CpG moiety can be modified easily if needed. Most importantly, quality control is quite easy compared to other pDNA complexes containing synthetic vehicles. However, it has been a very challenging problem to deliver naked pDNA into specific cells due to its large size and negative charge. *In vivo* delivery of naked pDNA is especially difficult with additional impediments of instability of pDNA in blood serum as well as in cellular sub-compartments such as cytosol, endosome, and nucleus after cell entry. Recently a hydrodynamic approach has been developed to drive efficient gene delivery into liver (Liu *et al.* 1999) and muscle (Danko *et al.* 1997) resulting in therapeutic levels of transgene expression in animal disease models, including hemophilia B (Miao *et al.* 2003) and others (Zhang *et al.* 2000). This method in its current form is not suitable for clinical use; however, notable recent advances have been made in large animal models (*vide, e.g.*, (Fabre *et al.* 2008; Kamimura *et al.* 2009; Suda *et al.* 2008). Alternatively, innovative US and MB technology to facilitate delivery of naked pDNA is a potential clinically feasible nonviral gene therapy approach (Miao *et al.* 2005; Shen *et al.* 2008; Song *et al.* 2011).

The potential for sonoporation to increase pDNA loading of cells was recognized years ago (Fechheimer *et al.* 1987). Sonoporation-enhanced transport of nanoparticles into cells is dependent on molecular size; uptake of particles ≤ 37 nm diameter was enhanced by sonoporation without gross damage, but particle uptake generally declined as particle size increased (Mehier-Humbert *et al.* 2005a). pDNA gene expression is faster with sonoporation than with liposome-based methods which depend on endocytosis (Mehier-Humbert *et al.* 2005), indicating that pDNA enters cells through transient pores. Moreover, due to the risk of pDNA degradation by serum nucleases and removal by phagocytes (Niidome & Huang 2002), injection of a nonviral gene transfer vector long before US treatment occurs is unlikely to be effective. With simultaneous delivery of pDNA and US exposure, UTMD can significantly facilitate the transfer of naked pDNA up to several hundred fold *in vitro* (Miller *et al.* 2002; Newman & Bettinger 2007) and up to several thousand fold *in vivo*, depending on different US and transfection systems (Chen *et al.* 2010; Miao *et al.* 2005; Shen *et al.* 2008; Song *et al.* 2011). Furthermore, with introduction of specific cis-acting elements in the gene transfer constructs, persistent expression of near-therapeutic levels of proteins can be achieved from episomal plasmids following UTMD-mediated gene therapy (Miao *et al.* 2005). These results demonstrate that development of UTMD has high potential to achieve a therapeutic effect for treating specific diseases.

4.2 Encapsulated or compacted DNA

Polyanionic solutes can be complexed to phospholipid polar head groups *via* $Ca2+$ bridges (Huster *et al.* 1999). Thus, anionic DNA molecules can be compacted onto cationic MB lipid shells. Polylysine has also been used to link naked DNA to phospholipid MB shells (Wang *et al.* 2009). It appears to be widely believed that enhanced efficiency of gene delivery can be obtained with pDNA in close proximity to MBs and any cell membrane 'defects' they may create, thereby increasing the probability that pDNA will be available to enter through these pores prior to their closure. Thus far, however, there has been inconsistent support for this idea. For example, pDNA coupled electrostatically with cationic MBs were used for local delivery of DNA to vascular muscle cells. One percent of cells were transduced with 40% of the cells remaining viable (Phillips *et al.* 2010). However, reporter plasmid bound to the cationic MB preparation MRX-225 was used to transfect canine myocardium which was exposed to diagnostic US. Reporter gene activity was only observed in the myocardium of those animals that received MB-linked DNA and were exposed to US but not in control untreated animals (Vannan *et al.* 2002). It is not clear if the pDNA-MB linkage influenced the experimental outcome.

Due to large size of DNA molecules and the concerns of enzymatic degradation of the injected pDNA as well as the low pDNA concentration in the vicinity of sonoporated cell membranes, polymer-coated MBs that can bind and protect the pDNA have been developed for UTMD-mediated gene delivery. Coating albumin-shelled MBs with poly(allylamine hydrochloride) (PAH) makes the surface charge of the MBs positive, but did not affect the size distribution of the MBs. The cationic coating allowed the MBs to bind to 100 fg of pDNA per MB and protected the bound DNA against nucleases. The PAH coating also significantly increased the lifetime of MBs (half-life ~7 h), making them more convenient for *in vivo* applications (Lentacker *et al.* 2006).

Another approach to compacting DNA onto MBs is to first incubate the pDNA with a cationic lipid such as GL67 or a cationic polymer such as polyethylenimine (PEI), followed by mixing with MBs. P To better understand the relationship between EI/DNA mixed with SonoVue MBs were injected intravenously in tumor–bearing mice. Following US exposure, reporter gene expression in tumor xenografts was significantly enhanced without causing any apparently adverse effect. Furthermore, with UTMD and PEI complex, vectors carrying a short hairpin RNA (shRNA) targeting human survivin were efficiently delivered into the tumor site, leading to inhibition of surviving gene expression and apoptosis of the tumor cells (Chen *et al.* 2010).

DNA loading of MB shells can be impressive, but may also be limiting to high yield gene therapies using UTMD. By first forming pDNA-Lipofectamine 2000 complexes, mixing these with home-made phospholipid mixtures, agitating under a perflutren atmosphere to generate MBs, and then repeatedly washing the bubbles to remove unassociated DNA, a per bubble loading of 50 x 10-15 gram/MB was achieved (Chen *et al.* 2006). Similar binding rates (approximately 100 fg/MB) were reported by others (Carson *et al.* 2011). Using a layer-by- layer approach, DNA loading of pre-existing cationic MBs was increased 10-fold by first bringing anionic DNA to the surface of the cationic bubble, followed by binding a coating of poly-cationic polylysine to the DNA, followed by more DNA, *etc.* Loading as high as ~2500 fg DNA/MB was achieved, without apparent impact on the bubbles' dynamical response to acoustic excitation (Borden *et al.* 2007). However, others have found that *in vitro* transfection rates are not enhanced by conjugation of reporter pDNA to MB shells (Tlaxca *et al.* 2010), as found in the *in vivo* canine myocardium example discussed earlier in this section.

4.3 Viral vectors

There have been many reports that US treatment alone, or in combination with MBs, can increase transgene uptake by cells (Miller *et al.* 2002; Newman & Bettinger 2007). Even with the difficulty of translating UTMD-mediated results obtained under tissue culture conditions (*e.g.*, infinite media for suspended cells *vs.* nearby noncompliant boundaries for monolayers) to the fully 3-D, viscoelastic intravascular environment *in vivo*, *in vitro* experiments often have the advantage of being better controlled and the results therefore more easily understood. This is not always the case, however. Zheng and colleagues exposed two different endothelial pigment cell lines to adenoviral gene vectors in combination with US or US and MBs (1 – 3 W/cm2; Pr appears to have been ~0.17 – 0.30 MPa at 1 MHz; SonoVue MBs were used when bubbles were employed). They found that treatment with US and MBs increased adenoviral gene transfer in human retinal pigment epithelium cells, but was without effect on rat retinal pigment epithelial cells under otherwise identical conditions (Zheng *et al.* 2009). The differences were speculated to arise in consequence to differing ability of the two cell lines to phagocytize the SonoVue MBs, to which the adenoviral vectors were believed to be attached.

It is difficult to limit the specificity of delivery of viral vectors, which are usually delivered systemically. By using retrovirus-loaded MBs, UTMD facilitated the delivery of viral vectors in a restricted area of cells exposed to Pr of 0.4 MPa or greater, despite uniform dispersion of the vector (Taylor *et al.* 2007). An envelope-deficient retroviral vector was combined with cationic MBs and added to target cells. Transduction efficiencies and sites can thus be controlled by means of US exposure. These results emphasize that UTMD can not only facilitate the delivery of nonviral vectors, but also has the potential to enhance efficiencies and restrict targeted sites of viral gene transfer.

4.4 Small RNAs and oligonucleotides

Small RNAs and oligonucleotides (ODN) have recently been developed as promising therapeutics to treat diseases like viral infections, cancer, and several genetic disorders. Among these, small inhibitory RNA (RNAi or siRNA) based therapeutics have been investigated for treating a number of different diseases, including viral infections (*e.g.*, hepatitis, HIV, influenza), cancer, Huntington's disease, and others. Other therapeutic agents including microRNA (miRNA), transfer RNA (tRNA), and antisense ODN are also being developed to regulate gene or cell functions as treatment regimens. Since these agents are small, it is expected that they will be good candidates for augmented delivery by UTMD.

The US contrast agent pioneer Thomas Porter recognized early that UTMD delivery of oligonucleotides had the potential to influence vascular tissue remodeling after injury. In a 2001 study, an oligonucleotide which inhibits vascular smooth muscle cell proliferation was bound to albumin-shelled MBs and UTMD effected by transcutaneous application of 20 kHz US to porcine carotid artery walls following balloon catheter injury. Thirty days after treatment, the percent area stenosis in UTMD-treated animals was half that in controls (Porter *et al.* 2001). The uptake of ODNs into intact *ex vivo* human saphenous veins and isolated smooth muscle cells from the veins was also potentiated by US (Kodama *et al.* 2005). In addition, UTMD facilitated the delivery of antisense ODN targeting the human androgen receptor (AR) in prostate tumor cells, resulting in 49% transfected cells, associated with a decrease in AR expression compared to untreated controls (Haag *et al.* 2006).

UTMD-mediated sonoporation (frequency: 1 MHz; intensity: 2 W/cm2; exposure time: 2 min) was capable of enhancing *in vivo* siRNA delivery into salivary gland of rats, leading to significant GAPDH gene silencing by 10-50% for 48 hours (Sakai *et al.* 2009). No gene silencing was observed with exposure to US only in the absence of Optison MBs. Intraventricular co-injection of siRNA-GFP and MB BR14 with concomitant ultrasonic exposure resulted in a substantial reduction in EGFP expression in the coronary artery in EGFP transgenic mice {Tsunoda *et al.* 2005). Liposomal MBs combined with US can efficiently delivery siRNA with only 10s of US exposure *in vitro*. siRNA was also efficiently delivered into the tibialis muscles using the same system and the gene-silencing effect could be sustained for more than 3 weeks {Negishi *et al.* 2008). These results demonstrate that UTMD-mediated delivery of siRNA can serve as a very useful tool for loss-of-function genetic engineering both *in vitro* and *in vivo*.

4.5 Transduced cell therapy

Cell therapy is a promising strategy for many applications, including genetic diseases, cancer, regenerative medicine, and others. However, it is very difficult to transfect certain cell types and maintain their viability following transfection, including hematopoietic and mesenchymal stem cells, T cells, and others which are important targets for cell therapy using the transfection methods currently available. UTMD has been demonstrated to facilitate the delivery of siRNA into mesenchymal stem cells (MSCs) {Otani *et al.* 2009), which knocked down mRNA expression of specific genes, leading to the improvement of cellular function and viability. The application of UTMD has high potential to facilitate the delivery of genetic materials into target cells and can be expanded for use in a variety of cell therapy protocols.

5. SELECTED APPLICATIONS OF ULTRASOUND-MEDIATED GENE DELIVERY

5.1 Cardiovascular

Recent reviews of cardiac applications of UTMD are provided elsewhere (Laing & McPherson 2009; Porter 2009). Desired endpoints may be improved vascular function (*e.g.*, following ischemia) or inhibition of vascularization to 'starve' tumors. The following discussion is intended only to provide an indication of some of the exciting avenues of research in the area.

Cardiovascular graft remodeling: Tissue remodeling after transplantation surgery is required for long term transplant success. It has been shown that US-mediated gene therapy can improve transplanted vessel patency after surgery. Carotid interposition saphenous vein grafts in pigs were treated *ex vivo* prior to transplantation with 1 MHz US at ~1.8 MPa Pa with both a MB contrast agent and a plasmid encoding for metalloproteinase 3 (TIMP-3; the enzyme inhibits post-graft vessel restriction) present during US exposure. At 4 weeks, luminal diameters in animals receiving the transfected grafts were significantly greater than in controls (Akowuah *et al.* 2005). Similarly, US treatment enhanced the delivery of an adenoviral vector to the aortic root, yielding a 2.5-fold enhancement in gene delivery (Beeri *et al.* 2002). A technical issue of note is that a balloon catheter was used to briefly occlude the aortic root above the sinuses to increase the dwell time of the injected MBs and adenoviral vectors.

Reperfusion therapies: US-mediated gene therapy to improve myocardial reperfusion following induced myocardial infarcts in mice was studied using Definity MBs, and UTMD achieved using high frequency (8 MHz), relatively high Pa (estimated ~4.5 MPa) US from a diagnostic US machine. Plasmids were either empty (controls) or encoded for Stem Cell Factor (SCF; expected to enhance reperfusion by recruitment of cells during tissue remodeling) or VEGF (expected to stimulate angiogenesis). At 21 days, UTMD with either VEGF or SCF-bearing pDNA increased the microvessel density and blood flow relative to controls (Fujii *et al.* 2009). Similarly, reperfusion of ischemic rat hind limbs was improved by US-mediated gene therapy. Cationic lipid-shelled MBs and a pDNA encoding for VEGF-165 were used; US frequency and amplitude were 1.3 MHz and ~2 MPa, respectively (Kobulnik *et al.* 2009). A murine cardiac infarct model treated with UTMD with pDNA encoding for either VEGF or stem cell factor (SCF) has been reported to increase reperfusion; the observation that SCF-encoding pDNA increased perfusion was interpreted as evidence for the recruitment of reparative cells into the area of infarction (Fujii *et al.* 2009).

Expression of a reporter plasmid gene delivered to the myocardium by UTMD methods was relatively brief (4 d), but was improved by retreatment (Bekeredjian *et al.* 2003). Under similar exposure conditions, damage to the heart was negligible (Bekeredjian *et al.* 2004). However, UTMD-enhanced gene delivery to the heart is often attended by at least minimal damage, which can include extravasation of large molecules and red cells (Hernot *et al.* 2010). It is noteworthy that under diagnostic US exposure conditions, premature ventricular contractions can occur with the use of US contrast agents (see, *e.g.* (Miller *et al.* 2005)), and in contrast with the results of (Hernot *et al.* 2010) these have been unambiguously correlated with cell killing of cardiomycetes. In rats, using 1.7 MHz ultrasound, premature complexes and cardiomycete death were observable at Pas of 2 MPa or greater (Miller *et al.* 2011). Premature contraction complexes appear to be related directly to extravascular cell killing, so even absent gross side effects, some side effects can be expected in gene therapies involving UTMD. This has important implications not only for safety, but also efficacy; *i.e.*, one hopes to transfect the target cells, not kill them. It seems unlikely that transfection to meaningful extents can be achieved without some cell killing, so attempts to optimize UTMD treatments must strive to achieve an acceptable balance between desired effect (transfection) and undesired effect (killing of the target cells).

When treating the heart using reporter genes, MBs and a diagnostic scanner as the acoustic source, superior results were obtained by moving the scan head about to 'paint' a larger volume of tissue (Geis *et al.* 2009). Another noteworthy finding was that moving the beam relative to the heart did not increase Evans blue dye extravasation, which suggests less microcirculation damage per unit transgene expression when the insonifying beam is moved relative to the target.

Inhibition of neovascularization: A substantial literature on therapies to inhibit neoangiogenesis exists. Here we mention one recent example: MBs and pDNA encoding for pigment epithelium derived factor, which inhibits neovascularization in the retina, were injected into the vitreous humor of rats having laser-induced choroidal injury, which leads to neovascularization. The eyes were treated immediately with 0.3 MHz US of Pa estimated to be in the range of 0.1 – 0.3 MPa. At 28 days post treatment, choroidal neovascularization was inhibited in the UTMD group relative to untreated controls (Zhou XY *et al.* 2009). Similarly, pDNA carrying a silencing sequence for the gene coding for survivin were introduced into implanted murine tumors using UTMD methods. Treated tumors were sonicated with 3 MHz US at an intensity of 2 W/cm^2 (estimated Pr: 0.2 - 0.5 MPa). Transgene expression was significantly increased in tumors treated with UTMB. It was proposed that the technique could be applied therapeutically to tumors to increase in tumor cell apoptosis *via* the silencing effect on survivin expression in transfected cells (Chen *et al.* 2010).

5.2 Skeletal muscle

Among the applications for gene therapy in skeletal muscle is the treatment of muscular dystrophy (see, e.g., (Alter *et al.* 2009) and references therein). As a specific recent example, Kodama and colleagues used various means to attempt delivery of PGL3 luciferase pDNA into murine skeletal muscle. Optison, human albumin MBs or acoustic liposomes (all formulated with C3F8 gas) were compared, using a 1 MHz US source; Pa was 0.2 MPa. Treatment with plasmid, US and either Optison or liposomes increased luciferase expression relative to controls. The ~200 nm diameter acoustic liposomes produced the highest expression rates, presumably because of their very high concentration (Kodama *et al.* 2010).

5.3 Pancreas

Pancreatic gene therapies have focused principally on treatments for diabetes types I and II. The feasibility of using pDNA and UTMD methods to deliver transgenes to the pancreas has been described by Grayburn and colleagues. Following injection of reporter gene pDNA attached to the MB phospholipid shells, the reporter gene was expressed with fairly high spatial specificity, declining after 4 days, but with measurable expression persisting for as long as 21 days ((Chen *et al.* 2006); see internal citations for a discussion of viral approaches). US of ~3 MHz was used, with an apparent Pa of ~1 MPa at the pancreas. When plasmid vectors carrying human insulin and hexokinase I gene were delivered using UTMD, transgene expression was obtained in islets and decreased circulating glucose levels were observed in treated rats. These data indicate that UTMD allows relatively noninvasive delivery of genes to pancreatic islets to modulate beta cell function. More recently, the same group reported that delivery of NeuroD1 gene driven by a rat insulin promoter (RAP3.1) into rats by UTMD-mediated gene delivery targeting pancreas *in vivo* promoted islet regeneration from surviving beta-cells, with normalization of glucose, insulin and C-peptide levels at 30 days (Chen *et al.* 2010).

5.4 Liver

Liver is a major organ for protein synthesis, and therefore represents an important target for gene therapy. We will discuss here UTMD-mediated gene therapy targeting liver for treating non-tumoral diseases.

Our group has extensively studied gene delivery of reporter and therapeutic genes into the liver. We have demonstrated that UTMD (1 MHz US) can significantly enhance gene transfer of naked pDNA into the mouse liver in the presence of either Optison (Miao 2005) or Definity MBs (Shen *et al.* 2008). Transgene expression was dependent exponentially on Pr, with an inflection point usually between 1 and 2.5 MPa followed by a plateau above 3 MPa (Song *et al.* 2011a), consistent with an inertial cavitation mechanism. More than thousand-fold enhancement of gene transfer efficiencies was obtained compared to control experiments in the absence of UTMD. Recently we have gained preliminary success in scaling up pDNA delivery in larger animal models, including rats (Song *et al.* 2011b) and dogs (Noble *et al.* 2011). Previously we have shown that near therapeutic levels of factor IX were achieved by UTMD-mediated gene delivery in mice. Technical improvements to further enhance gene transfer of factor VIII for treatment of hemophilia A and factor IX for treatment of hemophilia B are currently being pursued in small and large animal models.

5.5 Kidney

A number of kidney diseases could be potentially treated with gene therapies. Hydrodynamic approaches have met with some success. Xing *et al.* attempted to improve on these results by combining hydrodynamic and UTMD approaches to naked DNA reporter gene delivery to surgically-exposed rat kidneys. Combined US (unspecified frequency) from a Sonotron 2000 hand-held diagnostic US machine and hydrodynamic therapy together yielded better reporter gene transfection than hydrodynamic therapy alone, producing an approximately 4-fold increase in reported gene expression when the estimated Pa was in the range of 0.3 – 0.8 MPa and no MBs were used. The effect was intensity-dependent. When Optison MBs were injected with the naked DNA during hydrodynamic therapy and US exposure, the same effect (4-fold increase in gene expression) was observed at an intensity of only 1 Watt/cm^2 (estimated 0.2 – 0.5 MPa Pa) (Xing *et al.* 2009).

1.6 Skin (DNA vaccine)

The failure of wounds to heal in diabetic patients is a significant clinical problem. Gene therapies which promote angiogenesis represent a promising approach to this problem. VEGF-encoding gene vectors (either minicircle naked DNA; a supercoiled form with a molecular weight estimated as 331 g/mol, or naked DNA borne on the gene carrier branched polyethylinimine) were tested for efficacy in inducing circulating VEGF expression and accelerating wound healing in an induced diabetic mouse model. Wounds were treated by peripheral injection of gene vectors with or without exposure to US (1 MHz, 2 W/cm2, 20% duty cycle; estimated 0.25 - 0.5 MPa). In some treatments, SonoVue MBs were injected with the microcircle DNA prior to sonication. Markedly greater levels of circulating VEGF were observed in mice treated with [VEGF-encoding minicircle DNA + US + MBs] relative to controls, but not as high as those obtained using the polyethylinimine gene carrier. Nonetheless, the [minicircle DNA + US + MBs] treatment produced a significant improvement in healing rates of the treated skin wounds (Yoon *et al.* 2009).

5.7 Other solid organs

Brain: Much work has been done on 'opening' (*i.e.*, making more permeable) the blood- brain barrier, which so tightly regulates traffic between the vascular space and the brain that chemotherapeutic agents often cannot cross the barrier (Meairs & Alonso 2007). Much success with tracer molecules (*e.g.*, Evans blue dye, gadolinium MRI contrast agents, *etc.*) has been achieved using UTMD methods, principally in small animal models. However, low energy US applied through the temporal bone of swine produced short-term permeabilization of the blood-brain barrier with exogenous MBs (see (Xie *et al.* 2008) and citations within). However, we have found no reports of UTMD gene therapies attempted in the brain.

5.8 Endothelium

There are many papers which indicate that endothelial cells *in vitro* can express transgenes delivered by UTMD methods (see (Su *et al.* 2010) and internal references for recent examples). Su and colleagues found that treatment with 1 MHz US of 1 MPa Pr together with pGL3 pDNA and MBs yielded a 100-fold increase in luciferase expression relative to pDNA treatment alone. It remains to be determined if endothelial cells *in vivo* can be 'recruited' as effective synthesizers and secretors of therapeutic gene expression products.

5.9 Cell therapies: *Ex vivo* gene therapy

In this section, we discuss several examples of US-mediated gene delivery using cell cultures *in vitro*; these may have some mechanistic relevance to the *in vivo* condition. However, some of the results obtained using *in vitro* cells may have direct bearing on cell therapies.

Shock waves and MBs have been used *in vitro* to increase pDNA uptake in cultured HEK cells; the technique worked, but was associated with substantial cell killing (Bekeredjian *et al.* 2007) as might be expected (see, *e.g.*, (Brayman *et al.* 1999; Hwang *et al.* 2006; Hwang *et al.* 2003)). As a general rule, only a small fraction (~5 – 10%) of cells insonated with MBs *in vitro* undergo transient poration; these yields are often associated with cell killing rates of ~50% (see, *e.g.*

(AIUM 2000; AIUM 2008) and references therein). If those cells can be sorted (*e.g.*, by flow cytometry) and cultured to increase cell numbers, this may be acceptable for cell therapies – if the gene vector is integrated into the host genome. An obvious problem for cell therapies based on this approach with naked DNA, however, is that the pDNA does not replicate with the host cell, and so there would be little to be gained by increasing cell numbers of the transfected subpopulation of cells. However, using suspended KHT-C cells, and 3 V% Definity, high levels of *in vitro* permeabilization to 70 kDa fluorescein isothiocyanate (FITC) - dextran were achieved using 0.5 MHz US at a Pa of 0.57 MPa, 16 cycle pulses repeated at 3 kHz PRF. About 30% of the cells were permeabilized, with >95% retention of viability (Karshafian *et al.* 2009). If similar results can be obtained with naked DNA, the prospects for cell therapies using cells transfected using *in vitro* UTMD seem much brighter.

1.10 Gene therapies for tumoricidal activity

In this document, we aim to present an overview of the common types of microbubbles (MBs) used in gene therapies facilitated by the United States (US). We will discuss how the US can lead to the destruction of these MBs, either gradually or suddenly, as well as other methods of activation. Additionally, we will explore the mechanisms through which microvascular damage, extravasation, and the uptake of gene vectors by target cells can occur as a result of ultrasound-targeted microbubble destruction (UTMD). For a more thorough analysis of the physics, refer to the publication of Wu and Nyborg in 2008.

Generally, the targeting component of UTMD therapies relies on the fact that microbubbles (MBs) in tissues only respond to ultrasound (US) when the area is exposed to it. The inclusion of ligands in the MB shell can increase the concentration of the MBs and any cargo they carry in a specific area of interest. However, it is the targeted acoustic exposure that triggers the activation of the MBs. According to Emmer et al. (2007), when undergoing radial oscillations, the bubble moves forward abruptly, decelerating as it expands and accelerating as it collapses. The utilization of radiation force allows for the concentration of microbubbles (MBs) along a particular wall of an intracavitary space, as demonstrated by Horie et al. in 2010. This technique can also be applied to blood vessels.

The destruction of microbubbles (MB) in ultrasound-targeted microbubble destruction (UTMD) is triggered by acoustic exposure. This process can lead to the release of lipid or aqueous-phase drugs, as documented by Smith et al. in 2010. Additionally, it can cause minor damage to microvessels, enabling substances that are typically impermeable, such as drugs, plasmids, cells, or fragments of microbubbles, to exit the blood vessels and enter the interstitial space. MB activation is also linked to cellular permeabilization.

6. CONCISE SUMMARY

Considerable progress has been made in the field of ultrasound-mediated gene delivery, but, it is still in its early stages and there are few preclinical applications utilizing therapeutic genes. Several generic statements can be made: Typically, ultrasound-mediated gene transfer relies on or is significantly aided by external microbubbles, although there are exceptions. In situations when this is true, the sonic exposures typically cause the bubbles to exhibit nonlinear behavior. Considering the significant amount of effort being dedicated to this subject, it is quite likely that successful gene therapies will be accomplished in large animals and early human trials in the near future.

REFERENCES

AIUM. (2000) Mechanical Bioeffects from Diagnostic Ultrasound: AIUM Consensus Statements. Section 6 - Mechanical bioeffects in the presence of gas-carrier ultrasound contrast agents. *J Ultrasound Med* 19: 120 - 142.

AIUM. (2008) American Institute of Ultrasound in Medicine Consensus Report on Potential Bioeffects of Diagnostic Ultrasound: Executive Summary. *J Ultrasound Med* 27: 503- 515.

Akowuah EF, Gray C, Lawrie A, Sheridan PJ, Su CH, Bettinger T, Brisken AF, Gunn J, Crossman DC, Francis SE, Baker AH, Newman CM. (2005) Ultrasound-mediated delivery of TIMP-3 plasmid DNA into saphenous vein leads to increased lumen size in a porcine interposition graft model. *Gene Ther* 12: 1154-1157.

Alter J, Sennoga CA, Lopes DM, Eckersley RJ, Wells DJ. (2009) Microbubble stability is a major determinant of the efficiency of ultrasound and microbubble mediated in vivo gene transfer. *Ultrasound Med Biol* 35: 976-984.

Aoi A, Watanabe Y, Mori S, Takahashi M, Vassaux G, Kodama T. (2008) Herpes simplex virus thymidine kinase-mediated suicide gene therapy using nano/microbubbles and ultrasound. *Ultrasound Med Biol* 34: 425-434.

Apfel RE, Holland CK. (1991) Gauging the likelihood of cavitation from short-pulse, low duty cycle diagnostic ultrasound. *Ultrasound Med Biol* 17: 179 - 185.

Aronovich EL, McIvor RS, Hackett PB. (2011) The Sleeping Beauty transposon system - a non-viral vector for gene therapy. *Human Molec Genetics* Epub ahead of print.

Azuma H, Tomita N, Sakamoto T, Kiyama S, Inamoto T, Takahara K, Kotake Y, Segawa N, Morishita R, Takahara S, Hayasaki H, Otsuki Y, Horie S, Tanigawa N, Katsuoka Y. (2008) Marked regression of liver metastasis by combined therapy of ultrasound-mediated NF kappaB-decoy transfer and transportal injection of paclitaxel, in mouse. *Int J Cancer* 122: 1645-1656.

Beeri R, Guerrero JL, Supple G, Sullivan S, Levine RA, Hajjar RJ. (2002) New efficient catheter-based system for myocardial gene delivery. *Circulation* 106: 1756-1759.

Bekeredjian R, Bohris C, Hansen A, Katus HA, Kuecherer HF, Hardt SE. (2007) Impact of microbubbles on shock wave-mediated DNA uptake in cells in vitro. *Ultrasound Med Biol* 33: 743-750.

Bekeredjian R, Chen S, Frenkel PA, Grayburn PA, Shohet RV. (2003) Ultrasound-targeted microbubble destruction can repeatedly direct highly specific plasmid expression to the heart. *Circulation* 108: 1022-1026.

Bekeredjian R, Chen S, Pan W, Grayburn PA, Shohet RV. (2004) Effects of ultrasound- targeted microbubble destruction on cardiac gene expression. *Ultrasound Med Biol* 30: 539-543.

Bekeredjian R, Kroll RD, Fein E, Tinkov S, Coester C, Winter G, Katus HA, Kulaksiz H. (2007) Ultrasound targeted microbubble destruction increases capillary permeability in hepatomas. *Ultrasound Med Biol* 33: 1592-1598.

Blake JR, Gibson DC. (1987) Cavitation bubbles near boundaries. *Ann Rev Fluid Mechanics* 19: 99 - 123.

Borden MA, Caskey CF, Little E, Gillies RJ, Ferrara KW. (2007) DNA and polylysine adsorption and multilayer construction onto cationic lipid-coated microbubbles. *Langmuir* 23: 9401-9408.

Borden MA, Kruse DE, Caskey CF, Zhao S, Dayton PA, Ferrara KW. (2005) Influence of lipid shell physicochemical properties on ultrasound-induced microbubble destruction. *IEEE Trans UFFC* 52: 992-2002.

Brayman AA, Coppage ML, Vaidya S, Miller MW. (1999) Transient poration and cell surface receptor removal from human lymphocytes in vitro by 1 MHz ultrasound. *Ultrasound Med Biol* 25: 999-1008.

Carson AR, McTiernan CF, Lavery L, Hodnick A, Grata M, Leng X, Wang J, Chen X, Modzelewski RA, Villanueva FS. (2011) Gene therapy of carcinoma using ultrasound-targeted microbubble destruction. *Ultrasound Med Biol* 37: 393-402.

Carstensen EL, Gracewski S, Dalecki D. (2000) The search for cavitation in vivo. *Ultrasound Med Biol* 26: 1377-1385.

Caskey CF, Stieger SM, Qin S, Dayton PA, Ferrara KW. (2007) Direct observations of ultrasound microbubble contrast agent interaction with the microvessel wall. *J Acoust Soc America* 122: 1191-1200.

Chahine GL. (1977) Interaction between an oscillating bubble and a free-surface. *J Fluid Eng- T ASME* 99: 709-716.

Chen H, Brayman AA, Bailey MR, Matula TJ. (2010) Blood vessel rupture by cavitation. *Urological Res* 38: 321-326.

Chen H, Kreider W, Brayman AA, Bailey MR, Matula TJ. (2011) Blood vessel deformations on microsecond time scales by ultrasonic cavitation. *Phys Review Lett* 106: 034301- 034304.

Chen S, Ding JH, Bekeredjian R, Yang BZ, Shohet RV, Johnston SA, Hohmeier HE, Newgard CB, Grayburn PA. (2006) Efficient gene delivery to pancreatic islets with ultrasonic microbubble destruction technology. *Proc Nat Acad Sci USA* 103: 8469- 8474.

Chen S, Shimoda M, Wang MY, Ding J, Noguchi H, Matsumoto S, Grayburn PA. (2010) Regeneration of pancreatic islets in vivo by ultrasound-targeted gene therapy. *Gene Ther* 17: 1411-1420.

Chen W-S, Matula TJ, Brayman AA, Crum LA. (2003) A comparison of the fragmentation thresholds and inertial cavitation doses of different ultrasound contrast agents. *J Acoust Soc America* 113: 643 - 651.

Chen WS, Brayman AA, Matula TJ, Crum LA. (2003a) Inertial cavitation dose and hemolysis produced in vitro with or without Optison. *Ultrasound Med Biol* 29: 725- 737.

Chen WS, Brayman AA, Matula TJ, Crum LA, Miller MW. (2003b) The pulse length- dependence of inertial cavitation dose and hemolysis. *Ultrasound Med Biol* 29: 739- 748.

Chen ZY, Liang K, Qiu RX. (2010) Targeted gene delivery in tumor xenografts by the combination of ultrasound-targeted microbubble destruction and polyethylenimine to inhibit survivin gene expression and induce apoptosis. *J Exp Clin Cancer Res* 29: 152.

Chomas JE, Dayton P, Allen J, Morgan K, Ferrara KW. (2001) Mechanisms of contrast agent destruction. *IEEE Trans UFFC* 48: 232-248.

Chomas JE, Dayton P, Ferarra K. (2001) Threshold of fragmentation for ultrasonic contrast agents. *J Biomedical Optics* 6: 141 - 159.

Church CC, Carstensen EL. (2001) "Stable" inertial cavitation. *Ultrasound Med Biol* 27: 1435 - 1437.

Cosgrove D, Harvey C. (2009) Clinical uses of microbubbles in diagnosis and treatment. *Medical and Biological Engineering and Computing* 47: 813-826.

Daigeler A, Chromik AM, Haendschke K, Emmelmann S, Siepmann M, Hensel K, Schmitz G, Klein-Hitpass L, Steinau HU, Lehnhardt M, Hauser J. (2010) Synergistic effects of sonoporation and taurolidin/TRAIL on apoptosis in human fibrosarcoma. *Ultrasound Med Biol* 36: 1893-1906.

Danko I, Williams P, Herweijer H, Zhang G, Latendresse JS, Bock I, Wolff JA. (1997) High expression of naked plasmid DNA in muscles of young rodents. *Human Molec Genetics* 6: 1435-1443.

de Jong N, Emmer M, van Wamel A, Versluis M. (2009) Ultrasonic characterization of ultrasound contrast agents. *Medical and Biological Engineering and Computing* 47: 861-873.

Deng CX, Sieling F, Pan H, Cui J. (2004) Ultrasound-induced cell membrane porosity.*Ultrasound Med Biol* 30: 519-526.

Doinikov AA, Bouakaz A. (2010) Theoretical investigation of shear stress generated by a contrast microbubble on the cell membrane as a mechanism for sonoporation. *J Acoust Soc America* 128: 11-19.

Duvshani-Eshet M, Baruch L, Kesselman E, Shimoni E, Machluf M. (2006) Therapeutic ultrasound-mediated DNA to cell and nucleus: bioeffects revealed by confocal and atomic force microscopy. *Gene Ther* 13: 163-172.

Duvshani-Eshet M, Machluf M. (2005) Therapeutic ultrasound optimization for gene delivery: a key factor achieving nuclear DNA localization. *J Controlled Release* 108: 513-528.

Emmer M, van Wamel A, Goertz DE, de Jong N. (2007) The onset of microbubble vibration. *Ultrasound Med Biol* 33: 941-949.

Fabre JW, Grehan A, Whitehorne M, Sawyer GJ, Dong X, Salehi S, Eckley L, Zhang X, Seddon M, Shah AM, Davenport M, Rela M. (2008) Hydrodynamic gene delivery to the pig liver via an isolated segment of the inferior vena cava. *Gene Ther* 15: 452- 462.

Fechheimer M, Boylan JF, Parker S, Sisken JE, Patel GL, Zimmer SG. (1987) Transfection of mammalian cells with plasmid DNA by scrape loading and sonication loading. *Proc Nat Acad Sci USA* 84: 8463-8467.

Ferrara KW. (2008) Driving delivery vehicles with ultrasound. *Adv Drug Deliv Rev* 60: 1097- 1102.

Ferrara KW, Borden MA, Zhang H. (2009) Lipid-shelled vehicles: engineering for ultrasound molecular imaging and drug delivery. *Accounts Chem Res* 42: 881-892.

Fischer K, Gedroyc W, Jolesz FA. (2010) Focused ultrasound as a local therapy for liver cancer. *Cancer J* 16: 118-124.

Forbes MM, Steinberg RL, O'Brien WD, Jr. (2008) Examination of inertial cavitation of Optison in producing sonoporation of Chinese hamster ovary cells. *Ultrasound Med Biol* 34: 2009-2018.

Fujii H, Sun Z, Li SH, Wu J, Fazel S, Weisel RD, Rakowski H, Lindner J, Li RK. (2009) Ultrasound-targeted gene delivery induces angiogenesis after a myocardial infarction in mice. *JACC Cardiovascular Imaging* 2: 869-879.

Geis NA, Mayer CR, Kroll RD, Hardt SE, Katus HA, Bekeredjian R. (2009) Spatial distribution of ultrasound targeted microbubble destruction increases cardiac transgene expression but not capillary permeability. *Ultrasound Med Biol* 35: 1119- 1126.

Gross DR, Miller DL, Williams AR. (1985) A search for ultrasonic cavitation within the canine cardiovascular system. *Ultrasound Med Biol* 11: 85-97.

Haag P, Frauscher F, Gradl J, Seitz A, Schafer G, Lindner JR, Klibanov AL, Bartsch G, Klocker H, Eder IE. (2006) Microbubble-enhanced ultrasound to deliver an antisense oligodeoxynucleotide targeting the human androgen receptor into prostate tumours. *J Steroid Biochem and Molec Biol* 102: 103-113.

Hacein-Bey-Abina S, Le Deist F, Carlier F, Bouneaud C, Hue C, De Villartay JP, Thrasher AJ, Wulffraat N, Sorensen R, Dupuis-Girod S, Fischer A, Davies EG, Kuis W, Leiva L, Cavazzana-Calvo M. (2002) Sustained correction of X-linked severe combined immunodeficiency by ex vivo gene therapy. *New England J Med* 346: 1185-1193.

Hacein-Bey-Abina S, von Kalle C, Schmidt M, Le Deist F, Wulffraat N, McIntyre E, Radford I, Villeval JL, Fraser CC, Cavazzana-Calvo M, Fischer A. (2003) A serious adverse event after successful gene therapy for X-linked severe combined immunodeficiency. *New England J Med* 2003; 348: 255-256.

Hall TL, Hempel CR, Wojno K, Xu Z, Cain CA, Roberts WW. (2009) Histotripsy of the prostate: dose effects in a chronic canine model. *Urology* 74: 932-937.

Han YW, Ikegami A, Chung P, Zhang L, Deng CX. (2007) Sonoporation is an efficient tool for intracellular fluorescent dextran delivery and one-step double-crossover mutant construction in Fusobacterium nucleatum. *Appl Environ Microbiol* 73: 3677-3683.

Hernot S, Cosyns B, Droogmans S, Garbar C, Couck P, Vanhove C, Caveliers V, van Camp G, Bossuyt A, LaHoutte T. (2010) Effect of high-intensity ultrasound- targeted microbubble destruction of perfusion and function of the rat heart assessed by pinhole-gated SPECT. *Ultrasound Med Biol* 36: 158 - 165.

Horie S, Watanabe Y, Chen R, Mori S, Matsumura Y, Kodama T. (2010) Development of localized gene delivery using dual intensity ultrasound in the bladder. *Ultrasound Med Biol* 36: 1867 - 1875.

Huang SL. (2008) Liposomes in ultrasonic drug and gene delivery. *Adv Drug Deliv Rev* 60: 1167-1176.

Huster D, Paasche G, Dietrich U, Zschornig O, Gutberlet T, Gawrisch K, Arnold K. (1999) Investigation of phospholipid area compression induced by calcium-mediated dextran sulfate interaction. *Biophys J* 77: 879-887.

Hwang JH, Brayman AA, Reidy MA, Matula TJ, Kimmey MB, Crum LA. (2005) Vascular effects induced by combined 1-MHz ultrasound and microbubble contrast agent treatments in vivo. *Ultrasound Med Biol* 31: 553-564.

Hwang JH, Tu J, Brayman AA, Matula TJ, Crum LA. (2006) Correlation between inertial cavitation dose and endothelial cell damage in vivo. *Ultrasound Med Biol* 32: 1611- 1619.

Hwang JH, Vaezy S, Martin RW, Cho MY, Noble ML, Crum LA, Kimmey MB. (2003) High-intensity focused US: a potential new treatment for GI bleeding. *Gastrointest Endosc* 58: 111-115.

Juffermans LJ, Kamp O, Dijkmans PA, Visser CA, Musters RJ. (2008) Low-intensity ultrasound-exposed microbubbles provoke local hyperpolarization of the cell membrane via activation of BK(Ca) channels. *Ultrasound Med Biol* 34: 502-508.

Juffermans LJ, van Dijk A, Jongenelen CAM, Drukarch B, Reijerkerk AR, de Vries HE, Kamp O, Musters RJ. (2009) Ultrasound and microbubble-induced intra- and intercellular bioeffects in primary endothelial cells. *Ultrasound Med Biol* 35: 1917 - 1927.

Kamimura K, Suda T, Xu W, Zhang G, Liu D. (2009) Image-guided, lobe-specific hydrodynamic gene delivery to swine liver. *Molecular Ther* 17: 491-499.

Karshafian R, Bevan PD, Williams R, Samac S, Burns PN. (2009) Sonoporation by ultrasound-activated microbubble contrast agents: Effect of acoustic exposure parameters on cell membrane permeability and cell viability. *Ultrasound Med Biol* 35: 847 - 860.

Kay MA, Manno CS, Ragni MV, Larson PJ, Couto LB, McClelland A, Glader B, Chew AJ, Tai SJ, Herzog RW, Arruda V, Johnson F, Scallan C, Skarsgard E, Flake AW, High KA. (2000) Evidence for gene transfer and expression of factor IX in haemophilia B patients treated with an AAV vector. *Nature Genetics* 224: 257-261.

Kaya M, Gregory V TS, Dayton PA. (2009) Changes in lipid-encapsulated microbubble population during continuous infusion and methods to maintain consistency. *Ultrasound Med Biol* 35: 1748 - 1755.

Keravala A, Chavez CL, Hu G, Woodard LE, Monahan PE, Calos MP. (2011) Long-term phenotypic correction in factor IX knockout mice by using phiC31 integrase- mediated gene therapy. *Gene Ther* Epub ahead of print.

Kirn D, Niculescu-Duvaz I, Hallden G, Springer CJ. (2002) The emerging fields of suicide gene therapy and virotherapy. *Trends Molec Sci* 8: S68-73.

Kobulnik J, Kuliszewski MA, Stewart DJ, Lindner JR, Leong-Poi H. (2009) Comparison of gene delivery techniques for therapeutic angiogenesis: ultrasound-mediated destruction of carrier microbubbles versus direct intramuscular injection. *J Am Coll Cardiol* 54: 1735-1742.

Kodama T, Aoi A, Watanabe Y, Horie S, Kodama M, Li L, Chen R, Teramoto N, Morikawa H, Mori S, Fukumoto M. (2010) Evaluation of transfection efficiency in skeletal muscle using nano/microbubbles and ultrasound. *Ultrasound Med Biol* 36: 1196 - 1205.

Kodama T, Tan PH, Offiah I, Partridge T, Cook T, George AJ, Blomley MJ. (2005) Delivery of oligodeoxynucleotides into human saphenous veins and the adjunct effect of ultrasound and microbubbles. *Ultrasound Med Biol* 31: 1683-1691.

Kodama T, Tomita Y. (2000) Cavitation bubble behavior and bubble-shock wave interaction near a gelatin surface as a study of in vivo bubble dynamics. *Appl Phys B-Lasers* 70: 139-149.

Krupka TM, Solorio L, Wilson RE, Wu H, Azar N, Exner AA. (2009) Formulation and characterization of echogenic lipid-Pluronic nanobubbles. *Mol Pharm* 7: 49-59.

Lai CY, Wu CH, Chen CC, Li PC. (2006) Quantitative relations of acoustic inertial cavitation with sonoporation and cell viability. *Ultrasound Med Biol* 32: 1931-1941.

Laing ST, McPherson DD. (2009) Cardiovascular therapeutic uses of targeted ultrasound contrast agents. *Cardiovascular Res* 83: 626 - 634.

Lentacker I, De Geest BG, Vandenbroucke RE, Peeters L, Demeester J, De Smedt SC, Sanders NN. (2006) Ultrasound-responsive polymer-coated microbubbles that bind and protect DNA. *Langmuir* 22: 7273-7278.

Lentacker I, Wang N, Vandenbroucke RE, Demeester J, De Smedt SC, Sanders NN. (2009) Ultrasound exposure of lipoplex loaded microbubbles facilitates direct cytoplasmic entry of the lipoplexes. *Mol Pharmaceutics* 6: 457 - 467.

Li P, Armstrong WF, Miller DL. (2004) Impact of myocardial contrast echocardiography on vascular permeability: comparison of three different contrast agents. *Ultrasound Med Biol* 30: 83-91.

Liang HD, Lu QL, Xue SA, Halliwell M, Kodama T, Cosgrove DO, Stauss HJ, Partridge TA, Blomley MJ. (2004) Optimisation of ultrasound-mediated gene transfer (sonoporation) in skeletal muscle cells. *Ultrasound Med Biol* 30: 1523-1529.

Lindner JR. (2009) Contrast ultrasound molecular imaging: harnessing the power of bubbles. *Cardiovascular Res* 83: 615-616.

Liu F, Song Y, Liu D. (1999) Hydrodynamics-based transfection in animals by systemic administration of plasmid DNA. *Gene Ther* 6: 1258-1266.

Luo J, Zhou X, Diao L, Wang Z. (2010) Experimental research on wild-type p53 plasmid transfected into retinoblastoma cells and tissues using an ultrasound microbubble intensifier. *J Int Med Res* 38: 1005-1015.

Manno CS, Pierce GF, Arruda VR, Glader B, Ragni M, Rasko JJ, Ozelo MC, Hoots K, Blatt P, Konkle B, Dake M, Kaye R, Razavi M, Zajko A, Zehnder J, Rustagi PK, Nakai H, Chew A, Leonard D, Wright JF, Lessard RR, Sommer JM, Tigges M, Sabatino D, Luk A, Jiang H, Mingozzi F, Couto L, Ertl HC, High KA, Kay MA. (2006) Successful transduction of liver in hemophilia by AAV-Factor IX and limitations imposed by the host immune response. *Nature Med* 12: 342-347.

Marmottant P, de Jong N, Hilgenfeldt S, Lohse D. (2005) A model for large amplitude oscillations of coated bubbles accounting for buckling and rupture. *J Acoust Soc America* 118: 3499 - 3505.

Maruvada S, Hynynen K. (2004) Optical monitoring of ultrasound-induced bioeffects in glass catfish. *Ultrasound Med Biol* 30: 67-74.

Matula TJ. (2003) Bubble levitation and translation under single-bubble sonoluminescence conditions. *J Acoust Soc America* 114: 775-781.

Meairs S, Alonso A. (2007) Ultrasound, microbubbles and the blood-brain barrier. *Prog Biophys Mol Biol* 93: 354-362.

Mehier-Humbert S, Bettinger T, Yan F, Guy RH. (2005) Plasma membrane poration induced by ultrasound exposure: implication for drug delivery. *J Control Release* 104: 213-222.

Meijering BD, Juffermans LJ, van Wamel A, Henning RH, Zuhorn IS, Emmer M, Versteilen AM, Paulus WJ, van Gilst WH, Kooiman K, de Jong N, Musters RJ, Deelman LE, Kamp O. (2009) Ultrasound and microbubble-targeted delivery of macromolecules is regulated by induction of endocytosis and pore formation. *Circ Res* 104: 679-687.

Miao CH. (2005) A novel gene expression system: non-viral gene transfer for hemophilia as model systems. *Adv Genetics* 54: 143 - 177.

Miao CH, Brayman AA, Loeb KR, Ye P, Zhou L, Mourad P, Crum LA. (2005) Ultrasound enhances gene delivery of human factor IX plasmid. *Human Gene Ther* 16: 893-905.

Miao CH, Thompson AR, Loeb K, Ye X. (2001) Long-term and therapeutic-level hepatic gene expression of human factor IX after naked plasmid transfer in vivo. *Molecular Therapy* 3: 947-957.

Miao CH, Ye X, Thompson AR. (2003) High-level factor VIII gene expression in vivo achieved by nonviral liver-specific gene therapy vectors. *Human Gene Ther* 14: 1297- 1305.

Miao H, Gracewski SM, Dalecki D. (2008) Ultrasonic excitation of a bubble inside a deformable tube: implications for ultrasonically induced hemorrhage. *J Acoust Soc America* 124: 2374-2384.

Miller D, Li P, Armstrong WF. (2004) The effect of time and of vasoactive drugs on capillary leakage induced during myocardial contrast echocardiography. *Echocardiography* 21: 125-132.

Miller DL, Dou C, Wiggins RC. (2009) Glomerular capillary hemorrhage induced in rats by diagnostic ultrasound with gas-body contrast agent produces intratubular obstruction. *Ultrasound Med Biol* 35: 869-877.

Miller DL, Dou C, Wiggins RC. (2010a) Contrast-enhanced diagnostic ultrasound causes renal tissue damage in a porcine model. *J Ultrasound Med* 29: 1391-1401.

Miller DL, Dou C, Wiggins RC. (2010b) In vivo gas body efficacy for glomerular capillary hemorrhage induced by diagnostic ultrasound in rats. *IEEE Trans Biomed Eng* 57: 167-174.

Miller DL, Dou C, Wiggins RC, Wharram BL, Goyal M, Williams AR. (2007) An in vivo rat model simulating imaging of human kidney by diagnostic ultrasound with gas- body contrast agent. *Ultrasound Med Biol* 33: 129-135.

Miller DL, Dou H, Lucchesi BR. (2011) Are ECG premature complexes induced by ultrasonic cavitation electrophysiological responses to irreversible cardiomycete injury? *Ultrasound Med Biol* 37: 312 - 320.

Miller DL, Driscoll EM, Dou C, Armstrong WF, Lucchesi BR. (2006) Microvascular permeabilization and cardiomyocyte injury provoked by myocardial contrast echocardiography in a canine model. *J Am Coll Cardiol* 47: 1464-1468.

Miller DL, Gies RA. (1998a) Gas-body-based contrast agent enhances vascular bioeffects of 1.09 MHz ultrasound on mouse intestine. *Ultrasound Med Biol* 24: 1201-1208.

Miller DL, Gies RA. (1998b) The interaction of ultrasonic heating and cavitation in vascular bioeffects on mouse intestine. *Ultrasound Med Biol* 24: 123-128.

Miller DL, Gies RA. (2000) The influence of ultrasound frequency and gas-body composition on the contrast agent-mediated enhancement of vascular bioeffects in mouse intestine. *Ultrasound Med Biol* 26: 307-313.

Miller DL, Li P, Dou C, Gordon D, Edwards CA, Armstrong WF. (2005) Influence of contrast agent dose and ultrasound exposure on cardiomyocyte injury induced by myocardial contrast echocardiography in rats. *Radiology* 237: 137-143.

Miller DL, Li P, Gordon D, Armstrong WF. (2005) Histological characterization of microlesions induced by myocardial contrast echocardiography. *Echocardiography* 22: 25-34.

Miller DL, Pislaru SV, Greenleaf JE. (2002) Sonoporation: mechanical DNA delivery by ultrasonic cavitation. *Somatic Cell Molec Genetics* 27: 115-134.

Miller MW. (2000) Gene transfection and drug delivery. *Ultrasound Med Biol* 26 Suppl 1: S59- 62.

Mohan P, Rapoport N. (2010) Doxorubicin as a molecular nanotheranostic agent: effect of doxorubicin encapsulation in micelles or nanoemulsions on the ultrasound- mediated intracellular delivery and nuclear trafficking. *Molecular Pharmaceutics* 7: 1959-1973.

Negishi Y, Endo Y, Fukuyama T, Suzuki R, Takizawa T, Omata D, Maruyama K, Aramaki Y. (2008) Delivery of siRNA into the cytoplasm by liposomal bubbles and ultrasound. *J Control Release* 132: 124-130.

Newman CM, Bettinger T. (2007) Gene therapy progress and prospects: ultrasound for gene transfer. *Gene Ther* 14: 465-475.

Niidome T, Huang L.(2002) Gene therapy progress and prospects: nonviral vectors. *Gene Ther* 9: 1647-1652.

Noble ML, Brayman AA, Kuhr CS, Graves SS, Keilman GW, Chen L, Loeb KR, Storb RF, Miao CH. (2011) Effective gene transfer using microbubble-assisted ultrasound therapy in canine livers. *American Society of Gene and Cell Therapy 14th Annual Meeting 2011:* Oral abstract presentation.

Okada K, Kudo N, Hassan MA, Kondo T, Yamamoto K. (2009) Threshold curves obtained under various gaseous conditions for free radical generation by burst ultrasound - Effects of dissolved gas, microbubbles and gas transport from the air. *Ultrasonics Sonochem* 16: 512-518.

Otani K, Yamahara K, Ohnishi S, Obata H, Kitamura S, Nagaya N. (2009) Nonviral delivery of siRNA into mesenchymal stem cells by a combination of ultrasound and microbubbles. *J Control Release* 133: 146-153.

Overvelde M, Garbinb V, Sijl J, Dollet B, de Jong N, Lohse D, Versluis M. (2010) Nonlinear shell behavior of phospholipid-coated microbubbles. *Ultrasound Med Biol* 36: 2080 - 2092.

Phillips LC, Klibanov AL, Wamhoff BR, Hossack JA. (2010) Targeted gene transfection from microbubbles into vascular smooth muscle cells using focused, ultrasound mediated delivery. *Ultrasound Med Biol* 36: 1470-1480.

Pislaru SV, Pislaru C, Kinnick RR, Singh R, Gulati R, Greenleaf JF, Simari RD. (2003) Optimization of ultrasound-mediated gene transfer: comparison of contrast agents and ultrasound modalities. *Eur Heart J* 24: 1690-1698.

Plesset MS, Chapman RB. (1971) Collapse of an initially spherical vapour cavity in the neighbourhood of a solid boundary. *J Fluid Mech* 47: 283-290.

Porter TM, Smith DAB, Holland CK. (2006) Acoustic techniques for assessing the Optison destruction threshold. *J Ultrasound Med* 25: 1519-1529.

Porter TR. (2009) The utilization of ultrasound and microbubbles for therapy in acute coronary syndromes. *Cardiovasc Res* 83: 636-642.

Porter TR, Hiser WL, Kricsfeld D, Deligonul U, Xie F, Iversen P, Radio S. (2001) Inhibition of carotid artery neointimal formation with intravenous microbubbles. *Ultrasound Med Biol* 27: 259-265.

Postema M, Schmitz G. (2007) Ultrasonic bubbles in medicine: influence of the shell. *Ultrasonics Sonochem* 14: 438 - 444.

Postema M, van Wamel A, Lancee CT, de Jong N. (2004) Ultrasound-induced encapsulated microbubble phenomena. *Ultrasound Med Biol* 30: 827-840.

Price RJ, Skyba DM, Kaul S, Skalak TC. (1998) Delivery of colloidal particles and red blood cells to tissue through microvessel ruptures created by targeted microbubble destruction with ultrasound. *Circulation* 98: 1264-1267.

Qin S, Caskey CF, Ferrara KW. (2009) Ultrasound contrast microbubbles in imaging and therapy: physical principles and engineering. *Phys Med Biol* 54: R27-57.

Qiu Y, Luo Y, Zhang Y, Cui W, Zhang D, Wu J, Zhang J, Tu J. (2010) The correlation between acoustic cavitation and sonoporation involved in ultrasound-mediated DNA transfection with polyethylenimine (PEI) in vitro. *J Control Release* 145: 40-48.

Rahim AA, Taylor SL, Bush NL, ter Haar GR, Bamber JC, Porter CD. (2006) Spatial and acoustic pressure dependence of microbubble-mediated gene delivery targeted using focused ultrasound. *J Gene Med* 8: 1347-1357.

Robinson PB, Blake JR, Kodama T, Shima A, Tomita Y. (2001) Interaction of cavitation bubbles with a free surface. *J Appl Phys* 89: 8225-8237.

Sakai T, Kawaguchi M, Kosuge Y. (2009) siRNA-mediated gene silencing in the salivary gland using *in vivo* microbubble-enhanced sonoporation. *Oral Diseases* 15: 505-511.

Samuel S, Cooper MA, Bull JL, Fowlkes JB, Miller DL. (2009) An ex vivo study of the correlation between acoustic emission and microvascular damage. *Ultrasound Med Biol* 35: 1574-1586.

Sankin GN, Yuan F, Zhong P. (2010) Pulsating tandem microbubble for localized and directional single-cell membrane poration. *Physical Rev Lett* 105: 078101.

Sarkar K, Katiyar A, Jain P. (2009) Growth and dissolution of an encapsulated contrast microbubble: effects of encapsulation permeability. *Ultrasound Med Biol* 35: 1385- 1396.

Sarvazyan AP, Rudenko OV, Nyborg WL. (2010) Biomedical applications of radiation force of ultrasound: historical roots and physical basis. *Ultrasound Med Biol* 36: 1379-1394.

Sassaroli E, Hynynen K. (2007) Cavitation threshold of microbubbles in gel tunnels by focused ultrasound. *Ultrasound Med Biol* 33: 1651-1660.

Shen ZP, Brayman AA, Chen L, Miao CH. (2008) Ultrasound with microbubbles enhances gene expression of plasmid DNA in the liver via intraportal delivery. *Gene Ther* 15: 1147-1155.

Shima A, Tomita Y, Gibson DC, Blake JR. (1989) The growth and collapse of cavitation bubbles near composite surfaces. *J Fluid Mech* 203: 199 - 214.

Skyba DM, Price RJ, Linka AZ, Skalak TC, Kaul S. (1998) Direct in vivo visualization of intravascular destruction of microbubbles by ultrasound and its local effects on tissue. *Circulation* 98: 290-293.

Smith DA, Vaidya SS, Kopechek JA, Huang SL, Klegerman ME, McPherson DD, Holland CK. (2010) Ultrasound-triggered release of recombinant tissue-type plasminogen activator from echogenic liposomes. *Ultrasound Med Biol* 36: 145-157.

Snyder RO. (1999) Adeno-associated virus-mediated gene delivery. *J Gene Med* 1: 166-175.

Song J, Chappell JC, Qi M, VanGieson EJ, Kaul S, Price RJ. (2002) Influence of injection site, microvascular pressure and ultrasound variables on microbubble-mediated delivery of microspheres to muscle. *J Am Coll Cardiol* 39: 726-731.

Song S, Noble ML, Sun SX, Chen L, Brayman AA, Miao CH. (2011b) Optimization of high intensity therapeutic ultrasound and microbubbles for gene delivery into rat liver. *American Society of Gene and Cell Therapy 14th Annual Meeting:* Oral abstract presentation.

Song S, Shen Z, Chen L, Brayman AA, Miao AA. (2011a) Explorations of high-intensity therapuetic ultrasound and microbubble-mediated gene delivery in mouse liver. *Gene Therapy* Mar 31. [Epub ahead of print].

Stride E, Saffari N. (2003) On the destruction of microbubble ultrasound contrast agents. *Ultrasound Med Biol* 29: 563-573.

Stringham SB, Viskovska MA, Richardson ES, Ohmine S, Husseini GA, Murray BK, Pitt WG. (2009) Over-pressure suppresses ultrasonic-induced drug uptake. *Ultrasound Med Biol* 35: 409-415.

Su CH, Chang CY, Wang HH, Wu YJ, Bettinger T, Tsai CH, Yeh HI. (2010) Ultrasonic microbubble-mediated gene delivery causes phenotypic changes of human aortic endothelial cells. *Ultrasound Med Biol* 36: 449-458.

Suda T, Suda K, Liu D. (2008) Computer-assisted hydrodynamic gene delivery. *Molecular Therapy* 16: 1098-1104.

Talu E, Powell RL, Longo ML, Dayton PA. (2008) Needle size and injection rate impact microbubble contrast agent population. *Ultrasound Med Biol* 34: 1182-1185.

Taylor SL, Rahim AA, Bush NL, Bamber JC, Porter CD. (2007) Targeted retroviral gene delivery using ultrasound. *J Gene Med* 9: 77-87.

Tlaxca JL, Anderson CR, Klibanov AL, Lowrey B, Hossack JA, Alexander JS, Lawrence MB, Rychak JJ. (2010) Analysis of in vitro transfection by sonoporation using cationic and neutral microbubbles. *Ultrasound Med Biol* 36: 1907-1918.

Tsunoda S, Mazda O, Oda Y, Iida Y, Akabame S, Kishida T, Shin-Ya M, Asada H, Gojo S, Imanishi J, Matsubara H, Yoshikawa T. (2005) Sonoporation using microbubble BR14 promotes pDNA/siRNA transduction to murine heart. *Biochem Biophys Res Comm* 336: 118-127.

Vannan M, McCreery T, Li P, Han Z, Unger E, Kuersten B, Nabel E, Rajagopalan S. (2002) Ultrasound-mediated transfection of canine myocardium by intravenous administration of cationic microbubble-linked plasmid DNA. *J Am Soc Cardiogr* 15: 214-218.

Villanueva FS. (2009) Ultrasound mediated destruction of DNA-loaded microbubbles for enhancement of cell-based therapies: new promise amidst a confluence of uncertainties? *J Am Coll Cardiogr Cardiovasc Imaging* 2: 880-882.

Walton CB, Shohet RV. (2009) Tiny bubbles and endocytosis? *Circ Res* 104: 563-565.

Wang JF, Wu CJ, Zhang CM, Qiu QY, Zheng M. (2009) Ultrasound-mediated microbubble destruction facilitates gene transfection in rat C6 glioma cells. *Mol Biol Rep* 36: 1263- 1267.

Wang ZX, Wang ZG, Ran HT, Ren JL, Zhang Y, Li Q, Zhu YF, Ao M. (2009) The treatment of liver fibrosis induced by hepatocyte growth factor-directed, ultrasound-targeted microbubble destruction in rats. *Clin Imaging* 33: 454-461.

Wu J, Nyborg WL. (2008) Ultrasound, cavitation bubbles and their interaction with cells. *Advanced Drug Delivery Reviews* 60: 1103-1116.

Xie F, Boska MD, Lof J, Uberti MG, Tsutsui JM, Porter TR. (2008) Effects of transcranial ultrasound and intravenous microbubbles on blood brain barrier permeability in a large animal model. *Ultrasound Med Biol* 34: 2028-2034.

Xing Y, Pua EC, Lu X, Zhong P. (2009) Low-amplitude ultrasound enhances hydrodynamic-based gene delivery to rat kidney. *Biochem Biophys Res Commun* 386: 217-222.

Yant SR, Huang Y, Akache B, Kay MA. (2007) Site-directed transposon integration in human cells. *Nucleic Acids Res* 35: e50.

Ye X, Loeb KR, Stafford DW, Thompson AR, Miao CH. (2003) Complete and sustained phenotypic correction of hemophilia B in mice following hepatic gene transfer of a high-expressing human factor IX plasmid. *J Thrombosis Haemostasis* 1: 103-111.

Yoon CS, Jung HS, Kwon MJ, Lee SH, Kim CW, Kim MK, Lee M, Park JH. (2009) Sonoporation of the minicircle-VEGF(165) for wound healing of diabetic mice. *Pharm Res* 26: 794-801.

Zarnitsyn VG, Prausnitz MR. (2004) Physical parameters influencing optimization of ultrasound-mediated DNA transfection. *Ultrasound Med Biol* 30: 527-538.

Zhang G, Song YK, Liu D. (2000) Long-term expression of human alpha1-antitrypsin gene in mouse liver achieved by intravenous administration of plasmid DNA using a hydrodynamics-based procedure. *Gene Ther* 7: 1344-1349.

Zheng XZ, Li HL, Du LF, Wang HP, Gu Q. (2009) Comparative analysis of gene transfer to human and rat retinal pigment epithelium cell line by a combinatorial use of recombinant adeno- associated virus and ultrasound or/and microbubbles. *Bosn J Basic Med Sci* 9: 174-181.

Zhou S, Li S, Liu Z, Tang Y, Wang Z, Gong J, Liu C. (2010) Ultrasound-targeted microbubble destruction mediated herpes simplex virus-thymidine kinase gene treats hepatoma in mice. *J Exp Clin Cancer Res* 29: 170.

Zhou XY, Liao Q, Pu YM, Tang YQ, Gong X, Li J, Xu Y, Wang ZG. (2009) Ultrasound- mediated microbubble delivery of pigment epithelium-derived factor gene into retina inhibits choroidal neovascularization. *Chinese Medical J* 122: 2711-2717.

Zhou Y, Cui J, Deng CX. (2008) Dynamics of sonoporation correlated with acoustic cavitation activities. *Biophysical J* 94: L51-53.

Zhou Y, Kumon RE, Cui J, Deng CX. (2009) The size of sonoporation pores on the cell membrane. *Ultrasound Med Biol* 35: 1756 - 1760.

Zhou Y, Zhou XY, Wang ZG, Zhu YF, Li P. (2010) Elevation of plasma membrane permeability upon laser irradiation of extracellular microbubbles. *Lasers Med Sci* 25: 587-594.

11 Chapter

CHEMICAL STRUCTURE, MORPHOLOGY, AND GENE TRANSFER ACTIVITY OF POLYAMINE-LIPID CONJUGATES AS EFFECTIVE GENE CARRIERS

1. INTRODUCTION: AN OVERVIEW

One of the most difficult aspects of gene therapy is the development of more efficient and safer gene carriers using nonviral compounds (Kay, 1997; Lasic, 1997). In contrast to viral carrier systems, nonviral gene carrier systems offer the advantages of simplicity of use, lack of specific immune response, and ease of mass production due to the low cost of preparation. However, they are characterized by low transfection efficiency, which must be resolved (Miller, 1998; Li & Huang, 2000). To improve the efficiency of nonviral carriers, many synthetic organic compounds, including cationic lipids (Felgner & Ringold, 1989; MacDonald et al., 1999; Felgner et al., 1987; Behr et al., 1989; Meyer et al., 1998), polycations (Boussif et al., 1995; Petersen et al., 2002; Koide et al., 2006; Russ et al., 2008; Haensler & Szoka, 1993; Shim & Kwon, 2009), and combinations thereof (Guillot-Nieckowski et al., 2007; Wu et al., 2001; Ewert et al., 2006; Takahashi et al., 2007; Matsui et al., 2006; Mustapa et al., 2009; Kogure et al., 2008), have been developed as nonviral gene carriers (Mintzer, M. A. & Simanek, E. E., 2009; also references cited therein). Numerous studies have been conducted on the structure–activity relationships of cationic amphiphiles in relation to their hydrophobic and cationic components (Remy et al., 1994; Geall et al., 1999; Ewert et al., 2002; Byk et al., 1998; McGregor et al., 2001). In an aqueous phase, these amphiphiles form self-assembling micelles and liposomes. The structures of these structures have been analyzed using small-angle X-ray scattering (SAXS), transmission electron microscopy (TEM), and atomic force microscopy (AFM) to gain insight into the structure-activity relationship, particularly those involving ordered structures (lamellar, inverted hexagonal, and cubic phases) and their morphological changes (Koltover et al., 1998; Koynova, Wang & MacDonald, 2006), as well as the size of complexes (Aoyama et al., 2003).

It is most likely that the endosomal pathway is involved in the mechanism of gene delivery by these cationic carriers (Wrobel & Collins (1995)): (i) cellular uptake via endocytosis, (ii) DNA release from the endosome, and (iii) entry into the nucleus. Numerous researchers have developed cationic compounds that facilitate the process. For instance, ligand-conjugated molecules that target a receptor such as integrin (Mustapa et al., 2007; Varga, Wickham & Lauffenburger, 2000), pH-responsive or cleavable molecules that facilitate the escape of DNA from endosomes (Russ et al, 2008; Oupicky, Parker & Seymour, 2002; Dauty et al., 2001; Miyake et al., 2004; Anderson, Lynn & Lange, 2003), and the conjugation of nuclear localization signal peptides (NLS) (Zanta et al., 1999; Manickam & Oupicky, 2006) for steps (i)-(iii). A morphological transition from lamellar to inverted hexagonal phase in the acidic endosomal environment has been suggested to facilitate the endosomal release and escape of DNA for cationic lipids/DNA complexes (lipoplexes) (Bell et al., 2003; Xu & Szoka, 1996). Observation of the morphology and metamorphosis of polycation–DNA complexes (polyplexes and lipoplexes) is crucial for the development of novel synthetic carriers and for elucidating the mechanism of gene transfer, in addition to the investigation of intracellular trafficking (Koynova, Wang & MacDonald, 2006; Wan et al., 2008; Tarahovsky, Koynova & MacDonald, 2004).

We have previously reported that polycationic liposomes (PCL) containing cetylated polyethylenimine (cetyl-PEI) exhibit high gene transfer activity (Yamazaki et al., 2000; Oku et al., 2001; Matsuura et al., 2003). The hydrophobic cetyl portion anchors the cetyl-PEI molecule, which is then dispersed across the liposomal surface. In our previous report, we suggested a potential mechanism of PCL-mediated gene transfer in which PCL/DNA complexes are uptaken by the endosomal pathway. This was based on the tracking of fluorescence-labeled components, PCL lipid, cetyl-PEI, and DNA, which resulted in the release and transfer of the cetyl PEI/DNA complex into the nucleus via the cytosol (Sugiyama et al., 2004). DNA compaction is therefore essential, and the cetyl PEI / DNA complex's effective compaction is a result of both hydrophobic and electrostatic interactions.

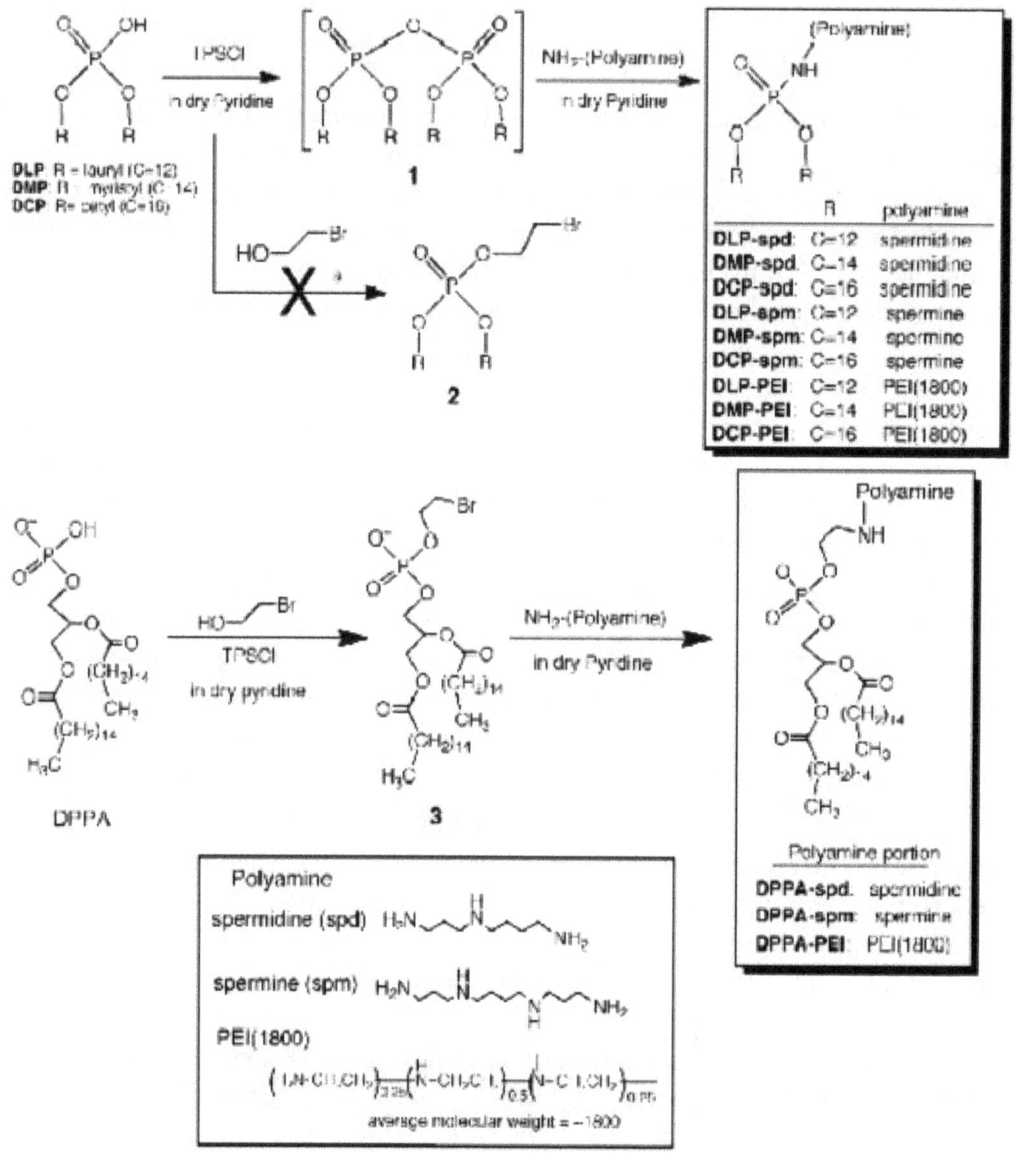

Scheme 1. Synthetic Strategy for Polyamine-lipid Conjugates.

PEI is employed as a gene transfer vector independently; however, it possesses inherent drawbacks, including polydispersity and cytotoxicity. We have previously reported the successful synthesis of a series of polyamine–dicetyl phosphate (DCP) conjugates through the reaction of a novel synthetic intermediate, dimerized DCP anhydride (compound 1 in Scheme 1), with a variety of polyamines, spermidine, spermine, and PEI (Scheme 1) (Dewa et al., 2004a). We anticipated minimal cytotoxicity due to the fact that spermidine and spermine are naturally occurring polyamines. Upon suspension in aqueous solution, they demonstrate moderate gene-transfer activity and form micellar aggregates. The extent of this activity is relatively insensitive to the modification of the polyamine portion. Additionally, we employed atomic force microscopy (AFM) to investigate the morphology of the conjugate/DNA complex. A concise discussion is provided regarding the correlation between the transfection efficiency of the conjugate/DNA and its assembling structure.

2. SYNTHETIC STRATEGY FOR A VARIETY OF POLYAMINE-LIPID CONJUGATES THROUGH FACILE SYNTHETIC ROUTES

In this report, we provide a concise description of (1) the simple synthetic strategy for polyamine-lipid conjugates and (2) evidence that the transgene activity is significantly increased when the conjugates are assembled into liposomes containing cholesterol and phospholipid. Additionally, we demonstrate that the activity is susceptible to chemical modification of the conjugate in both the polyamine and hydrophobic chain portions (Scheme 1). We also demonstrate that the gene transfer activity of the corresponding PCLs is significantly influenced by the type of polyamine in the conjugate. Specifically, there are significant differences between the lower molecular weight polyamines (spermidine and spermine) and the polymer type (PEI(1800)).

We also conducted an AFM analysis of the lipoplexes' morphology and deliberated on the correlation between their structure and their transfection efficiency. AFM analysis offers a significant advantage in the observation of lipoplex morphology, particularly for less ordered structures (Oberle et al., 2000).Nevertheless, there has been a lack of definitive evidence regarding the correlation between DNA release and morphological change. DNA release was demonstrated by AFM in this investigation as a consequence of the complex's disassociation (Dewa et al., 2010). We examine the relationships between morphology and activity using electrophoresis analysis, dynamic light scattering (DLS), and AFM observation. Furthermore, we present (5) our most recent endeavor to synthesize cleavable polyamine-lipid conjugates in a reductive cytosolic environment. In numerous instances, polycationic compounds have been synthesized through multistep reactions that consist of protection/deprotection reactions on polyamine moieties. Our polyamine-lipid compounds can be synthesized through two-step reactions that do not involve such protection/deprotection reactions. Scheme 1 illustrates the synthetic schemes, which encompass two categories of polyamine-lipid conjugates: dialkyl phosphate- and diacylphosphatidic acid-based compounds. In this section, we demonstrated a synthetic strategy for polyamine-lipid conjugates through straightforward routes (Dewa et al., 2004a,b, 2010).

2.1 Polyamine-dialkyl phosphate conjugates via a synthetic intermediate 1

Our preliminary idea for the syntheses of polyamine-dicetyl phosphate (DCP) conjugates was to use the bromoethylated compound, **2**, as a synthetic intermediate via condensation of DCP and 2-bromoethanol (Scheme 1) by using 1,3,5-triisopropylbenzenesulfonyl chloride, TPSCl (Uragami, Miyake & Regen, 2000; Tanaka, et al., 1987). Instead of the expected adduct, **2**, however, we obtained another product having high *Rf* value, 0.97 (*Rf* = 0.56 for DCP by eluting CHCl3/MeOH/H2O = 13/6/1, v/v/v). The resulting product is stable enough to be isolated by column chromatography. SIMS indicates almost double molecular mass, 1075.4 (546.9 for DCP). The observed IR absorption band at 952 cm-1 is assignable to the P–O–P stretching mode. The 31P NMR spectrum of the product, whose signals appear at δ = -12.27 and -0.13 ppm, is clearly distinguishable from that of DCP having a signal at δ =2.15 ppm. This evidence, taken together, leaves no doubt that the product is dimerized DCP in anhydrous form, **1**, [(C64H133O7P2)+ calcd 1075.9] connected via P–O–P bonding. There are some prior examples in the synthesis of pyrophosphate derivatives bearing small alkyl moieties, methyl, ethyl, propyl, and butyl (Cullis, Kaye & Trippett, 1987), however, the dimerized anhydride, **1**, bearing phospholipid-like long alkyl chains is an unprecedented compound. This compound is a convenient synthetic intermediate for formimg phosphoramidate bonds, as described below.

The anhydride **1** readily react with amines, *e.g.*, spermidine, spermine, and even polymer, PEI(1800), to form the phosphoramidate, P–N bond, providing the corresponding adducts shown in Scheme 1. The reactivity toward these nucleophiles indicates that the anhydrous compound **1** is potentially a good synthetic intermediate for making polyamine-dialkyl phosphate conjugates via the P–N bond. When reacted with spermine, for example, the adduct **DCP-spm** readily forms, concomitantly with the loss of DCP. The 31P NMR signal of compound **DCP-spm** is shifted downfield, to 9.76 ppm compared with that of DCP at 2.15 ppm, indicating the formation of the P–N bond in the compound. The down-field-shifted signal is attributed to the lower electronegativity of the nitrogen atom in spermine relative to the oxygen atom in DCP. Anhydride **1** is a useful synthetic intermediate because it is (1) stable yet reactive with amino groups, (2) very simple and easy to prepare, and (3) produced in high yield (~90 %). Furthermore, modification on the hydrophobic chain is easy when various dialkyl phosphates were used. These compounds could hence provide new category of polycationic lipids. Gall et al. have developed cationic lipid derivatives bearing P-N linkage, which possess gene transfer activity (Gall et al., 2010).

2.2 Polyamine-diacylphosphatidic acid conjugates via a synthetic intermediate 3

To diacylphosphatidic acid can be attached 2-bromoethanol with TPSCl, giving an intermediate **3**. When this reacted with an amino group of polyamine compounds via nucleophilic substitution reaction, a corresponding polyamine conjugate was formed as shown in Scheme 1 (Dewa et al., 2004b). Purified products can be afforded by column chromatography using an amino group-modified silicagel.

First, we describes trasnsfection efficacy of these polyamine-lipid conjugates. Figure 1A shows the efficacy of these carriers using β-galactosidase activity (milli-unit/well), in the absence of FBS. Compared with the commercially available transfection reagent, O-ethyl DOPC (E-DOPC) (MacDonald et al., 1999), which produced 0.1 milli-unit/well transgene expression, the compounds, **DCP-spd, DCP-spm, and DCP-PEI**, exhibit moderate efficiency, ~30 to 50 % of the efficiency of E-DOPC. The transfection efficiencies of the components themselves, *i.e.*, DCP, spermidine, spermine, and PEI(1800) were almost negligible. Thus, the transfection activity results from the conjugation of two moieties, a hydrophilic polyamine and a hydrophobic DCP. Compound **DCP-PEI** shows the highest efficiency at 3/1 (w/w) of **DCP-PEI** /DNA, whereas the efficiency of the other compounds are comparable and insensitive to the compound/DNA ratio within the range of error. In Figure 1B is shown the transfection efficiency in the presence of 20% FBS. The efficiency of these compounds was not influenced by the presence of 20% FBS, retaining 80–100 % of the activity (except compound **DCP-PEI** at 3/1 (w/w)). Such serum-resistant activity was also observed for the PCL gene transfection system previously reported (Matsuura et al., 2003). It is well known that serum often inhibits transfection; such inhibition is due to binding of negatively charged serum proteins to the cationic transfection reagents resulting in forming aggregates ineffective to the transfection. Although it is not clear why the polyamine-DCP conjugates are not influenced by the presence of the serum, the polyamine part may be assumed to efficiently interact with DNA via electrostatic interactions.

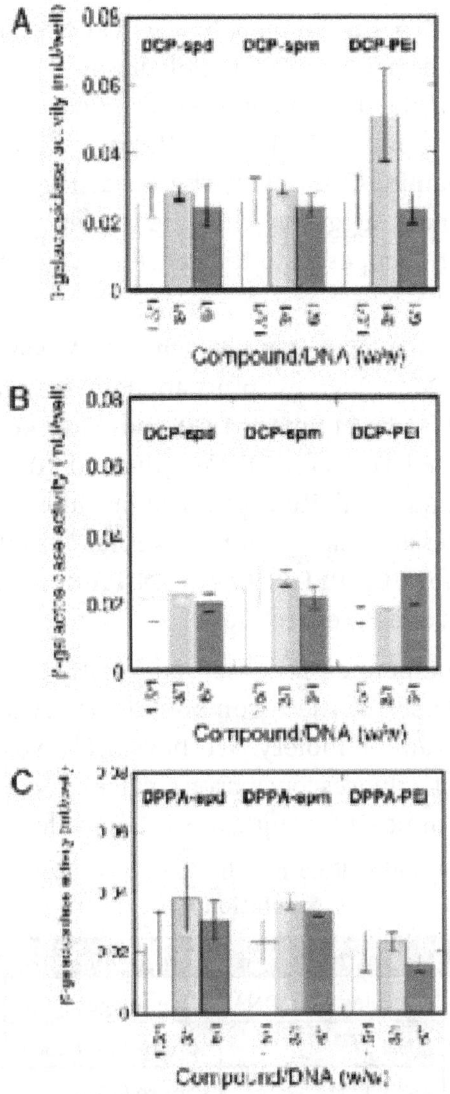

Fig. 1. Transfection efficacy of polyamine-DCP (A and B) and polyamine-DPPA conjugates (C) on VSMC, in the absence (A and C) and the presence of 20% of FBS (B). The ratios of compound/β-galactosidase plasmid DNA (w/w) were 1.5/1, 3/1, and 6/1, respectively.

3. TRANSFECTION EFFICACY OF VARIOUS POLYAMINE-LIPID CONJUGATES AS MICELLAR CARRIERS

Polyamine-DPPA derivatives also exhibited transfection activity, whose extent is almost comparable to the polyamine-DCP derivatives (Figure 1C). The tendency of the activity is **DPPA-spd** ≥ **DPPA-spm** > **DPPA-PEI.** In the following section, we will discuss on the relationship between gene transfer activity and morphology of polyamine-lipid complexes.

4. MORPHOLOGY OF THE MICELLAR AND COMPLEXES WITH DNA

It was found that DNA (ColE1 plasmid DNA) complex with various polyamines and polyamine-lipid conjugates by monitoring decrease of fluorescence from EtBr initially intercalated into DNA. With an increase in the cation (N) /anion (P) ratio, defined as N/P, the relative fluorescence intensity decreased as a result of complexation of DNA and polyamine. The polymeric molecule, PEI(1800) (N ~ 42 per molecule) most efficiently forms a complex with DNA; the complexation was almost complete at N/P = 3. Spermidine (N = 3 per molecule) and spermine (N = 4 per molecule) are less effective than PEI(1800), however, with complexation being complete at around N/P~7. For polyamine-DCP conjugates, the tendency for the fluorescence intensity to decrease is due to the complexation with DNA in the same way as that of the free amines, that is, **DCP-spd** < **DCP-spm** < **DCP-PEI.** DPPA-polyamine conjugates exhibited similar DNA condensation manner to DCP-polyamines, however, the tendency of DNA condensation was opposite way. The EtBr replacement experiment suggested that the tendency of DNA condensation is **DPPA-spd** > **DPPA-spm** > **DPPA-PEI**; the order is opposite to that of DCP-conjugates with regard to the polyamine portions.

In Table 1, the size of the micellar and complex with DNA was summarized. The mean particle diameters of various polyamine-DCP conjugates in an aqueous suspension, as given by dynamic light scattering is summarized in Table 1: 155 ± 54 nm for **DCP-spd**, 173 ± 46 nm for **DCP-spm**, 128 ± 38 nm for **DCP-PEI**, 90 ± 8 nm for **DPPA-spd**, 106 ± 41 nm for **DPPA- spm**, and 218 ± 35 nm for **DPPA-PEI.** AFM images for the suspension of these compounds exhibited spherical or ellipsoidal particles, whose sizes, defined according to diameter for the spheres and major × minor axes for the ellipsoid, were 132 nm for **DCP-spd**, 156 nm for **DCP-spm**, 209 × 145 nm for **DCP-PEI**, 96 nm for **DPPA-spd**, 111 nm for **DPPA-spm**, and 201 nm for **DPPA-PEI** (Figures 2A, B, C, D, E, and F, respectively).

The particle sizes for these compounds as observed by AFM show good agreement with those obtained by DLS. The molecular shapes of these compounds are regarded to be the "cone" type, due to the attachment of the large polyamine moiety. We have endeavored to make liposome from these compounds, however, entrapment of a fluorescence probe, calcein, was impossible. Therefore, we assume that the particle consisting of polyamine-DCP conjugates is a micelle- like aggregate.

Table 1. Estimated size and shape of polyamine-lipid micelles and complexes with ColE1 DNA suspended in aqueous solution [a)]

Compound	DLS (nm)		AFM/shape (size (nm)) and height (nm) [b)]			
	Micellar aggregate	Complex with DNA	Micelle aggregate		Complex with DNA	
DCP-spd	155 ± 54	409 ± 115	sphere (132)	21	aggregate (~1000 × 437)	16
DCP-spm	173 ± 46	237 ± 127	sphere (156)	24	aggregate (569 × 317)	20
DCP-PEI	128 ± 38	115 ± 40	ellipsoid (209 × 145)	12	sphere (120)	20
DPPA-spd	90 ± 8	140 ± 40	sphere (96)	24	sphere (129)	21
DPPA-spm	106 ± 41	109 ± 40	sphere (112)	16	sphere (172)	37
DPPA-PEI	218 ± 35	205 ± 39	aggregate (201)	12	aggregate (201)	17

The ratio of polyamine conjugate / DNA was 3/1 (w/w). Complexation of polyamine conjugate with DNA was carried out in water.

Sample solution was spread on a mica surface and dried. AFM images were obtained under dry condition. Width and height of the complexes were estimated from the AFM images in Figs 2 and 3.

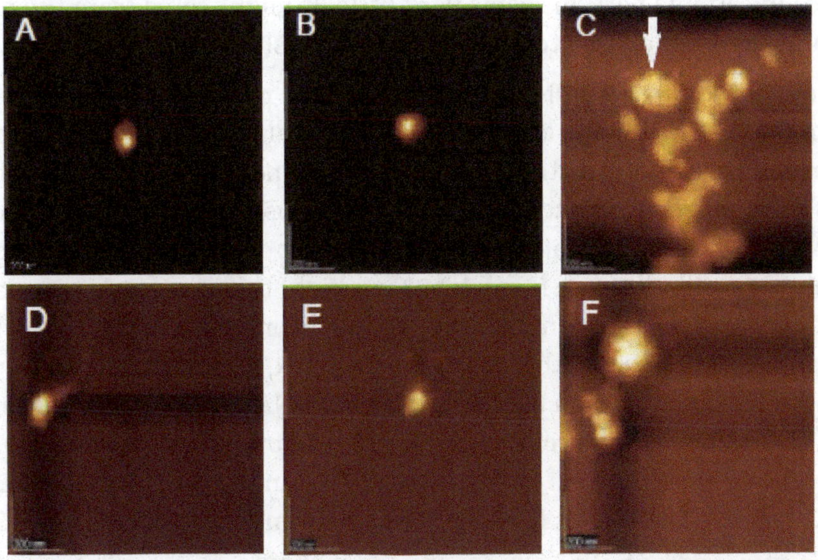

Fig. 2. AFM images of arrays of compounds, DCP-spd (A), DCP-spm (B), DCP-PEI (C), DPPA-spd (D), DPPA-spm (E), and DPPA-PEI (F). The compound was suspended in distilled water then dropped onto a mica surface by spin coating. All scale bars represent 200 nm. The object indicated by the arrow is discussed in the text.

Particle sizes of the conjugate/DNA (3/1: w/w) complex evaluated by DLS is 409 ± 115 nm for **DCP-spd**, 237 ± 127 nm for **DCP-spm**, 115 ± 40 nm for **DCP-PEI**, 140 ± 40 nm for **DPPA- spd**, 109 ± 40 nm for **DPPA-spm**, and 205 ± 39 nm for **DPPA-PEI**, respectively (Table 1). With increase in the size of polyamine portion in DCP-conjugates, the particle size significantly decreases, whereas opposite tendency was observed in DPPA-conjugates. AFM images support the tendency. Figure 3 shows an AFM image of DNA (A) and of the complexes it forms with various polyamines (B – I). Figure 3A reveals a clear image of partially-coiled ColE1 plasmid DNA (6646 bp), whose size is estimated to be 300~ 400 nm.

The height is ~ 2 nm, corresponding to the diameter of B-form DNA, ca. 2.4 nm. When DNA was complexed with compound **DCP-spd**, a "spider nest"-like structure was observed (Figure 3D). The size of the "quasi-ellipsoidal core region" is ~1000 × 437 nm and the height is ~16 nm. The height of the radiating peripheral "nest" region is ~ 2 nm, suggesting the nest region consists of free DNA. The height of the core region (~16 nm) is clearly larger than that of the DNA part, therefore, the core region of the structure must consist of the **DCP-spd**/DNA complex. Thus, complexation of compound **DCP-spd** and DNA gives rise to a segregated array, having complex and free DNA portions. The compound **DCP-spm** forms a similar but distinguishably different complex structure with DNA (Figure 3E), which resembles a "pearl necklace-like" aggregate (Yoshikawa et al., 1996), 569 × 317 nm in the plane of the substrate and 4~20 nm in height. The size of the pearl parts is 120 ~ 170 nm in diameter and 11 ~ 20 nm in height. These parts are connected by a region that is 4 ~ 5 nm in height. It appears from the image that the assembly consists of a tightly packed **DCP- spm**/DNA complex and regions of partially compacted DNA parts. Pronounced compaction of the complex was observed for the **DCP-PEI**/DNA, which formed a spherical cluster (120 nm in width and ~20 nm in height) (Figure 3F). The order of increasing compaction of DNA, **DCP-spd** < **DCP-spm** < **DCP-PEI**, is consistent with the extent of intercalation of EtBr as mentioned above, that is, the more tightly packed was the DNA complex, the lower was the extent of intercalation of EtBr. The tendency of size of the complex and the DNA condensation for DPPA-polyamine conjugates, that is, **DPPA-spd** > **DPPA-spm** (H) > **DPPA-PEI** (I). One may wonder why the morphologies of the complexes, DCP- and DPPA-based conjugates, as well as **DCP-spd** and **DCP-spm** with DNA are so different despite the structural difference in cationic portion seems very little. It might come from the difference in delicate balance of hydrophilic/hydrophobic factor and/or molecular shapes which reflect a packing parameter.

Upon comparison, of the observed DNA complex with free polyamines, spermine (Figure 3B) and PEI (Figure 3C), the complex size and shape are obviously distinguishable from the corresponding conjugate forms; much larger complexes form with these free polyamines. The dimension of the spermine/DNA complex (Figure 3B) is ~3 μm; the compaction of DNA is obviously incomplete, judging from the presence of the "nest" portion of DNA that are similar to peripheral part of the **DCP-spd**/DNA complex (Figure 3D). The PEI/DNA complex is smaller (416 × 218 nm by 22 nm high) than the former complex, suggesting that the greater cationic charge makes the complex smaller. When one considers the effect of the hydrophobic portion in the polyamine compound on the size of the complex, it is clear that hydrophobic alkyl parts in the conjugates play an important role in the compaction of DNA, which is most prominently observed in conjugate **DCP-PEI** (Figure 3C vs 3F). From these results, the prominent transfection efficiency of **DCP-PEI** likely results from the more efficient compaction of DNA in the complex **DCP-PEI**/DNA, whereas the lower activity of **DCP-spd** and **DCP-spm** is likely due to their much weaker compaction.

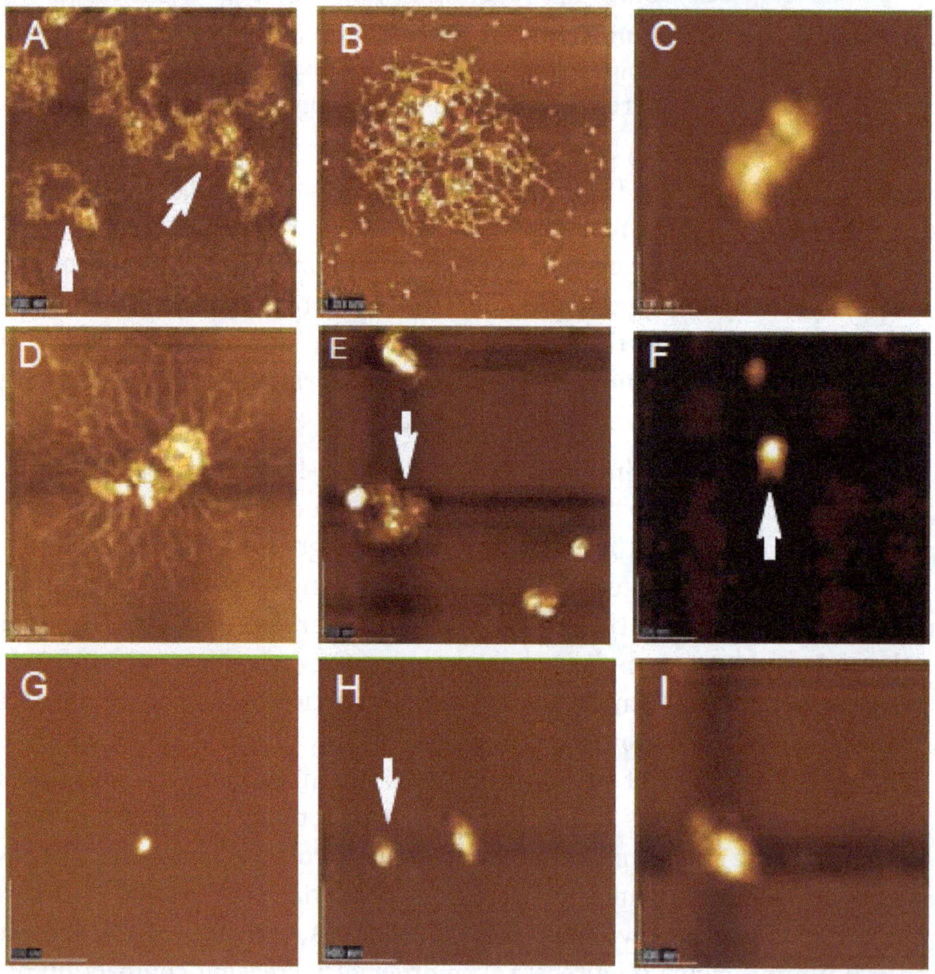

Fig. 3. AFM images of polyamine-ColE1 plasmid DNA complexes: (A), ColE1 plasmid DNA alone on polylysine-treated mica; (B), spermine/DNA; (C), PEI(1800)/DNA; (D), DCP- spd/DNA; (E), DCP-spm/ DNA; (F), DCP-PEI/DNA; (G), DPPA-spd/DNA; (H), DPPA-spm/DNA; (I), DPPA-PEI/DNA. The polyamine/ DNA ratio was 3/1 (w/w). Scale bars inserted in these images represent (A) 200, (B) 1000, (C) 200, (D) 500, (E) 500, (F) 200, (G) 500, 500, and (I) 200 nm, respectively. The objects indicated by arrow(s) are discussed in the text.

Although the precise mechanism remains to be clarified, the compaction by these polyamine compounds probably plays an important role because entry into the nucleus is thought to be a key step and a smaller complex is likely to be associated with more efficient transfection. The conjugates in this study possibly decompose into separate polyamine and DCP portions in endosome. We assume that two critically important factors are involved in the present transfection system; compaction of DNA by polyamine part and with hydrophobic dialkyl portion and decomposition of the complex in endosome so as to liberate the DNA/polyamine complex to be transported into nucleous. As previous report, DNA accompanied with cetyl-PEI enters into nucleus.

In this section, we described novel types of polyamine-dialkyl phosphate conjugates that have moderate gene transfection activity for β-galactosidase assay. These conjugates are easy to prepare via a novel synthetic intermediate, dimerized DCP anhydride, **1**. The synthetic approaches described herein are flexible and possess potential for the rationale design of highly efficient gene carriers with single or narrow ranges of molecular weight.

5. MORPHOLOGICAL EFFECT OF POLYAMINE-LIPID/DNA COMPLEXES ON THEIR TRANSFECTION ACTIVITY

As described above, when the polyamine-lipid conjugates were suspended in aqueous solution, they form micellar aggregates and exhibit moderate gene-transfer activity, the magnitude of which is relatively insensitive to the modification of the polyamine portion. Here we describe that preformed bilayer structure (as polycation liposomes, PCLs) significantly improves transfection efficacy compared with micellar aggregate form.

5.1 Preparation of Micellar Aggregates, PCL Vectors, and their complexes with DNA

Polycation liposome suspensions were typically prepared as follows: polyamine conjugate, phospholipid, 1,2-dioleoyl-*sn*-glycero-3-phosphoethanolamine (DOPE), and cholesterol (1/1/1 as a molar ratio) were dissolved in t-butyl alcohol. After removal of the solvent under reduced pressure, the residual solvent was removed by freeze-drying overnight. The lyophilized powder was hydrated with Tris-HCl buffer (20 mM, pH 8.0) followed by three freeze–thaw cycles and the resultant suspension was then subsequently extruded through polycarbonate membranes of 0.4, 0.2, and 0.1-μm pore diameter at room temperature. A suspension of the polyamine conjugate alone was prepared in the Tris-HCl buffer (20 mM, pH 8.0) by ultrasonication for 3 min. For convenience, particles so prepared are termed "micellar aggregates". Hereafter, polycationic liposomes (PCL) composed of the conjugates are described as "**conjugate**(PCL)", such as **DCP-spd**(PCL). Otherwise, micellar aggregates are described as just the name of conjugate, such as **DCP-spd**.

A plasmid encoding luciferase gene, pCAG-luc3 (6480 bp, a gift of DNAVEC Institute, Tsukuba, Japan), was amplified in *E. coli* JM109 (Nippon Gene, Toyama, Japan) and purified as described before (Matsuura et al., 2003). One microgram of the plasmid DNA in a TE buffer was added to a suspension of PCL containing 1 mM of polyamine conjugate so as to give the desired nitrogen/phosphate ratio, N/P. The mixture was incubated for 20 min at room temperature when used for transfection.

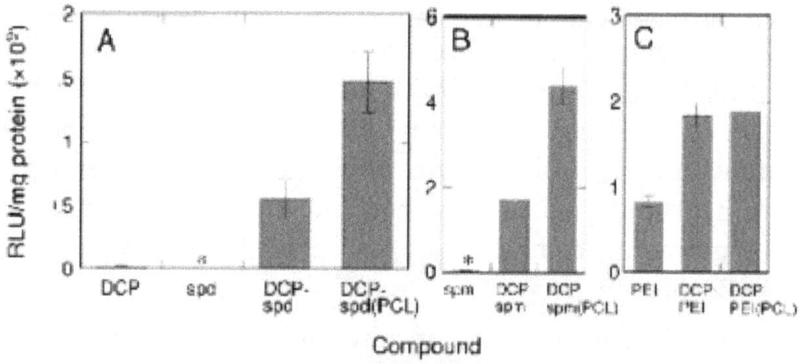

Fig. 4. Transfection efficacy of polyamine–lipid conjugates, DCP-spd (A), DCP-spm (B), and DCP-PEI (C), and their constituents, DCP (A) and polyamines, spermidine (spd, A), spermine (spm, B) and PEI1800 (PEI, C), on COS-1 cells. Efficacy was evaluated with the luciferase activity. The observed values for spermidine and spermine indicated by the asterisks (*) were apparently negligible on the activity scale shown. The conjugates DCP- spd, DCP-spm, and DCP-PEI represent their micellar aggregate forms, and DCP-spd(PCL), DCP-spm(PCL), and DCP-PEI(PCL) represent the conjugate-based PCLs (conjugate/DOPE/cholesterol = 1/1/1 (mol/mol/mol)). The nitrogen/phosphate (N/P) ratio was 16/1 for the polyamine conjugate/DNA complexes. For the monoanionic DCP, the molar ratio, DCP/nucleotide = 16/1, was applied as a negative control experiment. Transfection was conducted in the presence of 10% FBS.

5.2 Transfection procedure

COS-1 cells were cultured in Dulbecco's modified Eagle medium (DMEM) supplemented with 10% fetal bovine serum (FBS, Japan Bioserum Co. Ltd.) under a humidified atmosphere of 5% CO2 in air. One day before a transfection experiment, 1×105 COS-1 cells were seeded onto each of several 35-mm dishes and incubated overnight in a CO2 incubator. Then, the cells were washed twice with DMEM, and a suspension of lipoplex (1 μg DNA) was added to them in the presence of 10% FBS-DMEM. After 3 h incubation (37⌋C, 5% CO2), the cells were washed twice with DMEM and cultured for another 48 h in 10% FBS-DMEM. The cells in the 35-mm dishes were washed twice with phosphate-buffered saline at 37⌋C, and 200 μL of cell lysis buffer (LC-β, TOYO B-Net Co. Ltd., Tokyo) were added. After 15 min incubation, the cells were collected with a cell scraper, frozen at -80 ⌋C, and then thawed at room temperature. The lysate was centrifuged at 15,000 rpm for 5 min at 4 ⌋C. The supernatant was subjected to the luciferase assay (Pica Gene, TOYO B-Net Co. Ltd., Tokyo) using a luminophotometer (Luminescencer-PSN AB-2200, ATTO). The observed intensity in instrument light units was normalized to the amount of protein determined by BCA protein assay Kit (PIERCE) to give a relative light unit (RLU/mg protein).

5.3 Transfection efficacy of micellar aggregates and PCL vectors: Comparison of the dicetyl phosphate derivatives of spermidine, sperimine, and polyamine

Figure 4A shows the transfection efficacy of the polyamine conjugate, **DCP-spd**, and its constituent molecules, DCP and spermidine (spd). The conjugate (in this case an aqueous micellar suspension) shows greatly increased efficacy relative to the constituent molecules, DCP and spermidine. For these polyamine/DNA complexes, the nitrogen/phosphorous ratio (N/P) was 16. The data of the figure indicate that coupling the lipophilic and cationic portions is essential to obtain gene transfer. Such an effect of conjugating these two moieties was also observed for the spermine conjugate, **DCP-spm** (Figure 4B). When the conjugates were further formulated with DOPE and cholesterol (conjugate/DOPE/cholesterol = 1/1/1 (mol/mol/mol)) to generate polycationic liposomes (PCL) (**DCP-spd**(PCL) and **DCP- spm**(PCL)), efficacies were further enhanced by a factor of 2–3 relative to the micellar aggregate suspensions (**DCP-spd** and **DCP-spm**). The polycation PEI(1800) itself showed moderate activity and the conjugate, **DCP-PEI**, exhibited even greater activity. However, in contrast to the other conjugates, the activity of the liposomal form, **DCP-PEI**(PCL), was comparable to that of the micellar version, **DCP-PEI**. The cytotoxicity of the conjugates, both as micellar aggregates and PCLs, was low; the latter vectors, which exhibited higher activity, also had slightly higher toxicity than the former.

It was found that the DOPE-based PCL exhibits significantly greater activity than the DPPC- based compound for both of the **DCP-spd**(PCL) and the **DCP-spm**(PCL). This is indicative of a lipid-mediated gene transfer mechanism; DOPE is well-known as a "helper" lipid, which is believed to facilitate membrane fusion and endosomal escape of the DNA (Felgner et al., 1994). DOPE in the present PCL systems presumably also plays a role in the mechanism to expedite membrane fusion and destabilization of the endosomal membrane. DPPC, whose Tm is 41.5 °C, renders the PCL more stable and more rigid than does DOPE. Thus, it is likely that the fusogenic property of DOPE is responsible for the enhanced transfection activity of its complexes relative to those containing DPPC.

5.4 Optimal structure of the conjugate molecules for gene transfer

The facile synthetic route provides a variety of polycationic compounds that can be exploited to examine the effect of the polycationic and hydrophobic portions on transfection efficiency. We hence examined the effect on transfection activity of different polyamine conjugates incorporated into PCLs: C12, C14, or C16 alkyl chain in the lipophilic portion and spermidine, spermine, or PEI(1800) as the polycationic head group of the conjugate. The data on these compounds are shown in Figure 5A. This result reveals clear tendencies of longer length of the alkyl group and the lower molecular weight of the polyamines (spermidine, spermine) to enhance transfection. When compared with a commercial product, LipofectamineTM 2000, the **DCP-spd**(PCL) possessed 3.6-fold higher activity.

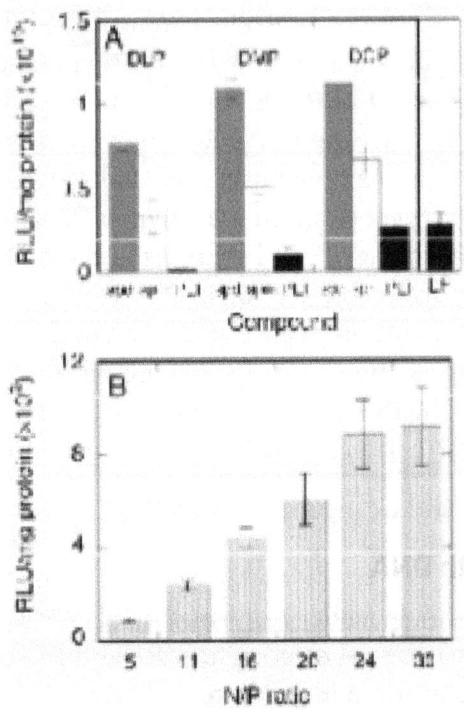

Fig. 5. (A) Effect of polyamine and hydrophobic portions on PCL-mediated gene transfer efficiency. PCL were composed of polyamine conjugate/DOPE/Cholesterol (1/1/1 mol/mol/mol). N/P ratio was 24. Transfection was done in the presence of 10% FBS. LF represents LipofectamineTM 2000 as a positive control experiment. (B) Effect of the N/P ratio on the transfection efficacy of DCP-spm(PCL) (DCP-spm/ DOPE/cholesterol = 1/1/1 (mol/mol/mol)). Transfection was in the presence of 10% FBS.

5.5 N/P-dependent efficacy and complexation of PCL with DNA

Figure 5B shows the dependence of **DCP-spm**(PCL) efficacy on the ratio of the number of nitrogen atoms in the conjugate to that of phosphate in the DNA (N/P). The efficacy increases with the N/P ratio essentially linearly up to 24. A similar N/P-dependence has been also observed for **DCP-spd**(PCL) (data not shown for clarity), indicating that excess polyamine relative to DNA is needed for effective gene transfer by PCLs. In contrast, the transfection activity of micellar aggregates **DCP-spd** and **DCP-spm** reaches plateau values in the N/P range of 11–16 (1~2 × 109 of RLU/mg protein). This tendency is consistent with our previous data obtained with the β-galactosidase expression system (Dewa et al., 2004a).

5.6 Chemical structure of the polyamine conjugates

The polyamine–dialkyl phosphate conjugates can be readily synthesized via a two-step reaction: (i) formation of dimerized dialkyl phosphate anhydride and (ii) its nucleophilic substitution with polyamines. The synthetic strategy gives access to a wide variety of polyamine–dialkyl phosphate derivatives. Conjugation of the polyamine and hydrophobic portions is required for an effective gene carrier (Figure 4). Such amphiphilicity is essential to condense DNA molecules, which requires both electrostatic and hydrophobic interactions (Yamazaki, 2000; Dewa et al., 2004a,b). We tested a number of combinations of dialkyl and polyamine portions for their activity in gene transfection. The longer alkyl chain exhibited higher efficiency (Figure 5A). The ζ-potential of the DOPE-based PCL increased with alkyl chain length, for example, 27.6, 33.4, and 37.1 mV for **DLP-spd**(PCL), **DMP-spd**(PCL), and **DCP-spd**(PCL), respectively, showing that the conjugate with the longer chain length provides the higher positive potential. The hydrophobic interaction results in stable incorporation of the conjugate into the PCL, consistent with transfection activity in the order of C16 ≥ C14 > C12.

We found that the micellar aggregate of **DCP-PEI** conjugate exhibited slightly higher transfection activity than the low-molecular weight amine conjugates, **DCP- spd** and **DCP-spm**, however, the activity of **DCP-PEI**(PCL) was marginal. The size of **DCP- PEI**(PCL)/DNA lipoplex is significantly greater than that of the other lipoplexes (Table 2, entry 14). The **DCP-PEI** conjugate is composed of DCP/PEI(1800) = ~1/1, judged by 1H- NMR (Dewa et al., 2004a). As previously reported, the cetyl-PEI, whose PCL possesses high transfection activity, consists of 10 cetyl portions in the polymer (Matsuura et al., 2003). The cetyl-PEI can attach to the PCL surface via the anchoring of cetyl portions in the lipid bilayer. However, that is not the case for **DCP-PEI**(PCL); the single hydrophobic portion in the conjugate is not enough to provide adequate covering of PEI over the PCL surface. This may cause "PEI-protrusion" from the surface, which gives rise to the large and heterogeneous aggregation seen upon combination with DNA molecules. This is likely the reason for the lower activity of the **DCP-PEI**(PCL). Taken together, these considerations suggest that a homogeneous positive charge distribution on the PCL surface is important to the transfection activity.

1.7 Morphological analysis of the complexes with DNA

5.7.1 Formation of complex with DNA

To elucidate the characteristics of N/P-dependence, formation of **DCP-spd**(PCL)/DNA and **DCP-spm**(PCL)/DNA lipoplexes was analyzed by agarose gel electrophoreses and DLS analysis. Electrophoretic analysis showed N/P ratio-dependent complexation; in the lower N/P range of 5–16, the open circular DNA band vanished and the supercoiled DNA band gradually faded for both of **DCP-spd**(PCL) and **DCP-spm**(PCL) and in the higher N/P range, 20–30, the latter DNA band totally disappeared. These observations indicate that the DNA molecules are completely entrapped within the lipoplex. Ethidium bromide (EtBr) replacement experiments also reveal condensation of DNA in the N/P range of 5–30 (Dewa et al., 2010).

The particle size of lipoplexes estimated by DLS analysis is summarized in Table 2. The diameters of the **DCP-spd**(PCL) and **DCP-spm**(PCL) alone are 158 ± 56, and 159 ± 30 nm, respectively (entries 1 and 7). Lipoplexes were larger and their size increased with increasing N/P up to 5 (entries 2 and 8 at N/P = 2 and entries 3 and 9 at N/P = 5). At N/P = 5, the PCL/DNA lipoplexes became larger with a broad distribution from 650 nm to over 1 μm (entries 3 and 9). In the higher N/P range (N/P = 16–24), sizes were reduced, converging at 261 ± 114 nm (**DCP-spd**(PCL), entry 5) and 256 ± 116 nm (**DCP-spm**(PCL), entry 11). ζ-potential measurements indicated the polarity of surface charge of lipoplexes inverts from negative in the lower N/P (5–11) to positive in the higher N/P (>16) regions.

AFM images revealed characteristic morphologies of lipoplexes in the both low and high N/P ranges (Figures 6A-D). When the **DCP-spd**(PCL)/DNA and **DCP-spm**(PCL)/DNA lipoplexes at N/P = 5 were put on PLL-treated mica (positively charged surface), large aggregates in the sub-~micrometer size range (600–1200 nm, Figures 6A and B) were observed. These structures resembled by bead-like aggregates (Yoshikawa et al., 1996) composed of small particles (80–120 nm in diameter, 8–20 nm in height) connected to one another. Such aggregates were not observed on a negatively-charged bare mica. This is understandable since the lipoplex at N/P = 5 is negatively charged, and the lipoplex must be adsorbed on the PLL-mica surface through electrostatic interaction to be imaged. The outer periphery of the large aggregates is rich in DNA molecules, presumably those were so loosely attached that they were liberated from the aggregates during electrophoresis.

Table 2. DLS analysis of various polyamine–dicetyl phosphate conjugate/DNA complexes

entry	compound	N/P ratio	pH	size (nm)	
				PCL	Micellar aggregate
1	DCP-spd	—	8	158 ± 56	249 ± 106
2		2	8	231 ± 89	
3		5	8	651 ± 501	1181 ± 1057
4		16	8	190 ± 83	
5		24	8	261 ± 114	1491 ± 1630
6		24	4	985 ± 1437	1624 ± 1566
7	DCP-spm	—	8	159 ± 30	225 ± 102
8		2	8	293 ± 112	
9		5	8	764 ± 378	321 ± 181
10		16	8	296 ± 168	
11		24	8	256 ± 116	1767 ± 1284
12		24	4	1033 ± 1060	1209 ± 949
13	DCP-PEI	—	8	214 ± 90	275 ± 184
14		24	8	1062 ± 783	317 ± 243

At the high N/P = 24, spherical complexes were observed for **DCP-spd**(PCL) and **DCP- spm**(PCL) complexes; their diameters were 200–400 nm for **DCP-spd**(PCL)/DNA (C) and 150–250 nm for **DCP-spm**(PCL)/DNA lipoplexes (D), and their heights were 12–30 and 27–60 nm, respectively. The size of the complexes is in good agreement with the values observed by DLS. The detailed topography of these **DCP-spd**(PCL)/DNA and **DCP- spm**(PCL)/DNA lipoplexes showed flat-topped spheres and spherical structures (line profiles (i)–(iii) for (C) and (iv)–(vi) for (D)). The profiles (ii) and (iii) are characteristically "step-like" (arrows). The heights indicated in (ii) are 6, 10, and 14 nm. Considering the thickness of lipid bilayer (4 nm) and the diameter of DNA (2 nm), these three values correspond to one bilayer (4 nm) + DNA (2 nm) = 6 nm, a double bilayer (8 nm) + DNA = 10 nm, and a triple bilayer (12 nm) + DNA = 14 nm, respectively. Thus, the step-like structure is reasonably indicative of a smectic lamellar assembly, where DNA molecules are laminated between bilayers.

Compared with **DCP-spd**(PCL) and **DCP-spm**(PCL) lipoplexes, the size of the **DCP- PEI**(PCL)/DNA lipoplex (N/P = 24) is large and has a broad distribution (1062 ± 783 nm, Table 2, entry 14). AFM images of the **DCP-PEI**(PCL)/DNA lipoplex show aggregates with heterogeneous and featureless shapes (Figure 6E and F). EtBr replacement experiments revealed DNA condensation in a similar manner to that of **DCP-spd**(PCL) and **DCP- spm**(PCL).

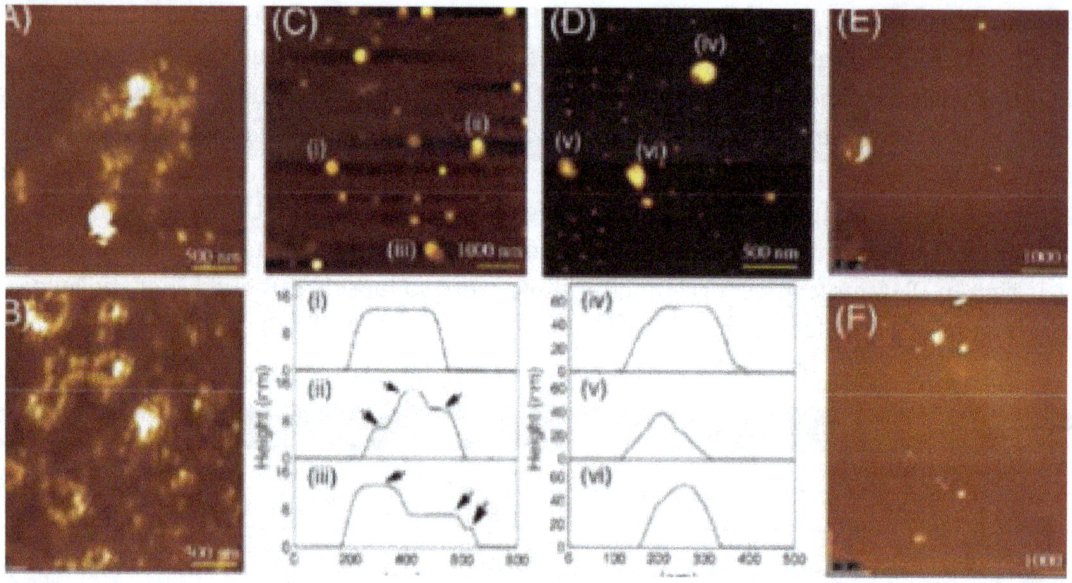

Fig. 6. AFM images of PCL/plasmid DNA (ColE1; 6646 bp) lipoplexes: (A), DCP- spd(PCL)/DNA (N/P = 5); (B), DCP-spm(PCL)/DNA (N/P = 5); (C), DCP- spd(PCL)/DNA (N/P = 24); (D), DCP-spm(PCL)/DNA (N/P = 24); (E and F), DCP- PEI(PCL)/DNA complex (N/P = 24). Scale bars shown in the images are 500 nm (A, B, and D) and 1000 nm (C, E, and F). The PCL/DNA complexes were applied to PLL-mica (A and B) and bare mica (C, D, E, F). All images were taken under an ambient air conditions. Height profiles of the objects (i)-(iii) in C and (iv)-(v) in D are shown below these images. Arrows in (ii) and (iii) indicate step-like profiles discussed in the text. Images E and F for DCP-PEI(PCL)/DNA were taken from a different area of the bare mica surface.

5.7.2 Effect of bilayer structure on the transfection activity

In the series of polyamines tested in this study, the low molecular weight polyamines were found to be more effective gene carriers when these conjugates were assembled into PCLs. The transfection activity of the PCL was ~3 times larger than as the corresponding micellar aggregate (Figure 4). The effect of the helper lipid, DOPE, was clearly substantial when compared with DPPC used instead of DOPE. DOPE, a predominantly non-lamellar lipid, is thought to facilitate fusion and destabilization of the endosomal membrane after uptake of cationic lipid/DNA complexes into a cell. In our previous report, we described intercellular trafficking of PCL (composed of cetyl-PEI and DOPE)–DNA complexes, which were taken up into cells by endosomal pathway (Sugiyama et al., 2004), followed by endosomal escape. In fact, transfection activity of the present PCLs was inhibited by nigericin, which is able to dissipate the pH gradient across the endosomal membranes, by 30-50%, suggesting that endosomal pathway is likely involved in the mechanism in the present lipoplex system. It appears therefore, that the mechanism of the lipofection by the compounds in this study may be similar to that of this and other agents known to be enhanced by DOPE. The lipoplexes made from the bilayer-structured PCLs evidently involve lamellar assemblies, given that AFM images reveal the presence of step-like profiles ((ii) and (iii) in Figure 6). The step-like profiles imply a lamellar complex, in which DNA rods (2 nm in diameter) are laminated between bilayers (4 nm thickness). Such an intrinsic bilayer structure may predispose lipoplexes to interact with cell and endosomal membranes. This is not the case for micellar aggregates, whose morphology is large spheres, in which polyamine conjugate and DNA molecules likely aggregate randomly (Table 2 entry 5). This may be one of the reasons for the higher activity of PCL-based lipoplex, whose size is more favorable to transfection.

The linear dependence of activity on N/P (Figure 5B) is related to the morphology of the lipoplex. The electrophoresis experiment and AFM images (Figures 6A-D) suggest a reasonable explanation of the dependence, namely, the following: In the low N/P range (~5), the PCLs inadequately condense DNA molecules, giving the bead-like structures (Figures 6A and B). The DNA molecules loosely packed in the complex are readily released during electrophoresis. Such a complex, whose ζ-potential is negative, is too large to be introduced into the cell membrane via endocytosis; therefore the transfection level is low. With increasing N/P ratio, the morphology of the lipoplex transforms from large bead-like structure into smaller particles, wherein DNA molecules are condensed more tightly (Figures 6C and D). The size of the lipoplexes, whose ζ-potential is positive, is 150–400 nm, is more favorable for cellular uptake via endocytosis (Koynova, Wang & MacDonald, 2006). Given that the lamellar assembly in the lipoplex is responsible for its effectiveness as a gene carrier (Koltover et al., 1998; Koynova, Wang & MacDonald, 2006), the population of active species for gene transfer would increase with increasing in the N/P. Although highly positive-charged carriers are generally toxic, the PCL described here exhibit low cytotoxicity, an advantage for in vitro and in vivo applications.

5.7.3 Disassembly of the lipoplexes and DNA release

When the **DCP-spd**(PCL)/DNA and **DCP-spm**(PCL)/DNA lipoplexes (N/P = 24) were incubated in acidic solution (down to pH 4), the particle sizes measured by DLS became significantly larger and exhibited broad distributions (Table 2, entries 6 and 12). AFM imaging revealed morphological transformation of the PCL/DNA complexes upon acidification. When the dispersion of **DCP-spd**(PCL)/DNA lipoplexes (N/P = 24) was acidified at pH 4 for 1 h by addition of acetic acid, deformed structures were observed on bare mica (Figure 7A). Relative to the original structure (Figure 6C), the complex is decisively deformed by the acid treatment. Although some of flat-topped sphere complexes remain, the predominant morphology is particles connected with strings. The height of the clusters is 25–53 nm and they are connected with string portions that are 6–10 nm high. When the acid-treated complex solution was put on PLL-mica, additional deformed objects appeared on the surface (Figure 7B). A "beads on a string" deformed structure is composed of very small particles (50~100 nm in diameter) and string parts (~70 nm in width and 2–5 nm in height). The beads on a string structure observed on the positively charged surface must consist of DNA-rich fragments associated with some lipid components. The **DCP- spm**(PCL)/DNA lipoplex maintains its spherical structure on bare mica (Figure 7C). On the PLL-mica, on the other hand, deformed structures were observed as in the case of **DCP- spd**(PCL)/DNA lipoplex (Figure 7D). Such morphological changes upon acidification result from disassembly of the lipoplexes and the accompanying DNA release. Gel electrophoretic analysis provided evidence for the DNA release; the released plasmid band increased with the acidification from pH 8 to pH 4. In sharp contrast, such a morphological change was not observed for the micellar aggregate **DCP-spd**. DLS analysis indicates an insensitivity of the micellar aggregates to acidification (Table 2, entries 6 and 12).

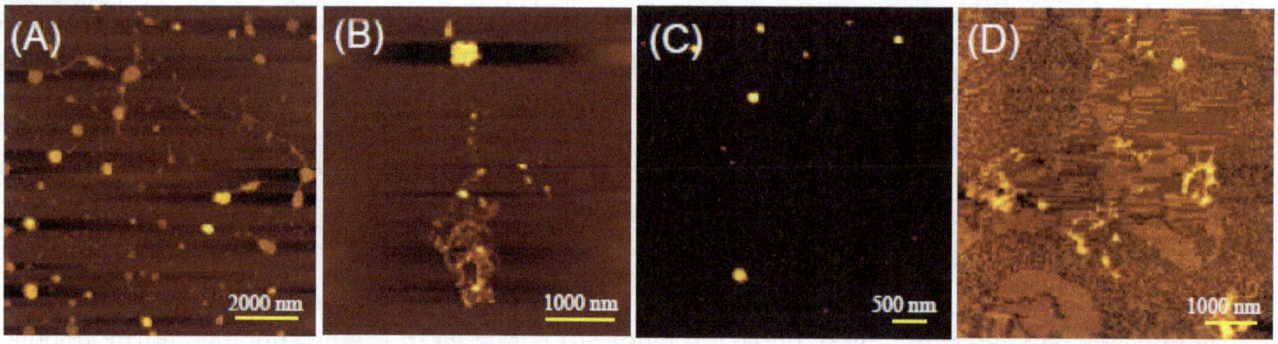

Fig. 7. AFM images of disassembled lipoplexes, DCP-spd(PCL)/DNA (A and B) and DCP- spm(PCL)/DNA (C and D) by acidification at pH 4. The acid-treated suspensions of complexes were put on bare mica (A, C) and PLL-mica (B and D). After removal of the solution the images were acquired. Scale bars: 2000 nm (A), 1000 nm (B), 500 nm (C), 1000 nm (D).

Facile escape from the acidic endosomal compartment is necessary for efficient gene transfer. Disassembly of the lipoplex associated with DNA release has been clearly observed by AFM imaging (Figure 7) and by electrophoretic analyses. Upon acidification (down to pH4), the extent of protonation of the polyamine portion is increased. Assuming that the lipoplex forms smectic lamella, electrostatic repulsion between layers must be increased with such protonation, resulting in the disruption of the lipoplex. For lipoplexes composed of **DCP-spd**(PCL) and **DCP-spm**(PCL), disruption accompanying DNA release has been confirmed. In sharp contrast, the size and shape of micellar aggregate/DNA complexes composed of **DCP-spd** and **DCP-spm** are insensitive to acidification (Table 2, entries 6 and 12). This would be due to their amorphous structure, which does not respond to pH change. Thus, the disassembly of the lipoplex composed of the bilayer-structured PCL is essential in effective gene transfer, especially in the process of endosomal escape, where lipid exchange and flip-flop are involved in disrupting the membrane and leading to DNA release (Xu & Szoka, 1996). Although the acidity of this experimental condition (pH4) seems higher than that in endosome (pH5.5), such a protonation process on the polyamines should be involved in the endosomal environment, because the protonation on the polyamines whose pKa values are > 8 may proceed rather gradually in the acidic region, especially in the self-assembled lamellar structure, where polyamines are densely packed. This finding suggests a strategy for molecular design—especially in the polyamine portion—in which a morphological transformation of lipoplexes is taken advantage of for new nonviral transfection strategies.

6. CLEAVABLE PEPTIDE-PHOSPHOLIPID CONJUGATES UNDER PHYSIOLOGICAL CONDITIONS

In an intracellular reductive environment, the release of DNA from complexes may be facilitated by detachable conjugates between cationic and hydrophobic lipid portions. This is a promising approach to enhance the efficacy of non-viral gene delivery. Disulfide-linked polymer (Oba et al., 2008) and gemini lipids (Behr) have been employed in the development of such methods. We have recently achieved the successful synthesis of oligo-arginine bearing phospholipids via disulfide linkage, DPPE-SS-Rn (Scheme 2).

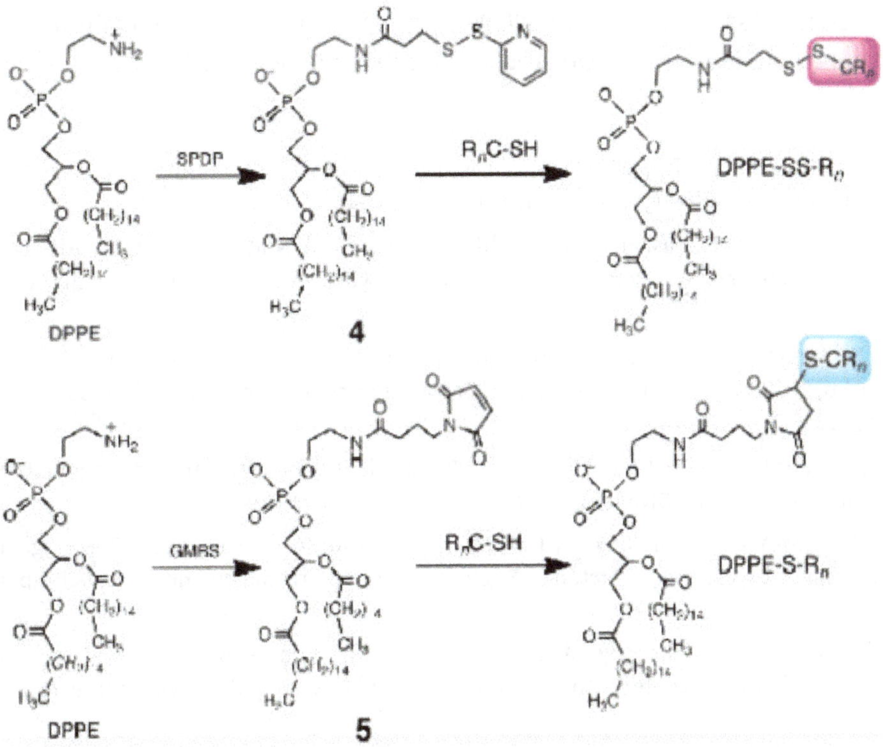

Scheme 2. Synthetic strategy for peptide-phospholipid conjugates

Oligo-arginines, e.g., TAT peptide, are well known for effective carriers across cell membranes (Futaki et al., 2001). The reaction scheme is shown in Scheme 2. In brief, 1,2-dipalmitoyl-*sn*-glycero-3-phospho-ethanolamine (DPPE) was reacted with a heterobifunctional coupling agent, *N*-succinimidyl-3-(2-pyridyldithio)-propionate (SPDP) to form compound **4**, followed by coupling with oligopeptides, (Arg) *n*Cys (R*n*C; *n* = 5, 8). Undetachable type of analogous conjugates (DPPE-S-R*n*) were also synthesized with a coupling agent, *N*-(4-maleimidobutyroxy)succinimide (GMBS) and DPPE to form compound **5**, followed by coupling with (Arg)nCys. When the disulfide-type conjugates were applied to gene delivery assay in vitro, 3-5 times higher trasnfection activity was observed compared with the corresponding undetachable conjugates. This result suggests that such a cleavable type of peptide-phospholipid conjugate is one of promising components to assemble effective non-viral vectors.

7. CONCISE SUMMARY

We have demonstrated that the gene transfer activity of PCL composed of the low-molecular-weight polyamine conjugates, DCP–spermidine (DCP-spd) and DCP–spermine (DCP-spm), is significantly higher than that of PEI(1800) conjugate-based DCP-PEI(PCL). The former compounds produce lipoplexes with a diameter of 150–400 nm, while the latter produces massive aggregates. The AFM images of the former compounds clearly demonstrate a morphological change upon acidification, which suggests that DNA is released from the lipoplexes. In contrast, the morphology of micellar aggregates is insensitive to pH change. The acidification-induced transformation may be significantly influenced by the chemical structure of the polyamine portion, as a pH-dependent transformation is essential for gene transfer. Additionally, we have discussed the correlation between the morphology of the lipoplexes and the N/P-dependence of transfection activity, as demonstrated by AFM. The morphological study conducted with AFM yielded valuable information that was instrumental in the development of strategies that would result in gene carriers that are optimally efficient and in the comprehension of the fundamental principles of lipoplexes with exceptional activity. Asai et al. (2001) have recently shown that DCP-polyaimie-based liposomes are efficacious for systemic siRNA delivery. Our recent efforts have provided evidence of this.

REFERENCES

Anderson, D. G., Lynn, D. M. & Lange, R. (2003) Semi-automated synthesis and screening of a large library of degradable cationic polymers for gene delivery. *Angew. Chem. Int. Ed.* 42, 3153–3158.

Aoyama, Y., Kanamori, T., Nakai, T., Sasaki, T., Horiuchi, S., Sando, S. & Niidome, T. (2003) Artificial viruses and their application to gene delivery. Size-controlled gene coating with glycocluster nanoparticles. *J. Am. Chem. Soc.* 125, 3455–3457.

Asai, T., Matsushita, S., Kenjo, E., Tsuzuku, T., Yonenaga, N., Koide, H., Hatanaka, K., Dewa, T., Nango, M., Maeda, N., Kikuchi, H. & Oku, N. (2011). Dicetyl Phosphate- Tetraethylenepentamine-Based Liposomes for Systemic siRNA Delivery. *Bioconjugate Chem.* 22, 429–435.

Behr, J.-P., Demeneix, B., Loeffler, J. P. & Perez-Mutul, J. (1989) Efficient gene transfer into mammalian primary endocrine cells with lipopolyamine-coated DNA. *Proc. Natl. Acad. Sci. U.S.A.* 86, 6982–6986.

Bell, P. C., Bergsma, M., Dolbnya, I. P., Bras, W., Stuart, M. C. A., Rowan, A. E., Feiters, M. C. & Engberts, J. B. F. N. (2003) Transfection mediated by gemini surfactants: Engineered escape from the endosomal compartment. *J. Am. Chem. Soc.* 125, 1551– 1558.

Boussif, O., Lezoulch, F., Zanta, M. A., Mergny, M. D., Scherman, D., Demeneix, B. & Behr, J.-P. (1995) A Versatile vector for gene and oligonucleotide transfer into cells in culture and in vivo: Polyethylenimine. *Proc. Natl. Acad. Sci. U.S.A.* 92, 7297–7301.

Byk, G., Dubertret, C., Escriou, V., Frederic, M., Jaslin, G., Rangara, R., Pitard, B., Crouzet, J., Wils, P., Schwartz, B. & Scherman, D. (1998) Synthesis, activity, and structure- activity relationship studies of novel cationic lipids for DNA transfer. *J. Med. Chem.* 41, 224–235.

Chittimalla, C., Zammut-Italiano, L., Zuber, G. & Jean-Paul Behr, J.-P. (2005). Monomolecular DNA Nanoparticles for Intravenous Delivery of Genes. *J. Am. Chem. Soc.*, 127, 11436-11441.

Cullis, P. M., Kaye, A. D. & Trippett, S. (1987). On the mechanism of reaction of activated phospholyl compounds with phosphoryl anions: a reassessment of the role of dioxadiphosphetanes. *J. Chem. Soc., Chem. Commun.* 1464 – 1466.

Ewert, K., Ahmad, A., Evans, H. M., Schmidt, H.-W., and Safinya, C. R. (2002) Efficient synthesis and cell-transfection properties of a new multivalent cationic lipid for nonviral gene delivery. *J. Med. Chem. 45*, 5023–5029.

Ewert, K. K., Evans, H. M., Zidovska, A., Bouxsein, N. F., Ahmad, A. & Safinya, C. R. (2006) A columnar phase of dendritic lipid-based cationic liposome-DNA complexes for gene delivery: Hexagonally ordered cylindrical micelles embedded in a DNA honeycomb lattice. *J. Am. Chem. Soc.* 128, 3998–4006.

Dauty, E., Remy, J.-S., Blessing, T. & Behr, J.-P. (2001) Dimerizable cationic detergents with a low cmc condense plasmid DNA into nanometric particles and transfect cells in culture. *J. Am. Chem. Soc.* 123, 9227–9234.

Dewa, T., Ieda, Y., Morita, K., Wang, L., MacDonald, R. C., Iida, K., Yamashita, K., Oku, N. & Nango, M. (2004a). Novel polyamine–dialkyl phosphate conjugates for gene carriers. Facile synthetic route via an unprecedented dialkyl phosphate. *Bioconjugate Chem.* 15, 824–830.

Dewa, T., Ieda, Y., Morita, K., Yoshida, K., Wang, L., MacDonald, R. C., Iida, K., Yamashita, K., Oku, N. & Nango, M. (2004b). Novel Polycationic phospholipid Conjugates for Gene Carriers. *Kobunshi Ronbunshu*, 61, 634-639.

Dewa, T., Asai, T., Tsunoda, Y., Kato, K., Baba, D., Uchida, M., Sumino, A., Niwata, K., Umemoto, T., Iida, K., Oku, N. & Nango, M. (2010). Liposomal Polyamine-Dialkyl Phosphate Conjugates as Effective Gene Carriers: Chemical Structure, Morphology, and Gene Transfer Activity. *Bioconjugate Chem.* 21, 844–852.

Felgner, P. L., Gadek, T. R., Holm, M., Roman, R., Chan, H. W., Wenz, M., Northrop, J. P., Ringold, G. M. & Danielsen, M. (1987) Lipofection: A highly efficient, lipid-mediated DNA-transfection procedure. *Proc. Natl. Acad. Sci. U.S.A.* 84, 7413– 7417.

Felgner, P. L. & Ringold, G. M. (1989) Cationic liposome-mediated transfection. *Nature* 337, 387–388.

Felgner, J. H., Kumar, R., Sridhar, C. N., Wheeler, C. J., Tsai, Y. J., Border, R., Ramsey, P., Martin, M. & Felgner, P. L. (1994). Enhanced gene delivery and mechanism studies with a novel series of cationic lipid formulations. *J. Biol. Chem.* 269, 2550–2561.

Futaki, S., Suzuki, T., Ohashi, W., Yagami, T.,Tanaka, S., Ueda, K. & Sugiura, Y. (2001). Arginine-rich Peptides. An Abundant Source of Membrane-permeable Peptides Having Potential as Carriers for Intracellular Protein delivery. *J. Biol. Chem.* 276, 5836–5840.

Gall, T. L., Loizeau, D., Picquet, E., Carmoy, N., Yaouanc, J.-J., Burel-Deschamps, L., Delépine, P., Giamarchi, P., Paul-Alain Jaffrées, P.-A., Lehn, P. & Montier, T. (2010). A Novel Cationic Lipophosphoramide with Diunsaturated Lipid Chains: Synthesis, Physicochemical Properties, and Transfection Activities. *J. Med. Chem.*53, 1496–1508.

Geall, A. J., Eaton, M. A. W., Baker, T., Catterall, C. & Blagbrough, I. S. (1999) The regiochemical distribution of positive charges along cholesterol polyamine carbamates plays significant roles in modulating DNA binding affinity and lipofection. *FEBS Lett.* 459, 337–342.

Guillot-Nieckowski, M., Joester, D., Stöhr, M., Losson, M., Adrian, M., Wagner, B., Kansy, M., Heinzelmann, H., Pugin, R., Diederich, F. & Gallani, J.-L. (2007) Self-assembly, DNA complexation, and pH response of amphiphilic dendrimers for gene transfection. *Langmuir* 23, 737–746.

Haensler, J.& Szoka, Jr. F. C. (1993) Polyamidoamine cascade polymers mediate efficient transfection of cells in culture. *Bioconjugate Chem.* 4, 372–379.

Kay, M. A., Liu, D. & Hoogerbrugge, P. M. (1997). Gene Therapy. *Proc. Natl. Acad. Sci. U.S.A.* 94, 12744–12746.

Kogure, K., Akita, H., Yamada, Y., and Harashima, H. (2008) Multifunctional envelope-type nano device (MEND) as a non-viral gene delivery system. *Adv. Drug Deliv. Rev.* 60, 559–571.

Koide, A., Kishimura, A., Osada, K., Jang, W.-D., Yamasaki, Y., Kataoka, K., Harada, A. & Nagasaki, Y. (2006) Semipermeable polymer vesicle (PICsome) self-assembled in aqueous medium from a pair of oppositely charged block copolymers: Physiologically stable micro-/nanocontainers of water-soluble macromolecules. *J. Am. Chem. Soc.* 128, 5988–5989.

Koltover, I., Salditt, T., Rädler, J. O. & Safinya, C. R. (1998) An inverted hexagonal phase of cationic liposome-DNA complexes related to DNA release and delivery. *Science, 281,* 78–81.

Koynova, R., Wang, L. & MacDonald, R. C. (2006) An intracellular lamellar-nonlamellar phase transition rationalizes the superior performance of some cationic lipid transfection agents. *Proc. Natl. Acad. Sci. U.S.A.* 103, 14373–14378.

Lasic, D. D. (1997) *Liposomes in Gene Derivary,* CRC Press, New York.

Li, S. & Huang, L. (2000) Nonviral gene therapy: promises and challenges, *Gene Therapy* 7, 31–34.

MacDonald, R.C., Ashley, G. W., Shida, M. M., Rakhmanova, V. A., Tarahovsky, Y. S., Pantazatos, D. P., Kennedy, M. T., Pozharski, E. V., Baker, K. A., Jones, R. D., Rosenzweig, H. S., Choi, K. L., Qiu, R. & McIntosh, T. J. (1999). Physical and biological properties of cationic triesters of phosphatidylcholine. *Biophys. J.* 77, 2612 – 2629.

Manickam D. S. & Oupicky, D. (2006) Multiblock reducible copolypeptides containing histidine-rich and nuclear localization sequences for gene delivery. *Bioconjugate Chem.* 17, 1395–1403.

Matsui, K., Sando, S., Sera, T., Aoyama, Y., Sasaki, Y., Komatsu, T., Terashima, T. & Kikuchi, (2006) Cerasome as an Infusible, cell-friendly, and serum-compatible transfection agent in a viral size. *J. Am. Chem. Soc.* 128, 3114–3115.

Matsuura, M., Yamazaki, Y., Sugiyama, M., Kondo, M., Ori, H., Nango, M. & Oku, N. (2003) Polycation liposome-mediated gene transfer in vivo. *Biochim. Biophys. Acta* 1612, 136 – 143.

McGregor, C., Perrin, C., Monck, M., Camilleri, P. & Kirby, A. J. (2001) Rational approaches to the design of cationic gemini surfactants for gene delivery. *J. Am. Chem. Soc.* 123, 6215–6220.

Meyer, O., Kirpotin, D., Hong, K., Sternberg, B., Park, J. W., Woodle, M. C. & Papahadjopoulos, D. (1998) Cationic liposomes coated with polyethylene glycol as carriers for oligonucleotides. *J. Biol. Chem.* 273, 15621–15627.

Miller, A. D. (1998) Cationic liposomes for gene therapy. *Angew. Chem. Int. Ed.* 37, 1768– 1785.

Mintzer, M. A. & Simanek, E. E. (2009) Nonviral vectors for gene delivery. *Chem. Rev. 109,* 259–302.

Miyata, K., Kakizawa, Y., Nishiyama, N., Harada, A., Yamasaki, Y., Koyama, H. & Kataoka, 1. (2004) Block catiomer polyplexes with regulated densities of charge and disulfide cross-linking directed to enhance gene expression. *J. Am. Chem. Soc.* 126, 2355–2361.

Mustapa, M. F. M., Bell, P. C., Hurley, C. A., Nicol, A., Guénin, E., Sarkar, S., Writer, M. J., Barker, S. E., Wong, J. B., Pilkington-Miksa, M. A., Papahadjopoulos-Sternberg, B., Shamlou, P. A., Hailes, H. C., Hart, S. L., Zicha, D. & Tabor, A. B. (2007) Biophysical characterization of an integrin-targeted lipopolyplex gene delivery vector. *Biochemistry* 46, 12930–12944.

Mustapa, M. F. M., Grosse, S., M., Kudsiova, L., Elbs, M., Raiber, E.-A., Wong, J. B., Brain, A. P. R., Armer, H. E. J., Warley, A., Keppler, M., Ng, T., Lawrence, M. J., Hart, S. L., Hailes, H. C., and Tabor, A. B. (2009) Stabilized integrin-targeting ternary LPD (Lipopolyplex) vectors for gene delivery designed to disassemble within the target cell. *Bioconjugate Chem. 20*, 518–532.

Oba, M., Aoyagi, K., Miyata, K., Matsumoto, Y., Itaka, K., Nishiyama, N., Yamasaki, Y., Koyama, H. & Kataoka, K. (2008). Polyplex Micelles with Cyclic RGD Peptide Ligands and Disulfide Cross-Links Directing to the Enhanced Transfection via Controlled Intracellular Trafficking *Mol. Pharmacuetics* 5, 1080–1092.

Oberle, V., Bakowsky, U., Zuhorn, I. S. & Hoekstra, D. (2000). Lipoplex formation under equilibrium conditions reveals a three-step mechanism. *Biophys. J. 79*, 1447– 1454.

Oku, N., Yamazaki, M., Matsuura, M., Sugiyama, M., Hasegawa, M. & Nango, M. (2001) A novel non-viral gene transfer system, polycation liposomes. *Adv. Drug Deliv. Rev.* 52, 209 – 218.

Oupicky, D., Parker, A. L. & Seymour, L. W. (2002) Laterally stabilized complexes of DNA with linear reducible polycations: Strategy for triggered intracellular activation of DNA delivery vectors. *J. Am. Chem. Soc.* 124, 8–9.

Petersen, H., Fechner, P. M., Martin, A. L., Kunath, K., Stolnik, S., Roberts, C. J., Fischer, D., Davies, M. C. & Kissel, T. (2002) Polyethylenimine-graft-poly(ethylene glycol) copolymers: influence of copolymer block structure on DNA complexation and biological activities as gene delivery system. *Bioconjugate Chem.* 13, 845– 854.

Remy, J.-S., Sirlin, C., Vierling, P. & Behr, J.-P. (1994) Gene transfer with a series of lipophilic DNA-binding molecules. *Bioconjugate Chem.* 5, 647–654.

Russ, V., Elfberg, H., Thoma, C., Kloeckner, J., Ogris, M. & Wagner, E. (2008) Novel degradable oligoethylenimine acrylate ester-based pseudodendrimers for in vitro and in vivo gene transfer. *Gene Therapy* 15, 18–29.

Shim, M. S. & Kwon, Y. J. (2009) Acid-responsive linear polyethylenimine for efficient, specific, and biocompatible siRNA delivery. *Bioconjugate Chem.* 20, 488–499.

Sugiyama, M., Matsuura, M., Takeuchi, Y., Kosaka, J., Nango, M. & Oku N. (2004) Possible mechanism of polycation liposome (PCL)-mediated gene transfer. *Biochim. Biophys. Acta* 1660, 24 – 30.

Takahashi, T., Hirose, J., Kojima, C., Harada, A. & Kono, K. (2007) Synthesis of poly(amidoamine) dendron-bearing lipids with poly(ethylene glycol) grafts and their use for stabilization of nonviral gene vectors. *Bioconjugate Chem.* 18, 1163– 1169.

Tanaka, T., Sakata, T., Fujimoto, K. & Ikehara, M. (1987). Synthesis of oligonucleotide with aliphatic amino or phosphate group at the 5' end by the phosphotriester method on a polystyrene support. *Nucleic Acids Res.* 15, 6209 – 6224.

Tarahovsky, Y. S., Koynova, R. & MacDonald, R. C. (2004) DNA release from lipoplexes by anionic lipids: Correlation with lipid mesomorphism, interfacial curvature, and membrane fusion. *Biophys. J.* 87, 1054–1064.

Uragami, M., Miyake, Y. & Regen, S. L. (2000). Influence of headgroup chirality on the mixing behavior of phosphatidylglycerol mimics in fluid bilayers. *Langmuir* 16, 3491 – 3496.

Varga, C. M., Wickham, T. J. & Lauffenburger, D. A. (2000) Receptor-mediated targeting of gene delivery vectors: Insights from molecular mechanisms for improved vehicle design. *Biotechnol. Bioeng.* 70, 593–605.

Wan, L., Manickam, D. S., Oupicky, D. & Mao, G. (2008) DNA release dynamics from reducible polyplexes by atomic force microscopy. *Langmuir* 24, 12474–12482.

Wrobel, I. & Collins, D. (1995) Fusion of cationic liposomes with mammalian cells occurs after endocytosis. *Biochim. Biophys. Acta* 1235, 296 – 304.

Wu, J., Lizarzaburu, M. E., Kurth, M. J., Liu, L., Wege, H., Zern, M. A. & Nantz, M. H. (2001) Cationic lipid polymerization as a novel approach for constructing new DNA delivery agents. *Bioconjugate Chem. 12*, 251–257.

Xu, Y. & Szoka, Jr. F. C. (1996) Mechanism of DNA release from cationic liposome/DNA complexes used in cell transfection. *Biochemistry* 35, 5616–5623.

Yamazaki, Y., Nango, M., Matsuura, M., Hasegawa, Y., Hasegawa, M. & Oku, N. (2000) Polyamine liposomes, a novel nonviral gene transfer system, constructed drom cetylated polyethyleninine. *Gene Therapy*, 7, 1148 – 1155.

Yoshikawa, Y., Emi, N., Kanbe, T., Yoshikawa, K. & Saito, H. (1996). Folding and aggregation of DNA chains induced by complexation with lipospermine: formation of a nucleosome-like structure and network assembly. *FEBS Lett.* 396, 71– 76.

Zanta, M. A., Beleguise-Valladier, P. & Behr, J.-P. (1999) Gene delivery: A single nuclear localization signal peptide is sufficient to carry DNA to the cell nucleus. *Proc. Natl. Acad. Sci. U.S.A.* 96, 91–96.

12 Chapter

FUNCTIONAL LIPOSOMES FOR GENE THERAPY: TECHNOLOGICAL ASPECTS OF SCALABLE PROCESSES

1. INTRODUCTION: AN OVERVIEW

Efficient transfection is essential for the success of gene delivery systems in in vivo or in vitro applications. Cationic liposomes continue to be a promising alternative for nonviral DNA carriers, primarily due to their ability to protect DNA from interstitial fluids and their ease of interaction with cells (Gregoriadis, 1993; Lasic, 1997). Nevertheless, in order to be effective in the immunological response, cationic liposomes must be functional and able to reach their specific target. Stability, reduced toxicity, specific targeting to the nucleus, and efficiency in delivering genes to cells are essential prerequisites for prophylactic and/or therapeutic performance. To satisfy these criteria, it is necessary to regulate critical physicochemical parameters in liposomes, including the zeta potential, size, polydispersity, concentration of the cationic lipid, DNA loading (as indicated by the R+/- molar charge ratio), and lipid functionality.

Cationic lipids have been employed in numerous laboratory experiments to investigate DNA vaccines. In 2000, Perrie and colleagues conducted a traditional investigation into the efficacy of cationic liposomes as DNA carriers, as well as the functionality and composition of lipids (Perrie & Gregoriadis, 2000; Perrie et al., 2001). The plasmid pRc/CMV HBS, which encodes the S (small) region of the hepatitis B surface antigen, was encapsulated in dehydrated-hydrated liposomes. These liposomes were composed of egg phosphatidylcholine (EPC, a bilayer-forming lipid), 1,2-dioleoyl-3-trimethylammonium-propane (DOTAP, a cationic lipid), and 1,2-dioleoyl-sn-glycero-3-phosphoethanolamine (DOPE, a helper lipid) in a 50:25:25 percent molar ratio. The authors demonstrated that the DNA vaccine is safeguarded from incubation with sodium dodecyl sulfate (SDS) by the encapsulation process, which is a result of DNA incorporation within the liposome lamella (Perrie & Gregoriadis, 2000). Additionally, cationic liposomes elicited a superior immunological response in comparison to naked DNA.

Various authors have reported that the size of nanoparticles is a critical parameter for the efficacy of transfection (Ma et al., 2007; Ogris et al., 1998; Rejman et al., 2004; Ross & Hui, 1999; Wiewrodt et al., 2002). This is in reference to polydispersity and size. In fact, the majority of the variability in transfection procedures is a result of non-viral gene delivery systems with high polydispersity index values. The particle size distribution's breadth is correlated with the polydispersity index. High polydispersity suggests that the optimal size range for cell transfection is not met by an increased number of population fractions (Hsieh et al., 2009).

This necessitates a technological process capable of efficiently constructing functional liposomes in conjunction with a rational design. In order to ensure the reproducibility of the formulation when additional liposomes are required for preclinical and clinical assays prior to industrial production, the technological processes must be scalable. In addition, a more precise control of the quality of the final product and a rational optimization of the process are facilitated by a deep comprehension of the phenomena involved and the correlation among operational variables and the physico-chemical parameters. Consequently, the desirable biological performance necessitates a strong connection between the process and the product.

The biological performance of cationic liposomes in gene therapy and vaccination is the subject of valuable results in the literature. Nevertheless, to the best of our knowledge, no study has yet been conducted that specifically examines the approach that connects the design, product, and production process. We achieved promising immunological responses and a prophylactic effect against tuberculosis (TB) by utilizing cationic dehydrated-hydrated liposomes (DRV) that were complexed to DNA containing the hsp65 gene on the external surface (Rosada et al., 2008). M. leprae is the source of Hsp65, a heat shock protein. For the prevention and treatment of tuberculosis, the biological effects of the hsp65 gene have been extensively investigated (de Paula et al., 2007; Silva et al., 2005). Formulations that are derived from the hsp65 gene have been employed in various immunization strategies and to assess protection against M. tuberculosis challenge (Souza et al., 2008).

We created a new and non-toxic formulation of cationic liposomes in which the hsp65 DNA vaccine was entrapped or complexed on the external surface of the cationic DRV. The formulation was employed to immunize rodents through intramuscular or intranasal routes. In the mouse model of tuberculosis, a single intranasal dose of the formulation elicited humoral and cellular immune responses that were as potent as those induced by four intramuscular doses of bare DNA. Additionally, the formulation permitted a 16-fold decrease in the quantity of DNA administered. Additionally, this formulation was shown to be biocompatible, stable, secure, and cost-effective to produce. We believe that this approach can be implemented in human vaccination against tuberculosis (TB) as either a single dose or in prime-boost protocols, which would have a significant effect on the management of this neglected disease (Rosada et al., 2008).

By imparting a net positive surface charge, DOTAP induces electrostatic binding to the cell membrane. The exchange of lipids with the endosomal membrane and the delivery of DNA into the cytoplasm are facilitated by the fusogenic properties of phosphatidylethanolamine (Xu and Szoka, 1996). Lastly, the presence of EPC enhances the stability of the liposomes, diminishes the cytotoxicity of the cationic lipids, and restricts the diameter of liposome construction, all of which affect macrophage phagocytosis. In comparison to encapsulated DNA, the complexation of DNA on the surface of liposomes resulted in increased cell accessibility (non-electrostatically bound DNA) and regulated antibody production in relation to DOTAP/DOPE lipoplexes for the same lipid composition (de la Torre et al., 2009). The liposomes were characterized based on their size, polydispersity, and zeta potential and were prepared using Bangham's method (Bangham et al., 1965). The external surface of liposomes was also examined to determine the availability and deposition of DNA (Rosada et al., 2008).

In order to assess the system's capacity for scale transposition, a scalable process was necessary to replicate the physico-chemical and biological characteristics of the liposomes that had been previously prepared on a laboratory scale for the subsequent pre-clinical and clinical assays.

This chapter discusses the top-down and bottom-up methodologies for the production of liposomes and the complexation of DNA with the objective of developing gene vaccines. Technological processes are categorized into strategies and the potential for scaling up. The properties of the liposomes produced and the physico-chemical aspects of the processes are interconnected, establishing a connection between the product and the process. Furthermore, we present our experience with the scalable production of cationic DNA-EPC/DOPE/DOTAP liposomes that were previously devised. This was achieved through a top-down process that involved the use of ultra-turrax and microchannel microfluidizer devices to homogenize and reduce the size of the liposomes.

2. LIPOSOMES: STRUCTURE, FORMATION, AND CHARACTERIZATION

In general, amphiphile molecules, when dispersed in a liquid medium, undergo internal self- organization, generating colloidal aggregates. Liposomes are formed through the aggregation of phospholipid molecules in an aqueous medium. Initially, a plain bilayer structure is formed when the relative volumes between the non-polar and polar parts of the phospholipid molecules are favorable (packing factor is close to 1) for vesiculation in closed structures. The timing of spontaneous self-aggregation is long, and in technological processes this time is reduced by promoting vesiculation via the addition of energy to the system and remotion of the organic solvent or detergent in which the lipids were initially dispersed (Gregoriadis, 1990; Lasic, 1993). Primary aggregation generates unilamellar liposomes, which undergo secondary aggregation, forming multilamellar liposomes (Figure 1).

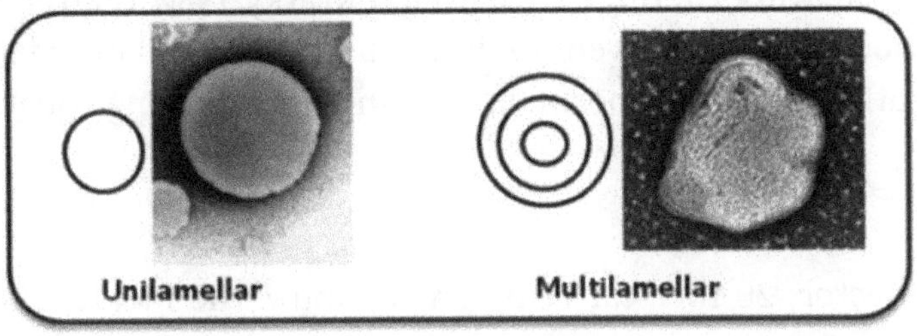

Fig. 1. Unilamellar and multilamellar liposomes.

The main physico-chemical characterization of cationic liposomes for gene vaccines includes average hydrodynamic diameter and size distribution, zeta potential, and morphology. These characterizations have been well described in the modern literature, including techniques, equipment, and the software used for analyzing data.

Regarding the average hydrodynamic diameter and size distribution, photon correlation spectroscopy (PCS) and dynamic laser light scattering using a Ne-He laser are generally used for measurements at various incidence angles. Particle diameter is calculated from the translational diffusion coefficient using the Stokes-Einstein equation:

$$d(H) = (kT)/(3\pi\eta D) \quad (1)$$

Where d(H) is the hydrodynamic diameter, D is the translational diffusion coefficient, k is Boltzmann's constant, T is the absolute temperature, and η is the viscosity. The mean diameter and size distribution are estimated by an adequate algorithm analysis. The results of the population distributions are expressed as the intensity of scattered light and automatically converted to number-weighted mean diameter and size distribution by adequate software. For more accurate size characterization, intensity and number-weighted mean diameters are considered. Although the whole range of diameters is shown in the intensity-weighted distribution, the proportionality to the sixth power of particle diameter underestimates small particles, which are only very weakly weighted (Egelhaaf et al., 1996). The corresponding number-weighted distribution converted using the Mie theory is in equivalent proportion to the first power of the diameter and determines the actual number of particles yielding the observed intensity in each size class (Hulst, 1969).

Several structural parameters involved in the process of liposome preparation can be studied and controlled using X-ray techniques. The simple piling of several polar lipid bilayers produces an X-ray diffraction pattern from which one can determine the periodicity, the type of lattice, and an estimate of water molecules in the interface. The colloidal size of self aggregates, such as the dispersion of liposomes, can be characterized by small angle X-ray scattering techniques directly in the liquid buffer media, complementing some indirect microscopy methods (Kratky 1988). Structural phase transition, swelling, crystallization, permeation, nucleation, and other properties will be reflected in a change in the scattering pattern. More than a few dimensions in polar lipid/water systems, such as, lattice periodicity and arrangement, aggregate size, and defects, are on the order of magnitude of the X-ray wavelength, a few Angstroms. The bigger the particle (up to a few hundred Angstroms), the smaller the scattering angle. The terms small angle X-ray scattering (SAXS) and wide angle X-ray scattering (WAXS) are applied. X-ray diffraction is a particular case of scattering for ordered systems.

2.1 Small angle X-ray scattering (SAXS)

1Here, we present an outline of the main features of the general X-ray techniques applied specifically to the case of liposome characterization. For better comprehension of the theory of X-ray scattering, the reader may like to consult the works of Glatter (Glatter & Kratky 1982, Glatter 1991) and Kratky (Kratky & Laggner 2001). For the notation, as a convention in the literature, and in order to avoid dependence on wavelength values, the scattering angle is normalized using the following equation:

$$q = |\mathbf{q}| = \frac{4\pi}{\lambda}\sin\theta \tag{2}$$

Where q is the wave vector, 2θ the scattering angle, and λ the incident wavelength. By using the wave vector instead of angles, one can directly compare experiments performed with different X-ray wavelengths. SAXS is the elastic scattering that occurs when X-rays strike a sample of given material. The electrons of each atom in the sample will re-emit the same energy isotropically. The total scattering amplitude, or amplitude "form factor" F(q), is a sum of all scattered waves in all directions as defined in Equation 3,

$$F(\bar{q}) = \int \rho(\bar{r})e^{-i\bar{q}\cdot\bar{r}}\,d\bar{r} \tag{3}$$

Where $\rho(\bar{r})e^{-i\bar{q}\cdot\bar{r}}$ is the amplitude of the wave scattered by an atom located at position r. The result is a maximum intensity in the direction of the incident beam decreasing smoothly as a function of the scattering angle. The intensity of the scattering will depend directly on the electronic density $\rho(\bar{r})$ of the sample, i.e., the number of electrons per unit of volume. More specifically, for our case, the higher the difference between the electronic densities of the scattering centers of the sample (e.g., liposomes) and the media (e.g., buffer solution), the higher the total scattering intensity. Lipid/water systems generally contain light atoms (low atomic number) and, consequently, very low electronic density contrast. A way to increase the scattering intensity is to use high flux X-ray sources, such as synchrotrons. The SAXS experiment will give information on the particle/system structure. Usually, the main goal of the SAXS analysis is the determination of the electron density $\rho(\bar{r})$. This is the so called inverse scattering problem since the information is given in reciprocal space. There are several procedures that can be applied to determine $\rho(\bar{r})$. One possible approach is the calculation of a theoretical intensity obtained for an assumed model object and the comparison to the experimental data. The more information one has about the sample, the more consistent the model will be and, consequently, the simpler the SAXS analysis. Nevertheless, some parameters can be determined directly from the scattering curve, for instance, the radius of gyration of a particle, Rg, which is the root mean square of the distances of all electrons from the center of gravity of the particle. The Rg is determined using the Guinier method, which approximates the scattering function to a simple distribution in the limit of very small angles (q~0.01Å-1). A graph plotting ln I(q) versus q^2 in the limit for very low angles might be characterized by a straight line with an angular coefficient proportional to Rg^2.

Considering that the particles are spheres with radius r, it yields:

$$R_g{}^2 = \frac{3}{5} r^2 \qquad (4)$$

The two requirements for using Guinier approximation are: the sample should be a monodisperse colloidal system, which implies no interaction between the scattering centers, and the experimental data must have a minimum q such that d<π/q, where d is the maximum particle dimension that can be studied (Glatter 1982). For systems of large unilamellar vesicles in which the dimensions of the particles are not inside the limit of SAXS experiments, other methods, such as visible light scattering, can be used (Glatter 2002). In some cases, determining the cross section distance would be interesting, such as in large unilamellar vesicle systems, where the thickness of the lamellae is inside the limits of the Guinier approximation. In this case, the graph of $I(q)q^2$ versus q, called the "thickness Guinier plot", is used to determine the radius of gyration of the thickness RT com $RT = T/\sqrt{12}$, were T is the bilayer thickness). Another approach is to obtain the electron density across the thickness of the vesicle, ρt(x). The latter approach can be applied both for unilamellar of multilamellar vesicles as will be discussed next.

1.2 X ray diffraction (XRD)

Diffraction can be viewed as a particular case of scattering: Due to the periodicity, d, of the system, the sum or integration of all scattered intensities turns into an interference pattern. In analogy, the function S(q) which is the result of the integration, is called the "structure factor". The scattering curve, or diffraction pattern, presents peaks for certain scattering angles, θ, which can satisfy Bragg's condition (Azaroff 1968):

$$n\lambda = 2d \sin\theta \qquad (5)$$

Where λ is the wavelength of the incident X-ray beam and n is the order of diffraction. We can observe a diffraction pattern, for example, in multilamellar lipid vesicles, which have a lipid bilayer periodicity on the order of 50 Å. The determination of the period of the bilayer is useful for characterizing the structural phase of the system and its transitions. If several orders of diffraction are observed, the structure factor function can be reconstructed and the electronic density profile of the lipid bilayer determined (Pachence & Blasie 1991). The incorporation of compounds inside the lipid bilayer can alter the density profile and be controlled (Sato et al. 2009). Also the matrix of acyl chains is a periodic arrangement with distances between chains on the order of 5 Å. The determination of this parameter is helpful for controlling the stability of the liposome. Any change in these distances will be followed by a dislocation of the diffraction peak.

2.3 Simultaneous determination of form factor and structure factor

Some systems present both form factor and structure factor. The scattering intensity from such a system can be written by the following expression:

$$I(q) = c\langle P(q)S(q)\rangle \qquad (6)$$

Where c is related to the concentration of particles in the system, P(q)=|F(q)|2 is the intensity form factor of the particles and carries information about particle shape and contrast and S(q) is the system structure factor, which carries information about possible interparticle interactions or arrangements. The brackets "⟨ ⟩" indicate that in the general case these two contributions have to be averaged together in the calculation. For highly anisotropic systems, like vesicles for example, where the transversal dimensions are much larger than the perpendicular dimension, the form factor can be considered as only corresponding to the perpendicular direction (1D function) and equation 6 can be rewritten as,

$$I(q) = \frac{c}{q^2}\langle P(q)S(q)\rangle \tag{7}$$

One usual approach is to use the decoupling approximation and treat the form factor and structure factor separately. For a lamellar system, the structure factor is well described by the MCT theory (Zhang et al. 1994)

$$S(q) = 1 + 2\sum_{n=1}^{N-1}\left(1 - \frac{n}{N}\right)\cos(nqd)\ e^{-(d/2\pi)^2 q^2 \eta\gamma}(\pi q)^{-(d/2\pi)^2 q^2 \eta} \tag{8}$$

The average number of coherent scattering bilayers in the stack is denoted as N, γ is Eulers' constant and d the separation between layers. The Caillé parameter η involves both the bending modulus K of lipid bilayers and the bulk modulus B for compression (Caille 1972, Zhang et al. 1994).

$$\eta = \frac{\pi k T}{2d^2\sqrt{BK}} \tag{9}$$

The description of the form factor can be done in several levels of detail. One of the simplest approximations is to consider a two step model given by (Nallet et al. 1993),

$$P(q) = \frac{4}{q^2}\left\{\Delta\rho_H \sin\left[q(\delta_H + \delta_T)\right] - \Delta\rho_H \sin(q\delta_T) + \Delta\rho_T \sin(q\delta_T)\right\}^2 \tag{10}$$

Where ΔρH and ΔρT are respectively the head group and tail group electron density contrasts and δH and δT are the sizes of the head group and tail group. Following the step model strategy, Glatter and co-workers developed the deconvolution square root method (Glatter & Kratky 1982, Fritz & Glatter 2006) where the electron density is described by several step functions and it is applied a constrained least squares fitting routine to obtain the step heights. Another strategy was proposed by Pabst et.al (Pabst et al. 2000), which models the bilayer using a two Gaussian system: a central Gaussian placed at the origin, which can model the central part of the tail groups, and is known to have negative contrast with respect to a water buffer (when using X-Rays), and a second Gaussian placed at a certain distance Z in such a way that it can model the position of the head group region. This approach and the two step model cannot describe high quality data, principally for high q values (Oliveira et al. Unpublished work). In a recent development, Oliveira and co-workers combined the advantages of the Gaussian description with the stability introduced by the Glatter method, enabling a more flexible and stable model procedure.

In this approach the profile is described by a symmetric sum of several equally spaced Gaussian functions. The amplitude of each Gaussian is smoothened by extra constraints. The constraints are used to ensure the numerical stability of the nonlinear least-square fit. By proper choice of the amplitude of each Gaussian it is possible to build a smooth profile that can describe more accurately the electron density of the bilayer. Usually, 4 Gaussians function are sufficient to describe a bilayer profile satisfactorily. The electron density is defined as

$$\rho(z) = \sum_{n=1}^{4} a_n\left[G_s(z, z_n, \sigma_n) + G_s(z, -z_n, \sigma_n)\right] / (1 + \delta_{i1}) \qquad \begin{matrix}\delta_{i1} = 1, i = 1 \\ \delta_{i1} = 0, i = 2,3,4\end{matrix} \tag{11}$$

$$G_s(z, z_n, \sigma) = \frac{1}{\sigma\sqrt{2\pi}}\exp\left[-(z - z_n)^2 / 2\sigma^2\right] \tag{12}$$

and the Kroninger delta function δi1 is used to avoid double counting for the central Gaussian. n is the order of Gaussian used. The profile is defined by the values of the amplitude factors an. Given a half bilayer thickness Z the centers of the Gaussians and the standard deviations (σ values) are defined as follows:

$$z_n = (n-1)2\sigma, \qquad \sigma = \frac{Z}{2n_{\max}\sqrt{2\ln 2}} \tag{13}$$

This choice of the centers and standard deviations gives a reasonable overlap between the Gaussian functions enabling the construction of smooth profiles. A typical profile built using equation 11 is shown in the figure 2 (left). One of the advantages of using a Gaussian set of functions for the representation of the profile is that, for the one-dimensional case of centro-symmetric bilayers, the integral in equation 3 has one analytical solution, given by:

$$F(q,n) = \sqrt{2\pi}\sigma \exp\left(-\frac{\sigma^2 q^2}{2}\right)\cos(q\, z_n) \tag{14}$$

The final scattering amplitude is just the addition of all the F(q,n) terms:

$$F(q) = \sum_{n=1}^{4} a_n F(q,n) \tag{15}$$

Regardless the methodology to calculate the form factor, this procedure gives the scattering contribution from the built shape of the electron density profile. For the multilamelar vesicles this profile will be a repeating unit for each vesicle layer. By using the structure factor given from the modified Caillé theory (equation 8), the final scattering intensity is then given by:

$$I_{MLV}(q) = \left(\left|F(q)\right|^2 S_{MCT}(q) + N_{diff}\left|F(q)\right|^2\right) \tag{16}$$

Where the second term gives rise to a diffusive scattering which might appear from the presence of single bilayers in the sample. Smearing effects, which are a consequence of the collimation of the camera pinholes, wavelength bandwidth, detector resolution, etc, can be taken into account by the use of the resolution function R(⟨q⟩,q), as described in the work of Pedersen (Pedersen et al. 1990). The final expression used to describe the data from multilamellar vesicles is given by:

$$I(q)_{FIT} = Sc_1 \frac{I_{MLV}(\langle q\rangle)}{q^2} + BG \tag{17}$$

Sc1 is an overall scale factor and BG is a constant background. Both are optimized during data fitting. Using the above mentioned procedure it is possible to perform a full curve fitting, retrieving simultaneously information about the form factor and structure factor. A simulated example and an application to real experimental data are shown in Figure 2(right).

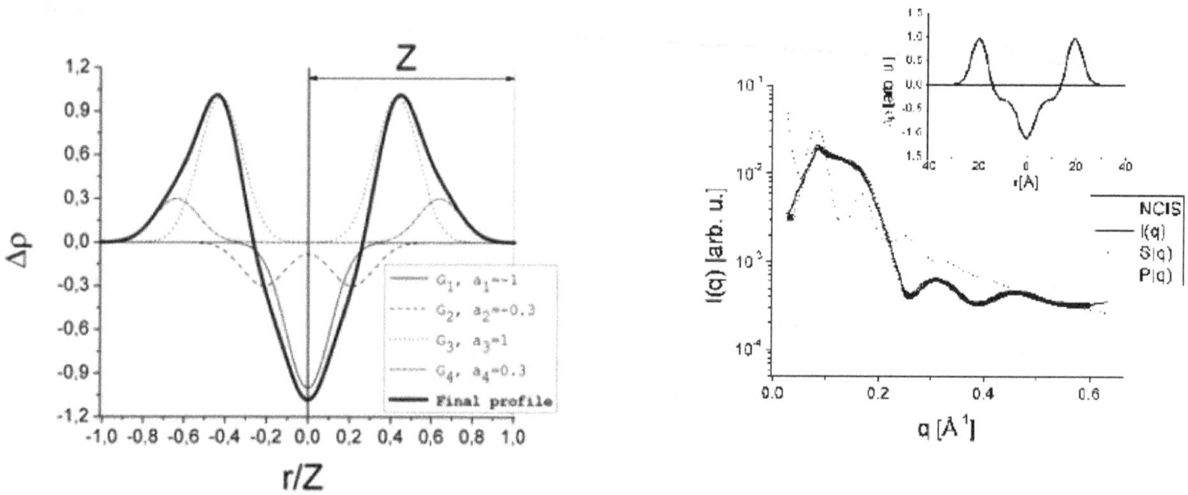

Fig. 2. Left: Construction of the electron density profile using four Gaussians (G1, G2, G3 and G4) with four different amplitudes (-1, -0.3, 1, 0.3). They are shown in four different lines. The final resultant electron density profile from the Gaussian model is shown in solid thick line. Right: Fitting of experimental data using the 4-Gaussians approach. Open circles: experimental data for Egg phosphatidylcholine (EPC) liposome system. Solid curve: Theoretical intensity I(q). Dotted line: structure factors S(q). Dashed line: intensity form factor P(q). The obtained number of layers was N=50, Interplanar distance d=73.5±0.2Å, Caillé parameter of η=0.13±0.01 and bilayer half size of Z=30.00±0.06Å. The obtained electron density profile is shown in the inset.

3. TOP-DOWN AND BOTTOM-UP APPROACHES IN PROCESSES

In general, the processes for producing nanomaterials can be characterized as two main approaches, top-down or bottom-up. The top-down approach starts with large particles that are comminuted to a nanometric size through the application of high-energy forces. This is the classical approach for the majority of nanoparticle production processes. Top-down approaches require highly precise control of the variables of the process in order to obtain the narrow particle size (Mijatovic et al., 2005). Lithography is the classical example. Other high-energy processes include grinding, high impact homogenization, ultrasound waves, and extrusion through nanoporous membranes (Sanguansri & Augustin, 2006).

Bottom-up approaches are based on the self-organization of molecules under thermodynamic control, generating nanostructures from atoms and molecules as a result of the effects of the chemical, physical, and process interventions on the balance of the intermolecular and intramolecular forces of the system components (Sanguansri & Augustin, 2006). Bottom-up approaches focus on the construction of functional materials, mimetizing the hierarchic organization of the molecules in live organisms, though the science still does not dominate the complex auto-aggregate structures in nature.

Regarding liposome production, top-down approaches comprise the high-energy comminution of a polydispersed population of multilamellar liposomes formed under non- controlled aggregation. The bottom-up approach manipulates the phospholipid molecules in controlled local aggregation in space and time, generating a monodispersed population of nano-sized unilamellar liposomes. Figure 3 illustrates top-down and bottom-up approaches for liposome production.

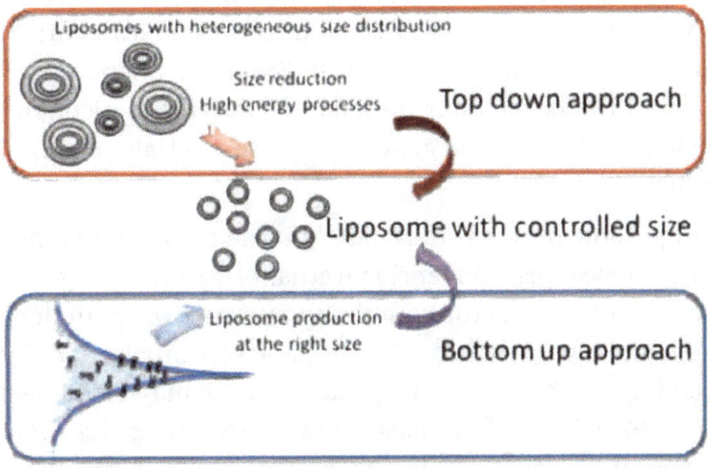

Fig. 3. Scheme of top-down and bottom-up approaches for liposome production.

4. TECHNOLOGICAL PROCESSES

Most of the conventional methods for liposome production require an additional unit operation for size reduction and polydispersity, as they are top-down approaches. In this approach, liposomes are produced from the hydration of a thin film of lipids using Bangham's method (Bangham et al., 1965), multitubular system (Torre et al., 2007; Tournier et al., 1999), detergent depletion or emulsion methods, ether/ethanol injection, and reverse phase evaporation (Lasic, 1993; New, 1990). All of these processes are discontinuous and only the ether/ethanol injection and multitubular system are scalable. Shearing or impact strategies are the general key for homogenation and reducing liposome size. Mechanical stirring, extrusion through orifices (French press), extrusion through membranes, high-pressure impactor, and microchannel microfluidizer are equipment used in high-energy processes.

Bottom-up processes are low energy processes inspired by biological systems and used for the development of functional nanomaterials, such as supramolecular structures, self- aggregated monolayers, Langmuir-Blodgett films, aggregated peptide nanotubes, and deposited polyelectrolites or proteins in multilayers (Mijatovic et al., 2005; Shimomura & Sawadaishi, 2001).

Microfluidic systems have been the main representatives of bottom-up processing for liposome production in continuous and scaled up processes. Different microfluidic systems can be applied to the production of liposomes and giant liposomes (Ota et al., 2009; Shum et al., 2008; Wagner et al., 2002). DNA complexation on the external surface of liposomes also constitute a promising technology for the production of gene vaccines.

4.1 Microfluidic systems for the production of cationic liposomes

Among different microfluidic geometries, hydrodynamic focusing (HF) is a promising technology and liposomes can be produced in sizes ranging from 50 to 500 nm (Jahn et al., 2004, 2007; Wagner et al., 2002). The HF consists of a device with four-microchannel intersection geometry (Figure 4A). The organic solvent, miscible in water (isopropanol, ethanol), containing dispersed lipids are injected in the middle stream and hydrodynamically compressed by two aqueous (or buffered) streams. The precise upstream flow rates are achieved using syringe pumps that control the position of the focused stream in the downstream channel and, consequently, the process quality parameters (Baldas & Caen, 2010). The use of stereo microscopy is recommended to monitor the process as presented in the schematic representation of the experimental apparatus in Figure 4B. The laminar flow rate allows the formation of a well defined region between the two miscible fluids. The interfacial forces between two miscible solvents (for example, water and isopropanol or water and ethanol) control the convective-diffusive process and liposome self-assembly (Jahn et al., 2010). From the phenomenological point of view, the continuous flow mode allows the continuous diffusion of water and alcohol, reducing lipid solubility, which causes thermodynamic instability and generates liposomes (Jahn et al., 2008). The continuous flow mode also increases productivity.

The HF geometry for liposome production is based on the conventional method of ethanol injection (bulk) adapted for microfluidic systems. Conventional ethanol injection consists of the controlled addition of an ethanol/lipid stream in a tank reactor containing buffered water under controlled agitation. The advantage is the use of ethanol; compared to other organic solvents (e.g., chloroform and methanol), it is less harmful and, depending on the lipid concentration, there may be no need for post-treatment size reduction (Kremer & Esker, 1977). Some of the disadvantages are low lipid concentration in ethanol (higher concentrations require post-treatment for size reduction) and a difficulty achieving reproducibility (Wagner et al., 2002).

The representative parameters for liposome size and polydispersity control in HF are channel geometry (deep and width), volumetric flow buffer/alcohol rate-ratio (FRR), and total volumetric flow rate (buffer+alcohol flow rates). Solvent diffusivity in water is another important parameter (Jahn et al., 2004, 2007, 2010). Basically, the narrower the microchannel width, the smaller the liposome size. For total volumetric flow rate, which promotes flow velocity at the channel, higher FRRs (which corresponds to an increased proportion of buffer to lipid/solvent flow) correspond to smaller liposome diameter (Janh et al., 2010). However, caution is required in this flow rate analysis as the final alcohol content in the liposome colloidal dispersion can change with FRR and as a consequence of particle size. Alcohol can be used to disrupt liposomes and, at low concentrations, liposome size can be increased due to alcohol incorporation into the bilayer. Precise analysis can be performed if the alcohol is removed after liposome processing. By decreasing the total volumetric flow rate, the residence time can be increased and a lower average vesicle diameter and narrower size distribution obtained. This behavior indicates that if increasing the total volumetric flow rate, the microchannel length must be longer to complete the alcohol diffusion, otherwise large particles will be obtained due to bulk mixing downstream of the channel (Janh et al., 2010).

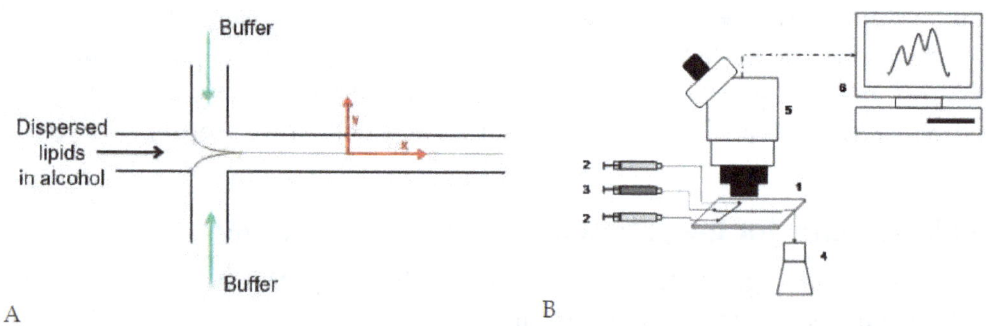

Fig. 4. (A) Schematic representation of hydrodynamic focusing (HF) in a microfluidic device with four-channel intersection geometry. The organic solvent (miscible in water) containing the dispersed lipids is injected in the middle stream and hydrodynamically compressed by two aqueous (or buffered) streams. The flow is in the x direction and alcohol diffusion is in the y direction. (B) Schematic representation of the experimental apparatus for liposome production using HF: (1) microfluidic device; (2) water syringe pumps; (3) syringe pump for lipids/ethanol stream; (4) collector flask; (5) stereo microscopy; (6) computer data acquisition (adapted from Zhang et al., 2008).

The lipid concentration in alcohol is also another important parameter for controlling size. The development of vaccines for *in vivo* applications requires the highest drug-loading capacity. In terms of gene vaccines, the DNA-loading capacity correlates with the cationic lipid content, defined by the molar charge rate (R+/-) between the cationic charges (from the cationic lipid) and negative charges (from the DNA). The R+/- and total lipid content are project parameters for scaling up (or scaling out) processes. As an example, the R+/- of the tuberculosis gene vaccine is 10 and the total lipid concentration 64 mM (Rosada et al., 2008). These parameters require cationic liposome production in a high lipid concentration under unusual microfluidic conditions. Aiming to explore the effect of lipid concentration on microfluidic HF processes, we investigated the influence of high lipid content (100 mM) and lipid composition in ethanol (EPC or EPC/DOTAP/DOPE) as a function of average size. We understand that the studied molar concentration is greater than the lipid solubility in ethanol (approximately 4 mM for EPC) (New, 1990) and, in this case, the lipid dispersion offers an additional barrier to ethanol diffusion into water. Based on this assumption, we simulated the required microchannel length for complete ethanol diffusion from a central stream (after HF) as a function of the lipid concentration in the ethanol stream considering the mass continuity equation (according to Figure 4A), without a chemical reaction (Equation 18). This simulation was performed based on a microfluidic glass device produced by the wet photolithographic process. The microchannels were etched with HF solution; upstream channels measured 140 ± 1 μm in width and downstream channels 200 ± 1 μm after the T junction. The microchannel was 50 ± 2 μm deep and the diffusion length 5 cm.

$$\frac{\partial C_E}{\partial t} + \left(v_x \frac{\partial C_E}{\partial x} + v_y \frac{\partial C_E}{\partial y} + v_z \frac{\partial C_E}{\partial z} \right) = D_{EW} \left(\frac{\partial^2 C_E}{\partial x^2} + \frac{\partial^2 C_E}{\partial y^2} + \frac{\partial^2 C_E}{\partial z^2} \right) \tag{18}$$

Where DEW is the ethanol diffusion coefficient in water and CE is the ethanol concentration. The Cartesian coordinates are used because the microchannel area presents rectangular geometry. Considering laminar flow in x direction and mass transfer along the y direction (Figure 4A), Equation 18 can be adapted for short times according to the Higbie penetration model (Higbie, 1935):

$$N_E = \sqrt{\frac{4.D_{EA}}{\pi.t}}.(C_{E_0} - C_{E\infty}) \tag{19}$$

Where *t* is the quotient between the difference in flow rates (water and ethanol) and distance x. The influence of lipid concentration was considered an additional barrier for ethanol diffusion due to the presence of lipid aggregates dispersed in the ethanol stream. In this case, the ethanol diffusion depends on the concentration gradient through ethanol/water streams, as well as its hydrophobic characteristics, with similar behavior as ethanol diffusion from aqueous solution to a phospholipid bilayer (Galindo-Rodriguez et al., 2004). The hydrophobic characteristics of the lipid can be expressed as the partition coefficient octanol/water (Po/w= 0,478). In this context, DEW was corrected by the partition coefficient according to Equation 20.

$$D_{EW}' = 0,478 * D_{EW} \tag{20}$$

Where DEW′ is the effective diffusion coefficient.

Figure 5 presents the simulation of the distance (x) for total ethanol diffusion as a function of ethanol flow rate (at a fixed FRR of 6.26 and 20). Increasing the lipid concentration in the ethanol stream, the distance x to complete ethanol diffusion will increase.

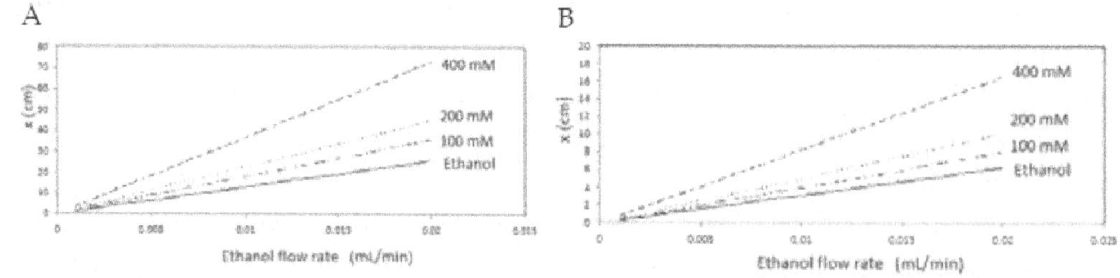

Fig. 5. Simulation of the distance for total ethanol diffusion into the water stream. The lipid (EPC) concentration was considered as an additional barrier to ethanol diffusion. (A) FRR = 6.26. (B) FRR = 20.

According to the simulation, ethanol dispersion requires the lowest length for ethanol diffusion for a lipid concentration of 100 mM (among the simulated concentrations). Experimental evaluations with this lipid concentration were performed at different total flow velocities (total flow rate) and lipid compositions (EPC and EPC/DOTAP/DOPE – cationic liposomes) as presented in Table 1. The average diameter is strongly influenced by the total flow velocity and lipid composition. Decreasing the flow velocity decreases the average size, and this behavior is not dependent on the FRR. This parameter is associated with the residence time inside the microchannel, suggesting that higher residence time is required to control liposome size. Another interesting parameter is the lipid composition. The viscosity of the ethanol dispersion was 1.07 and 1.21 cP for EPC/DOTAP/DOPE and EPC, respectively. Lower viscosity probably offers lower mass transfer resistence, reflecting smaller diameters (and standard deviations) and polydispersity index values. This difference reflected in HF is confirmed by stereomicroscopic observation along the microchannel. A high polydispersity index (Table 1) can also indicate the possibility for further process optimization in terms of flow velocity, FRR, and lipid concentration.

Table 1. Average liposome size (Z-average) obtained at different lipid compositions and flow rates.

Lipid composition	Total flow velocity (cm/min)	FRR	Z-average (nm)	Polidispersity Index
EPC	*5.6*	*6*	*342.8 ± 134.5*	*0.427 ± 0.140*
	4	*7*	*271.6 ± 37.5*	*0.408 ± 0.145*
EPC/DOTAP/DOPE	*8.1*	*10.6*	*162.1 ± 88.13*	*0.606 ± 0.157*
	6.1	*7.7*	*92.92 ± 10.14*	*0.441 ± 0.028*
FRR: volumetric flow buffer/alcohol rate-ratio.				
Total lipid concentration of 100 mM in ethanol stream.				

4.2 Microfluidic systems for electrostatic complexation of DNA

The electrostatic interactions between DNA and cationic liposomes produce particles with different sizes and morphology (Mannisto et al., 2002; Oberle et al., 2000) that depend on R+/-, buffer ionic strength, order of component addition, reaction conditions, and the type of lipids (Mount et al., 2003; Zelphati et al., 1998). In this context, HF can also be used to control the diffusion process for DNA compaction, producing well organized aggregates. The flow velocity is the major parameter controlling the aggregation process (Dootz et al., 2006).

Otten et al. (2005) investigated the HF microfluidic device to produce cationic liposome- DNA complexes. The DNA solution is introduced in the central stream and the cationic liposome stream introduced at a lateral position. The average liposome size was 200 nm (composed of 1:1 DOTAP and DOPC with a lipid concentration of 25 mg.mL^{-1}), and the DNA was calf thymus (5 mg.mL^{-1}). The flow velocity was

100 mm.s-1, varying with vLiposome = 13vDNA and vLipossomas = 130vDNA, where v is the flow velocity. The authors concluded that the complex is formed in two steps. The first step relates to the formation of a multilamellar complex, followed by the second step in which DNA is organized inside the lamellae. The central stream can be focused according the FRR, and reducing the diffusional length allows faster mixing (Knight et al., 1998).

A micromixer (Jellema et al., 2010) and multi-inlet microfluidic HF (MF) system (Koh et al., 2010) were recently investigated as alternatives for producing the liposome-DNA complexes. These studies point out new and promising alternatives aiming to control the aggregation process between DNA and cationic liposomes.

5. PHYSICO-CHEMICAL ASPECTS

The processes that use top-down strategies for liposome production differ according to the remotion of solvent and/or application of shear. The remotion of solvent is carried out by evaporation (Bangham´s method or multitubular system), spray drying or evaporation in reverse phase, sublimation (dehydration–hydration), or solubilization in water by ether/ethanol injection in diluted or concentrated phases. The incomplete remotion of solvent has a direct impact on liposome size.

We have observed that the presence of organic solvent in spray-dried lipid structures generates more amorphous or more crystalline domains due to the packing of lipids in the bilayers (Alves and Santana, 2004). The operational conditions interfere with the drying rate, and mass transfer limitations result in the complete evaporation of ethanol. The higher the evaporation rate, the higher the amount of ethanol remaining inside the particles because a shield of packed lipids close to the surface and more amorphous structures are formed. At a slower rate of evaporation, it is more controlled, generating more crystalline structures. The subsequent hydration of the solid particles with different levels of crystallinity influences the size and polydispersity of the generated liposomes.

Similar mass transfer limitations are present when the organic solvent is removed by diffusion in water. A compromise between the diffusion and vesiculation rates controls particle size. The opposite rates of diffusion and hydration depend on the barrier created by the phospholipids, which is a function of its concentration and depends on the interaction among lipids. In previous studies, we characterized the packing of pseudo-ternary mixed Langmuir monolayers of EPC/DOTAP/DOPE using surface–molecular area curves. The interactions and miscibility behavior were inferred from the curves by calculating the excess free energy of the mixture (ΔGExc). The deviation from ideal showed dependence on the lipid polar head type and monolayer composition (Rigoletto et al., 2011).

The rates of vesiculation and solvent diffusion are comparable in magnitude only at very low lipid concentrations, generating small liposomes in an excess of water. Under other conditions, the barrier created by the extension of the primary aggregation of lipids delays solvent remotion, generating large multilamellar liposomes. Otherwise, the rate of vesiculation is associated with the hydrophobicity of the non-polar groups. Therefore, the vesiculation rate in the primary aggregate controls liposome size. When cationic lipids are used, such as in the DOTAP/DOPE/EPC system, the electrostatic repulsion among the molecules benefits solvent diffusion and hydration, generating smaller liposomes. Because of the softness of the primary aggregate and the bilayer fluidity imparted by the phase transition temperature, liposomes are prone to fusion due to the non-ordered Brownian movement of the colloids in the medium, and multilamellar vesicles are generated in a broad range of sizes. The remaining solvent generates polydispersed liposomes, which are more favorable for fusion. However, charged lipids stabilize liposomes. Therefore, mass transfer is critical in discontinuous processing, controlling liposome size and polydispersity. Despite limitations in mass transfer and interactions, discontinuous processes are still preferable due to the simplicity in carrying out massive liposome production and scaling up processes. In this context, the development of technologies that promote the production of cationic liposomes with controlled size and low polydispersity index, with low energy consumption and the elimination of organic solvents, is still a challenge.

Continuous processes in microchannels with HF reduce the limitations due to mass transfer in the interface of the primary aggregate, and the continuous operation decreases interactions among particles due to Brownian movement. Therefore, particle size is controlled by the relative flow rates between the phases and by microchannel length.

6. A CASE STUDY OF THE PRODUCTION OF CATIONIC LIPOSOMES AND GENE VACCINES IN SCALABLE TOP-DOWN PROCESSING

Although top-down strategies are important and high shear processing has various uses, no systematic studies have been carried out on the effects of liposome comminution on the physico-chemical and surface properties of liposomes. The data in the literature are sparse in regards to the kind of impellers, comminution equipment, or type of lipids used.

Aiming to produce the gene vaccine composed of EPC/DOTAP/DOPE liposomes with DNA complexed on their external surface, we initially studied the significance of the process variables for the properties of liposomes composed of egg lecithin. Mechanical forces were used for homogenation and comminution in Caules type stirrer, Ultra-Turrax, and microchannel microfluidizer equipment. A main variable was selected and its effects on the physico-chemical properties of the liposomes characterized. Finally, a scalable discontinuous process was established and EPC/DOTAP/DOPE liposomes produced and complexed with DNA. The physico-chemical and biological properties of the gene vaccine were compared with our previous gene vaccine prepared using Bangham's method.

1.1 Effects of the process variables on liposome properties

6.1.1 Statistical analysis

We studied the effects of the shear rate using multi-factorial statistical experimental planning (Montgomery 2008) in order to delineate the relative importance and influence of the shear rate and feed flow rate on the mean diameter, polydispersity, zeta potential, and viscosity of egg lecithin liposomes. The liposomes were prepared with high lipid concentration (300 mM), aiming for applications in scaling up processes. Food grade egg lecithin (60% phosphatidylcholine content) from Degussa (GmbH Germany) and ethanol 99- 100% from Labsynth ltda. (São Paulo- Brazil) were used as the lipid and solvent, respectively.

Figure 6 shows the experimental outline for liposome preparation. The 300 mM ethanol- lipid suspension in a beacker (1) was fed at a previously defined flow rate through a peristaltic pump (2) into the bottom of a 150 mL beaker with four fins containing pure Milli Q water (3). Continuous mechanical stirring was provided by a Caules type stirrer or Ultra- Turrax® IKA T25 (Ika Works) (4). The final lipid concentration was 50 mM. After the feeding was complete, stirring was maintained for an additional 15 min. Comminution was also carried out in a microchannel microfluidizer (Microfluidizer® M-110P) with 100 mL of a liposome dispersion pre-processed through an Ultra-Turrax® at 5000 to 21000 s-1 shear rate and 0.09 to 0.96 mL.s-1. The microfluidizer worked in the pressure range of 200 to 1500 bar and for various passages. All liposome dispersions were stored at 8°C for 12 hours before characterization.

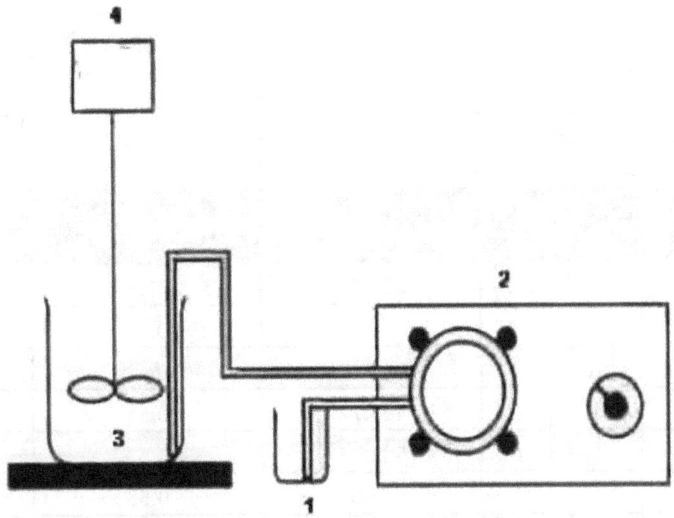

Fig. 6. Scheme of the experimental set-up used for liposome preparation using (A) Ultra- Turrax® or Caules mechanical stirrer. 1) Tank containing the lipid dispersion in ethanol; 2) peristaltic pump; 3) tank with four fins containing (150 mL) water; 4) Ultra-Turrax® or Caules mechanical stirrer. (B) Microchannel microfluidizer system processed the samples from the Ultra-Turrax®.

6.1.2 Significance of shear and feed flow rates

The effects of the shear rate and feed flow rate on the mean diameter, polydispersity, and zeta potential are presented in Table 2 and in terms of the significance of the independent variables and their interaction in the Pareto graphics in Figure 8. For the Caules type stirrer, the effects of shear rate and feed flow rate on the mean diameter of liposomes were not significant (Figure 7A1). The shear rate was significant for polydispersity (Figure 7A2), and both variables were significant for zeta potential (Figure 7A3). The interactions between the variables were not significant for the three response variables.

Table 2. (A,B) Physico-chemical properties of liposomes as a function of the operational variables.

Table-A

Factors			–	0	+
1: Shear rate (s^{-1})			2, 860	12,140	21,430
2: Inlet lipid solution flow rate (mL/s)			0.09	0.54	0.96
Experiment	**1**	**2**	**Diameter (nm)**	**Polydispersity**	**Zeta Potential (mV)**
1	-1	-1	386.10	0.37	-53.73
2	1	-1	254.73	0.31	-56.33
3	-1	1	550.20	0.58	-53.73
4	1	-1	246.50	0.35	-48.57
5	0	0	374.83	0.47	-55.90
6	0	0	359.00	0.36	-54.60
7	0	0	358.90	0.45	-54.90

Table-B

Factors			–	0	+
1 Shear rate (s^{-1})			1,000	1,750	2,530
2 Inlet lipid solution flow rate (mL/s)			0.09	0.54	0.96
Experiment	**1**	**2**	**Diameter (nm)**	**Polydispersity**	**Zeta Potential (mV)**
1	-1	-1	626.17	0.65	-55.00
2	1	-1	414.87	0.51	-71.27
3	-1	1	616.47	0.68	-57.50
4	1	-1	446.77	0.51	-52.60
5	0	0	582.77	0.65	-57.27
6	0	0	642.67	0.58	-55.80
7	0	0	567.40	0.62	-56.00

Data obtained using ultra-turrax showed a significant influence of the shear and feed flow rates, as well as their interactions, on the mean diameter and polydispersity (Figure 7B1,B2). However, these variables had no significant effect on the zeta potential (Figure 7B3). The mean diameter was influenced more by shear rate, whereas both shear rate and feed flow rate were significant for polydispersity.

The results show that the shear rate range of the Caules stirrer did not comminute the aggregates less than 600 nm in size, but it destroyed the progressive aggregations, influencing the polydispersity and zeta potential.

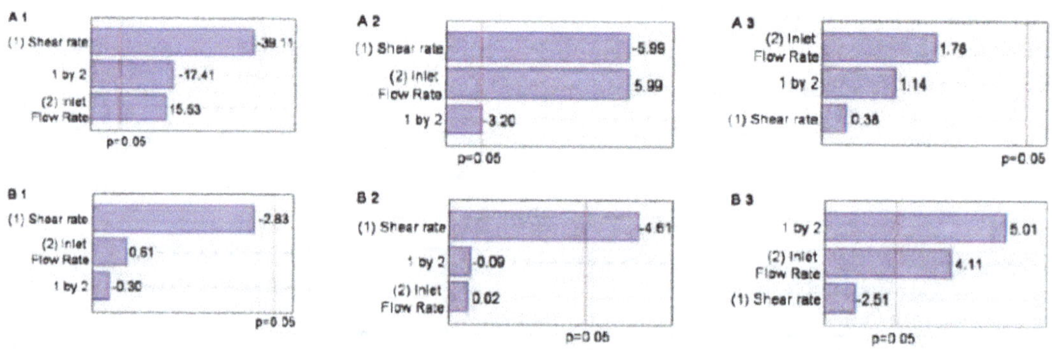

Fig. 7. Pareto´s graphic for liposome production using (A) Caules and (B) Ultra-Turrax® stirrers. (A1,B1) mean diameter hidrodinâmico médio cumulativo, (A2,B2) polydispersity, (A3,B3) zeta potential.

However, the higher shear rates provided by the ultra-turrax better control phospholipid aggregation, generating smaller liposomes. Control is strongly dependent on the intensity of shear, and sizes in the range of 200 nm were obtained at the superior shear rate limit (21530 s-1) provided by the equipment. Feed flow rate influence mainly occurs at the inferior shear rate limit (2860 s-1). Though the shear rate range for ultra-turrax interferes with primary liposome aggregation, the zeta potential is not influenced. Therefore, for egg lecithin (300 mM concentration) liposomes, the shear rate range between 1000 and 21430 s-1 delimits the liposome size from approximately 600 to 200 nm.

1.13 Effects of the shear rate on the liposome properties

Mean diameter - Additional data allowed the construction of the curve presented in Figure 8A. The curve shows a clear relationship between liposome comminution and shear rate, with the mean diameter exponentially decaying with applied shear rate. The error bars are higher at lower shear rates due to the poor homogenization of liposomes provided by the lower pumping capabilities of the mechanical systems.

The microchannel microfluidizer, working in the pressure range of 200 to 1500 bars, provided shear rates in the range of 2×105 to 6×105 s-1 .The data from the microchannel microfluidizer were obtained for pre-treated liposomes using Ultra-Turrax at shear and feed flow rates of 5600 to 24000 s-1 and 0.09 to 0.96 mL.s-1, respectively. The results show that the pre-formed liposomes reached the nanometric range (100 nm) in only one passage using Ultra-Turrax and the high shear rate range of the microchannel microfluidizer.

This comminution behavior agrees with the results reported by Diat et al. (1993a). Through a balance between elastic and viscous forces in the liposomes, the mean diameter is reduced according to the square root of the applied shear rate (Equation 21).

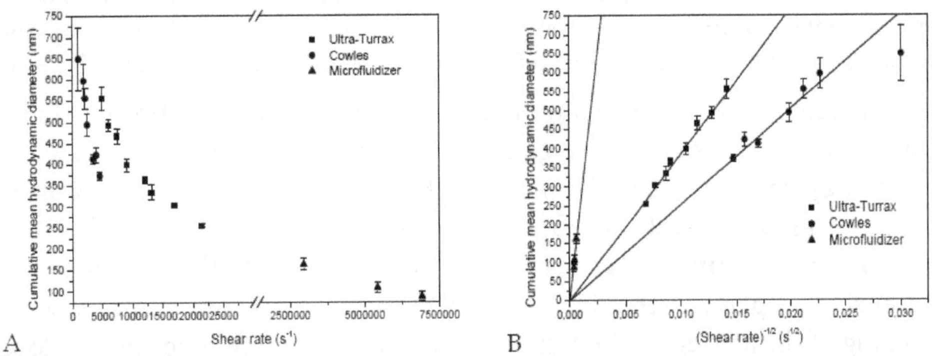

Fig. 8. Cumulative hydrodynamic mean diameter as a function of the shear rate provided by Caules stirrer, Ultra-Turrax®, or microchannel microfluidizer. (A) Z-average values obtained from light scattering measurements. (B) Linear relationship from Equation 21 proposed by Diat et al. (1993a).

$$R = \sqrt{\frac{4.\pi.\left(2.k + \bar{k}\right)}{\mu.d.\dot{\gamma}}} \qquad (21)$$

Where R is the liposome radius at equilibrium, k and k are the average and Gaussian elastic constant of the membrane, respectively, μ is the viscosity of the liposome dispersion, y' is the shear rate, and d is the interlamellar distance. The different slopes for the straight lines in Figure 9B were obtained for the shear rate ranges of the devices used, agreeing with the mass balance between elastic and viscous forces in the liposomes described by Diat et al. (1993b). Higher shear rates produce higher slopes, indicating the presence of liposomes with higher elastic constant, lower viscosity, and shorter interlamellar distance.

6.1.3.1 Viscosity, surface tension, and zeta potential

The shear rate also reduced the viscosity of the liposome dispersion from 5 to 2 mPa, but no significant changes were observed in the surface tension as a consequence of the reduction in size. However, the reduction in size also resulted in rearrangement of the lipids in the external layer, changing the zeta potential from -50 to -40 mV.

6.1.3.2 Lamellar packing

Small angle X-ray scattering (SAXS) characterization showed changes in the interlamellar distances for the applied shear rate ranges (Figure 9). The distance decreased with higher shear rates. The decreasing interlamellar distance was a consequence of the loss of interlamellar water, calculated to be up to 12% for the highest level of shear rate, as well as of lipid packing in the liposome, found to be less than 2%. The water layer is calculated as the difference between the full period and the bilayer thickness (see Table 4).

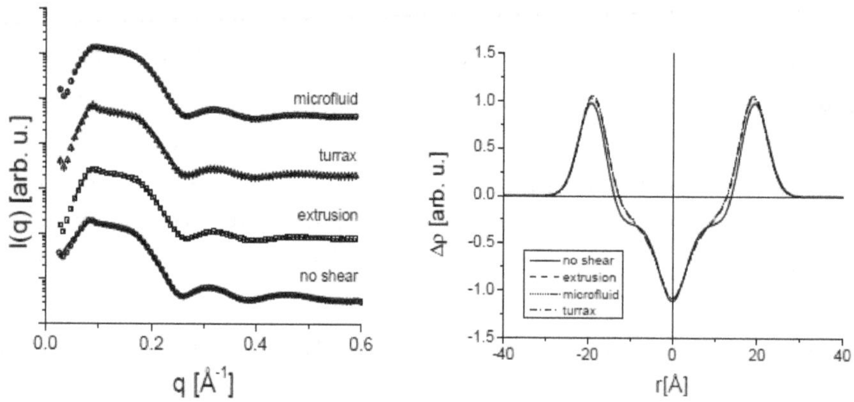

Fig. 9. Left: Small angle X-ray scattering (SAXS) profiles determined for the liposomes treated under moderate shear (extruded through polycarbonate membranes) and high shear produced by Ultra-Turrax® (21430 s^{-1}) and microchannel microfluidizer devices (6 961 000 s^{-1}). Right: Electron density profile across the bilayer obtained from the fitting (see Table 3).

These factors may explain the observed changes in zeta potential and the different slopes of the straight lines obtained for the studied shear rate ranges. In addition to the interlamellar distance and packing, the elasticity of the bilayer may also be influenced by the shear rate and changes in the elastic constants determined. In general, elastic modulus is not the same as stiffness. Elastic modulus is a property of the constituent material, whereas stiffness is a property of the structure. In the case of liposomes, the elastic modulus is an intensive property of the lipids, and stiffness is an extensive property dependent on the structure and packing of the aggregate. Therefore, we associated the changes in the slope of the straight lines in Figure 8B with changes in the elastic constants as a consequence of the changes in packing in the liposomes.

Table 3. Structural parameters obtained from full curve modeling of SAXS data.

Sample	Period[Å]	Bilayer thickness[Å]	N	η
no shear	76.7±0.2	59,9±0.1	3.2±0.1	0.18±0.02
Extrusion	73.4±0.1	59,3±0.1	3.4±0.1	0.26±0.01
Turrax	72.5±0.2	59,1±0.1	4.1±0.1	0.26±0.01
Microfluid	72.3±0.1	59,2±0.2	2.3±0.1	0.09±0.01

N is the average number of bilayers and η is the Caillé parameter.

6.1.3.3Morphlogy

Figure 10 presents transmission electronic microscope images of the liposomes under various shear treatments compared to control (Bangham's method).

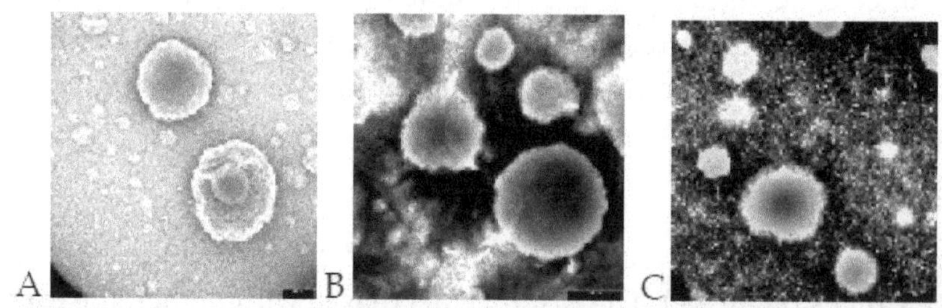

Fig. 10. Transmission electronic microscopy of liposomes obtained using (A) Bangham's method followed by extrusion in membranes, (B) ultra-turrax under a shear rate of 21430 s^{-1}, (C) microchannel microfluidizer under a shear rate 6 961 000 s^{-1}. Bars :200nm

Under low shear, the liposome surface looks thicker (Figure 10A), whereas it looks more packed with ultra-turrax treatment (Figure 10B). However, in both cases the liposome morphology is spherical. Under the highest shear, the morphology of the liposomes was not spherical, and irregularities in the surface reflect stretching of the aggregated structure (Figure 8C). These results reflect the observed change in packing and zeta potential as a consequence of the shear rate.

6.1.4 Scalable top-down process for liposome production

The scalable process described in Figure 11 was established based on these results and validated for the production of EPC/DOPE/DOTAP liposomes. Previous assays (data not shown) validated the changes in the purity of EPC from 99-100% (analytical grade) to 96- 98% (commercial grade) and the lipid concentration from 16 mM (used in the Bangham's method) to 64 mM.

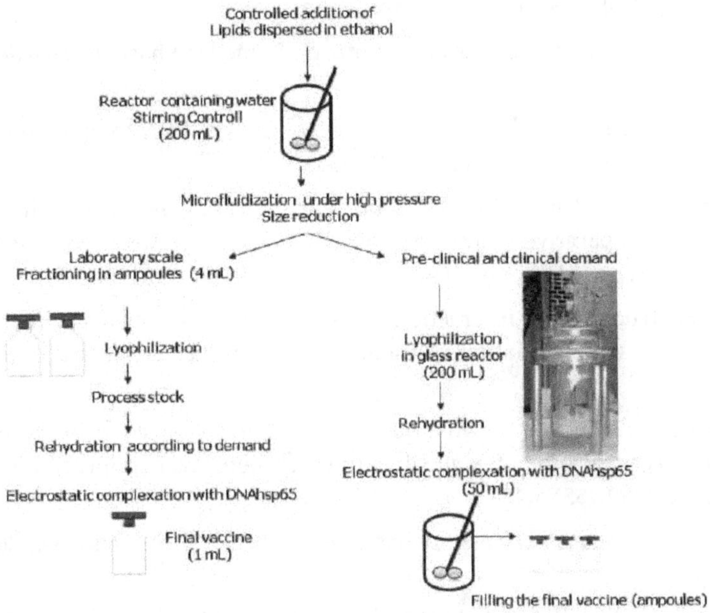

Fig. 11. Schematic diagram of the scalable process for cationic liposome production.

The injection of ethanol was used modified for the higher lipid concentration. Lipid concentration in the ethanol phase was previously optimized with an aim to minimize the concentration of ethanol in the liposome formulation, as well as prevent obstructions of the microchannels in the microfluidizer.

7. CONCISE SUMMARY

The top-down and bottom-up methodologies are advantageous for liposome production. By comprehending the physico-chemical characteristics and behavior of variables involved in the process, we can establish a conscious scalable process for liposome production and establish a connection between the final properties of the gene vaccine as a product, the performance of the production process, and the functional liposomes. The results of this chapter present opportunities for the development of novel gene or non-gene vaccines in future research through the use of scalable top-down and bottom-up processes.

REFERENCES

Alves, G. P., and Santana, M. H. A. (2004) Phospholipid dry powders produced by spray drying processing: Structural, thermodynamic and physical properties, *Powder Technology*, Vol. 145, pp 139-148, ISSN 0032-5910.

Azaroff, L. V. (1968) *Elements of x-ray crystallography*, McGraw Hill Book, New York.

Baldas, L., and Caen, R. (2010) Experimental methods *in: Microfluidics*, Vol. Edited by Stéphane Colin, pp 330-348.

Bangham, A. D., Standish, M. M., and Watkins, J. C. (1965) Diffusion of univalent ions across lamellae of swollen phospholipids, *Journal of Molecular Biology*, Vol. 13, p 238, ISSN 0022-2836.

Caille, A. (1972) *C. R. Seances Acad. Sci. B*, Vol. 274, p 891, ISSN 0249-6313.

Ciccariello, S., and Sobry, R. (1999) Small-angle scattering intensity behaviours of cylindrical, spherical and planar lamellae, *Journal of Applied Crystallography*, Vol. 32, pp 892-901, ISSN 0021-8898.

Cullity, B. D. (1978) *Elements of x ray diffraction*, 2nd ed., Addison Wesley, USA.

de Paula, L., Silva, C. L., Carlos, D., Matias-Peres, C., Sorgi, C. A., Soares, E. G., Souza, P. R., Blades, C. R., Galleti, F. C., Bonato, V. L., Goncalves, E. D., Silva, E. V., and Faccioli,

L. H. (2007) Comparison of different delivery systems of DNA vaccination for the induction of protection against tuberculosis in mice and guinea pigs, *Genet Vaccines Ther*, Vol. 5, p 2, ISSN 1479-0556.

Diat, O., and Roux, D. (1993a) Preparation of monodisperse multilayer vesicles of controlled size and high encapsulation ratio, *Journal De Physique II*, Vol. 3, pp 9-14, ISSN 1155- 4312.

Diat, O.; Roux, D.; Nallet, F. (1993b) Effect of Shear on a Lyotropic Lamellar Phase. *Journal de Physique II*, Vol.3, pp. 1427-1452.

Dootz, R., Otten, A., Koster, S., Struth, B., and Pfohl, T. (2006) Evolution of DNA compaction in microchannels, *Journal of Physics-Condensed Matter*, Vol. 18, pp S639-S652, ISSN 0953-8984.

Egelhaaf, S. U., Wehrli, E., Muller, M., Adrian, M., and Schurtenberger, P. (1996) Determination of the size distribution of lecithin liposomes: A comparative study using freeze fracture, cryoelectron microscopy and dynamic light scattering, *Journal of Microscopy-Oxford*, Vol. 184, pp 214-228, ISSN 0022-2720.

Fritz, G., and Glatter, O. (2006) Structure and interaction in dense colloidal systems: Evaluation of scattering data by the generalized indirect fourier transformation method, *Journal of Physics: Condensed Matter*, Vol. 18, pp S2403-S2419, ISSN 0953- 8984.

Galindo-Rodriguez, S., Allemann, E., Fessi, H., and Doelker, E. (2004) Physicochemical parameters associated with nanoparticle formation in the salting-out, emulsification-diffusion, and nanoprecipitation methods, *Pharmaceutical Research*, Vol. 21, pp 1428-1439, ISSN 0724-8741.

Glatter, O. (1982) Data treatment, In *Small angle x ray scattering* (Glatter, O. & Kratky, O., Eds.), p 137, Academic Press, London.

Glatter, O. (1991) Scattering studies on colloids of biological interest (amphiphilic systems), In *Trends in colloid and interface science v* (Corti, M. & Mallamace, F., Eds.), pp 46-54, Springer Berlin / Heidelberg.

Glatter, O. (2002) Neutrons, x-rays and light: Scattering methods applied to soft condensed matter, In *Static light scattering of large systems* (Lindner, P. & Zemb, T., Eds.), pp 171-201, Elsevier, ISBN 0-444-51122-9, Amsterdam.

Glatter, O., and Kratky, O., (Eds.) (1982) *Small angle x-ray scattering*, Academic Press, London.

Gregoriadis, G. (1993) *Liposome technology*, CRC Press, Inc, ISBN 0-8493-6709-3.

Higbie, R. (1935) The rate of absorption of a pure gas into a still liquid during short periods of exposure, *Transactions of the American Institute of Chemical Engineers*, Vol. 31, pp 365-389, ISSN 0096-7408.

Hsieh, A. T. H., Hori, N., Massoudi, R., Pan, P. J. H., Sasaki, H., Lin, Y. A., and Lee, A. P. (2009) Nonviral gene vector formation in monodispersed picolitre incubator for consistent gene delivery, *Lab on a Chip*, Vol. 9, pp 2638-2643, ISSN 1473-0197.

Hulst, H. C. v. d. (1957) *Light scattering by small particles*, Wiley, New York.

Jahn, A., Reiner, J. E., Vreeland, W. N., DeVoe, D. L., Locascio, L. E., and Gaitan, M. (2008) Preparation of nanoparticles by continuous-flow microfluidics, *Journal of Nanoparticle Research*, Vol. 10, pp 925-934, ISSN 1388-0764.

Jahn, A., Stavis, S. M., Hong, J. S., Vreeland, W. N., Devoe, D. L., and Gaitan, M. (2010) Microfluidic mixing and the formation of nanoscale lipid vesicles, *Acs Nano*, Vol. 4, pp 2077-2087, ISSN 1936-0851.

Jahn, A., Vreeland, W. N., DeVoe, D. L., Locascio, L. E., and Gaitan, M. (2007) Microfluidic directed formation of liposomes of controlled size, *Langmuir*, Vol. 23, pp 6289-6293, ISSN 0743-7463.

Jahn, A., Vreeland, W. N., Gaitan, M., and Locascio, L. E. (2004) Controlled vesicle self- assembly in microfluidic channels with hydrodynamic focusing, *Journal of the American Chemical Society*, Vol. 126, pp 2674-2675, ISSN 0002-7863.

Jellema, R. K., Bomans, P., Deckers, N., Ungethum, L., Reutelingsperger, C. P. M., Hofstra, L., and Frederik, P. M. (2010) Transfection efficiency of lipoplexes for site- directed delivery, *Journal of Liposome Research*, Vol. 20, pp 258-267, ISSN 0898- 2104.

Knight, J. B., Vishwanath, A., Brody, J. P., and Austin, R. H. (1998) Hydrodynamic focusing on a silicon chip: Mixing nanoliters in microseconds, *Physical Review Letters*, Vol. 80, pp 3863-3866, ISSN 0031-9007.

Koh, C. G., Zhang, X. L., Liu, S. J., Golan, S., Yu, B., Yang, X. J., Guan, J. J., Jin, Y., Talmon, Y., Muthusamy, N., Chan, K. K., Byrd, J. C., Lee, R. J., Marcucci, G., and Lee, L. J. (2010) Delivery of antisense oligodeoxyribonucleotide lipopolyplex nanoparticles assembled by microfluidic hydrodynamic focusing, *J. Control. Release*, Vol. 141, pp 62-69, ISSN 0168-3659.

Kratky, O. (1988) The importance of x-ray small-angle scattering in colloid research, In *Dispersed systems* (Hummel, K. & Schurz, J., Eds.), pp 1-14, Springer Berlin / Heidelberg.

Kratky, O., and Laggner, P. (2001) X-ray small-angle scattering, In *Encyclopedia of physical science and technology* (Robert, A. M., Ed.), pp 939-988, Academic Press, ISBN 978-0- 12-227410-7, New York.

Kremer, J. M. H., Esker, M. W. J., Pathmamanoharan, C., and Wiersema, P. H. (1977) Vesicles of variable diameter prepared by a modified injection method, *Biochemistry*, Vol. 16, pp 3932-3935, ISSN 0006-2960.

Lasic, D. D. (1993) *Liposomes: From physics to applications*, Amsterdam:Elsevier Science Publishers B.V.

Mannisto, M., Vanderkerken, S., Toncheva, V., Elomaa, M., Ruponen, M., Schacht, E., and Urtti, A. (2002) Structure-activity relationships of poly(l-lysines): Effects of pegylation and molecular shape on physicochemical and biological properties in gene delivery, *J. Control. Release*, Vol. 83, pp 169-182, ISSN 0168- 3659.

Mijatovic, D., Eijkel, J. C. T., and van den Berg, A. (2005) Technologies for nanofluidic systems: Top-down vs. Bottom-up - a review, *Lab on a Chip*, Vol. 5, pp 492-500, ISSN 1473-0189.

Montgomery, D. C. (2008) *Design and analysis of experiments*, John Wiley & Sons.

Mount, C. N., Lee, L. K., Yasin, A., Scott, A., Fearn, T., and Shamlou, P. A. (2003) The influence of physico-chemical and process conditions on the physical stability of plasmid DNA complexes using response surface methodology, *Biotechnology and Applied Biochemistry*, Vol. 37, pp 225-234, ISSN 0885-4513.

Nagle, J. F., Zhang, R., Tristram-Nagle, S., Sun, W., Petrache, H. I., and Suter, R. M. (1996) X- ray structure determination of fully hydrated l alpha phase dipalmitoylphosphatidylcholine bilayers, *Biophys J*, Vol. 70, pp 1419-1431, ISSN 0006-3495.

Nallet, F., Laversanne, R., and Roux, D. (1993) Modelling x-ray or neutron scattering spectra of lyotropic lamellar phases : Interplay between form and structure factors, *J. Phys. II France*, Vol. 3, pp 487-502.

New, R. R. C. (1990) *Liposomes: A practical approach (the practical approach series)*.

Oberle, V., Bakowsky, U., Zuhorn, I. S., and Hoekstra, D. (2000) Lipoplex formation under equilibrium conditions reveals a three-step mechanism, *Biophysical Journal*, Vol. 79, pp 1447-1454, ISSN 0006-3495.

Ogris, M., Steinlein, P., Kursa, M., Mechtler, K., Kircheis, R., and Wagner, E. (1998) The size of DNA/transferrin-pei complexes is an important factor for gene expression in cultured cells, *Gene Therapy*, Vol. 5, pp 1425-1433, ISSN 0969- 7128.

Oliveira, C. L. P., Sankar R., Kloesgen, M. B., and Pedersen, J. S. (Unpublished work).

Ota, S., Yoshizawa, S., and Takeuchi, S. (2009) Microfluidic formation of monodisperse, cell- sized, and unilamellar vesicles, *Angewandte Chemie-International Edition*, Vol. 48, pp 6533-6537, ISSN 1433-7851.

Otten, A., Koster, S., Struth, B., Snigirev, A., and Pfohl, T. (2005) Microfluidics of soft matter investigated by small-angle x-ray scattering, *Journal of Synchrotron Radiation*, Vol. 12, pp 745-750, ISSN 0909-0495.

Pabst, G., Rappolt, M., Amenitsch, H., and Laggner, P. (2000) Structural information from multilamellar liposomes at full hydration: Full q-range fitting with high quality x- ray data, *Physical Review E*, Vol. 62, p 4000.

Pachence, J. M., and Blasie, J. K. (1991) Structural investigation of the covalent and electrostatic binding of yeast cytochrome c to the surface of various ultrathin lipid multilayers using x-ray diffraction, *Biophysical journal*, Vol. 59, pp 894-900, ISSN 0006-3495.

Pedersen, J. S., Posselt, D., and Mortensen, K. (1990) Analytical treatment of the resolution function for small-angle scattering, *Journal of Applied Crystallography*, Vol. 23, pp 321-333, ISSN 0021-8898.

Perrie, Y., Frederik, P. M., and Gregoriadis, G. (2001) Liposome-mediated DNA vaccination: The effect of vesicle composition, *Vaccine*, Vol. 19, pp 3301-3310, ISSN 0264- 410X.

Perrie, Y., and Gregoriadis, G. (2000) Liposome-entrapped plasmid DNA: Characterisation studies, *Biochimica Et Biophysica Acta-General Subjects*, Vol. 1475, pp 125-132, ISSN 0304-4165.

Rejman, J., Oberle, V., Zuhorn, I. S., and Hoekstra, D. (2004) Size-dependent internalization of particles via the pathways of clathrin-and caveolae-mediated endocytosis, *Biochemical Journal*, Vol. 377, pp 159-169, ISSN 0264-6021.

Rigoletto, T. D., Zaniquelli, M. E. D., Santana, M. H. A., and de la Torre, L. G. (2011) Surface miscibility of epc/dotap/dope in binary and ternary mixed monolayers, *Colloids and Surfaces B-Biointerfaces*, Vol. 83, pp 260-269, ISSN 0927-7765.

Rosada, R. S., de la Torre, L. G., Frantz, F. G., Trombone, A. P. F., Zarate-Blades, C. R., Fonseca, D. M., Souza, P. R. M., Brandao, I. T., Masson, A. P., Soares, E. G., Ramos,

S. G., Faccioli, L. H., Silva, C. L., Santana, M. H. A., and Coelho-Castelo, A. A. M. (2008) Protection against tuberculosis by a single intranasal administration of DNA-hsp65 vaccine complexed with cationic liposomes, *BMC Immunol.*, Vol. 9, p 13, ISSN 1471-2172.

Ross, P. C., and Hui, S. W. (1999) Lipoplex size is a major determinant of in vitro lipofection efficiency, *Gene Therapy*, Vol. 6, pp 651-659, ISSN 0969-7128.

Sato, T., Sakai, H., Sou, K., Medebach, M., Glatter, O., and Tsuchida, E. (2009) Static structures and dynamics of hemoglobin vesicle (hbv) developed as a transfusion alternative, *J Phys Chem B*, Vol. 113, pp 8418-8428, ISSN 1520-6106.

Shimomura, M., and Sawadaishi, T. (2001) Bottom-up strategy of materials fabrication: A new trend in nanotechnology of soft materials, *Current Opinion in Colloid & Interface Science*, Vol. 6, pp 11-16, ISSN 1359-0294.

Shum, H. C., Lee, D., Yoon, I., Kodger, T., and Weitz, D. A. (2008) Double emulsion templated monodisperse phospholipid vesicles, *Langmuir*, Vol. 24, pp 7651-7653, ISSN 0743-7463.

Silva, C. L., Bonato, V. L. D., Coelho-Castelo, A. A. M., De Souza, A. O., Santos, S. A., Lima, K. M., Faccioli, L. H., and Rodrigues, J. M. (2005) Immunotherapy with plasmid DNA encoding mycobacterial hsp65 in association with chemotherapy is a more rapid and efficient form of treatment for tuberculosis in mice, *Gene Therapy*, Vol. 12, pp 281-287, ISSN 0969-7128.

Torre, L. G., Carneiro, A. L., Rosada, R. S., Silva, C. L., and Santana, M. H. A. (2007) A mathematical model describing the kinetic of cationic liposome production from dried lipid films adsorbed in a multitubular system, *Brazilian Journal of Chemical Engineering*, Vol. 24, pp 477-486, ISSN 0104-6632.

Tournier, V. H., Trionex, M. S., and Chable-Beaumont, C. G. L. (1999) Liposomes with enhanced entrapment capacity and their use in imaging, In *US Patent* (Bracco Reserach S.A, Ed.).

Wagner, A., Vorauer-Uhl, K., Kreismayr, G., and Katinger, H. (2002) The crossflow injection technique: An improvement of the ethanol injection method, *Journal of Liposome Research*, Vol. 12, pp 259-270, ISSN 0898-2104.

Wiewrodt, R., Thomas, A. P., Cipelletti, L., Christofidou-Solomidou, M., Weitz, D. A., Feinstein, S. I., Schaffer, D., Albelda, S. M., Koval, M., and Muzykantov, V. R. (2002) Size-dependent intracellular immunotargeting of therapeutic cargoes into endothelial cells, *Blood*, Vol. 99, pp 912-922, ISSN 0006-4971.

Xu, Y., Szoka, F.C.J. (1996) Mechanism of DNA release from cationic liposome/DNA complexes used in cell transfection, *Biochemistry*, Vol. 35, pp. 5616-5623, ISSN: 0006- 2960.

Zelphati, O., Nguyen, C., Ferrari, M., Felgner, J., Tsai, Y., and Felgner, P. L. (1998) Stable and monodisperse lipoplex formulations for gene delivery, *Gene Therapy*, Vol. 5, pp 1272-1282, ISSN 0969-7128.

Zhang, R., Suter, R. M., and Nagle, J. F. (1994) Theory of the structure factor of lipid bilayers, *Phys Rev E Stat Phys Plasmas Fluids Relat Interdiscip Topics*, Vol. 50, pp 5047-5060, ISSN 1063-651X.

Zhang, R., Tristram-Nagle, S., Sun, W., Headrick, R. L., Irving, T. C., Suter, R. M., and Nagle, J. F. (1996) Small-angle x-ray scattering from lipid bilayers is well described by modified caille theory but not by paracrystalline theory, *Biophys J*, Vol. 70, pp 349- 357, ISSN 0006-3495.

Zhang, S. B., Zhao, B., Jiang, H. M., Wang, B., and Ma, B. C. (2007) Cationic lipids and polymers mediated vectors for delivery of sirna, *J. Control. Release*, Vol. 123, pp 1- 10, ISSN 0168-3659.

Zhang, S. H., Shen, S. C., Chen, Z., Yun, J. X., Yao, K. J., Chen, B. B., and Chen, J. Z. (2008) Preparation of solid lipid nanoparticles in co-flowing microchannels, *Chemical Engineering Journal*, Vol. 144, pp 324-328, ISSN 1385-8947.

13 Chapter

CATIONIC LIPOSOMES IN VARIOUS STRUCTURAL CONFIGURATIONS FOR GENE DELIVERY

1. INTRODUCTION: AN OVERVIEW

In recent decades, gene therapy has emerged as a valuable experimental tool for the investigation of gene function and regulation. This strategy is particularly promising for the treatment of numerous refractory diseases, including inherited diseases (Martin-Rendon & Blake, 2003) and acquired immunodeficiency syndrome (AIDS) (Fanning et al., 2003). The objective of gene therapy is to enable the expression of the protein coded in the target cells, which in turn treats the disease through the secretion of the protein from the transfected cells (El-Aneed, 2004).

The success of gene therapy is contingent upon the use of appropriate transfection vectors, which possess a favorable safety profile and a high efficiency of gene transfer to target cells. Generally, these vectors are categorized into two primary groups: viral and non-viral (Liu & Huang, 2002). At present, viral vectors that are constructed from a variety of viruses, including retroviruses and adenoviruses, have demonstrated some success in the field of cancer gene therapy (Williams et al., 2010). The performance and pathogenicity of these vectors have been assessed in animal models. In order to completely realize their clinical potential, it is necessary to address a number of issues, such as the limited opportunity for repeated administrations due to the acute inflammatory response, the difficulty in production, and the delayed humeral or cellular immune responses (Love et al., 2010). Consequently, it is imperative to create a more adaptable and efficient security system in the realm of gene delivery vectors. Synthetic non-viral vectors are potential alternatives to viral vectors and may contribute to the resolution of certain issues (Zhang et al., 2010). The most extensively studied of these compounds are cationic compounds, which are primarily cationic lipids and cationic polymers, and are believed to cause fewer safety issues due to their relative simplicity.

Numerous cationic lipids have been synthesized and employed for the delivery of nucleic acids into cells over the past two decades, following the initial description of effective in vitro transfection with cationic lipids by Felgner et al. in 1987 (Felgner et al., 1987). The efficacy of gene delivery has been enhanced by cationic liposomes, which are comprised of lipid constituents and possess a typical bilayer structure. Dioleylphosphatidyl choline (DOPC) and dioleylphosphatidyl ethanolamine (DOPE), which are neutral lipids (Zuhorn et al., 2005), are frequently used in conjunction with cationic lipids. They are crucial in the formation of lipoplexes by combining cationic liposomes and genes, as they can influence the morphology of lipoplexes.

A particulate complex, polyplex, is formed by cationic polymers, which condense DNA through ionic interaction at physiological pH. This complex is capable of gene transfer into the targeted cells (El-Aneed, 2004). They have the ability to condense the negatively charged DNA to a relatively small size and decrease its susceptibility to nucleases, which may make them advantageous for enhancing transfection proficiency. The synergistic effect on transfection efficiency and a promising solution to the problem that frustrates us may be achieved by incorporating polycations (such as poly-L-lysine and protamine) into cationic liposomes as co-polymers (Gao & Huang, 1996; Li & Huang, 1997). In this chapter, we summarize the impact of cationic liposomes on gene delivery at various structural levels and contribute to the elucidation of the mechanism that controls the formation and behavior of cationic liposomes in gene delivery. The initial phase of research on the structure–activity relationship of cationic lipids involves the synthesis of new vectors and efforts to enhance transfection efficiencies and reduce cytotoxicity by modifying the hydrophilic, hydrophobic, and linker domains. The second study examines the morphologies of lipoplexes in relation to the effects of helper lipids and particle diameters. The final method involves the hybridized use of non-viral vectors, such as the complexes of cationic lipsomes and polymers, conjugates of lipids and peptides, and targeting moieties and lipids.

2. THE CHEMICAL COMPOSITION OF CATIONIC COMPOUNDS

A vast array of cationic lipids has been created in order to comprehend the correlation between gene delivery and chemical structure. In general, a cationic lipid that is employed in gene therapy is composed of three fundamental domains: a hydrophilic cationic headgroup, a hydrophobic domain, and a linker bond that connects the hydrophilic and hydrophobic regions (Fig. 1) (Gao & Hui, 2001). The physical parameters of the liposome are determined by the chemical structure of cationic lipids, which is a critical factor in both transfection activity and cytotoxicity levels. The impact of chemical structure modifications on gene delivery is thoroughly examined by examining a substantial body of literature and is subsequently illustrated.

2.1 Headgroup domain

Since the first description of successful *in vitro* transfection with cationic lipid—DOTMA (Fig. 1.) by Felgner et al. in 1987 (Felgner et al., 1987), progress has been made in the design and analysis of each domain. The effect of transfection efficiency and cytotoxicity is associated with the cationic nature of the vectors, which is mainly determined by the structure of its hydrophilic headgroup. In general, the different types of headgroups fall into the following categories: primary, secondary, tertiary amines or quaternary ammonium salts (of which polyamines have often seen the most success) and guanidinium, amidine, as well as heterocyclic ring (Heyes et al., 2002). In addition, other charged groups that have been shown capable of binding plasmids have since been employed. For example, Floch et al. (Floch et al., 2000) showed that cationic lipids (Fig. 2.) characterized by a cationic charge carried by a phosphorus or arsenic atom instead of a nitrogen atom, led to an increased transfection efficiency (up to 7 times according to the cell lines tested) and reduced toxicity.

In hydrophilic headgroup, tertiary amine or quaternary ammonium groups are the most frequently used in many of the established cationic lipids (Felgner et al., 1987; Gao & Huang, 1991; Leventis & Silvius, 1990). A typical example that modified to the chemical structure of headgroup was described by Felgner et al. (Felgner et al., 1994). They synthesized a series of 2,3-dialkyloxy quaternary ammonium compounds containing a hydroxyl moiety (Fig. 3.), which were more efficient in transfection compared with DOTMA that lacks a hydroxyl group on the quaternary amine. In our previous work, we also synthesized a series of cationic lipids containing a hydroxyl moiety on the quaternary amine for liposome-mediated gene delivery (Fig. 4.). Several cationic liposomes with relatively higher transfection efficiency were selected after *in vitro* transfection studies of the prepared cationic liposomes, whose biological performance was superior or parallel to that of the commercial transfection agents, Lipofectamine2000 and DOTAP. It is suggested that the headgroup hydration can be decreased by the incorporation of a hydroxyalkyl chain capable of hydrogen bonding to neighbouring headgroups while it improves the compaction of DNA by several mechanisms, for example, DNA can form hydrogen bonds with the lipid, and the hydroxyl group can enhance the membrane hydration. Accordingly, several groups have synthesised cationic lipids that varied the chain length of the hydroxyalkyl moiety, while keeping the remaining structure unchanged, and observed that the activity of lipid increased with the decrease in the hydroxyalkyl chain length. It shows that a decrease in the number of carbon atoms in the hydroxyalkyl chain, providing more rigidity to the terminal hydroxyl group, are very efficient in compacting DNA, which is responsible for the more observed transfection activity (Felgner et al., 1994; Bennett et al.,1997).

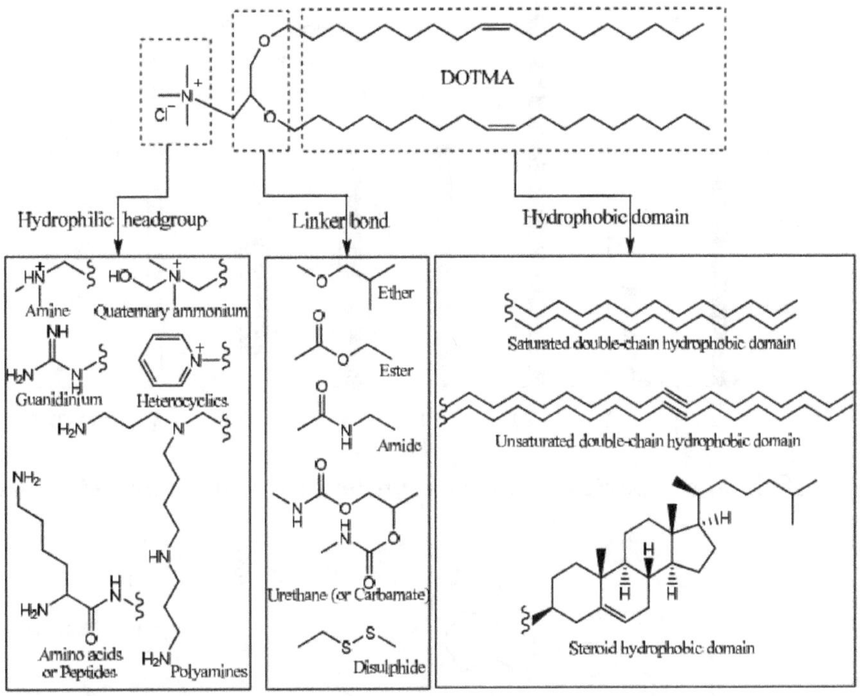

Fig. 1. Representative structure of the cationic lipid DOTMA. Examples of cationic lipid structural components: hydrophilic headgroup, linker bond, and hydrophobic domain.

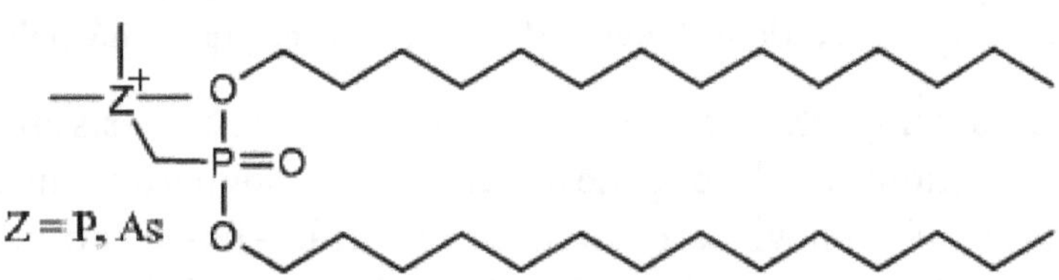

Fig. 2. Alternative cations in the lipid headgroup.

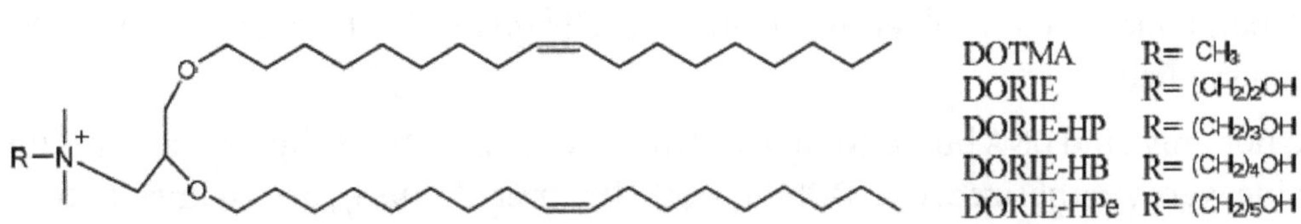

Fig. 3. Double-chain cationic lipids containing a hydroxyl moiety on the quaternary amine.

A number of cationic lipids that spread the positive charge of the cationic head by delocalizing it into a heterocyclic ring have been designed in an attempt to improve transfection efficacy and to lower toxicity. Heterocyclic cationic lipids with morpholine or piperazine polar heads conjugated to cholesterol directly or via a spacer (Fig. 5.) have been reported to compare favourably with linear polyamine head groups with similar or greater number of charges (Gao & Hui, 2001). And then, the use of imidazole or pyridine rings have been also reported to display higher transfection efficiency and reduced cytotoxicity when compared with classical transfection systems (Ilies et al., 2003; Medvedeva et al., 2009).

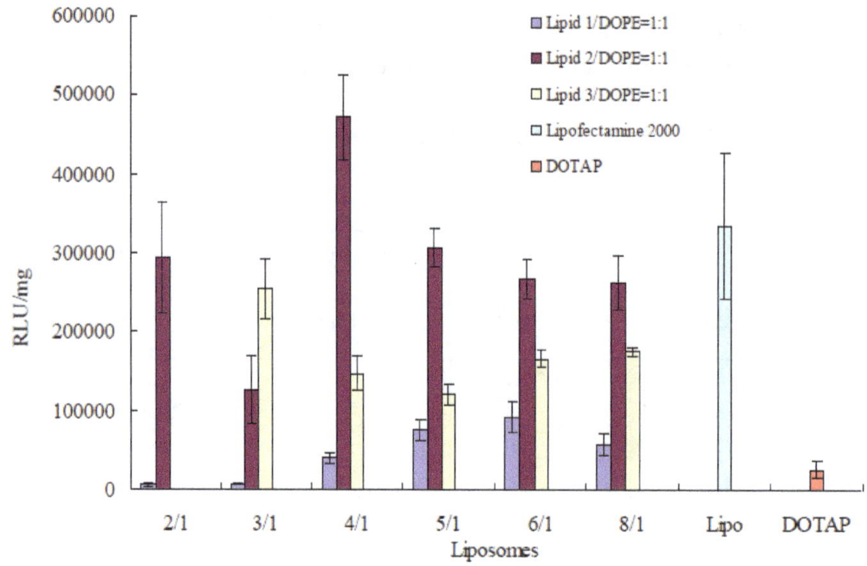

Fig. 4. Transfection comparisons of lipids synthesized and commercial reagents.

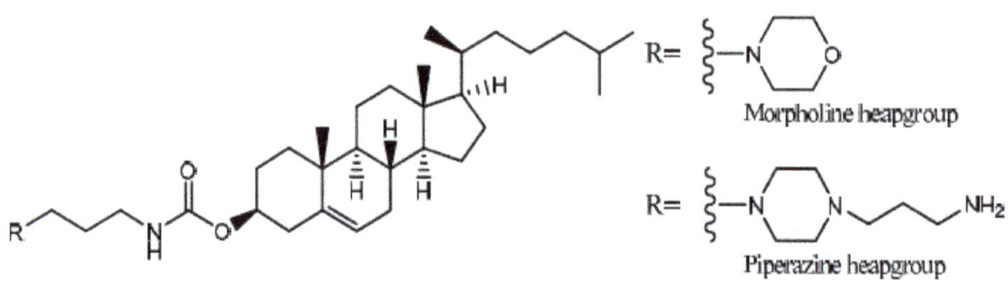

Fig. 5. Heterocyclic cationic lipids with morpholine or piperazine polar heads.

Guanidine and its salts are important intermediates for organic synthesis and medicine; they have also been proposed for making cationic lipids for gene delivery. Yingyongnarongkul et al. (Yingyongnarongkul et al., 2004) studied a library of aminoglycerol–diamine conjugate- based cationic lipids with urea linkage between varying length of diamines and hydrophobic chains, and found two compounds with bis-guanidinium and one tail had transfection activity superior to that of the commercial lipid transfection reagent effective and merit further investigation. In addition, several cationic lipids containing guanidine (Floch et al., 2000), amidine (Lensink et al., 2009), or cyclic guanidine (Frederic at al., 2000) in the head group have been studied.

As a way to not only bind DNA but to compact DNA, polyvalent cationic lipids known as lipopolyamines and lipospermine were synthesized, e.g., DOSPA, DOGS (Behr et al., 1989), are claimed to be more efficient than single-charged lipids such as DOTMA, DOTAP, DC-Chol, DMRIE (Ferrari et al., 1998). The choice of headgroup may depend on the desired and specific function of the cationic lipid, but for gene transfection it is essential that the headgroups of cationic lipid interact strongly with the minor groove of DNA via its multivalent headgroup, and has the ability to efficiently condense DNA. In addition, a few parameters, such as the shape (linear, T-shaped, globular, branched) and the length of the multivalent headgroup, and the distance between two consecutive nitrogen atoms in the polyamine, are also important in transfection efficiency and cytotoxicity (Byk et al., 1998; Fujiwara et al., 2000).

In summary, the choice of cationic headgroups has expanded into the use of natural architectures and functional groups with recognized DNA binding modes. Multivalent cationic lipids are more likely to be the most efficient in compacting DNA; however, they are prone to formation of micelles typically contributing to increased toxicity (Pedroso de Lima et al., 2003). The incorporation of a heteroatomic group as the substitution of the liner amine headgroup, such as pyridinium and guanidine, can improve the compaction of DNA by the positive charge of the cationic head, and then transfection efficiency is increased and toxicity is decreased significantly.

2.2 Hydrophobic domain

The hydrophobic domains represent the non-polar hydrocarbon moieties of cationic lipids and are usually made of two types of hydrophobic moieties—aliphatic chains, steroid domain. Transfection efficiency and toxicity of cationic lipids can be affected by structural variations in the hydrophobic domain such as length, the specific type of chemical bonds, and the relative position of the hydrocarbon chains (Zhi at al., 2010).

Cationic lipids with aliphatic chains have been very thoroughly researched. The chains are either linear and saturated or linear and mono-unsaturated and used in liposomal vectors ranging from C5:0 to C18:1, but oleyl, lauryl, myristyl, palmityl and stearyl, have been the most researched ones (Niculescu-Duvaz et al., 2003). A common variation is the use of branched (Ferrari et al., 2002), acetylenic (Fletcher et al., 2006) chains and *cis*- monounsaturated alkyl chains (Bennett at al., 1997).

It is commonly believed that cationic lipids have one to four hydrocarbon chains. Several studies have also shown that incorporating aliphatic chains with different numbers can improve transfection efficiency potentially by promoting endosomal escape (Felgner et al., 1987; Tang & Hughes, 1999a; Gaucheron at al., 2002; Zhi et al., 2010). Cationic lipids with double-chain hydrocarbons in the hydrophobic domain represent the majority of cationic lipids synthesized so far. Cationic lipids containing two aliphatic chains such as DOTMA and DOTAP, are among the most active lipids for systemic gene delivery. However, Tang et al. (Tang & Hughes, 1999a) demonstrated that 6-lauroxyhexyl ornithinate (LHON) with one tail was more efficient and of lower cytotoxicity compared to DOTAP. Generally speaking, for aliphatic chains, single-tailed and three-tailed cationic lipids are better known as surfactants because of their ability to form micelles in solution, but they are more toxic and less efficient than their double-tailed counterparts. Usually, cationic lipids with double- tailed hydrocarbons are capable of forming liposomes by themselves or with a helper phospholipid.

Therefore, most of the aliphatic chains in the cationic lipids are double-tailed. It is generally agreed that the length and saturation of the aliphatic chains incorporated into cationic lipids significantly affect their transfection efficiency. In order to gain chain length– activity correlation, Felgner et al. (Felgner et al., 1994) studied a series of hydroxyethyl quaternary ammonium lipids with myristoyl (diC14:0, DMRIE), palmitoyl (diC16:0, DPRIE), stearoyl (diC18:0, DSRIE), and oleoyl (diC18:1, DORIE) chains. They observed that a comparison of vectors based solely on the lengths of the two aliphatic chains led to identify the order C14:0 > C18:1 > C16:0 > C18:0. Our study on double-chain cationic lipids also showed increasing transfection efficiency with decrease of the chain length (Liu et al., 2008). It was therefore proposed that cationic lipids with shorter chain length (for saturated chains) were generally important for acquiring high transfection efficiency, since they are responsible for membrane fluidity and good lipid mixing within the bilayer. Beyond that, the best chains in terms of benefit to transfection are frequently the unsaturated ones. The overwhelming majority of results showed that the unsaturated C18:1 oleyl was the optimal aliphatic chain, which was frequently the best choice for good transfection (Fletcher et al., 2006).

Some new vectors were designed to covalently connect some special moieties in the hydrophobic chains, in order to get the relationship between hydrophobic chains and transfection efficiency. Jacopin et al. (Jacopin et al., 2001) synthesized a glycosylated analogue (Fig. 6.) of the dialkylamidoglycylcarboxyspermines, which formed stable particles at low charge ratio and was efficient for gene delivery. Many groups also reported a few other glycosylated cationic bolaamphiphiles similar to the compound (Fabio et al., 2003; Brunelle et al., 2009). In addition, the fluorinated part of the hydrophobic chain can also influence the transfection efficiency of cationic lipids *in vivo* and *in vitro*. Many varieties of fluorinated cationic lipids have been developed as transfecting agents, which are very efficient in compacting DNA and delivering genes into cells *in vivo* and *in vitro* (Gaucheron et al., 2001a, 2001b, 2001c, 2001d).

In the steroid groups, cholesterol is by far the most frequently encountered and used as an alternative to aliphatic chains because of its rigidity, as well as its endogenous biodegradability and fusion activity. An example is cationic lipid 'GL-67' (Fig. 7.), which has been found to be particularly efficient for gene transfer to cultured cells and in murine lungs

(Lee et al., 1996). Other steroid compounds used as hydrophobic moieties for cationic lipids include vitamin D (Ren et al., 2000), bile acids (Randazzo et al., 2009), antibiotic (Kichler et al., 2005), cholestane and litocholic acid (Fujiwara et al., 2000).

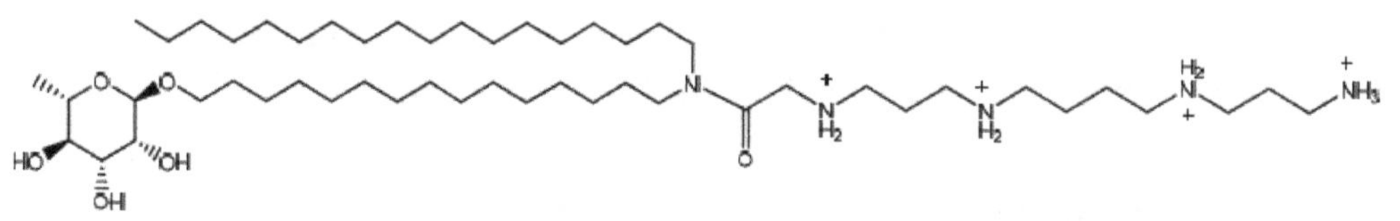

Fig. 6. Glycosylated analogue of lipopolyamines.

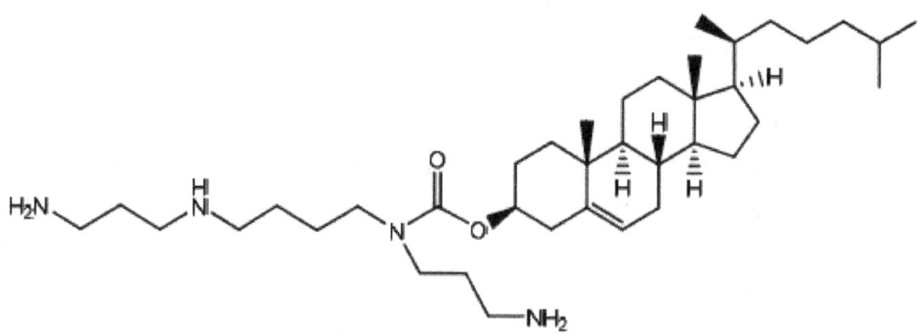

Fig. 7. Chemical structure of GL-67.

To summarise, the hydrophobic domain of cationic lipids mainly includes aliphatic chains and steroid domain, which determines the phase transition temperature and the fluidity of the bilayer, and influences the stability and toxicity of liposomes, the DNA protection from nucleases, and the DNA release from complex.

1.3 Linker bond

For lipids without a backbone, the linker bond that acts as a connector between the hydrophobic and cationic headgroup domains can affect the transfection efficiency, biodegradability and stability of cationic lipids. Linker bonds are commonly ether, ester (Leventis & Silvius, 1990), amide (Behr et al., 1989) or urethane (or carbamate) (Lee et al., 1996) groups (Koynova & Tenchov, 2010), but other groups such as redox-sensitive disulphide (Byk et al., 1998, 2000) have also been employed (Fig. 1.). Cationic lipids with ether bonds such as DOTMA in the linker domain generally render better transfection efficiency, but they are too stable to be biodegraded thus may cause higher toxicity. Compared with ether bonds, although cationic lipids with ester bonds such as DOTAP are more biodegradable and associated with less cytotoxicity in cultured cells (Leventis & Silvius, 1990; Choi et al., 2001), those with ester may also decrease the stability of liposomes in systemic circulation.

The chemistry of the linker has most often been of the carbamate or amide variety, both of which are chemically stable and biodegradable, and cationic lipids with these linkers could be used as efficient gene delivery carriers (Ren et al., 2001; Liu et al., 2005a, 2005b, 2008). A typical example of cationic lipid with carbamate linker is DC-Chol, which was the first lipid used in clinical trials because of its combined properties of transfection efficiency, stability, and low toxicity (Gao & Huang, 1995). As well known, when incorporating a carbamate group into the linker, it may therefore be hoped that the pH drop will act as a trigger, disconnecting the hydrophobic and hydrophilic portions of the lipoplex, and thereby to release DNA after entering endosomes in cell because of the pH decreasing (Liu et al., 2005a, 2008). We synthesized a series of carbamate-linked cationic lipids for liposome-mediated gene delivery, which proved to have good gene transfection properties (Fig. 4.).

It is familiar to chemists that compounds comprising redox-sensitive disulphide bonds is stable chemically as long as no reducing agents, and it is expected that these disulphide- linked lipids can keep stable in the circulation system while decomposing to release DNA after entering endosomes in cells (in a similar manner as the pH-sensitive systems) (Tang & Hughes, 1999b). Byk et al. (Byk et al., 1998, 2000) prepared a series of lipopolyamines that harbor a disulfide bridge within different positions in the backbone of the lipids as biosensitive function. They found that an early release of DNA during or after penetration into cells, probably promoted by reduction of a disulfide bridge placed between the polyamine and the lipid, implied a total loss of transfection efficiency.

In addition, structural variations at the linker region such as length, the specific type of chemical bonds and the relative position of the hydrocarbon chains can affect the transfection efficiency, biodegradability and stability of cationic lipids (Fujiwara et al., 2000). The level of hydration and toxicity of the lipid can also be determined by the length of the linker (Floch et al., 2000). In a word, the use of linkers incorporating functional groups that are cleavable on shorter time scales and under specific stimuli is however of emerging interest, as DNA release may here be facilitated by a triggered decomplexation mechanism.

From the chemistry point of view, the structure of cationic compounds is an important factor for their transfection activity and toxicity. Some common conclusions can be achieved by comparing the different structures and their transfection activity in the same family or different families of lipids. The transfection efficiency is not only determined by one domain of cationic lipids, but also depends on the combination of them (Tang & Hughes, 1999a). In general, it seems when researchers design cationic compounds for gene delivery, the balances between the opposite factors including fluidity and rigidity, symmetry and asymmetry, saturation and unsaturation, linearity and branching, short chain and long chain, hydrophilicity and lipophilicity of compounds should be taken into serious consideration.

3. HELPER LIPIDS AND MORPHOLOGY OF LIPOPLEXES

Neutrally charged helper lipids such as DOPE, DOPC (Fig. 8.), are often employed with cationic lipids in order to gain high transfection efficiency (Felgner & Ringold, 1989). When cationic liposomes are mixed with DNA, lipoplexes are formed with heterogeneous morphologies including beads on a string structure (Felgner & Ringold, 1989), spaghettis or meatballs structure (Sternberg et al., 1994), multilamellar structure LCα, inverted hexagonal phase structure HCII (Koltover et al., 1998), a map-pin structure (Sternberg et al., 1998) and a sliding columnar phase (O'Hern & Lubensky, 1998). Helper lipids play very important role during the formation of lipoplexes by combining cationic liposomes and genes, as they could determine the morphologies of lipoplexes. It has shown lipoplex size is very important for gene transfer to actively endocytosing cells (Ross & Hui, 1999), as such the influences on transfection efficiency: DNA ratio, types of liposomes, incubation time in polyanion containing media, and time of serum addition, are channeled mostly through their influences on lipoplex size.

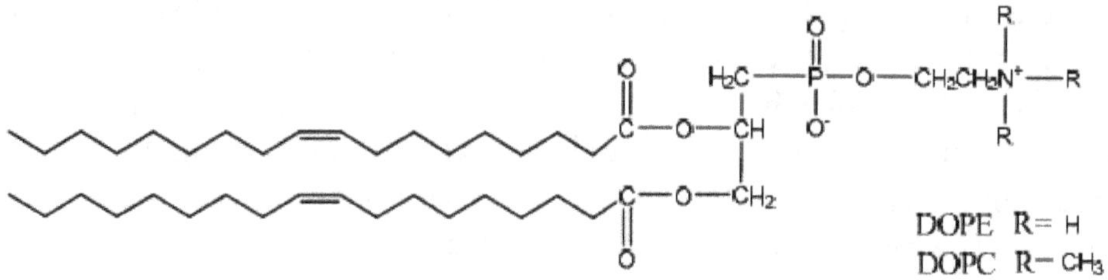

Fig. 8. Chemical structure of DOPE and DOPC.

3.1 DOPE

DOPE often presents a super synergistic effect when used in cationic liposomes, because DOPE destabilized lipid bilayers, and it was believed to be involved in endosomal disruption (Litzinger & Huang, 1992), allowing the release of DNA into the cytosol (Farhood et al., 1995) and leading to mixed bilayers (Scarzello et al., 2005). Most studies have shown that lipoplexes containing the non-bilayer-phase-preferring lipid DOPE or cholesterol would promote HCII organization (Zuhorn et al., 2005). A transition from the LCα phase to the HCII phase could be expected by increasing weight fraction of DOPE, via controlling the spontaneous radius of curvature "Ro" of the lipid layers, favored by the elastic free energy (Safinya, 2001). Another helper lipid, cholesterol, could also promote HII^c organization as DOPE. It has been proved that *in vivo* applications cholesterol was a more effective helper lipid than DOPE (Lasic, 1997).

Koltover et al. (Koltover et al., 1998) disclosed the reason in the level of phase transition through synchrotron small-angle X-ray scattering (SAXS) and optical microscopy to show the phase transition from Lα to HII induced by DOPE via controlling the spontaneous curvature $Co = 1/Ro$ of the lipid monolayer. It has been concluded that DOPE facilitates endosomal escape by forming an unstable inverted hexagonal phase at the endosomal pH that destabilizes both the complex and the endosomal membrane. But in a recent study (Leal et al., 2010), they developed CL-siRNA complexes with a novel cubic phase nanostructure exhibiting efficient silencing at low toxicity by using glycerol monooleate other than DOPE as the helper lipid. The inverse bicontinuous gyroid cubic nanostructure was unequivocally established from synchrotron X-ray scattering data, while fluorescence microscopy revealed colocalization of lipid and siRNA in complexes.

Tubes of lipoplexes containing DOTAP/MOG, DOTAP or DOTAP/PC, and DOTAP/DOPE were observed in freeze-fracture electron micrographs. The tubes were extremely short and appeared bead-like in lipoplexes containing DOTAP/MOG, slightly longer in those containing DOTAP or DOTAP/PC, and extensively elongated in DOTAP/DOPE lipoplexes (Xu et al., 1999). The spaghetti-like structures, occurring at DNA: lipid concentrations which were used during transfection and their diameter came closest to the diameter of the nuclear pores, may be the active cationic lipoplexes (Zhdanov et al., 2002). In the study of the structure and morphology of DC-Chol–DOPE/DNA complexes it was found the existence of cluster-like aggregates made of multilamellar DNA/lipid domains coexisting with other multilamellar lipoplexes or, alternatively, with DNA-coated vesicles (Amenitsch et al., 2010). The further study showed that DC-Chol-DOPE/DNA lipoplexes preferentially used a raft mediated endocytosis, while DOTAP-DOPC/DNA systems were mainly internalized by not specific fluid phase macropinocitosys. Most efficient multicomponent lipoplexes, incorporating different lipid species in their lipid bilayer, can use multiple endocytic pathways to enter cells. Their data demonstrated that efficiency of endocytosis was regulated by shape coupling between lipoplex and membrane lipids to suggest that such a shape-dependent coupling regulated efficient formation of endocytic vesicles thus determining the success of internalization (Marchini et al., 2010).

Kato et al. (Kato et al., 2010) observed the effect of phase separation of the membrane by changing PE from DOPE to dipalmitoylethanolamine (DPPE), which corresponded to a change from a homogeneous single phase to two segregated phases of liquid-ordered and liquid-disordered states on the membrane. This study further proved that helper lipids could change the morphologies of lipoplexes through the mutual interaction with DNA based on their chemical structures. Several helper lipids such as dilauroylphosphatidylethanolamine (C12:0), dimyristoyl phosphatidyl ethanolamine (C14:0), dipalmitoylphosphatidylethanolamine (C16:0), diphytanoyl phosphatidylethanolamine (C16:0, branched), distearoylphosphatidylethanolamine (DSPE, C18:0) were compared with DOPE (C18:1) to show that the branched and unsaturated species combined with cationic lipids acted in physical synergism to increase transfection efficiency (Heinze et al., 2010).

1.4 DOPC

Ewert et al. (Ewert et al., 2004) demonstrated that σM, the average membrane charge density of the CL-vector, was a key universal parameter that governed the transfection behavior of LαC complexes in cells. DOPC favors the formation of LαC type of lipoplexes, in which, a system of DOPC/DOTAP-DNA lipoplex showed a strong dependence on the molar fraction of neutral lipid DOPC (ΦDOPC) and therefore membrane charge density σM. The transfection efficiency started low for 0.5 < ΦDOPC < 0.7 and increased dramatically to a similar value, at ΦDOPC = 0.2, with HIIC lipoplex achieved by the DOPE/DOTAP-DNA. In contrast to LαC complexes, HIIC complexes containing DOPE exhibited no dependence on σM. The transfection efficiency increased exponentially with a linear increase of σM for the MVL5/DOPC/DNA lipoplex bearing LαC (Ewert et al., 2002). And then, they found that the curve of transfection efficiency versus σM assumed a bell-shape with increasing σM using MVL type of cationic lipids (Ahmad et al., 2005). Ewert et al. (Ewert et al., 2006) also found that hexagonally arranged tubular lipid micelles (HIC) surrounded by DNA rods were formed though DOPC was used in the dendritic lipid-based cationic liposome.

Later it has been proved that the enhanced transfection efficiency was supported by a mesoscale computer modeling of cationic lipid lipoplexes in LαC phase at high concentrations of cationic lipid (Farago et al., 2006). Recently, a study (Kedika & Srilakshmi, 2011) showed that DOPC was a more efficacious colipid than DOPE. The difference in the transfection efficiencies of lipoplexes in the presence of colipids DOPE and DOPC was explained as the uptake of the lipoplexes in the presence of DOPE took place mainly from the fusion of the lipoplexes with the plasma membrane, whereas "endocytosis" facilitated uptake in the presence of DOPC. Many researchers have agreed membrane charge density σM is a universal parameter governing the transfection efficiency of LCα lipoplexs (Ewert et al., 2005a, 2005b; Lin, 2003). But for the question, which morphology among LαC governed main by DOPC and HIIC governed mainly by DOPE is favored in terms of transfection efficiency, we still need to carry out more research.

1.3 Lipoplex sizes

Another parameter of morphologies affecting transfection efficiency is lipoplex sizes, for the important role of lipoplex sizes in determining the nature of the entry pathway by endocytosis (Wasungu & Hoekstra, 2006). Though it is difficult to unify the size effect of lipoplexs on the transfection activity so far, most transfection complexes fall within an average size range of 100–300nm. The lipoplex particles can be categorized as small (≤100nm), medium (100–200nm), large (200–1000nm) or giant (≥1000nm). (Donkuru et al., 2010).

Some times large lipoplexes sizes could be more efficient to transfer genes because large particles lead to maximum contact with cells (Kennedy et al., 2000), the formation of large intracellular vesicles which are more easily disrupted, thus releasing DNA into the cytoplasm (Escriou et al., 1998), phagocytic activity accompanied by endosomal escape (Xu et al., 1999) and faster sedimentation and better cellular trafficking (Lee et al., 2003). At the same time, some reports supported that particles with smaller size would gain high transfection efficiency (Pitard et al., 1997; Kneuer et al., 2006). The requirement for efficient transfection may be different *in vivo* and *in vitro*. Compared with *in vitro* delivery, small particles tend to have high transfection efficiency *in vivo* because of the ability of small particles to traverse narrow capillary networks. Large particles typically have low *in vivo* transfection efficiencies, while 200-400nm is the optimal size for lipoplexes *in vitro* (Zhdanov et al., 2002; Kennedy et al., 2000). Measurement of the endosomal uptake of fluorescent dextran beads of various sizes clarified that particles smaller than 200 nm were predominantly taken up by means of clathrin mediated endocytosis; with increasing the size, a shift to another mechanism occurred, so that particles larger than 500 nm were taken up predominantly by caveolae mediated pathways (Rejman et al., 2004).

Carriere et al. (Carriere et al., 2002) have proved that lipofection inhibition by serum was largely due to the serum inhibition of lipoplex size growth, and may be overcome by using large, stable lipoplexes. Lipoplexes of over 700 nm mean diameter induced efficient transfection in the presence or absence of serum (Turek et al., 2000), but lipoplexes of less than 250 nm in size showed efficient transfection only in the absence of serum. It was reported that the particle sizes may be one of the factors that were contributed to serum resistance of EDL (ethanol-dried lipid-DNA) lipoplexes, and the large cationic lipoplexes may delay the dissociation of DNA with lipid, thereby enhancing DNA transfection efficiency (Lian & Ho, 2003).

Although a general rule is not obtained until now, there is no doubt that high transfection would be gained from large lipoplexes when endocytosis is dominant, because large particles facilitate membrane contact and fusion. When cells are not actively endocytosing, either small particles may have high transfection efficiency, or lipoplex sizes don't correlate with lipofection efficiency. The possibility of a final agreement on the lipoplexes size effect may be very small, as the other conditions of every transfection case could be different. The controllable assembly of lipoplexes may provide a solution to this problem.

4. HYBRID VECTORS BASED ON CATIONIC LIPIDS

The hybridized utilization of non-viral vectors also provides an alternative solution to the delivery of genes. We could hybridize cationic lipsomes and polymers; introduce peptides and targeting moieties into lipids for approaching the requirements of gene therapy (Zhang et al., 2010).

1.1 Hybrids of cationic liposomes and polymers

Cationic polymers could combine with DNA to form a particulate complex, polyplex, capable of gene transfer into the targeted cells (El-Aneed, 2004), because most of them are completely soluble in water. Therefore, they have the obvious advantage of compressing DNA molecules to a relatively small size

(Gershon et al., 1993; Ruponen et al., 1999). But they do not contain a hydrophobic moiety (Elouahabi & Ruysschaert, 2005), this may hinder the transfection efficiency and cause cytotoxicity to some degree.

Liposome-mediated gene transfer could be improved by natural polycations such as protamine sulfate (PS), poly(L-lysine) (PLL), and spermine (Li & Huang, 1997; Cheng et al., 2009). The addition of poly(L-lysine) and protamine dramatically reduced the particle size of the complex formed between DNA and cationic liposomes and rendered DNA resistant to the nucleases (Gao & Huang, 1996). These polycations could form a complex with DNA and condense DNA from extended conformation to highly compact structure into 30-100 nm in size. A type of hybrid vectors were developed by Huang et al. (Gao & Huang, 1996; Lee & Huang, 1996) in which poly(lysine)-condensed DNA was entrapped into folate-targeted cationic liposomes (LPD). They found LPD vectors to be more efficient and less cytotoxic compared to conventional cationic liposomal vectors. Later, they modified LPDs through different cationic liposomes wherein LPDs were used to deliver antisense oligodeoxynucleotide and siRNA (Li & Huang, 2006; Chen et al., 2009; Gao & Huang, 2009). As a cationic polymer, PEI is commonly used for the delivery of genes. The hybrid usage with cationic liposomes provides a promising way to the field of gene transfer. In a study, the combination of PEI and DOTAP-Chol caused more than 10-fold increase in the transfection efficiency and less toxicity in many cells compared with using polymer or liposome alone(Lee et al., 2003). Nearly at the same time, PEI2K-DNA-Dosper complexes showed much more cellular uptake of DNA than PEI2K-DNA complexes and two times higher transfection than Dosper-DNA complexes. It has been hypothesized that Dosper improved the cellular uptake of PEI2K-DNA complexes and PEI2K improved a transfer of the complexes from lysosomes to nucleus (Lampela et al., 2003).

In recent years, chitosan-based carriers have become one of the non-viral vectors that have gained increasing interest as a safer and cost-effective delivery system for gene materials, as they have beneficial qualities such as low toxicity, low immunogenicity, excellent biocompatibility as well as a high positive charge density (Zhang et al., 2007; Mao et al., 2010). Katas et al. (Katas & Alpar, 2006) may be the first group to investigate the use of chitosan to deliver siRNA *in vitro*. Two types of cell lines, CHO K1 and HEK 293 were used to reveal that preparation method of siRNA association to the chitosan played an important role on the silencing effect. Chitosan-TPP nanoparticles with entrapped siRNA were shown to be better vectors as siRNA delivery vehicles compared to chitosan-siRNA complexes possibly due to their high binding capacity and loading efficiency. We have combined chitosans and cationic liposomes to form a ternary lipopolyplex which could facilitate the delivery of genes into cells more efficiently than the utilization of a lipoplex or a polyplex alone (Fig. 9.). The confocal microscopy method proved that exogenous DNA molecules entered the nucleus through the nuclear membrane other than via the NPC. The results explored novel ways based on the hybrid vectors to enhance the pDNA delivery with chitosans and with further suitable intracellular mechanism, allowing the development of nonviral gene delivery and may provide the most exciting solution for hybrid biomaterials design used for beneficial candidates for gene therapy.

4.2 Conjugates of peptides and lipids

One of the most challenge things for gene delivery by cationic liposome method is the toxicity of cationic lipids originated in the cationic nature. The replacement of cationic head groups has been a major trend with other more biocompatible groups, such as peptides, in recent years, as the cationic lipids with quaternary ammonium head groups can become cytotoxic by interacting with critical enzymes such as PKC (Bottega & Epand, 1992). The commonly used headgroups are peptides consisting of amino acids, such as lysine, arginine, histine, ornigine and tryptophan. Therefore, the conjugates of peptides and lipids are much less toxic, whilst keeping the same transfection efficiency (Behr et al., 1989; Ahn et al., 2004).

The first polypeptide cationic liposome prepared by a polycondensation reaction was described by Folda et al. (Folda et al., 1982). A subset of lipitoids with a repeated side chain trimer motif conjugated with dimyristoyl phosphatidyl-ethanolamine (DMPE) mediated DNA were also found to transfer cells with high efficiency (Huang et al., 1998). A compound which contained cholesterol and a dipeptide consisting of glycine and sterically protected arginine has been proved to be suitable for *in vitro* transfection in the presence of 10% sera more efficiently than other cholesterol derivatives (Sochanik et al., 2000). Peptide-based gemini surfactants GS could lead to an increase in levels of gene expression *in vitro* compared to well-established non-viral reagents (McGregor et al., 2001).

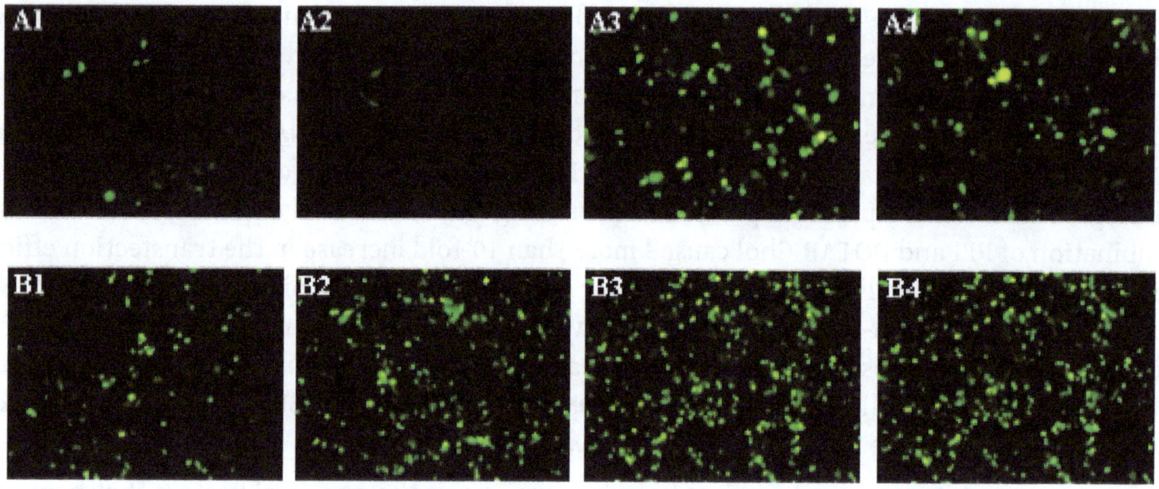

Fig. 9. The role of chitosans on transfection efficiency of cationic liposomes. (A in Hela cells, A1: lipoplex, A2: polyplex, A3: lipopolyplex 1, A4: lipopolyplex 2; B in Hep2 cells, B1: lipoplex, B2: lipopolyplex 1, B3: lipopolyplex 2, B4: lipopolyplex 3)

Obata et al. (Obata et al., 2008) have proved that the lysine- or arginine-type lipids exhibited higher gene expression efficiencies than that of Lipofectamine2000, with COS-7 cells. A series of new lipophilic peptides possessing a cationic tripeptide headgroup were effective non-viral vectors for gene delivery. Then, they also synthesized a series of cationic amino acid-based lipids having a spacer between the cationic head group and hydrophobic moieties and examined the influence of the spacer on a liposome gene delivery system. (Obata et al., 2009). At present, they are investigating the ability of cationic liposomes composed of 1,5-dihexadecyl *N*-arginyl-L-glutamate (Arg-Glu2C16) to carry nucleic acids into neuronal cells. Arg-Glu2C16, as a model cationic amino acid-based lipid, had a high capability as a gene carrier, even for neuronal transfection (Obata et al., 2010). Coles et al. (Coles et al., 2010) has synthesized positively charged peptide-based carriers which could interact with DNA improved by performing isothermal titration calorimetry and particle size and zeta potential experiments. The particle sizes of the carrier/DNA complexes varied over the different charge ratios from 200-800nm. The utilization of lipophilic carriers is a promising approach to improve the bioavailability of gene delivery.

Some peptide head groups could endow additional functions to the lipids, such as membrane-disturbing ability. In a hybrid molecule, the covalent coupling of an amphipathic and membrane-disturbing peptide to a lipid moiety might create a stable and efficient peptide-based gene transfer system. The luciferase activity induced by the dioleoylmelittin/DNA complex was 5-500-fold higher than that induced by a cationic lipid/DNA complex, depending on the cationic lipid and the cell-line (Legendre et al., 1997). Later, a membrane-disrupting peptide derived from the influenza virus was covalently linked to different polymethacrylates using N-succinimidyl 3-(2-pyridyldithio)propionate (SPDP) as a coupling agent to increase the transfection efficiency of polyplexes based on these polymers. *In vitro* transfection and toxicity were tested in COS-7 cells, and these experiments showed that the polyplexes with grafted peptides had a substantially higher transfection activity than the control polyplexes, while the toxicity remained unchanged (Funhoff et al., 2005).

1.3 Conjugates of targeting moieties and lipids

For a number of gene therapy applications, targeted gene delivery systems have attracted great attention due to their potential in directing the therapeutic genes to the target cells, and it may help minimize adverse effects such as cytotoxicity or immune reactions, as well as maximizing the efficacy of the therapeutic response. The targeted delivery of the lipoplexes may be achieved through the addition of targeting moieties (e.g., ligands) into liposomes by direct formulation, with no covalent bond to any lipid (Seol et al., 2000); conjugated to helper lipid (Dauty et al., 2002), or connected directly to the cationic lipids (Kawakami et al., 2000a; Gaucheron et al., 2001c). For targeted lipoplexes, modified with a targeting moiety such as folate (Dauty et al., 2002), galactose(Kawakami et al., 2000a; Gaucheron et al., 2001c), mannose (Kawakami et al., 2000b), antibodies (Duan et al., 2008) and transferrin (Seol et al., 2000; Sakaguchi et al., 2008), the uptake can be receptor mediated and enhanced (Zhang et al., 2010).

Targeting of the folate receptor (FR) had received much attention in recent years, since the folate receptor is a tumor marker over expressed in large numbers of cancer cells, including cancers of the ovary, kidney, uterus, testis, brain, colon, and in addition, folic acid is a relatively small molecule (MW 441 Da), therefore, it has the advantages of being stable and nonimmunogenic compared to monoclonal antibodies (Kane et al., 1986), and still having a relatively high receptor affinity. In a study, the folate moiety was attached to a lipid membrane anchor via a cysteinyl-PEG3400 spacer, which greatly increased specific cellular uptake to FR overexpressing cancer cells in comparison with unmodified cationic liposome and can significantly improve the transfection efficiency of a cationic liposomal formulation (Reddy et al., 2002). Recently, Yoshizawa et al. (Yoshizawa et al., 2008) developed a folate-linked nanoparticle (NP-F), which was composed of cholesteryl-3β- carboxyamidoethylene-N-hydroxyethylamine, Tween 80 and folate-poly(ethylene glycol)- distearoyl-phosphatidylethanolamine conjugate (f-PEG2000-DSPE), and was delivered synthetic siRNA with high transfection efficiency and selectivity into nasopharyngeal tumor KB cells.

The asialoglycoprotein receptor (ASPGR), present at the surface of hepatocytes, could recognize and bind to β-D-galactoside terminated glycoproteins for the targeting of cationic vector-based gene delivery systems (Ashwell & Harford, 1982). Kawakami et al. (Kawakami et al., 1998) have studied liposomal with asialoglycoprotein receptor gene carrier systems for gene delivery to hepatocytes, which was a novel galactosylated cholesterol derivatives, cholesten-5-yloxy-N-(4-((1-imino-2-β-D-thiogalactosylethyl) amino)alkyl) formamide. In human hepatoma cells (HepG2), the liposomes containing this galactolipid showed higher transfection activities than DC-Chol liposomes based on a receptor-mediated mechanism.

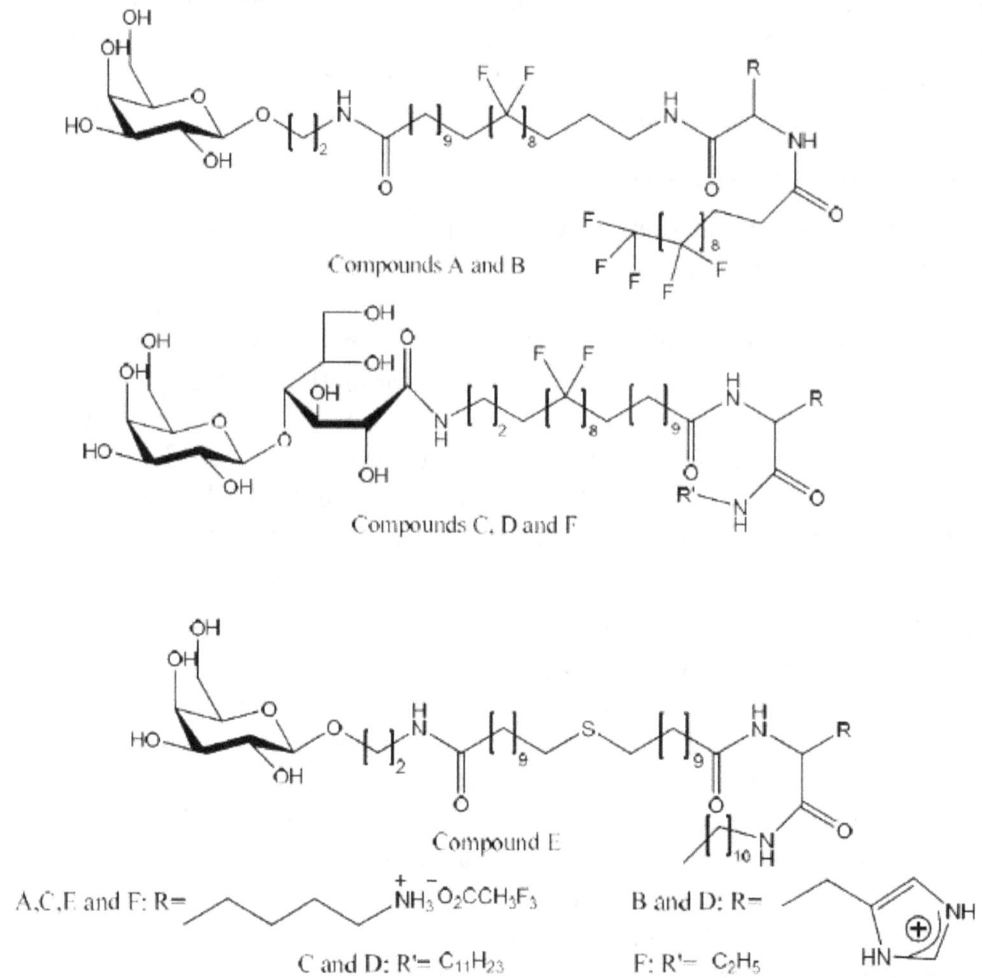

Fig. 10. A set of dissymmetric hemifluorocarbon bolaamphiphiles.

Later, they used galactosylated cationic liposomes to target liver cell asialoglycoprotein receptors *in vivo* (Kawakami et al., 2000a). Many groups also reported a few other glycosylated cationic bolaamphiphiles similar to the compound (Letrou-Bonneval et al., 2008). Brunelle et al. (Brunelle et al., 2009) synthesized a new series of dissymmetric hemifluorocarbon bolaamphiphiles (Fig. 10.), and the dissymmetric functionalization of diiodoperfluorooctane led to bolaamphiphile molecules composed of a partially fluorocarbon core end-capped with a glycoside and an ammonium salt. They found that the incorporation of two fluorinated segments in the molecular structure of the bolaamphiphiles is detrimental for an efficient DNA condensation.

The mannose receptor (MR) is found on the surface of macrophages and dendritic cells can recognize complex carbohydrates that are located on glycoproteins that are a part of many different biological processes, and bind terminal mannoses found on vector. Kawakami et al. (Kawakami et al., 2000b) have developed a novel mannosylated cholesterol derivative, Man- C4-Chol, consisting of modified cationic liposomes with mannose moieties for NPC- selective gene delivery via mannose receptors on NPC. The mannosylated cationic liposomes can deliver pDNA to liver nonparenchymal cells (Kawakami et al., 2000b) and splenic DCs and improve immune activation in DNA vaccines (Hattori et al., 2004).

Since theirs early use for the targeting to erythroblasts (Zhang et al., 2010), targeting proteins such as antibodies or transferrin have become one of the most widely used ligands for targeting of synthetic vectors, and been used in conjunction with cationic lipoplexes (Rao, 2010) to mediate uptake of plasmid DNA and antisense oligonucleotides. Another important class of cell proteins with promise for targeted gene delivery is growth factor receptor, which is a most commonly single pass transmembrane protein with an extracellular ligand binding domain and an intracellular region with enzymatic activity, usually a tyrosine kinase domain, which transmits a growth factor signal from the cell's environment to its interior. It has been used to target polylysine complexes (Cristiano & Roth, 1996), liposomes (Kikuchi, 1996), PEI polyplexes (Cristiano & Roth, 1996), and adenovirus-derived peptides (Medina-Kauwe et al., 2001) to receptor-positive cells.

Targeting provides a generic strategy to improve the specificity of a pharmaceutical formulation independently of the specificity of the drug or gene itself, primarily through a modulation of the carriers' biodistribution, so that a dose differential is created between healthy and diseased tissue.

5. CONCISE SUMMARY

Cationic lipisomes-mediated gene transfer has been demonstrated to be a safe and effective method for gene transfer in gene therapy, and it is expected to be implemented in the near future. Many of the traditional cationic lipids that have been used in gene transfer for a long time are protected by trademarks, including lipofectin, lipofectamine, and transfectam. Nevertheless, these cationic lipids may not be sufficient for clinical trials; conversely, they have certain limitations in a specific application field. Subsequently, numerous novel cationic lipids with chemical structures that satisfy gene therapy specifications have been investigated. The formulations of cationic liposomes should be the next level of development, in addition to the lipid structure. This is because other ingredients, such as heper lipids, could be selected to reduce cytotoxicity and enhance transfection efficacy. Different formulations may result in a variety of lipoplex morphologies that are formed by the combination of cationic liposomes and genes. It demonstrates that the efficacy of transfection can be enhanced through the intentional design of morphologies. Another approach to gene delivery is the hybrid utilization of non-viral vectors that are based on cationic liposomes. The promising combination of cationic lipsomes and polymers, as well as the favorable conjugation between lipids and peptides and between targeting moieties and lipids, has been demonstrated in a significant amount of research.

REFERENCES

Ahmad, A.; Evans, H.M.; Ewert, K.; George, C.X.; Samuel, C.E. & Safinya1, C.R. (2005). New Multivalent Cationic Lipids Reveal Bell Curve for Transfection Efficiency Versus Membrane Charge Density: Lipid–DNA Complexes for Gene Delivery. *J. Gene. Med.*, Vol.7, No.6, pp. 739-748, ISSN 1099-498X

Ahn, C.H.; Chae, S.Y.; Bae, Y.H. & Kim, S.W. (2004). Synthesis of Biodegradable Multi-block Copolymers of Poly(L-lysine) and Poly(ethylene glycol) As a Non-viral Gene Carrier. *J. Controlled Release*, Vol.97, No.3, pp. 567–574, ISSN 0168-3659

Amenitsch, H.; Caracciolo, G.; Foglia, P.; Fuscoletti, V.; Giansanti, P.; Marianecci, C.; Pozzi, D. & Laganà, A. (2011). Existence of Hybrid Structures in Cationic Liposome/DNA Complexes Revealed by Their Interaction with Plasma Proteins. *Colloids Surf., B*, Vol.82, No.1, pp. 141-146, ISSN 0927-7765

Ashwell, G. & Harford, J. (1982). Carbohydrate-specific Receptors of the Liver. *Annu. ReV. Biochem.*, Vol.51, pp. 531–554, ISSN 0066-4154

Behr, J.P.; Demeneix, B.; Loeffler, J.P. & Perez-Mutul, J. (1989). Efficient Gene-transfer into Mammalian Primary Endocrine-cells with Lipopolyamine-coated DNA. *Proc. Natl. Acad. Sci. U.S.A.*, Vol.86, No.18, pp. 6982–6986, ISSN 0027-8424

Bennett, M.J.; Aberle, A.M.; Balasubramaniam, R.P.; Malone, J.G.; Malone, R.W. & Nantz, M.H. (1997). Cationic Lipid-mediated Gene Delivery to Murine Lung: Correlation of Lipid Hydration with in Vivo Transfection Activity. *J. Med. Chem.*, Vol.40, No.25, pp. 4069-4078, ISSN 0022-2623

Bottega, R. & Epand, R.M. (1992). Inhibition of Protein Kinase C by Cationic Amphiphiles. *Biochemistry*, Vol.31, No.37, pp. 9025-9030, ISSN 0006-2960

Brunelle, M.; Polidori, A.; Denoyelle, S.; Fabiano, A.S.; Vuillaume, P.Y.; Laurent- Lewandowski, S. & Pucci, B. (2009). A Structure-activity Investigation of Hemifluorinated Bifunctional Bolaamphiphiles Designed for Gene Delivery. *C. R. Chimie.*, Vol.12, No.1-2, pp. 188-208, ISSN 1631-0748

Byk, G.; Dubertret, C.; Escriou, V.; Frederic, M.; Jaslin, G.; Rangara, R.; Pitard, B.; Crouzet, J.; Wils, P.; Schwartz, B. & Scherman, D. (1998). Synthesis, Activity, and Structure- Activity Relationship Studies of Novel Cationic Lipids for DNA Transfer. *J. Med. Chem.*, Vol.41, No.2, pp. 224-235, ISSN 0022-2623

Byk, G.; Wetzer, B.; Frederic, M.; Dubertret, C.; Pitard, B.; Jaslin, G. & Scherman, D. (2000). Reduction-sensitive Lipopolyamines as a Novel Nonviral Gene Delivery System for Modulated Release of DNA with Improved Transgene Expression. *J. Med. Chem.*, Vol.43, No.23, pp. 4377-4387, ISSN 0022-2623

Carriere, M.; Tranchant, I.; Niore, P.A.; Byk, G.; Mignet, N.; Escriou, V.; Scherman, D. & Herscovici, J. (2002). Optimization of Cationic Lipid Mediated Gene Transfer: Structure-function, Physico-chemical, and Cellular Studies. *J. Liposome. Res.*, Vol.12, No.1-2, pp. 95-106, ISSN 0898-2104

Chen, Y.C.; Sen, J.; Bathula, S.R.; Yang, Q.; Fittipaldi, R. & Huang, L. (2009). Novel Cationic Lipid That Delivers siRNA and Enhances Therapeutic Effect in Lung Cancer Cells. *Mol. Pharm.*, Vol.6, No.3, pp. 696-705, ISSN 1543-8384

Cheng, H.; Zhu, J.L.; Zeng, X.; Jing, Y.; Zhang, X.Z. & Zhuo, R.X. (2009). Targeted Gene Delivery Dediated by Folate-polyethylenimine-block-poly(ethylene glycol) with Receptor Selectivity. *Bioconjugate Chem.*, Vol.20, No.3, pp. 481-487, ISSN 1043-1082

Choi, J.S.; Lee, E.J.; Jang, H.S. & Park, J.S. (2001). New Cationic Liposomes for Gene Transfer into Mammalian Cells with High Efficiency and Low Toxicity. *Bioconjugate Chem.*, Vol.12, No.1, pp. 108-113, ISSN 1043-1082

Coles, D.J.; Esposito, A.; Chuah, H.T. & Toth, I. (2010). The Synthesis and Characterization of Lipophilic Peptide-based Carriers for Gene Delivery.*Tetrahedron*, Vol.66, No.29, pp. 5435-5441, ISSN 0040-4020

Cristiano, R.J. & Roth, J.A. (1996). Epidermal Growth Factor Mediated DNA Delivery into Lung Cancer Cells via the Epidermal Growth Factor Receptor. *Cancer Gene Ther.*, Vol.3, No.1, pp. 4-10, ISSN 0929-1903

Dauty, E.; Remy, J. S.; Zuber, G. & Behr, J.-P. (2002). Intracellular Delivery of Nanometric DNA Particles via the Folate Receptor. *Bioconjugate Chem.*, Vol.13, No.4, pp. 831- 839, ISSN 1043-1082

Donkuru, M.D.; Badea, I.; Wettig, S.; Verrall, R.; Elsabahy, M. & Foldvari, M. (2010). Advancing Nonviral Gene Delivery: Lipid- and Surfactant-based Nanoparticle Design Strategies. *Nanomedicine (Lond)*, Vol.5, No.7, pp. 1103-1127, ISSN 1743-5889

Duan, Y.J.; Zheng, J.N.; Han, S.F.; Wu, Y.; Wang, Y.M.; Li, D.J.; Kong, D.L. & Yu, Y.T. (2008). A Tumor Targeted Gene Vector Modified with G250 Monoclonal Antibody for Gene Therapy. *J. Controlled Release*, Vol.127, No.2, pp. 173-179, ISSN 0168-3659

El-Aneed, A. (2004). An Overview of Current Delivery Systems in Cancer Gene Therapy. *J. Controlled Release*, Vol.94, No.1, pp. 1-14, ISSN 0168-3659

Elouahabi, A. & Ruysschaert, J.M. (2005). Formation and Intracellular Trafficking of Lipoplexes and Polyplexes. *Mol. Ther.*, Vol.11, No.3, pp. 336–347, ISSN 1525-0016

Escriou, V.; Ciolina, C.; Lacroix, F.; Byk, G.; Scherman, D. & Wils, P. (1998). Cationic Lipid- mediated Gene Transfer: Effect of Serum on Cellular Uptake and Intracellular Fate of Lipopolyamine/DNA Complexes. *Biochim. Biophys. Acta.*, Vol.1368, No.2, pp. 276-288, ISSN 0005-2736

Ewert, K.; Ahmad, A.; Evans, H.M. & Safinya, C.R. (2005a). Cationic Lipid-DNA Complexes for Non-viral Gene Therapy: Relating Supramolecular Structures to Cellular Pathways. *Expert Opin. Biol. Ther.*, Vol.5, No.1, pp. 33-53, ISSN 1471-2598

Ewert, K.; Ahmad, A.; Evans, H.M.; Schmidt, H.W. & Safinya, C.R. (2002). Efficient Synthesis and Cell-transfection Properties of a New Nultivalent Cationic Lipid for Nonviral Gene Delivery. *J. Med. Chem.*, Vol.45, No.23, pp. 5023-5029, ISSN 0022-2623

Ewert, K.; Evans, H.M.; Ahmad, A.N.; Slack, L.; Lin, A.J.; Martin-Herranz, A. & Safinya, C.R. (2005b). Lipoplex Structures and Their Distinct Cellular Pathways. *Adv. Genet.*, Vol.53, No.8, pp. 119-155, ISBN 13:978-0-12-374621-4

Ewert, K.; Evans, H.M.; Zidovska, A.; Bouxsein, N.F.; Ahmad, A. & Safinya, C.R. (2006). A Columnar Phase of Dendritic Lipid-based Cationic Liposome-DNA Complexes for Gene Delivery: Hexagonally Ordered Cylindrical Micelles Embedded in A DNA Honeycomb Lattice. *J. Am. Chem. Soc.*, Vol.128, No.12, pp. 3998-4006, ISSN 0002- 7863

Ewert, K.; Slack, N.L.; Ahmad, A.; Evans, H.M.; Lin, A.J.; Samuel, C.E. & Safinya, C.R. (2004). Cationic Lipid-DNA Complexes for Gene Therapy: Understanding the Relationship Between Complex Structure and Gene Delivery Pathways at The Molecular Level. *Curr. Med. Chem.*, Vol.11, No.2, pp. 133-149, ISSN 1568-0118

Fabio, K.; Gaucheron, J.; Giorgio, C.D. & Vierling P. (2003). Novel Galactosylated Polyamine Bolaamphiphiles for Gene Delivery. *Bioconjugate Chem.*, Vol.14, No.2, pp. 358-367, ISSN 1043-1082

Fanning, G.; Amado, R. & Symonds, G.J. (2003). Gene Therapy for HIV/AIDS: The Potential for a New Therapeutic Regimen. *J Gene Med*, Vol.5, No.8, pp. 645-653, ISSN 1521- 2254

Farago, O.; Gronbech-Jensen, N. & Pincus, P. (2006). Mesoscale Computer Modeling of Lipid-DNA Complexes for Gene Therapy. *Phys. Rev. Lett.*, Vol.96, No.4, pp. 018102- 018106, ISSN 0031-9007

Farhood, H.; Serbina, N. & Huang, L. (1995). The Role of Dioleoyl Phosphatidylethanolamine in Cationic Liposome Mediated Gene Transfer. *Biochim. Biophys. Act.*, Vol.1235, No.2, pp. 289-295, ISSN 0005-2736

Felgner, J.H.; Kumar, R.; Sridhar, C.N.; Wheeler, C.J.; Tsai, Y.J.; Border, R.; Ramsey, P.; Martin, M. & Felgner, P.L. (1994). Enhanced Gene Delivery and Mechanism Studies with a Novel Series of Cationic Lipid Formulations. *J. Biol. Chem.*, Vol.269, No.4, pp. 2550-2561, ISSN 0021-9258

Felgner, P.L.; Gadek, T.R.; Holm, M.; Roman, R.; Chan, H.S.; Wenz, M.; Northrop, J.P.; Ringgold, G.M. & Danielson, M. (1987). Lipofection: A Highly Efficient, Lipid- Mediated DNA-transfection Procedure. *Proc. Natl. Acad. Sci. U.S.A.*, Vol.84, No.21, pp. 7413-7417, ISSN 0027-8424

Felgner, P.L. & Ringold, G.M. (1989). Cationic Liposome-mediated Transfection. *Nature*, Vol.11, No.2, pp. 387-388, ISSN 0028-0836

Ferrari, M.E.; Nguyen, C.M.; Zelphati, O.; Tsai, Y. & Felgner, P.L. (1998). Analytical Methods for the Characterization of Cationic Lipid Nucleic Acid Complexes. *Hum. Gene Ther.*, Vol.9, No.3, pp. 341-351, ISSN 1043-0342

Ferrari, M.E.; Rusalov, D.; Enas, J. & Wheeler, C.J. (2002). Synergy between Cationic Lipid and Co-lipid Determines the Macroscopic Structure and Transfection Activity of Lipoplexes. *Nucleic Acids Res.*, Vol.30, No.8, pp. 1808-1816, ISSN 0305-1048

Fletcher, S.; Ahmad, A.; Perouzel, E.; Heron, A.; Miller, A.D. & Jorgensen, M.R. (2006). In Vivo Studies of Dialkynoyl Analogues of DOTAP Demonstrate Improved Gene Transfer Efficiency of Cationic Liposomes in Mouse Lung. *J. Med. Chem.*, Vol.49, No.1, pp. 349-357, ISSN 0022-2623

Floch, V.; Loisel, S.; Guénin, E.; Hervé, A.C.; Yaouanc, J.J.; Clément, J.C.; Férec, C. & Des Abbayes, H. (2000). Cation Substitution in Cationic Phosphonolipids: A New Concept to Improve Transfection Activity and Decrease Cellular Toxicity. *J. Med. Chem.*, Vol.43, No.23, pp. 4617-4628, ISSN 0022-2623

Folda, T.; Gros, L. & Ringsdorf, H. (1982). Polyreactions in Oriented Systems, 29a). Formation of Oriented Polypeptides and Polyamides in Monolayers and Liposomes. *Makromol Chem Rapid Commun*, Vol.3, No.3, pp. 167–174, ISSN 0173-2803

Frederic, M.; Scherman, D. & Byk, G. (2000). Introduction of Cyclic Guanidines into Cationic Lipids for Non-viral Gene Delivery. *Tetrahedron Lett.*, Vol.41, No.5, pp. 675-679, ISSN 0040-4039

Fujiwara, T.; Hasegawa, S.; Hirashima, N.; Nakanishi, M. & Ohwada, T. (2000). Gene Transfection Activities of Amphiphilic Steroid–polyamine Conjugates. *Biochim. Biophys. Acta, Biomembr.*, Vol.1468, No.1-2, pp. 396-402, ISSN 0005-2736

Funhoff, A.M.; van Nostrum, C.F.; Lok, M.C.; Kruijtzer, J.A.; Crommelin, D.J. & Hennink, W.E. (2005). Cationic Polymethacrylates with Covalently Linked Membrane Destabilizing Peptides as Gene Delivery Vectors. *J. Controlled Release*, Vol.101, No.1- 3, pp. 233–246, ISSN 0168-3659

Gao, H. & Hui, K.M. (2001). Synthesis of a Novel Series of Cationic Lipids that Can Act As Efficient Gene Delivery Vehicles Through Systematic Heterocyclic Substitution of Cholesterol Derivatives. *Gene Ther*, Vol.8, No.11, pp. 855-863, ISSN 0969-7128

Gao, K. & Huang, L. (2009). Nonviral Methods for siRNA Delivery. *Mol. Pharm.*, Vol.6, No.3, pp. 651-658, ISSN 1543-8384

Gao, X. & Huang, L. (1991). A Novel Cationic Liposome Reagent for Efficient Transfection of Mammalian Cells. *Biochem. Biophys. Res. Commun.*, Vol.179, No.1, pp. 280-285, ISSN 0006-291X

Gao, X. & Huang, L. (1995). Cationic Liposome-mediated Gene Transfer. *Gene Ther*, Vol.2, No.10, pp. 710-722, ISSN 0969-7128

Gao, X. & Huang, L. (1996). Potentiation of Cationic Liposomemediated Gene Delivery by Polycations. *Biochemistry*, Vol.35, No.3, pp. 1027-1036, ISSN 0006-2960

Gaucheron, J.; Boulanger, C.; Santaella, C.; Sbirrazzuoli, N.; Boussif, O. & Vierling, P. (2001a). In Vitro Cationic Lipid-mediated Gene Delivery with Fluorinated Glycerophosphoethanolamine Helper Lipids. *Bioconjugate Chem.*, Vol.12, No.6, pp. 949-963, ISSN 1043-1082

Gaucheron, J.; Santaella, C. & Vierling, P. (2001b). Highly Fluorinated Lipospermines for Gene Transfer: Synthesis and Evaluation of Their in Vitro Transfection Efficiency. *Bioconjugate Chem.*, Vol.12, No.1, pp. 114-128, ISSN 1043-1082

Gaucheron, J.; Santaella, C. & Vierling, P. (2001c). *In Vitro* Gene Transfer with a Novel Galactosylated Spermine Bolaamphiphile. *Bioconjugate Chem.*, Vol.12, No.4, pp. 569- 575, ISSN 1043-1082

Gaucheron, J.; Santaella, C. & Vierling, P. (2001d). Improved in Vitro Gene Transfer Mediated by Fluorinated Lipoplexes in the Presence of a Bile Salt Surfactant. *J. Med. Chem.*, Vol.3, No.4, pp. 338-344, ISSN 0022-2623

Gaucheron, J.; Wong, T.; Wong, K.F.; Maurer, N. & Cullis, P.R. (2002). Synthesis and Properties of Novel Tetraalkyl Cationic Lipids. *Bioconjugate Chem.*, Vol.13, No.3, pp. 671-675, ISSN 1043-1082

Gershon, H.; Ghirlando, R.; Guttman, S.B. & Minsky, A. (1993). Mode of Formation and Structural Features of DNA-Cationic Liposome Complexes Used for Transfection. *Biochemistry*, Vol.32, No.28, pp. 7143–7151, ISSN 0006-2960

Hattori, Y.; Kawakami, S.; Suzuki, S.; Yamashita, F. & Hashida, M. (2004). Enhancement of Immune Responses by DNA Vaccination Through Targeted Gene Delivery Using Mannosylated Cationic Liposome Formulations Following Intravenous Administration in Mice. *Biochem. Biophys. Res. Commun.*, Vol.317, No.4, pp. 992–999, ISSN 0006-291X

Heinze, M.; Brezesinski, G. & Dobner, B. (2010). LangnerA: Novel Cationic Lipids Based on Malonic Acid Amides Backbone: Transfection Efficacy and Cell Toxicity Properties. *Bioconjugate Chem.*, Vol.21, No. 4, pp. 696-708, ISSN 1043-1802

Heyes, J.A.; Niculescu-Duvaz, D.; Cooper, R.G. & Springer, C.J. (2002). Synthesis of Novel Cationic Lipids: Effect of Structural Modification on the Efficiency of Gene Transfer. *J. Med. Chem.*, Vol.45, No.1, pp. 99–114, ISSN 0022-2623

Huang, C.Y.; Uno, T.; Murphy, J.E.; Lee, S.; Hamer, J.D.; Escobedo, J.A.; Cohen, F.E.; Radhakrishnan, R.; Dwarki, V. & Zuckermann R.N. (1998). Lipitoids-novel Cationic Lipids for Cellular Delivery of Plasmid DNA *in Vitro*. *Chem Biol.*, Vol.5, No.6, pp. 345–354, ISSN 1074-5521

Legendre, J.Y.; Trzeciak, A.; Bohrmann, B.; Deuschle, U.; Kitas, E. & Supersaxo, A. (1997). Dioleoylmelittin as a Novel Serum-Insensitive Reagent for Efficient Transfection of Mammalian Cells. *Bioconjugate Chem.*, Vol.8, No.1, pp. 57–63, ISSN 1043-1082

Ilies, M.A.; Seitz, W.A.; Caproiu, M.T.; Wentz, M.; Garfield, R.E. & Balaban, A.T. (2003). Pyridinium-based Cationic Lipids As Gene-transfer Agents. *Eur. J. Org. Chem.*, Vol.2003, No.14, pp. 2645–2655, ISSN 1099-0690

Jacopin, C.; Hofland, H.; Scherman, D. & Herscovici, J. (2001). Synthesis and Transfecting Properties of a Glycosylated Polycationic DNA Vector. *Bioorg. Med. Chem. Lett.*, Vol.11, No.3, pp. 419-422, ISSN 0960-894X

Kane, M.A.; Elwood, P.C.; Portillo, R.M.; Antony, A.C. & Kolhouse, J.F. (1986). The Interrelationship of the Soluble and Membrane-associated Folate-binding Proteins in Human KB cells. *J. Biol. Chem.*, Vol.261, No.33, pp. 15625–15631, ISSN 0021-9258

Katas, H. & Alpar, H.O. (2006). Development and Characterisation of Chitosan Nanoparticles for siRNA Delivery. *J. Controlled Release*, Vol.115, No.2, pp. 216-225, ISSN 0168-3659

Kato, A.; Tsuji, A.; Yanagisawa, M.; Saeki, D.; Juni, K.; Morimoto, Y. & Yoshikawa, K. (2010). Phase Separation on a Phospholipid Membrane Inducing a Characteristic Localization of DNA Accompanied by Its Structural Transition. *J. Phys. Chem. Lett.*, Vol.1, No. 23, pp. 3391-3395, ISSN 1948-7185

Kawakami, S.; Fumoto, S.; Nishikawa, M.; Yamashita, F. & Hashida, M. (2000a). In Vivo Gene Delivery to the Liver Using Novel Galactosylated Cationic Liposomes. *Pharm. Res.*, Vol.17, No.3, pp. 306–313, ISSN 0724-8741

Kawakami, S.; Sato, A.; Nishikawa, M.; Yamashita, F. & Hashida, M. (2000b). Mannose Receptor-mediated Gene Transfer into Macrophages Using Novel Mannosylated Cationic Liposomes. *Gene Ther*, Vol.7, No.4, pp. 292–299, ISSN 0969-7128

Kawakami, S.; Yamashita, F.; Nishikawa, M.; Takakura, Y. & Hashida, M. (1998). Asialoglycoprotein Receptor-mediated Gene Transfer Using Novel Galactosylated Cationic Liposomes. *Biochem. Biophys. Res. Commun.*, Vol.252, No.1, pp. 78-83, ISSN 0006-291X

Kedika, B. & Patri, S.V. (2011). Design, Synthesis, and in Vitro Transfection Biology of Novel Tocopherol Based Monocationic Lipids: A Structure-Activity Investigation. *J. Med. Chem.*, Vol.54, No.2, pp. 548–561, ISSN 0022-2623

Kennedy, M.T.; Pozharski, E.V.; Rakhmanova, V.A. & MacDonald, R.C. (2000). Factors Governing the Assembly of Cationic Phospholipid-DNA Complexes. *Biophys. J.*, Vol.78, No.3, pp. 1620-1633, ISSN 0006-3495

Kichler, A.; Leborgne, C.; Savage, P.B. & Danos, O. (2005). Cationic Steroid Antibiotics Demonstrate DNA Delivery Properties. *J. Controlled Release*, Vol.107, No.1, pp. 174- 182, ISSN 0168-3659

Kikuchi, A.; Sugaya, S.; Ueda, H.; Tanaka, K.; Aramaki, Y.; Hara, T.; Arima, H.; Tsuchiya, S.

& Fuwa, T. (1996). Efficient Gene Transfer to EGF Receptor Overexpressing Cancer Cells by Means of EGF-labeled Cationic Liposomes. *Biochem. Biophys. Res. Commun.*, Vol.227, No.3, pp. 666–671, ISSN 0006-291X

Kneuer, C.; Ehrhardt, C.; Bakowsky, H.; Kumar, M.N.; Oberle, V.; Lehr, C.M.; Hoekstra, D. & Bakowsky, U. (2006). The Influence of Physicochemical Parameters on the Efficacy of Non-viral DNA Transfection Complexes: a Comparative Study. *J. Nanosci Nanotechnol.*, Vol.6, No.9-10, pp. 2776-2782, ISSN 1550-7033

Koltover, I.; Salditt, T.; Radler, J.O. & Safinya, C.R. (1998). An Inverted Hexagonal Phase of Cationic Liposome-DNA Complexes Related to DNA Release and Delivery. *Science*, Vol.281, No.5373, pp. 78-81, ISSN 0036-8075

Koynova, R. & Tenchov, B. (2010). Cationic Lipids: Nolecular Structure/Transfection Activity Relationships and Interactions with Biomembranes. *Top Curr Chem*, Vol.296, No.2010, pp. 51-93, ISSN 0340-1022

Lampela, P.; Elomaa, M.; Ruponen, M.; Urtti, A.; Mannisto, P.T. & Raasmaja, A. (2003). Different Synergistic Roles of Small Polyethylenimine and Dosper in Gene Delivery. *J. Controlled Release*, Vol.88, No.1, pp. 173-183, ISSN 0168-3659

Lasic, D.D. (1997). Recent Developments in Medical Applications of Liposomes: Sterically Stabilized Liposomes in Cancer Therapy and Gene Delivery *in Vivo*. *J. Controlled Release*, Vol.48, No.2-3, pp. 203-222, ISSN 0168-3659

Leal, C.; Bouxsein, N. F.; Ewert, K.K. & Safinya C.R. (2010). Highly Efficient Gene Silencing Activity of siRNA Embedded in a Nanostructured Gyroid Cubic Lipid Matrix. *J. Am. Chem. Soc.*, Vol.132, No.47, pp. 16841-16847, ISSN 0002-7863

Lee, C.H.; Ni, Y.H.; Chen, C.C.; Chou, C.K. & Chang, F.H. (2003). Synergistic Effect of Polyethylenimine and Cationic Liposomes in Nucleic acid Delivery to Human Cancer Cells. *Biochim. Biophys. Acta, Biomembr.*, Vol.1611, No.1-2, pp. 55-62, ISSN 0005-2736

Lee, E.R.; Marshall, J.; Siegel, C.S.; Jiang, C.; Yew, N.S.; Nichols, M.R.; Nietupski, J.B.; Ziegler, R.J.; Lane, M.B.; Wang, K.X.; Wan, N.C.; Scheule, R.K.; Harris, D.J.; Smith, A.E. & Cheng, S.H. (1996). Detailed Analysis of Structures and Formulations of Cationic Lipids for Efficient Gene Transfer to the Lung. *Hum. Gene Ther.*, Vol.7, No.14, pp. 1701-1717, ISSN 1043-0342

Lee, L.K.; Siapati, E.K.; Jenkins, R.G.; McAnulty, R.J.; Hart, S.L. & Shamlou, P.A. (2003). Biophysical Characterization of an Integrin-targeted Non-viral Vector. *Med. Sci. Monit.*, Vol.9, No.1, pp. 54-61, ISSN 1234-1010

Lee, R.J. & Huang, L. (1996). Folate-targeted, Anionic Liposome-entrapped Polylysine- condensed DNA for Tumor Cell-specific Gene Transfer. *J. Biol. Chem.*, Vol.271, No.14, pp. 8481-8487, ISSN 0021-9258

Letrou-Bonneval, E.; Chèvre, R.; Lambert, O.; Costet, P.; André, C.; Tellier, C. & Pitard, B. (2008). Galactosylated Multimodular Lipoplexes for Specific Gene Transfer into Primary Hepatocytes. *J Gene Med*, Vol.10, No.11, pp. 1198-1209, ISSN 1521-2254

Lensink, M.F.; Lonez, C.; Ruysschaert, J.M. & Vandenbranden, M. (2009). Characterization of the Cationic DiC14-amidine Bilayer by Mixed DMPC/DiC14-amidine Molecular Dynamics Simulations Shows an Interdigitated Nonlamellar Bilayer Phase. *Langmuir*, Vol.25, No.9, pp. 5230-5238, ISSN 0743-7463

Leventis, R. & Silvius, J.R. (1990). Interactions of Mammalian Cells with Lipid Dispersions Containing Novel Metabolizable Cationic Amphiphiles. *Biochim. Biophys. Acta, Biomembr.*, Vol.1023, No.1, pp. 124-132, ISSN 0005-2736

Lian, T. & Ho, R.J. (2003). Design and Characterization of a Novel Lipid-DNA Complex That Resists serum-induced Destabilization. *J. Pharm. Sci.*, Vol.92, No.12, pp. 2373-2385, ISSN 0022-3549

Lin, A.J.; Slack, N.L.; Ahmad, A.C.; George, X.; Samuel, C.E. & Safinya, C.R. (2003). Three- dimensional Imaging of Lipid Gene-carriers: Membrane Charge Density Controls Universal Transfection Behavior in Lamellar Cationic Liposome-DNA Complexes. *Biophys. J.*, Vol.84, No. 5, pp. 3307-3316, ISSN 0006-3495

Litzinger, D. & Huang, L. (1992). Phosphatidylethanolamine Liposomes: Drug Delivery, Gene Transfer and Immunodiagnostic Applications. *Biochim.Biophys.Act.*, Vol.1113, No.2, pp. 201-227, ISSN 0005-2736

Liu, D.L.; Hu, J.J.; Qiao, W.H.; Li, Z.S.; Zhang, S.B.; Cheng, L.B. (2005a). Synthesis of Carbamate-linked Lipids for Gene Delivery. *Bioorg. Med. Chem. Lett.*, Vol.15, No.12, pp. 3147–3150, ISSN 0960-894X

Liu, D.L.; Hu, J.J.; Qiao, W.H.; Li, Z.S.; Zhang, S.B.; Cheng, L.B. (2005b). Synthesis and Characterization of a Series of Carbamate-linked Cationic Lipids for Gene Delivery. *Lipids*, Vol.40, No.8, pp. 839–848, ISSN 0024-4201

Liu, D.L.; Qiao, W.H.; Li, Z.S.; Chen, Y.; Cui, X.; Li, K.; Yu, L.; Yan, K.; Zhu, L.; Guo, Y. & Cheng, L.B. (2008). Structure–function Relationship Research of Glycerol Backbone- based Cationic Lipids for Gene Delivery. *Chem. Biol. Drug Des.*, Vol.71, No.4, pp. 336–344, ISSN 1747-0277

Liu, F. & Huang, L. (2002). Development of Non-viral Vectors for Systemic Gene Delivery. *J. Controlled Release*, Vol.78, No.1-3, pp. 259-266, ISSN 0168-3659

Li, S. & Huang, L. (1997). In Vivo Gene Transfer via Intravenous Administration of Cationic Lipid-protamine-DNA (LPD) Complexes. *Gene Ther*, Vol.4, No.9, pp. 891-900, ISSN 0969-7128

Li, S.D. & Huang, L. (2006). Targeted Delivery of Antisense Oligodeoxynucleotide and Small Interference RNA into Lung Cancer Cells. *Mol. Pharm.*, Vol.3, No.5, pp. 579–588, ISSN 1543-8384

Love, K.T.; Mahon, K.P.; Levins, C.G.; Whitehead, K.A.; Querbes, W.; Dorkin, J.R.; Qin, J. & Anderson, D.G. (2010). Lipid-like Materials for Low-dose, in Vivo Gene Silencing. *Proc. Natl. Acad. Sci. U.S.A.*, Vol.107, No.5, pp. 1864–1869, ISSN 0027-8424

Mao, S.; Sun, W. & Kissel, T. (2010). Chitosan-based Formulations for Delivery of DNA and siRNA. *Adv Drug Deliv Rev.*, Vol.62, No.1, pp. 12-27, ISSN 0169-409X

Marchini, C.; Pozzi, D.; Montani, M.; Alfonsi, C.; Amici, A.; Amenitsch, H.; Sanctis, S.C.D. & Caracciolo, G. (2010). Tailoring Lipoplex Composition to the Lipid Composition of Plasma Membrane: A Trojan Horse for Cell Entry. *Langmuir*, Vol.26, No.17, pp. 13867-13873, ISSN 0743-7463

Martin-Rendon, E. & Blake, D.J. (2003). Protein Glycosylation in Disease: New Insights into the Congenital Muscular Dystrophies. *Trends Pharmacol Sci*, Vol.24, No.4, pp. 178- 183, ISSN 0165-6147

McGregor, C.; Perrin, C.; Monck, M.; Camilleri, P. & Kirby, A.J. (2001). Rational Approaches to the Design of Cationic Gemini Surfactants for Gene Delivery. *J. Am. Chem. Soc.*, Vol.123, No.26, pp. 6215-6220, ISSN 0002-7863

Medina-Kauwe, L.K.; Maguire, M.; Kasahara, N. & Kedes, L. (2001). Nonviral Gene Delivery to Human Breast Cancer Cells by Targeted Ad5 Penton Proteins. *Gene Ther*, Vol.8, No.23, pp. 1753–1761, ISSN 0969-7128

Medvedeva, D.A.; Maslov, M.A.; Serikov, R.N.; Morozova, N.G.; Serebrenikova, G.A.; Sheglov, D.V.; Latyshev, A.V.; Vlassov, V.V.; Zenkova, M.A. (2009). Novel Cholesterol-Based Cationic Lipids for Gene Delivery. *J. Med. Chem.*, Vol.52, No.21, pp. 6558-6568, ISSN 0022-2623

Niculescu-Duvaz, D.; Heyes, J. & Springer, C. J. (2003). Structure-activity Relationship in Cationic Lipid Mediated Gene Transfection. *Curr. Med. Chem.*, Vol.10, No.14, pp. 1233-1261, ISSN 0929-8673

Obata, Y.; Ciofani, G.; Raffa, V.; Cuschieri, A. & Takeoka. S. (2010). Evaluation of Cationic Liposomes Composed of an Amino Acid-based Lipid for Neuronal Transfection. *Nanomed: Nanotech, Biol. Med.*, Vol.6, No.1, pp. 70-77, ISSN 1549-9634

Obata, Y.; Saito, S.; Takeda, N. & Takeoka, S. (2009). Plasmid DNA-encapsulating Liposomes: Effect of a Spacer between the Cationic Head Group and Hydrophobic Moieties of the Lipids on Gene Expression Efficiency. *Biochim. Biophys. Acta, Biomembr.*, Vol.1788, No.5, pp. 1148-1158, ISSN 0005-2736

Obata, Y.; Suzuki, D. & Takeoka, S. (2008). Evaluation of Cationic Assemblies Constructed with Amino Acid Based Lipids for Plasmid DNA Delivery. *Bioconjugate Chem.*, Vol.19, No.5, pp. 1055–1063, ISSN 1043-1082

O'Hern, C.S. & Lubensky, T.C. (1998). Sliding Columnar Phase of DNA-Lipid Complexes. *Phys. Rev. Lett.*, Vol.80, No.19, pp. 4345-4348, ISSN 0031-9007

Randazzo, R.A.S.; Bucki , R.; Janmey, P.A. & Diamond, S.L. (2009). A Series of Cationic Sterol Lipids with Gene Transfer and Bactericidal Activity. *Bioorg Med Chem.*, Vol.17, No.9, pp. 3257-3265, ISSN 0968-0896

Rao, N.M. (2010). Cationic Lipid-mediated Nucleic Acid Delivery: Beyond Being Cationic. *Chem* Phys Lipids, Vol.163, No.3, pp. 245-252, ISSN 0009-3084

Reddy, J.A.; Abburi, C.; Hofland. H.; Howard, S.J.; Vlahov, I.; Wils, P. & Leamon C.P. (2002). Folate-targeted, Cationic Liposome-mediated Gene Transfer into Disseminated Peritoneal Tumors. *Gene Ther*, Vol.9, No.22, pp. 1542–1550, ISSN 0969-7128

Pedroso de Lima, M.C.; Neves, S.; Filipe, A; Düzgünes, N. & Simoes, S. (2003). Cationic Liposomes for Gene Delivery: From Biophysics to Biological Applications. *Curr. Med. Chem.*, Vol.10, No.14, pp. 1221-1231, ISSN 0929-8673

Pitard, B.; Aguerre, O.; Airiau, M.; Lachagès, A.M.; Boukhnikachvili, T.; Byk, G.; Dubertret, C.; Herviou, C.; Scherman, D.; Mayaux, J.F. & Crouzet, J. (1997). Virus-sized Self- assembling Lamellar Complexes between Plasmid DNA and Cationic Micelles Promote Gene Transfer. *Proc. Natl. Acad. Sci. U. S. A.*, Vol.94, No.26, pp. 14412- 14417, ISSN 0027-8424

Rejman, J.; Oberle, V.; & Zuhorn, I.S. (2004). Size-dependent Internalization of Particles via the Pathways of Clathrin-and Caveolae-mediated Endocytosis. *Biochem. J.*, Vol.377, No.1, pp. 159-169, ISSN 0264-6021

Ren, T.; Zhang, G.; Liu, F. & Liu, D. (2000). Synthesis and Evaluation of Vitamin D-based Cationic Lipids for Gene Delivery in Vitro. *Bioorg. Med. Chem. Lett.*, Vol.10, No.9, pp. 891-894, ISSN 0960-894X

Ren, T.; Zhang, G.S. & Liu, D.X. (2001). Synthesis of Bifunctional Cationic Compound for Gene Delivery. *Tetrahedron Lett.*, Vol.42, No.6, pp. 1007–1010, ISSN 0040-4039

Ross, P.C. & Hui, S.W. (1999). Lipoplex Size Is a Major Determinant of in Vitro Lipofection Efficiency. *Gene. Ther.*, Vol.6, No.4, pp. 651-659, ISSN 0969-7128

Ruponen, M.; Ylä-Herttuala, S. & Urtti, A. (1999). Interactions of Polymeric and Liposomal Gene Delivery Systems with Extracellular Glycosaminoglycans: Physicochemical and Transfection Studies. *Biochim. Biophys. Acta, Biomembr.*, Vol.1415, No.2, pp. 331- 341, ISSN 0005-2736

Safinya, C.R. (2001). Structures of Lipid–DNA Complexes: Supramolecular Assembly and Gene Delivery. *Curr. Opin. Struct. Biol.* Vol.11, No.4, pp. 440- 448 ISSN 0959-440X

Sakaguchi, N.; Kojima, C.; Harada, A.; Koiwai, K.; Shimizu, K.; Emi, N. & Kono, K. (2008). Generation of Highly Potent Nonviral Gene Vectors by Complexation of Lipoplexes and Transferrin-bearing Fusogenic Polymer-modified Liposomes in Aqueous Glucose Solution. *Biomaterials*, Vol.29, No.9, pp. 1262-1272, ISSN 0142-9612

Scarzello, M.J.; Smisterova, A.; Wagenaar, M.C.; Stuart, D.; Hoekstra, J.B. & Engberts, R. (2005). Sunfish Cationic Amphiphiles: Toward an Adaptative Lipoplex Morphology. *J. Am. Chem. Soc.* Vol.127, No.29, pp. 10420-10429, ISSN 0002-7863

Seol, J.G., Heo, D.S., Kim, H.K., Yoon, J.-H., Choi, B.I., Lee, H.-S., Kim, N.K. & Kim, C.Y. (2000). Selective Gene Expression in Hepatic Tumor with Trans-arterial Delivery of DNA/Liposome/Transferrin Complex. *In Vivo*, Vol.14, No.3, pp. 513–518, ISSN 0258-851X

Sochanik, A.; Kaida, I.; Mitrus, I.; Rajca, A. & Szala, S. (2000). A New Cholesterol Derivative Suitable for Transfecting Certain Type of Cells in the Presence of 10% Serum. *Cancer Gene Ther.*, Vol.7, No.4, pp. 513–520, ISSN 0929-1903

Sternberg, B.; Sorgib, F.L. & Huang, L. (1994). New Structures in Complex Formation between DNA and Cationic Liposomes Visualized by Freeze-fracture Electron Microscopy. *FEBS. Lett.*, Vol.356, No.2-3, pp. 361-366, ISSN 0014-5793

Sternberg, B.; Hong, K.; Zheng, W. & Papahadjopoulos, D. (1998). Ultrastructural Characterization of Cationic Liposome-DNA Complexes Showing Enhanced Stability in Serum and High Transfection Activity in Vivo. *Biochim. Biophys. Act.*, Vol.1375, No.1-2, pp. 23-35, ISSN 0005-2736

Tang, F. & Hughes, J.A. (1999a). Synthesis of a Single-tailed Cationic Lipid and Investigation of Its Transfection. *J. Controlled Release*, Vol.62, No.3, pp. 345-358, ISSN 0168-3659

Tang, F. & Hughes, J.A. (1999b). Use of Dithiodiglycolic Acid As a Tether for Cationic Lipids Decreases the Cytotoxicity and Increases Transgene Expression of Plasmid DNA in Vitro. *Bioconjugate Chem.*, Vol.10, No.5, pp. 791–796, ISSN 1043-1082

Turek, J.; Dubertret, C.; Jaslin, G.; Antonakis, K.; Scherman, D. & Pitard, B. (2000). Formulations Which Increase the Size of Lipoplexes Prevent Serum-associated Inhibition of Transfection. *J. Gene. Med.*, Vol.2, No.1, pp. 32-40, ISSN 1099-498X

Wasungu, L. & Hoekstra, D. (2006). Cationic Lipids, Lipoplexes and Intracellular Delivery of Genes. *J. Controlled Release,* Vol.116, No.2, pp. 255-264, ISSN 0168-3659

Williams, P.D.; Ranjzad, P.; Kakar, S.J. & Kingston, P.A. (2010). Development of Viral Vectors for Use in Cardiovascular Gene Therapy. *Viruses*, Vol.2, No.2, pp. 334-371, ISSN 1999-4915

Xu, Y.; Hui, S.W.; Frederik, P. & Szoka, F.C.J. (1999). Physicochemical Characterization and Purification of Cationic Lipoplexes. *Biophys. J.*, Vol.77, No.1, pp. 341-353, ISSN 0006- 3495

Yingyongnarongkul, B.E.; Howarth, M.; Elliott, T. & Bradley, M. (2004). Solid Phase Synthesis of 89 Polyamine-based Cationic Lipids for DNA Delivery to Mammalian Cells. *Chem. Eur. J.*, Vol.10, No.2, pp. 463–473, ISSN 1521-3765

Yoshizawa, T.; Hattori, Y.; Hakoshima, M.; Koga, K. & Maitani, Y. (2008). Folate-linked Lipid-based Nanoparticles for Synthetic siRNA Delivery in KB Tumor xenografts. *Eur J Pharm Biopharm.*, Vol.70, No.3, pp. 718-725, ISSN 0939-6411

Zuhorn, I.S.; Bakowsky, U.; Polushkin, E.; Visser, W.H.; Stuart, M.C.; Engberts, J.B. & Hoekstra, D. (2005). Nonbilayer Phase of Lipoplex-membrane Mixture Determines Endosomal Escape of Genetic Cargo and Transfection Efficiency. *Mol. Ther.*, Vol.11, No.5, pp. 801-810, ISSN 1525-0016

Zhang, Z.; Yang, C.; Duan, Y.; Wang, Y.; Liu, J.; Wang, L. & Kong, D. (2010). Poly(ethylene glycol) Analogs Grafted with Low Molecular Weight Poly(ethylene imine) As Non- viral Gene Vectors. *Acta Biomater.*, Vol.6, No.7, pp. 2650-2657, ISSN 1742-7061

Zhang, S.B.; Zhao, B.D.; Jiang, H.M.; Wang, B. & Ma, B.C. (2007). Cationic Lipids and Polymers Mediated Vectors for Delivery of siRNA. *J. Controlled Release*, Vol.123, No.1, pp. 1-10, ISSN 0168-3659

Zhang, S.B.; Zhao, Y.N.; Zhao, B.D. & Wang, B. (2010). Hybrids of Nonviral Vectors for Gene Delivery. *Bioconjugate Chem.*, Vol.21, No.6, pp. 1003-1009, ISSN 1043-1082

Zhdanov, R.I.; Podobed, O.V. & Vlassov, V.V. (2002). Cationic Lipid–DNA Complexes- lipoplexes-for Gene Transfer and Therapy. *Bioelectrochemistry.* Vol.58, No.1, pp. 53- 64, ISSN 1567-5394

Zhi, D.F.; Zhang, S.B.; Wang, B., Zhao, Y.N., Yang B.L. & Yu S.J. (2010). Transfection Efficiency of Cationic Lipids with Different Hydrophobic Domains in Gene Delivery. *Bioconjugate Chem.*, Vol.21, No.4, pp. 563-577, ISSN 1043-1082

14 Chapter

DNA TRANSFECTION AND NEUTRAL LIPOSOMES

1. INTRODUCTION: AN OVERVIEW

Today, one of the most extensive disciplines of chemical, biological, and medical research is non-viral gene transfer vectors for human gene therapy (HGT) applications. The future of a safe and efficient gene therapy will be contingent upon the development of appropriate synthetic vectors of genetic material, rather than on viruses, according to certain authors (Kostaleros & Miller, 2005).

The rationale for this preference is that viruses, despite their high transfection efficiency of genetic material, may be subject to significant drawbacks, including a high cost of transferring systems preparation, immune response (Marshall, 2000), and potential oncogenic activity (Hacein-Bey-Abina et al., 2003). In contrast, synthetic vectors possess numerous potential benefits, including the absence of immunogenicity and oncogenicity, the absence of restrictions on the size of nucleic acids that can be transported within cells (Harrington et al., 1997; Roush, 1997; Willard, 2000), and the simplicity and affordability of the preparation process. However, the low efficiency of the transfections in vivo is an inconvenient problem that accompanies their use. The cationic liposomes (CLs) are the synthetic vectors that have been the most extensively studied by researchers.

This was the case after Felgner and colleagues (Felgner et al., 1997) described the synthesis of the first cation lipid, N-[1- (2,3-dioleoyloxy)propyl]-N,N,N-trimethylammonium chloride (DOTMA), and demonstrated its ability to bind DNA and transfect it in both in vitro and *in vivo* experiments. Other cationic lipids, including the commercially widely used N-[1- (2,3-dioleoyloxy)propyl]-N,N,N-trimethylammonium methylsulfate (DOTAP), 1,2-dioleoyl- 3-dimethylammonium-propane (DODAP), dimethyldioctadecylammonium bromide (DDAB), and 3β-[N-(N',N'-dimethylaminoethane)-carbamoyl] cholesterol hydrochloride (DC- Chol) were synthesized in the past and are still being synthesized. Several years following the discovery of Felgner, scientists also developed an interest in cationic polymers (Wu, G.Y. & Wu; C.H., 1987). These polymers gained popularity following the discovery of polyethylenimines (Boussif, 1995) and continue to be a subject of great interest. The interest in both classes of cationic compounds is derived from the formation of stable complexes with DNA, known as lipoplexes and polyplexes, respectively. These complexes are formed through an electrostatic interaction between the cationic head of lipids and the negative phosphates of DNA. The significant milestone of 20% of the ongoing clinical trials being conducted with synthetic vectors has been achieved as a result of the substantial amount of data and experiments reported in the literature with cationic vectors, as well as some encouraging results in in vivo transfection experiments (Edelstein et al., 2004).

Nevertheless, the objective of achieving a higher efficiency remains unresolved. The long-term objective of research in this field, as recently asserted by Safinya et al. (2006), is to establish a general fundamental theory that can facilitate the design and implementation of specialized vector synthesis to achieve the utmost efficiency in a variety of *in vivo* applications. Therefore, the prevailing view is that additional research and development will be necessary before the efficiency of this technology can be compared to that of viral vectors. In reality, the experience has demonstrated that cationic carriers also have some significant drawbacks that affect their transfection efficiency in a more or less negative manner. These drawbacks include inherent cytotoxicity (Filion & Phillips, 1998; Lv et al., 2006) that causes negative effects on cells, such as the inhibition of protein kinase C (PKC) and the limited stability of their complexes with plasmid DNA in serum (Foradada et al., 2000). These drawbacks are the reason for the current restriction of a generalized and extensive use.

The concept of employing neutral liposomes as DNA carriers appears to be intriguing and has the potential to provide promising results in this context. It is widely recognized that neutral liposomes are generally nontoxic (Koiv et al., 1995) and relatively stable in serum (Tardi et al., 1996), which renders them potentially intriguing gene transfer vectors. Neutral liposomes (NLs) have not yet garnered widespread attention in the context of the HGT, despite the fact that the researchers' initial interest in DNA entrapment was redirected to neutral liposomes (Budker et al., 1978). This is despite the strategic characteristics of NLs. There are two likely causes for this situation: the first, and more significant, is the absence of positive charge, which renders it virtually impossible to establish a stable interaction between NLs and DNA. The second is the significant leading role that the cationic carriers have assumed, which has polarized the researchers' interest and pushed aside other alternatives.

This chapter will address the two distinct trajectories that NLs follow, which are determined by the distinct functions they perform in HGT applications. The former will address their function as DNA transfection aids when combined with cationic liposomes, while the latter will discuss their success and potential as autonomous and independent DNA carriers. Despite the fact that the literature reviewed thus far is not particularly extensive, it is possible to anticipate intriguing opportunities for NLs as prospective synthetic vectors of genetic material. It is possible that they will be viewed as a viable alternative to cationic vectors in the near future, rather than a provocative challenge. This forecast is substantiated by the findings of numerous studies. It is obvious that further investigation is required to establish a framework that is suitable for the proper treatment of the numerous facets of the transfection process and to identify more effective experimental conditions that will ensure high transfection efficiency, particularly *in vivo*. The numerous studies that have been conducted to date and the vast amount of data that has been accumulated in the field of cationic liposomes will assist in the development of this framework and the exploration of the various aspects of the specific transfection with NLs.

2. NEUTRAL LIPIDS AS HELPERS OF CATIONIC LIPOSOMES IN DNA TRANSFECTION

Since the discovery of the cationic liposomes as DNA carriers for gene transfer applications it has been clear that higher efficiencies could be obtained by adding a neutral lipid to the lipoplexes with the role of transfection helper. The approach followed in this section is not to evaluate all the issues related to DNA transfection with lipoplexes mediated by neutral lipids, but rather discuss the specific role of neutral lipids in determining the lipoplex transfection efficiency. The ultimate purpose is to highlight the factors that cause such increase of efficiency and check how they may help in designing the best experimental conditions for the transfections procedures with independent NLs.

Cationic systems for gene therapy are generally prepared by mixing a cationic lipid with DNA and a neutral helper co-lipid; such non toxic helper induces the dual result to reduce the amount of the toxic cationic component and, more important, to alter the physical properties of the delivery vehicle, in a way that favours some of the most complex steps of the whole mechanism of transfection: the consequence is that both actions affect the quality of the transfection. The most widely used neutral helpers in DNA transfection experiments are the 1,2-dioleoyl-*sn*-glycero-3-phosphocholine (DOPC) and the 1,2-dioleoyl-*sn*-glycero-3- phosphoethanolamine (DOPE), sometimes in combination with cholesterol: the main difference between the two phospholipids is that DOPC induces a lamellar LCα phase in the lipoplexes, while DOPE induces an inverted hexagonal C HII phase. Earlier studies led to propose that DOPE is responsible for more efficient transfections because its hexagonal phase is able to fuse readily with anionic vesicles (Koltover et al., 1998) and destabilize the bilayer membranes, making easier the DNA escape from the endosomes, once the lipoplex has entered the cells. Therefore it was believed that DOPE was more useful than DOPC to realize the most efficient transfections. In support of this claim it was also suggested (Mui et al., 2000) that the inverted hexagonal phase promoted by DOPE has a higher ability to disrupt the membrane integrity than the lamellar one induced by DOPC. It turned soon evident that the situation was not so simple and the interpretation was not unambiguous: indeed, if it is verified that the DNA complex with cationic DOTAP, in which DOPE is present as helper in an amount of 70%, transfects better than the corresponding complex including the same amount of DOPC (Koltover et al., 1998), lamellar LCα complexes, showing similar transfection efficiency, were also synthesized (Lin et al., 2003). More generally, many literature data show the difficulty in finding a general correlation between vector formulations and transfection efficiency.

Trying to explain these apparently contradictory results, some researchers have recently suggested the opportunity to consider the possibility that complexes can bear a structural modification within the phase of the interaction of the systems DNA-carrier with each individual cell, the transition LCα to CHII phase being an example among others (Safinya, 2001). Starting from the observation that a large number of efficient complexes are assembled in the LCα phase, it was suggested (Caracciolo & Caminiti, 2005) that perhaps a compelling correlation between the structure of a complex and its transfection efficiency does not simply exist and that the lower transfection efficiency of the synthetic carriers of DNA with respect to virus depends on a poor understanding of the supramolecular structures of the complexes, on the mechanism of their interaction with cells and of the release of DNA within the nucleus. If these statements are true, and there are reasons to confirm them, a more exhaustive knowledge of the subject is particularly important also with reference to the use of NLs as independent carriers of DNA, because it is likely that similar problems will arise and the need to overcome them will be even more important. Of course, understanding the mechanism is crucial for any successful design of a non viral gene delivery, whatever path has been chosen. Besides some early studies, that assumed a fusion between liposomes and cell membranes as the initial step of the process, it is today recognized that the uptake into an endocytic pathway is required for fusion to occur (Wrobel & Collins, 1995) and the whole aspect has been the object of deep attention (Liu & Huang, 2002) leading to suggest some main steps, namely non specific interaction with the cell surface, endocytosis into endocytic vesicles, trafficking and release of the DNA from endosomal compartment, nuclear uptake and transgene expression.

Special attention was also devoted to the intracellular trafficking of cationic vectors and the role of neutral helpers (Elouahabi & Ruysschaert, 2005): six steps of the whole transfection process were identified and analyzed, namely (i) interaction between vectors and nucleic acids with formation of complexes, (ii) binding of a complex to the cell, (iii) effect of serum, uptake of the complex by the cell, (v) escape from endosomes and dissociation of the complex and finally (vi) nuclear entry of DNA.

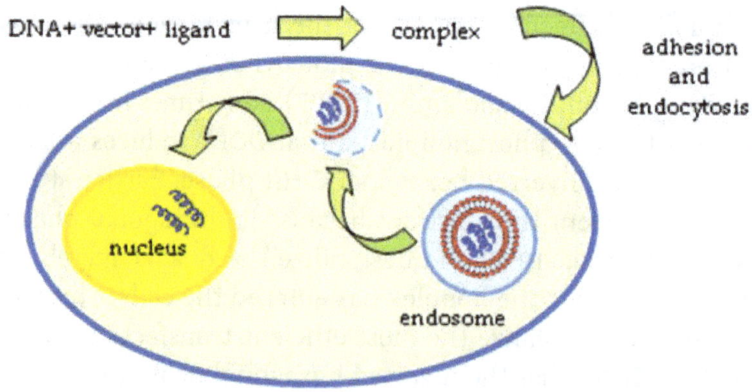

Fig. 1. A schematic representation of the steps involved in the liposome mediated DNA transfection.

We will consider here only the aspects likely to have significant involvements on the transfection process operated by NLs as unique carriers of genetic material. Neutral helpers play a not marginal role in the formation of the complexes: meaningful, though in some way contradictory, is the role of the helpers in determining the effect of serum on the uptake of lipoplexes, an issue that has implications in the *in vivo* transfections. No significant inhibition of serum was observed in transfecting COS-7 cells with cationic pyridinium- derived lipids (SAINT) in DOPE/DNA lipoplexes (Zuhorn et al., 2002), while a strong inhibitory effect operates in lipoplexes obtained by polycationic lipids like DOGS.

CATIONIC LIPIDS

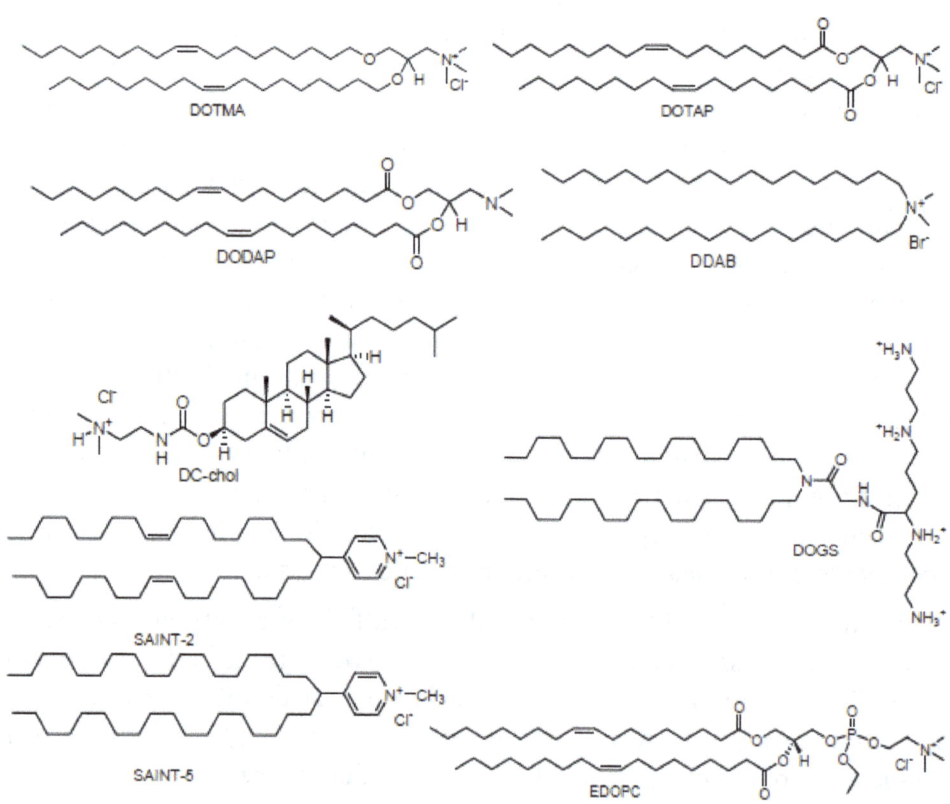

However this negative effect may be avoided if one operates at slight alkaline pH that favours a lamellar organisation of the lipoplexes (Boukhnikachvili et al., 1997). Likewise DOTAP/DOPE and DC-Chol/

DOPE prepared at high +/- charge ratio are not sensitive to the inhibitory effect of serum and indeed, at some ratios, are even more efficient (Yang & Huang, 1997). A different perspective on the role of helpers is offered by other researchers: it was observed (Fasbender & al., 1997) that when DOPE is incorporated into the complexes of DNA with three different cationics, its effect on the gene expression in COS-1 cells is different, depending on each cationic lipid. When the cationic lipids and DOPE were formulated separately and then complexed with DNA, no difference in activity was observed over that obtained with cationic lipids alone. Finally (Felgner et al., 1994) unsaturated PE co-lipids enhance lipoplexes activity while saturated PE and PC have no enhancing effect or even have an inhibitory effect.

The escape of DNA from the endosomes is strictly depending on the nature of the neutral co-lipid: it was found that the fusion with endosomes is likely the way for the release of DNA into the cytoplasm. This mechanism was supported by the finding that efficient transfections require the fusogenic lipid DOPE, which is able to promote a transition from bilayers to hexagonal structures, the latter being known to catalyze the fusion process (Koltover et al., 1998; Mok & Cullis, 1997). The ability of liposomes to fuse with endosomal membranes was also proved by several studies (Koltover et al., 1998; Farhood et al., 1995; Mok & Cullis, 1997) and some evidence was found that the helper lipids can adapt their actions depending on the cationic lipid and the target cells (Fasbender et al., 1997).

In order to better clear up the situation, an interesting study was made a few years ago (Zuhorn et al., 2005). It provides deeper insight into the involvement of helper lipids in the liposomes mediated gene delivery. Two different helpers, DOPE, which has a propensity to adopt an inverted hexagonal phase, and the lamellar phase forming dipalmitoylphosphatidylethanolamine (DPPE), have been compared as neutral co-lipids in lipoplexes formed with SAINT-2 and plasmid DNA, with the specific aim of studying the endosomal escape of the genetic cargo in the cytosol for transport to the nucleus. As usual, it was found that the helper determines the *in vitro* transfection efficiency (COS-7 cells were used), DPPE inducing a significantly lower efficiency (≅ 25% of cell transfected) than DOPE (≅ 75%), despite an equal interaction of both SAINT-2/DNA/DOPE and SAINT- 2/DNA/DPPE with cells. Assuming that the translocation of the nucleic acids through the endosomal membrane is the crucial step of the overall process, a mimic membrane consisting of phosphatidylserine (PS): phosphatidylcholine (PC): phosphatidylethanolamine (PE) anionic vesicle was used to simulate this step. Without helper lipids, a limited fraction of DNA was released from SAINT-2 lipoplex and no effect was promoted by the inclusion of DOPC. On the contrary, inclusion of DOPE significantly enhanced the amount of DNA released and, more interesting, a comparable effect was induced by DPPE (40% release for DPPE versus 50% for DOPE). This result can find an explanation since the *x*-ray diffraction analysis after incubation of SAINT-2/DPPE with the anionic PS:PC:PE (1:1:2) revealed the presence of a mixed lamellar-hexagonal phase. The lower efficiency of the DPPE containing complex is consistent with this partial transition from the lamellar to the hexagonal phase. These results confirm that the limiting step in the overall transfection pathway depends on the level of DNA translocation through the endosomal membrane. The results obtained in the transfections with two different lipoplexes containing SAINT with different tails, namely SAINT-2 (C18:1) and SAINT-5 (C18:0) lead (Zuhorn et al., 2002) to analogous conclusions. Both amphiphiles may make transfection and DOPE strongly promotes the SAINT-2 mediated one, but not the SAINT-5. The relatively rigid SAINT-5 membrane forms structurally deformed lipoplexes hampering the plasmid translocation through endosomal and/or nuclear membranes.

What it has been said so far makes clear that there are still many aspects concerning the DNA transfection process that require a better investigation: among them, the need of a comprehensive knowledge of what really happens when a lipid-DNA complex interacts with a cell, an issue extremely important also in the neutral lipid mediated DNA transfections. In this connection a new perspective has gained recently ground: it identifies as one of the most critical factors of the transfection process the evolution of the structure of the lipoplexes that occurs when they interact with cells. That means to introduce the idea that the differences in the transfection efficiency, often observed and not always unambiguously interpreted, may depend on each particular cellular variety and emphasize the importance to consider both lipid composition of lipoplexes and target membranes (Koynova et al., 2005). Ancestors of this new perspective are some studies that demonstrated the ability of anionic lipids to promote the release of DNA from lipoplexes, by neutralizing their positive charge (Szoka et al. 1996; Zelphati & Szoka, 1996; McDonald et al., 1999). It was observed that the DNA release from complexes with the cationic lipids o-ethyldioleoylphosphatidylcholinium (EDOPC) or DOTAP, after mixing them with some different negatively charged lipids, depends on both lipoplexes and negative lipids. Most significant, the transfection efficiencies of DNA complexes with two very similar cationic phospholipids, bearing only a minimal structural difference in one of the two hydrocarbon tails, the carbon-carbon double bond bearing oleoyldecanoyl-ethylphosphatidylcholine (C18:1/C10-EPC) and the completely saturated stearoyldecanoyl-ethylphosphatidylcholine (C18:0/C10-EPC) were compared (Koynova et al., 2006). The former complex shows a 50-fold higher transfection efficiency than the latter in human umbilical artery endothelial cells. A reasonable explanation of this different behaviour lies in the great difference of these lipids in the phase evolution found in mixing with biomembrane mimicking lipid mixtures (DOPC/DOPE/DOPS/chol). The C18:1 lipoplex underwent a transition to the fusogenic non lamellar cubic phase, whereas the C18:0 did not. All these new perspectives must be seriously considered by all aiming at studying the DNA transfection with NLs, since an analogous behaviour will be probably characterize those experimental setups.

Before ending these considerations on the role of the helper co-lipids in the processes of non viral DNA transfection, it is advisable to say something about a neglected aspect of the topic. As a matter of fact it is surprising that the continuous growing of number and features of cationic lipids, in the search for most suitable vectors of genetic material, no analogous interest has been reserved, for many years, for new co-lipids. The idea that improved transfections could be realized also with the aid of new and more appropriate helpers has developed only in the last ten years. After the discovery that high transfection efficiency could be obtained with fluorinated double chain lipospermines, forming fluorinated lipoplexes (Gaucheron et al., 2001a, 2001b), a partially fluorinated analogue of DOPE, identified as [F8E11][C16]OPE from the number of fluorine atoms, was synthesized and compared with DOPE as helper of fluorinated lipoplexes (Boussif et al., 2001): this compound, inactive itself in promoting transfection, increased the *in vitro* and *in vivo* gene transfer of the lipoplex obtained from the pentacationic pcTG90 to a larger extent than DOPE. The synthesis was then extended to more fluorinated glycerophosphoethanolamines (Gaucheron et al., 2001) confirming that lipoplexes formulated with fluorinated helper lipids are attractive candidates for gene delivery both *in vitro* and *in vivo*. Several reasons were identified to explain these results: fluorinated co- lipids have a larger ability to preserve the integrity of complexed DNA in a biological environment and a larger propensity to promote fusion with endosomes and subsequent destabilisation, allowing more efficient DNA release in the cytosol; their high hydrophobic and lipophobic character can preserve the lipoplexes from the effect of the interactions with lipophilic and hydrophilic biocompounds; finally fluorinated DOPE compounds are expected to have a greater tendency to promote a lamellar to a an inverted hexagonal phase transition with the consequence of a higher effectiveness in disrupting membranes than DOPE.

It is commonly accepted that one of the main features of DOPE as helper depends on its polymorphism under various concentration and temperature conditions. Its ability to enhance transfection efficiency is related to its preference for the fusogenic HII phase, which can promote fusion with cellular membranes, especially the endosomal ones, thereby facilitating the escape of the genetic material; however, the low Lα/HII phase transition temperature (Th = 10 °C) makes cationic liposomes too unstable in the *in vivo* environment. An approach to solve the problem might be to synthesize analogues of DOPE in which the phase transition is near the physiological temperature. Some molecules having these characteristics have been synthesized (Fletcher et al., 2006) and correspond to a series of dialkynoyl analogues of DOPE where the cis-double bond in the two oleoyl fatty acid chains is replaced by a triple bond located in different positions of the hydrocarbon tails. With this modified geometry a new intermolecular packing is realized, able to induce an increase of the phase transition at the physiological conditions.

The achievements just shown were based on the concept to modify DOPE: a different approach to the search for more efficient helpers has been realized by synthesizing completely new lipids characterized by the presence of an imidazole polar head (Mével et al., 2008). These lipophosphoramides are neutral at physiological pH: the protonation occurring in the acidic compartments of the cell, namely the endosomes, induces fusion of liposomes with endosomes and structural changes that favour the release of DNA in the cytosol. Three phosphoramidates with a cationic polar head derived from natural aminoesters or a methylimidazolium salt were also synthesized and these cationic lipids were formulated with each one of the two new helpers and with cholesterol or DOPE for a comparison of transfections; it is worth noticing that the new helper lipids can improve the transfection by a factor of 100 compared with DOPE.

3. LIPOSOMES AND MEMBRANE FUSION MIMICKING

A remote introduction to the processes of gene transfer is represented by a series of studies aiming at mimicking the process of cell fusion by using model membranes mainly composed of phosphatidylcholine (PC) and phosphatidylserine (PS). It was demonstrated that addition of liposomes containing these lipids fuse with plasma membranes in the presence of Ca^{2+} and Mg^{2+} (Papahadjopoulos et al., 1976). Considering that cellular fusion is one of the most fundamental processes in life and that its role has been recognized in the interaction of liposomes with endosomes within the processes of DNA transfection, such studies are highly significant for the knowledge of this particular step of the non viral GT and in designing lipids with the highest characteristics of fusion. The study of the effect of divalent metal cations in the interaction and mixing of membrane components in vesicles prepared from phospholipids led to find that low amounts of Ca^{2+} and Mg2+ induce extensive mixing of vesicle membrane components and important structural rearrangements to form new membrane structures. The result is a true fusion rather than a simple mixing of vesicles that occurs in the absence of cations. Some evidence was found that fusion of vesicle membranes by Ca^{2+} and Mg^{2+} is not simply due to electrostatic charge neutralization, but rather to changes in molecular packing. It is possible to see here an anticipation of the phase transitions that many years after was demonstrated to occur in DNA transfection with co-lipid added lipoplexes, as reported in the previous section. These results have been confirmed by finding that Ca^{2+} and Mg^{2+} produce structurally different complexes with PE (Newton et al., 1978). A different behaviour in fusion induced by these two cations was found in mixture of PS/PC, PS/PE and PS/PC/PE (Düzgünes et al. 1981). The extent of fusion by Ca^{2+} in mixed PE/PC was lower compared to that of pure PS vesicles and was completely inhibited when PC reached 50% in the mixture; rapid fusion was instead obtained in mixtures PS/PE. Mg^{2+} can fuse PS only in the presence of PE. The fusogenic capacity of Mg^{2+} was instead completely absent in mixtures PS/PC/PE with 10% of PC. These results show clearly a marked difference between Ca^{2+} and Mg^{2+} against fusion: as we will see later, this difference will appear also in some processes concerning the formation of complexes of DNA with neutral liposomes.

A rational mechanism to interpret these results was tried out: kinetics of the interaction between PS vesicles in the presence of Ca^{2+} (Portis et al., 1979) show the formation of two different complexes: the former develops when the cations bind only to individual vesicles, the latter, which seems correlated with the beginning of membrane fusion, when the vesicles come to close apposition; the former complex is obtained also with Mg^{2+}. Its characteristics led to suggest that this complex is formed when the divalent cations bind to PS head groups on one bilayer only (cis complex). The latter, obtained only with Ca^{2+}, shows different characteristics and seems to involve a polydentate chelation of Ca^{2+} with the head groups of PS from apposed membranes (trans complex). The formation of this PS/Ca^{2+} complex is of crucial importance for the fusion of the vesicles. Apart from the names used to identify these complexes, it is worth noting that the structure suggested for the latter complex agrees with the one found in the complex between the neutral lipid DPPC and DNA, promoted by divalent metal cations, showing a L^C_α phase. A more detailed study (Wilschut et al., 1980) allowed to obtain further information about the process: it was demonstrated that fusion is one of the earliest events during the Ca2+ induced aggregation of SUVs (small unilamellar vesicles) of PS and occurs at a similar time scale, which means that fusion doesn't require initial rupture of the vesicles. The close contact between the vesicles induced by Ca^{2+} is sufficient to trigger the immediate fusion of the two membranes and the mixing of the internal volumes with a relative low leakiness of their content: which makes the Ca^{2+}/PS system an almost ideal model for membrane fusion. This model has been later deeply developed and is the basis to explain the processes which occur in the cytosol when the complexes liposomes/DNA encounter the endosomes and release the DNA. With these last findings the route to the DNA delivery to cells by means of liposomes was opened.

4. THE NEUTRAL LIPOSOMES AS INDEPENDENT DNA TRANSFECTION AGENTS

After the Bangham's work (Bangham et al., 1965) liposomes were extensively used as models of biological membranes (Sessa & Weissmann, 1968) on the basis of their lamellar structure. It was seen that they are able to discriminate ions as natural membranes do, and that it is easy to vary their surface charge, in order to modulate the diffusion of a large amount of cations and anions. It was proved that it is possible to incorporate proteins in their lamellar structure and that their composition can be modified to mimic the properties of a large variety of natural membranes. Basically, it was recognized that liposomes are a valuable instrument to study many problems concerning natural membrane structure and function. What's more, it was assumed that, if liposomes were able to incorporate proteins, enzymes, drugs or nucleic acids, an important step towards a true *in vitro* replica of the membranes of living systems would be obtained.

4.1 Liposomes and polynucleotide entrapment

Soon these foreseen opportunities began to turn into actual tasks: liposomes started being applied as carriers of different molecules into target cells (Dimitriadis, 1979; Tyrrell et al., 1976; Finkelstein & Weissmann, 1978) or of enzymes in enzyme replaced therapy (Gregoriadis & Buckland, 1973). It was in those years that the entrapment of synthetic polynucleotides (Magee et al., 1976), as well as natural ones (Hoffman et al., 1978; Lurquin, 1979), was undertaken. Large unilamellar liposomes were obtained (Dimitriadis, 1978) by adding ribonucleic acid (globin mRNA) to PS and it was demonstrated that mRNA is really entrapped and not simply adhering to the surface. A different experiment was realized with the aim of clearing up the mechanism of crossing the hydrophobic barriers formed by protein-lipid membranes and the nature of bonds, providing adsorption of polynucleotides in the membranes (a mechanism unknown at that time). It was demonstrated (Budker et al., 1978) that polynucleotides are adsorbed by liposomes of PC forming stable complexes in the presence of Mg^{2+} or Ca^{2+} ions, but not in the absence of these ions. This result suggested that this interaction is due to the action of bivalent cations, which crosslink phosphate groups of polynucleotides with the ones of PC. It was also found that the complexes obtained are stable, but that the addition of monovalent cations reduces the extent of complexation.

It has been said already that one of the main reasons of the success of cationic liposomes resides in their positive charge that enables attractive interactions both with the negative phosphate groups of the polynucleotides and the negatively charged cell wall. Encapsulation of DNA into a vector is of course the first and irreplaceable step to realize a synthetic vector driven gene therapy and must be solved within neutral liposomes, where the absence of charge does not allow the formation of a stable aggregation with negative DNA. However gene transfer and gene therapy need an efficient encapsulation of plasmid DNA into neutral liposomes and an attractive interaction with the negative cell wall which is the necessary step for the endocytic internalisation of the construct.

The methods generally used to realize a stable entrapment can be schematically indicated in three main classes: reverse phase evaporation, dehydration-rehydration and freeze-thawing. According to the first method the nucleic acids are dissolved in water and the solution is added to lipids dispersed in an organic solvent, then evaporated to induce vesiculation (Szoka & Papahadjopoulos, 1978). In the second procedure the nucleic acids are added to a dispersion of SULs (small unilamellar liposomes) and the mixture is dehydrated until almost dryness; afterwards the material is rehydrated and vortexed to induce the formation of liposomal aggregates (Deamer & Barchfeld, 1982). The third method involves the addition of nucleic acids to a dispersion of SULs followed by numerous freeze-thawing operations and by a final extrusion to obtain homogeneously sized vesicles (Chapman et al., 1990). The last method was applied to encapsulate a 3368 base pair DNA (Monnard et al., 1997), using liposomes prepared with 1-palmitoyl-2-oleoyl-*sn*-glycero-3-phosphocholine (POPC) mixed with a little amount of the negative PS or the cationic didodecyl-methylammonium bromide (DDAB) as cosurfatants. The yields of entrapment calculated over the amount of the initial material were 27% in pure POPC, 26% in POPC/PS 9:1 and 50% in POPC/DDAB 99:1. While the addition of PS has evidently no influence on the entrapment, the one of traces of the positive DDAB, doubles the percentage of entrapped DNA.

An important contribute to the entrapment of nucleic acids in neutral liposomes was done some years later (Bayley & Sullivan, 2000). Plasmid DNAs (pDNA) were trapped into pure DOPC, DOPC/DOPE 1:1 and DOPC/DOPE/Chol 1:1:1 by simply adding CaCl2 and ethanol to the initial mixture DNA/lipids. With optimized amounts of ethanol and calcium the entrapment percentages were 65-70% for DOPC, 70-80% for DOPC/DOPE 1:1 and only 35-40% for DOPC/DOPE/Chol 1:1:1. Most important the neutral liposome complexes obtained from DOPC and DOPC/DOPE are stable for at least two weeks in PBS (phosphate buffered saline) at 4 °C.

1.2 Some early experiments of DNA delivery to cells by means of liposomes

The need of modifying the expression of the eukaryotic genome to study the protein synthesis led to encapsulate a functional rabbit globin mRNA in lecithin liposomes, made by neutral PC and PE, to realize its selective insertion into differentiated eukaryotic cells in vitro and express a globin-like protein (Ostro et al., 1978). The authors claimed this result as the first successful attempt to entrap and deliver high molecular weight RNA with a liposome. New applications followed and led to improve the technique. Poliovirus RNA was encapsulated in liposomes of PS and delivered efficiently to cells in an infectious form (Wilson et al., 1979). A comparison between large unilamellar vesicles (LUVs) and multilamellar vesicles (MLVs) of PS indicates that LUVs deliver their content to cell cytoplasm much more efficiently than MLVs and that LUV-entrapped poliovirus RNA produces infection titers 10 to 100 fold higher than when delivered with other techniques. Likewise, DNA isolated from simian virus 40 (SV40) was encapsulated in LUVs of PS and delivered to a monkey cell line (Fraley et al., 1980).

The infectivity realized with this method was enhanced at least 100 fold over that of free naked DNA. This process was then used as a probe to study liposome-cell interactions and determine conditions favouring the intracellular delivery of liposome content to cells (Fraley et al., 1981). The efficiency of DNA delivery was found dependent both on size of vesicles and the resistance of liposomes to cell induced leakage of content. Acidic phospholipids are much more effective in both binding and delivery, and PS was found to be the best in both events. Inclusion of cholesterol in liposomes reduces the cell-induced leakage of vesicle content and enhances the delivery of DNA to cells. A brief exposure of cells to glycerol solutions enhances infectivity of the SV40 DNA when encapsulated into the negatively charged liposome of PS, but not in neutral and positively charged liposomes. Morphological studies indicate that the glycerol treatment stimulates membrane vacuolisation and suggest that the enhanced uptake of liposomes occurs by an endocytic-like process. As it was said in a previous paragraph, endocytosis is the mechanism followed in the phase of the internalisation of the complexes liposome/DNA into cells. Additional attempts to transfect DNA to cells by means of bivalent cations mediated complexes of neutral liposomes are reported by the literature (Kovalenko et al., 1996).

4.3 The problem of the uptake of liposomes by the reticuloendothelial system

It is known that many liposomes are removed in liver and spleen from the blood circulation within minutes: this property, beneficial when they are employed to carry drugs for treating intracellular infections of the reticuloendothelial system (RES), has limited their use as delivery carriers of material to sites beyond the RES. Designing liposomes with prolonged circulation time requires a reduction of the rate of their clearance by the RES and of the leakage of liposome cargo in blood stream. The search for a solution of the problem has led to the discovery of the so called stealth liposomes, sterically stabilized by the presence of bulky groups: the so-called PEG-liposomes, namely polyethyleneglycol functionalized liposomes, are the most important tools in ensuring a prolonged circulation time in blood. The incorporation of PEG into conventional liposomes provides a steric barrier at the liposome surface that inhibits opsonisation, therefore extending the persistence time of liposomes in the blood. Positive effects are a prolonged circulation lifetime of lipoplexes and a reduced formation of aggregates (Klibanov et al., 1990; Papahadjopoulos et al., 1991). The incorporation of 1,2-dioleoyl-N- (methoxy-polyethyleneglycol-succinyl-)phosphatidylethanolamine (PEG-PE) in liposomes composed of egg phosphatidylcholine-cholesterol, exposed to human serum at 37 °C increases the blood circulation half-life ten times the one of simple phosphatidylcholine- cholesterol liposome (Klibanov et al., 1990). While PEG-lipids play an active role in limiting excessive inhibition and fusion during the self-assembling phase when cationic lipids associate with anionic DNA protecting DNA from nuclease degradation in plasma, the steric barrier introduced by PEG is expected to inhibit also the process of fusion with the endosomes and by consequence reduce the transfection activity. Conflicting results have been obtained so far. It has been found (Song et al., 2002) that, owing to the presence of PEG lipids with long acyl chains (< 14 carbons), the contact between complexes and endosomal membranes doesn't allow membrane disruption. In general it seems that the biological and physicochemical characteristics of the DNA/copolymer complexes, including PEG, are influenced by the copolymer architecture (Deshpande et al., 2004) and that the transfection efficiency is strongly correlated with the level of cellular association and uptake of the DNA/copolymer complexes.

1.4 The structure of the complexes of DNA with zwitterionic liposomes

In previous paragraphs we pointed out that the structure of the lipoplexes represents a fundamental tool in understanding and planning DNA transfection systems. The same remarks are valid for complexes of

DNA with neutral (zwitterionic) liposomes in order to evaluate correctly their behaviour and, in case, design the necessary developments to achieve better results in DNA transfection experiments both *in vitro* and *in vivo*. These complexes were initially studied in order to understand the influence of DNA structural transition of neutral lipids: DSC thermograms of the DPPC/DNA/Ca^{2+} complex (Tarahovsky et al., 1996) reveal a distinct maximum at the temperature of 316.3 K in addition to the main maximum at 314.6 K. Since a direct relationship was observed between the molar proportion of DNA in samples and the value of the height of the second peak, it was hence assumed that the higher temperature transition corresponds to the formation of the complex. In another work (Kharakoz et al., 1999) it was demonstrated that DPPC/DNA complexes could be obtained by simply mixing the DNA solutions with an aqueous lipid dispersion in the presence of Ca^{2+} and that their formation could be obtained with both MLVs and ULVs. The stoichiometry was determined in 4.5 to 5 strongly bound lipid molecules per molecule of nucleotide, depending on the method used in a temperature-scanning ultrasonic study.

From this result and the ones obtained in a small angle x-ray scattering experiment (SAXS), a model was proposed (McManus et al., 2003) for the interaction of DNA and DPPC in the presence of $CaCl_2$. The lamellar repeat distance in complexes with MLVs at 298 K increases slightly as Ca2+ concentration increases, but it drops to a minimum at a Ca^{2+} concentration equal to 5 mM. At this concentration a special compact structural arrangement is observed, indicative of increased order. Combining this finding with the above result on the ratio of 4.5 to 5 lipid molecules per molecule of nucleotide, it was inferred that roughly one $CaCl_2$ binds two DPPC molecules and a model was proposed where every Ca^{2+} bridges two adjacent DPPC molecules through their phosphate groups. A different possibility was formerly considered (Bruni et al., 1997) in a study on the interactions of bivalent metal cations with double-stranded polynucleotides or DNA and egg yolk PC. Scatchard plots of PC/DNA/Mn^{2+} and DNA/Mn^{2+} complexes, combined with data of elemental analysis, support an arrangement where each Mn2+ bridges two DNA phosphates with three PC molecules. One more schematic model for interpreting the DNA-lipid interaction mediated by Ca^{2+} and Mg^{2+} has been working on the zwitterionic 1,2-dimyristoylphopsphoetyhanolamine (DMPE).

Following this suggestion (Gromelski & Brezesinski, 2006) the divalent cations bridge the negative part of the zwitterionic phospholipid headgroups, thereby making the lipid monolayers positive. Divalent cations also interact with the negative DNA phosphate moieties, condensing the DNA and leading to an ordered alignment of the DNA strands. If not all charges are screened by the divalent cations, the DNA aggregate remains partially negative and can interact either via divalent cations with the lipid phosphate groups or directly with the positively charged ethanolamine groups of DMPE when the lipid phosphate groups are bridged by divalent cations.

This new class of complexes consists of ternary systems NLs/DNA/M2+ where M refers mostly to Ca, Mg, a choice consistent with the previously reported experiments of membrane fusion, and sometimes Mn. The formation of the ternary complexes is the result of a self assembling process in which the driving force is represented by the release of the counter-ion entropy upon neutralization of DNA phosphate groups by metal cations (Cl^- in the examples discussed). Studies on the structure of these ternary complexes were undertaken mainly by means of x-ray diffraction technique.

NEUTRAL LIPIDS

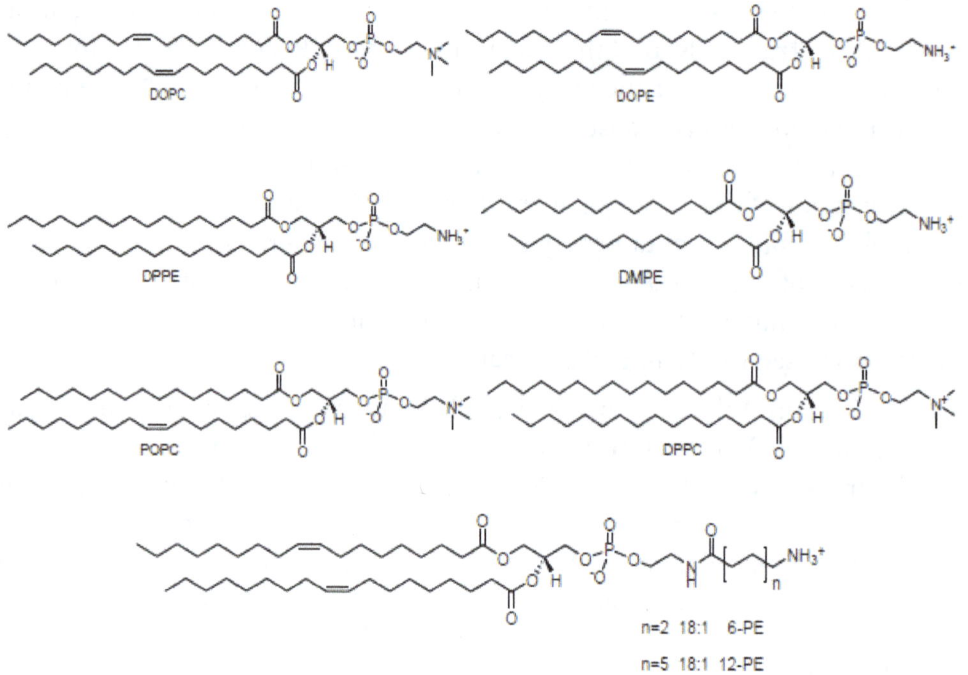

In all the experiments performed by the authors of this review, XRD measurements were carried out at the high brilliance beamline ID02 of the European Synchrotron Radiation Facility (Grenoble, France). The energy of the incident beam was 12.5 keV (λ= 0.995 Å), the beam size 100x100 μm2, and the sample-to-detector distance 1.2 m. The 2D diffraction patterns were collected by a CCD detector. The small angle q range from qmin = 0.1 nm^{-1} to qmax = 4 nm-1 with a resolution of 5 x 10^{-3} nm^{-1} (fwhm) was investigated: the samples were held in a 1 mm-sized glass capillary. To avoid radiation damage, each sample was exposed to radiation for 3 sec/frame. To calculate the electron density maps, the integrated intensities of the diffraction peaks were determined by fitting the data with series of Lorentz functions, using a nonlinear baseline. The Lorentz correction was performed multiplying each integrated intensity by sin θ and the intensities were then calibrated dividing by the multiplicity of the reflection (Harper et al., 2001; Francescangeli et al., 1996). The square root of the corrected peak was finally used to determine the modulus of the form factor *F* of each respective reflection. The electron density profile Δρ along the normal to the bilayers was calculated by Fourier sum,

$$\Delta\rho = \frac{\rho(z)-\langle\rho\rangle}{\left[\langle\rho^2(z)\rangle-\langle\rho\rangle^2\right]^{1/2}} = \sum_{l=1}^{N} F_l \cos\left(2\pi\, l\frac{z}{d}\right)$$

where ρ(*z*) is the electron density, ⟨ρ⟩ its average value, *N* the highest order of fundamental reflection observed in the SAXS pattern; *Fl* is the form factor of the *(00l)* reflection, *d* the thickness of the repeating unit and the origin of the *z* axis is chosen in the middle of the lipid bilayers. The phase problem was solved by means of a pattern recognition approach based on the histogram of the electron density map (Tristram-Nagle et al., 1998) and the results were found to be in agreement with those obtained with different approaches.

In a first example (Francescangeli et al., 2003), a DOPC liposome was mixed with calf thymus DNA in hepes buffered aqueous solutions of divalent cations and simultaneous small (SAXS) and wide (WAXS) angle x-ray scattering measurements were carried out.

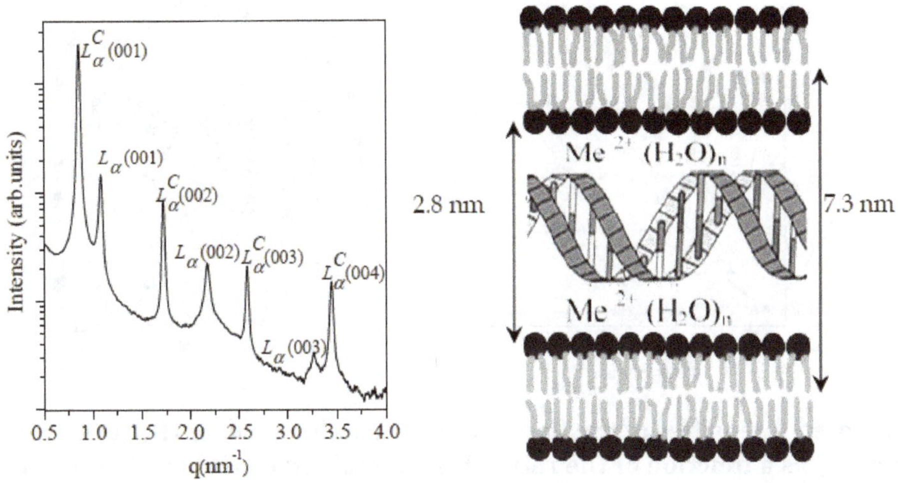

Fig. 2. Left: synchrotron SAXS pattern of DOPC/DNA/Mn^{2+} complex at molar ratio 3:4:12.

(Reprinted from MROC, 2011, 8, 38) Right: the model proposed for the LCa phase of the ternary complex, reporting the main structural parameters.

In a typical experiment the mole ratio DOPC:DNA:Mn2+ was 3:4:12 and the corresponding synchrotron x-ray diffraction (XRD) pattern (Figure 2, left) at 298 K is reported. Two series of spacings are present in the x-ray pattern: the one indicated with the symbol C Lα (d= 7.34 nm), independent of the concentration of the cation, has been attributed to the ternary complex and the one indicated with the symbol Lα (d= 5.88 nm) to the complex DOPC/Mn2+. The ternary complex is characterized by the lamellar symmetry of the C Lα (Luzzati, 1968), consisting of an ordered multilamellar assembly where the hydrated DNA helices are sandwiched between the liposome bilayers. This structure is similar to that found in CLs/ DNA complexes (Rädler et al., 1997; Podgornik et al. 1989). A pictorial representation (Figure 3, left) of the ternary complex DOPC/DNA/Mn2+ has been proposed: the DNA strands are sandwiched between the lipid bilayers and bound together through the hydrated metal ions: the value of 2.8 nm between two lipid bilayers is sufficient to accommodate a hydrated double strand of DNA (Figure 2, right).

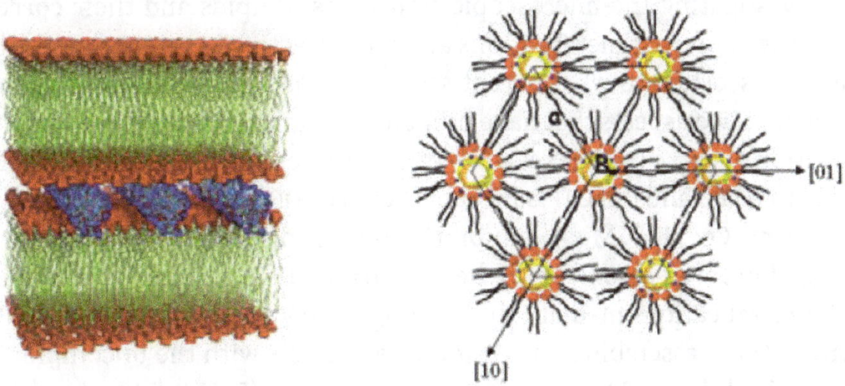

Fig. 3. A pictorial image of the DOPC/DNA/Mn2+ (left) and of the DOPE/DNA/M2+ (right) ternary complexes.

The simultaneous presence of two lamellar structures, confirmed by an analogous XRD study (Uhrikova et al., 2005), was interpreted by plotting (Figure 4, left) the integrated intensities of the first order diffraction peaks of the DNA complex and of the DOPC liposome as a function of the ratio of the metal ion concentration versus the one of the DNA phosphate groups.

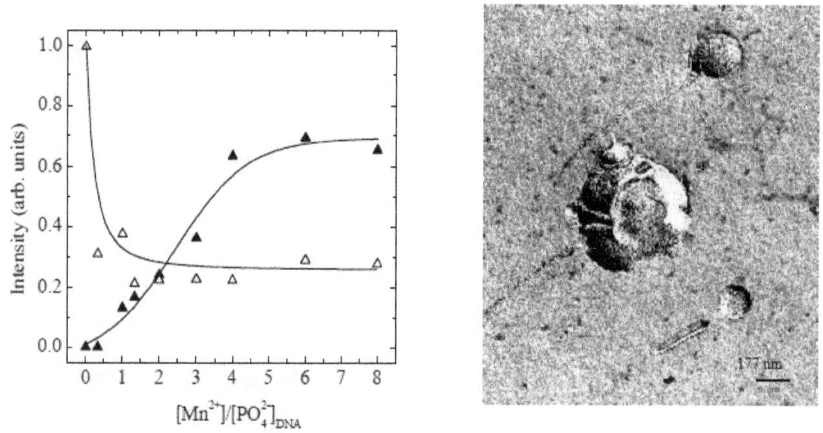

Fig. 4. Left: integrated intensities of the first order diffraction peak of the ternary complex (▲) and of DOPC liposome (△) as a function of the ratio of the metal ion concentration to the concentration of DNA phosphate groups.
(Reprinted from Phys. Rev. E, 2003, 67, 11904).

Right: freeze-fracture EM micrograph of the DOPC/DNA/Mn2+ complex.

An increase of the Mn2+ concentration favours the formation of the ternary complex accompanied to a complementary reduction of the DOPC: the saturation is reached at a ratio [Mn2+]:[PO42-] ≅ 6, corresponding to a constant volume fraction of the two structures (~ 70% to ~ 30% respectively). A freeze-fracture EM micrograph of the ternary complex has also been made and reported in figure 4 (right). An analogous study was made with the neutral liposome 1,2-dipalmitoylphosphatidylcholine (DPPC), bearing completely saturated hydrocarbon tails (McManus et al., 2003) : as for DOPC two coexisting phases are present and the ternary complex shows a lamellar structure, the DNA layers being embedded in the DPPC layers. The repeat distance is 7.84 nm at 298 K. At this temperature the DPPC/DNA/Ca2+ complex is in the gel thermotropic phase (Lβ').

These works have been followed by an extended approach to ternary complexes based on NLs, bearing unsaturated (DOPC, DLPC and DOPE) or saturated (DPPC) hydrocarbon tails. A twofold goal has been pursued when investigating the microscopic structures of lipids and their corresponding ternary complexes: to test whether different metal cations are equally active in promoting the DNA condensation with different lipids and ascertain to what extent structure and phase symmetry of the lipids affect the structure of the complexes; two aspects that have fundamental implications in view of an approach to gene delivery based application of these complexes. Different varieties of DNA (calf-thymus, salmon sperm and plasmid) complexed with DOPC and DLPC (Pisani et al., 2005) or DPPC (Pisani et al. 2006), in the presence of different cations (Ca^{2+}, Mn^{2+}, Co^{2+}, Fe^{2+}, Mg^{2+}), exhibit the already discussed multilamellar liquid-crystalline C Lα phase, consisting of ordered assemblies, where hydrated DNA helices are sandwiched between the lipid bilayers, and the metal cations mediate the binding of the phosphate groups of DNA with the lipid polar heads. Also within these assemblies the C Lα phase coexists with the uncomplexed Lα phase of the parent lipid. A systematic series of SAXS measurements in DOPC/DNAct/M^{2+} complexes, prepared with different metal cations, was performed as a function of the number of metal ion moles (n).

The results obtained from these spectra are reported in figure 5: a remarkable constancy of the lamellar spacings of the C Lα accompanied by a slight decrease of the lamellar repeat distance of the uncomplexed Lα , reported in the figure agrees with the model proposed in figure 3. As an example we report (Figure 6) the analysis of the ternary complex with Mg^{2+}: again two sets of peaks (each one including fundamental and high-order harmonics) related to distinct lamellar structures C Lα and Lα , with layer spacings d1= 7.52 nm and d2= 5.9 nm respectively, are present. The SAXS pattern (A), and the relative electron density profile (B) are shown: in the latter, the two peaks with the maximum of electron density correspond to phospholipids' polar headgroups, while the minimum correlates with the terminal hydrocarbon chain region.

The distance between the centres of the density maxima gives a good approximation of the bilayer thickness (dPP = 4.51 nm): it follows that the water-layer thickness can be calculated as dW = d1 – dPP = 7.52 – 4.51 = 3.01 nm sufficient to accommodate a double stranded DNA helix surrounded by one water hydration layer plus two thin layers of hydrated metal ions. Likewise, it was calculated a water layer thicknesses dW in the range of 2.8-3.0 nm in the complexes with DLPC and in the range of 2.9-3.2 nm in the ones with DPPC, depending on metal cation.

The SAXS pattern of the DPPC/DNA/Ca2+ (Figure 7, left) shows a correlation peak, marked as DNA in the figure, corresponding to the DNA-DNA interaction, indicative of a higher organization of the DNA chains between the liposome layers. The thermotropic phase behaviour in a temperature range between 303 K and 328 K, well above the main transition temperature of the pure lipid (Tm = 314 K) was studied (Figure 7, right), leading to the important conclusion that coexistence of complexed and uncomplexed phases persists over the whole explored thermal range. A further relevant effect is observed: while the uncomplexed lipid exhibits the same thermotropic phase behaviour as pure DPPC, i.e. *Lβ' -Pβ' -Lα* , the mesomorphic behaviour of the bound lipid in the complex is partially altered.

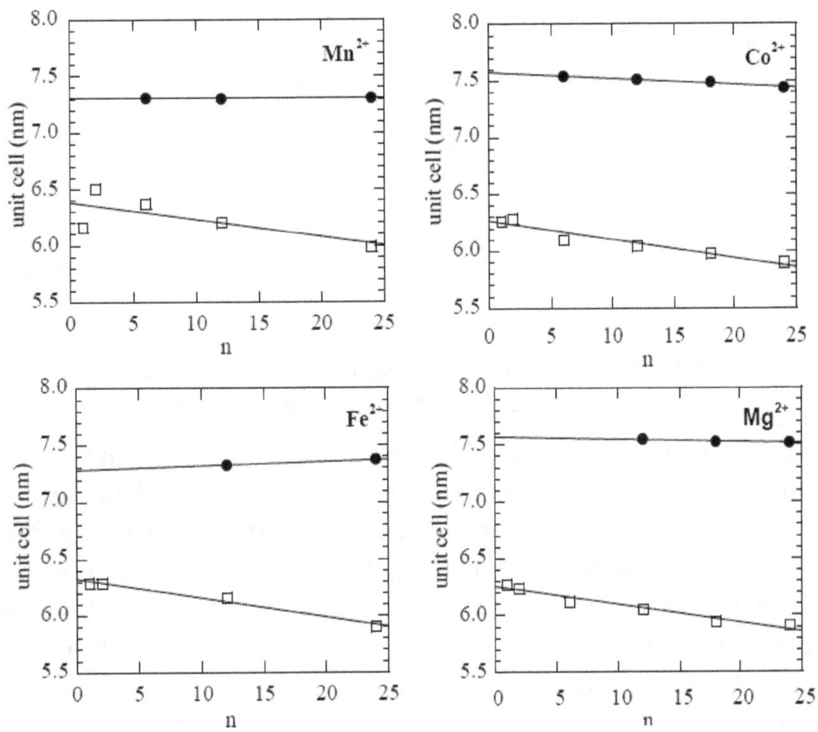

Fig. 5. Lamellar d-spacings of the L^{C}_{α} of ternary complex DOPC/DNAct/M^{2+} (•) and of the Lαphase of DOPC (∀) as a function of the metal mole number n in the DOPC/DNAct/M2+ complexes at molar ratios 3:4:n. (Reprinted from: M. Pisani, P. Bruni, C. Conti, E. Giorgini, O. Francescangeli. Self-Assembled Liposome-DNA-Metal Complexes Related to DNA Delivery. Mol. Cryst. Liq. Cryst., 2005, 434, 643).

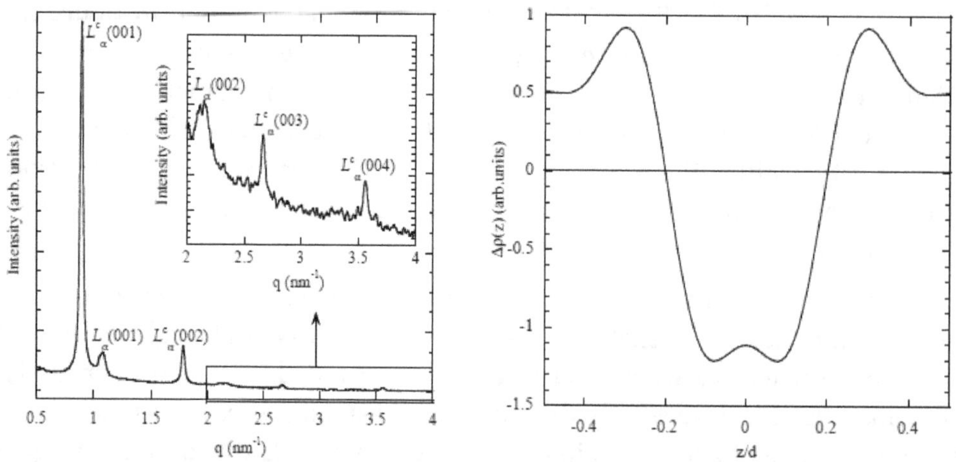

Fig. 6. Left: synchrotron SAXS pattern of the ternary complex DLPC/DNAct/Mn2+ at 3:4:12 molar ratio. Right: electron density profile along the normal to the bilayers in the LC$_a$ phase. (Reprinted from: M. Pisani, P. Bruni, C. Conti, E. Giorgini, O. Francescangeli. Self- Assembled Liposome-DNA-Metal Complexes Related to DNA Delivery. Mol. Cryst. Liq. Cryst., 2005, 434, 643).

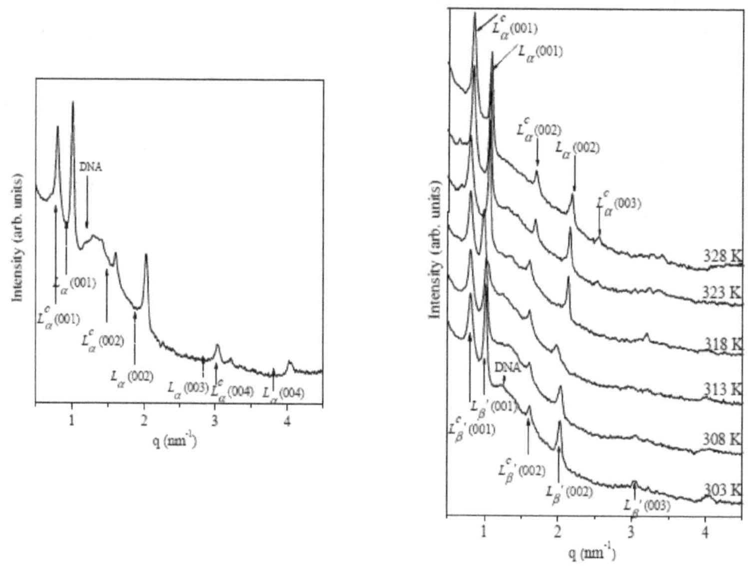

Fig. 7. Left: SAXS pattern of DPPC/DNA/Ca2+ complex at molar ratio 3:4:24. Right: synchrotron XRD patterns as a function of temperature.

This is highlighted by the disappearance of the ripple phase and the remarkable increase of the main transition temperature: the observed thermotropic phase sequence of the complex goes directly from *Lcβ* to *c La* . (Pisani et al. 2006). In addition the effect of the temperature on the formation of the DOPC/ DNA/Mn2+ complex has been determined (Francescangeli et al., 2003): figure 8 shows the temperature evolution of the XRD patterns of this complex in the range 290 to 320 K (left) and the temperature dependence (right) of the integrated intensities of the two small-angle reflections for the L$_\alpha$ phase for the ternary complex and the Lα phase of DOPC, respectively.

This is highlighted by the disappearance of the ripple phase and the remarkable increase of the main transition temperature: the observed thermotropic phase sequence of the complex goes directly from *Lcβ* to *c La* . (Pisani et al. 2006). In addition the effect of the temperature on the formation of the DOPC/ DNA/Mn2+ complex has been determined (Francescangeli et al., 2003): figure 8 shows the temperature evolution of the XRD patterns of this complex in the range 290 to 320 K (left) and the temperature dependence (right) of the integrated intensities of the two small-angle reflections for the L_α phase for the ternary complex and the Lα phase of DOPC, respectively.

The evolution of the equilibrium concentrations of the two phases clearly shows that the increase of the temperature favours the formation of the complex, the relative concentrations of the lamellar phases of pure lipids lowering in favour of the one of the ternary complex.

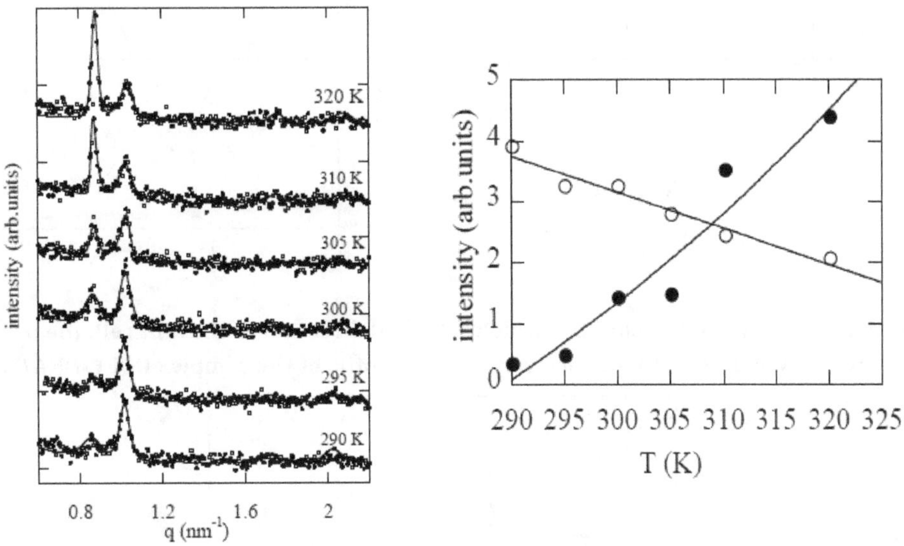

Fig. 8. Left: temperature evolution of the XRD patterns of DOPC/DNA/Mn2+ in the range 290 to 320 K. Right: temperature dependence of the integrated intensities of the two small-angle reflections for the LC_a and the Lα phases. (Reprinted from Rec. Res. Devel. In Macromol., 2003, 7, 247).

The complex DOPC/DNA/Mn2+ has also been studied in a solid supported phase (Caracciolo et al., 2004) by Energy Dispersion X-ray Diffraction (Caminiti & Rossi Albertini, 1999; Caracciolo et al., 2002) and it has been found that its structure is essentially identical to that in aqueous solution. The effect of hydration on the structural features of these multilamellar systems has also been explored (Caminiti et al., 2005), considering that adsorbed water plays a major role in the effectiveness of lipid drug delivery systems where lipid-cell interactions are involved. The hydration kinetics of oriented DOPC shows that the long-range order in a multilamellar lipid system strictly depends on the hydration level: adsorbed water molecules first promote a spatial coherence along the normal to the lipid bilayers, then penetrate the interbilayer region and behave as bulk water, producing disorder. The existence of a correlation between the degree of hydration of lipid bilayers and the structure of interbilayer water (Ge & Freed, 2003; Zhou et al., 1999) has been confirmed. We have already reported that DOPE induces a structural transformation of the lipoplexes when added as a co-lipid: the equilibrium phase of pure DOPE in excess water consists of an inverted hexagonal HII lattice (Turner & Gruner, 1992), whose structure elements are infinitely long rigid rods, all identical and cristallographically equivalent, regularly packed in a 2D hexagonal lattice. The cylinders are filled by water and dispersed in the continuous medium of the hydrocarbon chains, whereas the polar groups are located at the water- hydrocarbon interface. The SAXS pattern of pure DOPE allows to calculate a unit cell spacing a = 7.44 nm (Francescangeli et al., 2004) and its electron density profile calculated along the [10] direction (Figure 9, left) shows an average diameter of the water core dW =3.02 nm. DOPE and divalent metal cations Mn2+, Mg2+, Co2+ and Fe2+ in water solution condense DNA into ternary complexes DOPE/DNA/M2+ characterized by an inverted- hexagonal phase $H_{II}C$. Also in this case two different sets of peaks with different unit cell spacings, namely a = 7.45

nm and aC= 6.87 nm respectively, have been observed. The former corresponds to the phase HII of pure DOPE, the latter is instead consistent with the 2D columnar inverted hexagonal phase $H_{II}C$ of the DOPE/DNA/Fe2+ complexes.

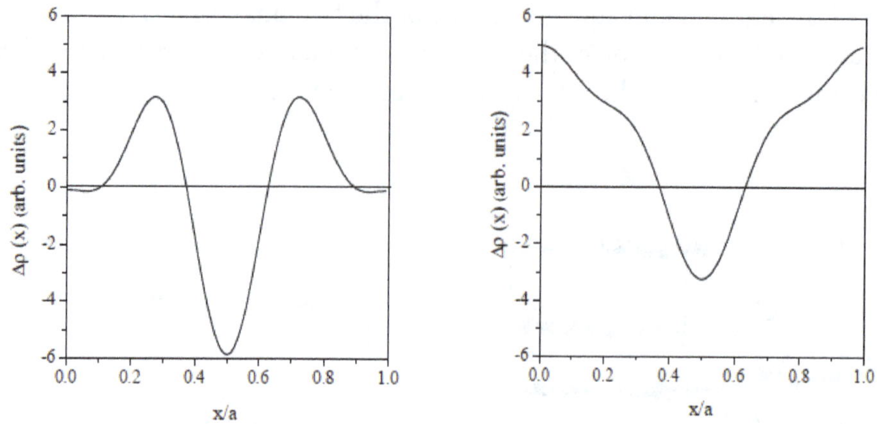

Fig. 9. Left: electron density profile of the pure DOPE along the [10] direction of the unit cell: the origin corresponds to the centre of water core. Right: electron density profile of the complex DOPE/DNA/Fe^{2+}.

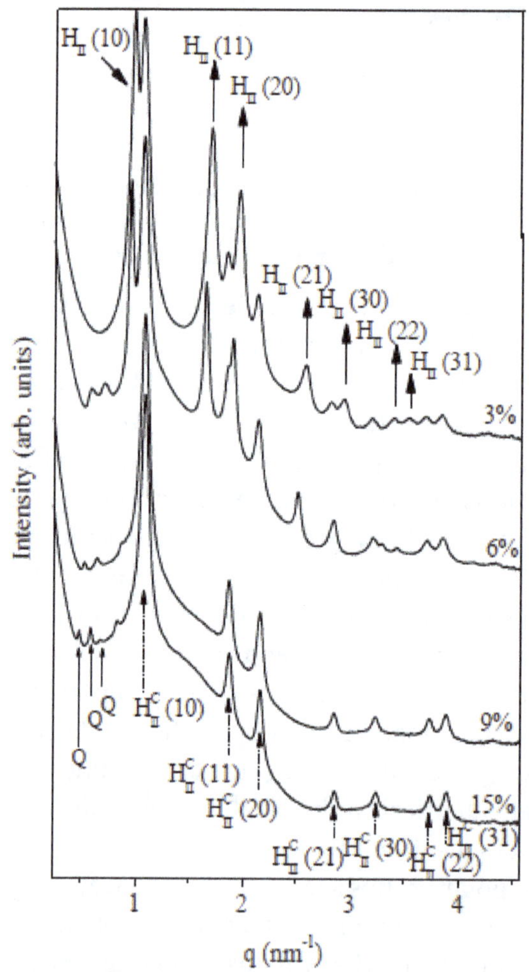

Fig. 10. Synchrotron XRD patterns of the DOPE/DOPE-PEG(350)/DNA/Mn^{2+} complex as a function of different concentrations of the DOPE/PEG component in the lipid mixture. The pattern of the cubic phase is clearly visible at 6%, 9% and 15% concentration of the pegilated DOPE.

In this structure DNA strands are supposed to fill the water gap inside the cylinders of pure DOPE, as it is supported by the electron density profiles (Figure 9 right) calculated along the [10] direction. The two shoulders, at z/d ~ 0.26 and 0.73 respectively, correspond to phosphate groups and are used to localize the centres of the polar head. From the structural data, values of dPP = 3.26 nm and dL = 4.36 nm were calculated, leading to a water layer thickness dW = 2.51 nm, large enough to accommodate a double-stranded DNA molecule surrounded by a hydration layer (Podgornik et al., 1989). A pictorial representation of this structure is reported in Figure 3 (right). Unlike complexes organized in the C Lα the ratio between HII and $H^{C}II$ depends also on the incubation time: after 48 hours, phase $H^{C}II$ disappears completely and is transformed into the HICI of ternary complex.

The effect of pegylation on NLs has also been studied (Pisani at al., 2008, 2009) in mixed complexes DOPE/DOPE-PEG(350)/DNA/M2+ (M = Ca, Mg, Mn). XRD investigation on the complex with Mn^{2+} shows that with 3% of DOPE-PEG, the two phases HICI and HII coexist as usual: the former being attributed to DOPE/DOPE-PEG(350)/DNA/Mn^{2+}, the latter to DOPE/DOPE-PEG(350)/Mn^{2+}. Interestingly a new phase, indexed in the SAXS pattern as Q (Figure 10), appears at higher concentrations of DOPE-PEG (6, 9 and 15%): the corresponding peaks are spaced in the ratios $\sqrt{2};\sqrt{3};\sqrt{4};\sqrt{6};\sqrt{8};\sqrt{9};\sqrt{10}$ consistent with a cubic Q224 phase with the space group *Pn3m*. A transition HII → Q_{II} has been found in different contexts (Koynova et al., 1997) : this ability, together with the well known fusogenic property of DOPE and its destabilizing effect on targeted endosomal membranes makes the complexes DOPE/DOPE-PEG(350)/DNA/Mn2+ extremely interesting for application in HGT.

1.5 DNA transfection experiments in vitro

A first attempt of *in vitro* transfection was made with a couple of complexes DOPC/pDNA/M^{2+}, with M = Ca or Mn, on a mouse fibroplast NIH 3T3 cell line (Bruni et al., 2006): using standard methods the green fluorescent protein was expressed by both complexes. Figure 11 reports an improved result obtained later, using a 15 mM concentration of Ca2+ in the complex.

Other attempts of *in vitro* transfections made in our laboratories are compared in a series of histograms (Figure 12) which show that the low efficiency of pure DOPC can be increased by addition of both 1, 2-dioleoyl-*sn*-glycero-3-phosphoethanolamine-N-hexanoylamine (6PE) or 1,2-dioleoyl-*sn*-glycero-3-phosphoethanolamine-N-dodecanoylamine (12PE) to DOPC/DNA/M2+. This result confirms that the transfection efficiency is strongly dependent on an appropriate mixutures of liposomes as DNA carriers.

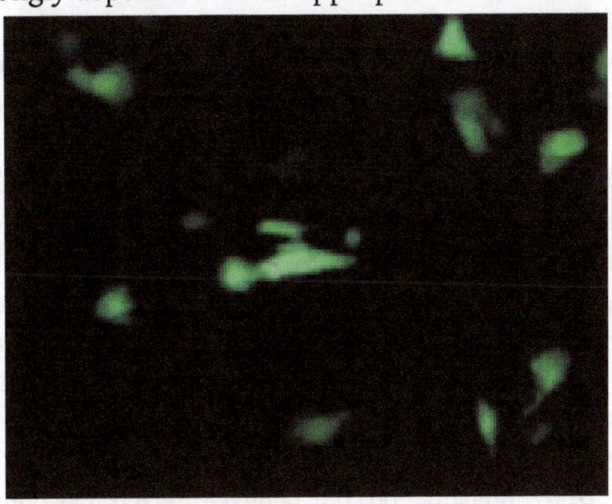

Fig. 11. Fluorescence micrograph of mouse fibroplast NIH 3T3 cells transfected with pGreen Lantern complexed with DOPC liposome in presence of Ca^{2+}.

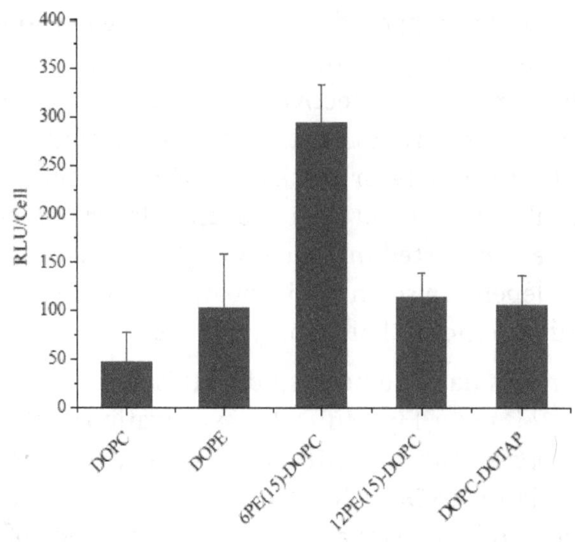

Fig. 12. Luciferase expression following 6hrs transfection with different complexes in NIH3T3 cell line. Expression efficiency is expressed as Relative Luminometric Units per cell (RLU/cell).

In vitro transfection efficiencies of DOPE/DNA complexes mediated by cations bearing varying charges, including K^+, Mg^{2+}, Ca^{2+}, and La^{3+}, have been the subject of an intriguing comparison (Tresset et al., 2007). Pure DOPE has a modest negative charge at physiological pH, which is not affected by K^+ due to its low density of binding sites. In contrast, the charge density of Mg^{2+} and Ca^{2+} has been measured to be high, with La^{3+} being particularly noteworthy, as it is 100 times more dense than the two bivalent cations. The normal presence of the two HII and HCII phases has been confirmed with the other cations, while SAXS of the corresponding ternary complexes demonstrates the absence of any ordered structure induced by K^+. The transfection efficacy of the two cell lines U87 and hepG2 has been assessed. The absence of a complexation by K^+, as evidenced by x-rays, has resulted in the absence of transfection. Rather, the efficacy increases in the following order: $Mg^{2+} < Ca^{2+} < La^{3+}$, with the latter being 2.6-fold more efficient than the lipoplex DNA/DOTAP/DOPE. It is also of great significance that the highest efficiency measured with the La3+ complex was achieved with an ion concentration three orders of magnitude lower than that of Ca^{2+}. This result is exceedingly favorable in terms of toxicity, as has been demonstrated.

5. CONCISE SUMMARY

At present, the utilization of NLs as autonomous carriers of genetic material in human gene therapy is regarded as an opportunity that necessitates extensive investigation in order to serve as a genuine alternative to CLs. Given the numerous constraints that the latter still encounter, particularly in in vivo applications, and their sluggish progress, it is crucial to consider the parallel path of NLs as potential autonomous carriers. Their high stability in serum and lack of toxicity are significant attributes that favor them. Some of the findings presented here are noteworthy developments. The structure and symmetry of the parent lipids are reflected in the complexes, and the various bivalent metal cations are equally active in promoting the condensation of DNA into the ternary complexes. These findings will provide a structure-composition correlation that can be utilized to optimize the design of these materials as non-viral DNA carriers in HGT. Our laboratories are currently conducting additional research on the application of pegylated NLs in the treatment of brain-related maladies, a field in which CLs are currently being tested (Zhang et al., 2002; Pardridge, 2007; Boardo, 2007). It is possible that NLs could yield superior results due to their capacity to decrease opsonization. The recent interest in the development of intelligent carriers on CLs (Voinea & Simionescu, 2002; Shi et al., 2002; Alvarez-Lorenzo et al., 2009) may also present an intriguing opportunity for NLs. The transfection is likely to be influenced by a variety of factors, including the structural knowledge of DNA complexes with NLs. Numerous additional factors, including efficient DNA entrapment, complex size, and Z-potential values, are crucial acquisitions that must be attained. The introduction of NLs into the field of HGT and the subsequent opportunity to compare their properties and activity with those of cationic and anionic liposomes will facilitate a more comprehensive comprehension of these processes. In this context, it is reassuring to cite the perspective of Rädler, one of the most distinguished experts in cationic lipids: "the resources allocated to the development of less toxic cationic-DNA complexes may, in the future, be balanced by research that explores the potential for the creation of comparable complexes from entirely non-toxic components, such as the NLs/DNA/M2+ complexes."

REFERENCES

Alvarez-Lorenzo, C.; Bromberg, L. & Concheiro, A. (2009). Light-sensitive intelligent drug delivery systems. *Photochem. Photobiol.*, 85, 848–860, 0031-8655.

Bailay, A.L. & Sullivan, S.M. (2000). Efficient encapsulation of DNA plasmids in small neutral liposomes induced by ethanol and calcium. *Biochim. Biophys. Acta*, 1468, 239-252, 005-2736.

Bangham, A.D., Standish, M.M. & Weissmann, G. (1965). The action of steroids and streptolysin S on the permeability of phospholipids structures to cations. *J. Mol. Biol.*, 13, 253-259, 0022-2836.

Boado R.J. (2007). Blood-brain barrier transport of non-viral gene and RNAi therapeutics. *Pharm. Res.*, 24, 1772-1787, 0724-8741.

Boukhnikachvili. T., Aguerre-Chariol, O., Airlau, M., Lesieur, S., Ollivon, M: & Vacus, J. (1997). Structure of in-serum transfecting DNA-cationic lipid complexes. *FEBS Lett.*, 409, 188-194, 0014-5793.

Boussif, O., Lezoualch, F., Zanta, M. A.,. Mergny, M. D., Scherman, D., Demeneix, B.J. & Behr P. (1995). A versatile vector for gene and oligonucleotide transfer into cells in culture and in vivo: polyethylenimine *Proc. Natl. Acad. Sci. USA*, *92*, 7297-7301, 0027-8424.

Boussif, O., Gaucheron, J., Boulanger, C., Santaella, C., Kolbe, H.V.J. & Vierling, P. (2001). Enhanced in vitro and in vivo cationic lipid-mediated gene delivery with a fluorinated glycerophospho-ethanolamine. *J.Gen. Med.*, 3, 109-114, 1099-498X.

Bruni, P., Gobbi, G., Morgganti, G., Iacussi, M. &Maurelli E. (1997). Use and activity of metals in biological systems. The interaction of bivalent metal cations with double- stranded polynucleotides and phospholipids. *Gazz. Chim. Ital.*, 127, 513-517, 0016- 5603.

Bruni, P.; Pisani, M.; Amici, A.; Marchini, C.; Montani, M. & Francescangeli, O. (2006). Self- assembled ternary complexes of neutral liposomes, deoxyribonucleic acid and bivalent metal cations. Promising vectors for gene transfer? *Applied Physics Letters*, 88, 73901, 1-3, 0003-6951.

Budker, V. G., Kazatchkov, Y. A. & Naumova, L. P. (1978). Polynucleotides adsorb on mitochondrial and model lipid membranes in the presence of bivalent cations. *FEBS Lett.*, 95, 143-146, 0014-5793.

Caminiti, R. & Rossi Albertini, V. (1999). The kinetics of phase transitions observed by energy-dispersive X-ray diffraction. *Int. Rev. Phys. Chem.*, 18, 263-299, 0144-235X.

Caminiti, R.; Caracciolo, G.; Pisani, M. & Bruni, P. (2005). Effect of hydration on the long- range order of lipid multilayers investigated by in situ time-resolved energy dispersive X-ray diffraction. *Chem. Phys. Lett.*, 409, 331-336, 0009-2614.

Caracciolo, G.; Caminiti, R.; Pozzi, D.; Friello, M.; Boffi, F. & Congiu Castellano A. (2002). Self-assembly of cationic liposomes-DNA complexes: a structural and thermodynamic study by EDXD. *Chem. Phys. Lett.*, 351, 222-228, 0009-2614.

Caracciolo, G.; Sadun, C.; Caminiti, R.; Pisani, M.; Bruni, P. & Francescangeli, O. (2004). Structure of solid-supported lipid-DNA-metal complexes investigated by energy dispersive X-ray diffraction. *Chem. Phys. Lett.*, 397, 138-143, 0009-2614.

Caracciolo, G. & Caminiti, R. (2005). Do DC-Chol/DOPE-DNA complexes really form an inverted hexagonal phase? *Chem. Phys. Lett.*, 411, 327-332, 0009-2614.

Chapman, C.J., Erdahl, W.L., Taylor, R.W. & Pfeiffer, D.R. (1990). Factors affecting solute entrapment in phospholipid vesicles prepared by the freeze-thaw extrusion method: a possible general method for improving the efficiency of entrapment. *Chem. Phys. Lipids*, 55, 73-83, 0009-3084.

Deamer, D.W. & Barchfield, G.L. (1982). Encapsulation of macromolecules by lipid vesicles under simulated prebiotic conditions. *J.Mol. Evol.*, 18, 203-206, 0022-2844.

Deshpande, M.C.; Davies, M.C.; Garnett, M.C.; Williams, P.M.; Armitage, D.; Bailey, L.; Vamvakaki, M.; Armes, S.P. & Stolnik, S. (2004): The effect of poly(ethylene glycol) molecular architecture on cellular interaction and uptake of DNA complexes. *J. Control. Release*, 97, 143–156, 0168-3659.

Dimitriadis, G.J. (1978). Entrapment of ribonucleic acid in liposomes. *FEBS Lett.*, 86, 289-293, 0014-5793.

Dimitriadis, G.J. & Butters, T.D. (1979). Liposome-mediated ricin toxicity in ricin-resistant cells. *FEBS Lett.*, 98, 33-36, 0014-5793.

Düzgünes, N., Wilschut, J., Fraley, R. & Papahadjopoulos, D. 1981. Studies on the mechanism of membrane fusion. Role of head-group composition in calcium- and magnesium-induced fusion of mixed phospholipid vesicles. *Biochim. Biophys. Acta*, 642, 182-195, 0005-2736.

Edelstein, M.J., Abedi, M.R., Wicson, J. & Edelstein, M.R. (2004): Gene therapy clinical trials worldwide 1989-2004. *J. Genet. Med.*, 6, 597-602, 1226-1769.

Elouahabi, A. & Ruysschaert, J.M. (2005). Formation and intracellular trafficking of lipoplexes and polyplexes. *Mol. Ther.*, 11, 336-347, 1525-0016.

Farhood, H., Serbina, N. & Huang, L. (1995), The role of dioleyl phosphatidylethaloamine in cationic liposome- mediated gene transfer. *Biochim. Biophys. Acta*, 1235, 289-295, 0005-2736.

Fasbender, A., Marshall, J., Moninger, T.O., Grust, T., Cheng, S. & Welsh, M.J. (1997). Effect of co-lipids in enhancing cationic-lipid mediated gene transfer in vitro and in vivo. *Gen. Ther.*, 4, 716-725, 0969-7128.

Felgner, P.L., Gadek,T.R., Holm, M., Roman, R., Chan, H.W., Wenz, M., Northrop, J.P., Ringold, G.M. & Danielsen, M. (1987). Lipofection: A highly efficient, lipid- mediated DNA-transfection procedure. *Proc. Natl. Acad. Sci. USA*, 84, 7413-7417, 0027-8424.

Felgner, J.H., R Kumar, C N Sridhar, C J Wheeler, Y J Tsai, R Border, P Ramsey, M Martin, & P L Felgner. (1994). Enhanced gene delivery and mechanism studies with a novel series of cationic lipid formulations. *J. Biol. Chem.*, 269, 2550-2561, 0021-9258.

Filion, M.C., Phillips, N.C. (1998). Major limitations in the use of cationic liposomes for DANN delivery. *Int. J. Pharm.*, 162, 159-170, 0378-5143.

Finkelstein, M. & Weissmann, G. (1978). The introduction of enzymes into cells by means of liposomes. *J. Lip. Res.*, 19, 289-303, 0022-2275.

Fletcher S., Ahmad, A., Perouzel, E., Jorgensen & M.R., Miller, A.D. (2006). A dialkynoyl analog of DOPE improves gene transfer of lower-charged cationic lipoplexes. *Org. Biomol. Chem.*, 4, 196-199, 1477-0520.

Foradada, M.; Pujol, M.D.; Bermudez, J. & Esterlich, J. (2000). Chemical degradation of liposomes by serum components detected by NMR. *Chem. Phys. Lipids*, 104, 133-148, 0009-3084.

Fraley, R., Subramani, S., Berg, P. & Papahadjopoulos, D. (1980). Introduction of liposome- encapsulated SV40 DNA into cells. *J. Biol. Chem.*, 255, 10431-10435, 0021-9258.

Fraley, R., Straubinger, R.M., Rule, G., Springer, A.L. & Papahadjopoulos, D. (1981). Liposome-mediated delivery of deoxyribonucleic acid to cells: enhanced efficiency of delivery related to lipid composition and incubation conditions. *Biochemistry*, 20, 6978-6987, 0006-2960.

Francescangeli, O.; Rinaldi, D.; Laus, M.; Galli, G. & Gallot, B.J. (1996). An X-ray study of a smectic C and smectic A liquid crystal polyacrylate. *J. Phys. II*, 6, 77-89.

Francescangeli, O., Pisani, M., Stanic, V., Bruni, P. & Iacussi, M. (2003). Supramolecular ordering of self-assembled liposome-DNA- metal complexes. *Recent Res. Dev. Macromol. Res.*, 7, 247-263, 81-271-0025-0.

Francescangeli, O., Stanic, V., Gobbi, L., Bruni, P., Iacussi, M., Tosi, G. & Bernstorff, S. (2003). Structure of self-assembled liposome-DNA-metal complexes. *Phys. Rev. E*, 67, 11904 (1-11), 1539-3755.

Francescangeli, O.; Pisani, M.; Stanic, V.; Bruni, P. & Weiss, T.M. (2004). Evidence of an inverted hexagonal phase in self-assembled phospholipid-DNA-metal complexes. *Europhysics Letters*, 67, 669-675, 0295-5075.

Gaucheron, J., Santaella, C. & Vierling, P. (2001a). Higly fluorinated lipospermines fo gene transfer: synthesis and evaluation of their in vitro transfection efficiency. *Bioconj. Chem.*, 12, 114-128, 1043-1802.

Gaucheron, J., Santaella, C. & Vierling, P. (2001b). Improved in vitro gene transfer mediated by fluorinated lipoplexes in the presence of a bile salt surfactant. *J.Gene Med.*, 3, 338- 344, 1029-498X.

Gaucheron, J., Boulanger, C., Santaella, C., Sbirrazzuoli, N., Boussif. O. & Vierling, P. (2001c). In vitro cationic lipid-mediated gene delivery with fluorinated glycerophosphoethanolamine helper lipids. *Bioconj. Chem.*, 12, 949-963, 1043-1802.

Ge, M. & Freed J.H. (2003). Hydration, structure and molecular interactions in the headgroup region of dioleoylphosphatidylcholine bilayers: an electron spin resonance study. *Biophys. J.*, 85, 4023-4040, 006-3495.

Gregoriadis, G. & Buckland R. A. (1973). Enzyme-containing Liposomes alleviate a Model for Storage Disease. *Nature*, 244, 170-172, 0038-0836.

Gromelski, S. & Brezezinski, G. (2006). DNA condensation and interaction with zwittrionic phospolipids mediated by divalent cations. *Langmuir*, 22, 6293-6301, 0743-7463.

Hacein-Bey-Abina, S. et al. (2003). LMO2-Associated clonal T cell proliferation in two patients after gene therapy for SCID-X1. *Science*, 302, 415-419, 0036-8075.

Harper, P. E.; Mannock, D.A.; Lewis, R.N.A.H.; McElhaney, R.N. & Gruner, S.M. (2001). X- Ray diffraction structures of some phosphatidylethanolamine lamellar and inverted hexagonal phases. *Biophys. J.*, 81, 2693-2706, 0006-3495.

Harrington, J.J., Van Bokkelen, G., Mays, R.W., Gustashaw, K. & Williard, H.F. (1997). Formation of *de novo* centromeres and construction of first-generation human artificial microchromosome. *Nat. Genet.*, 15, 345-355, 1061-4036.

Hoffman, R.M., Margolis, R.B. & Bergelson, L.D. (1978). Binding and entrapment of high molecular weight DNA by lecithin liposomes. *FEBS Lett.*, 93, 365-368, 0014-5793.

Kharakoz, D.P., Khusainova, R.S., Gorelov, A.V. & Dawson, K.A. (1999). Stoichiometry of dipalmitoylphosphatidylcholine-DNA interaction in the presence of Ca2+: a temperature-scanning ultrasonic study. *FEBS Lett.*, 466, 27-29, 0014-5793.

Klibanov, A.L., Maruyama, K:, Torchilin, V.P. & Huang, L. (1990). Amphipatic polyethylenglycos effectively prolong the circulation time of liposomes. *FEBS*, 268, 235-237, 0014-5793.

Koiv, A., Palvimo, J. & Kinnunen, P.K. (1995). Evidence for ternary complex by histone H1, DNA and liposomes. *Biochemistry*, 34, 8018-8027, 0006-2960.

Koltover, L., Salditt, T., Rädler, J.O. & Safinya, C.R. (1998). An inverted hexagonal phase of cationic liposome-DNA complexes related to DNA release and delivery. *Science*, 281, 78-81, 0036-8075.

Kostarelos, K. & Miller, A.D. (2005). Synthetic, self-assembly ABCD nanoparticles; a structural paradigm for synthetic non-viral vectors. *Chem. Soc. Rev.*, 34, 970-994, 0306-0012.

Kovalenko D.V.; Shafei, R.A.; Zelenina, I.A.; Semenova, M.L.; Samuilova, O.V. & Zhdanov, R.I. (1996). Metallo nucleoliposome complexes as a vehicle for gene delivery to mouse skeletal muscles in vivo. *Genetika, 32*, 1299-1301, 1022-7954.

Koynova, R.; Tenchov, B. & Rapp, G. (1997). Low amounts of PEG-lipid induce cubic phase in phosphatidylethanolamine dispersions. *Biochim. Biophys. Acta*, 1326, 167-170, 0005-2736.

Koynova, R., Wang, L., Tarahovsky, W. & McDonald, R.C. (2005). Lipid phase control of DNA delivery. *Bioconj. Chem.*, 16, 1335-1339, 1043-1802.

Koynova, R., Wang, Li. & MacDonald, R.C. (2006). An intracellular lamellar-non lamellar phase transition rationalizes the superior performance of some cationic lipid transfection agents. *Proc. Natl. Acad. Sci. USA*, 103, 14373-14378, 0027-8424.

Lin, A.J., Slack, N.L., Ahmad, A., George, C.X., Samuel, C.E. & Safinya, C.R. (2003). Three- dimensional imaging of lipid gene-carriers: membrane charge density controls universal transfection behavior in lamellar cationic liposome-DNA complexes *Biophys. J.*, 84, 3307-3316, 0006-3495.

Liu, F. & Huang, L. (2002). Development of non-viral vectors for systemic gene delivery. *J. Control. Rel.*, 78, 259-266, 0168-3659.

Lurquin, P.F. (1979). Entrapment of plasmid DNA by liposomes and their interactions with plant protoplast. *Nucleic Acid Res.*, 6, 3773-3784, 0305-1048.

Luzzati, V. (1968). *Biological Membranes*, Academic Press, New York, London.

Lv, H., Zhang, S., Wang,B., Cui, S. & Yan, J. (2006). Toxicity of cationic lipids and cationic polymers in gene delivery. *J.Control. Rel.*, 114, 100-109, 0168-5659.

Magee, W., Talcott, M., Straub, S. & Vriend, D. (1976). A comparison of negatively and positively charged liposomes containing entrapped polynosinic polycytidylic acid for interferon induction in mice. *Biochim. Biophys. Acta*, 451,610-618, 0304-4165.

Marshall, E. (2000). Gene Therapy on Trial. *Science*, 288, 951-957, 0036-8075.

McDonald, R.C., Ashley, G.W., Shida M.M., Rakhmanova, V.A., Tarahovsky, Y.S., Pantazatos, P.P., Kennedy, M.T., Pozharski, E.V., Baker, K.A., Jones, R.D., Rosenzweig, H.S., Choi, K.L., Qiu, R.Z. & McIntosh T. J. (1999). Physical and Biological Properties of cationic triesters of phosphatidylcholine. *Biophys. J.*, 77, 2612-2629, 0006-3495.

McManus, J.J., Rädler, J.O. & Dawson, K.A. (2003). Does calcium turn a zwitterionic lipid cationic? *J. Phys. Chem. B*, 107, 9869-9875, 1520-6106.

McManus, J.J., Rädler, J.O. & Dawson, K.A. (2003). Phase behavior of DPPC in a DNA- calcium-zwitterionic lipid complex studied by small-angle x-ray scattering. *Langmuir*, 19, 9630-9637, 0743-7463.

Mével, M., Neveu, C., Goncalves, C., Yaouanc, J.J., Pichon, C., Jaffrès, P.A. & Midoux, P. (2008). Novel neutral imidazole-lipophosphoramides for transfection assays. *Chem Comm.*, 3124-3126, 1359-7345.

Mok, K.W. & Cullis, P.R. (1997). Structural and fusogenic properties of cationic liposomes in the presence of plasmid DNA. *Biophys. J.*, 73, 2534-2545, 0006-3495.

Monnard, P.A., Oberholzer, T. & Luisi, P.L. (1997). Entrapment of nucleic acids in liposomes. *Biochim. Biophys. Acta*, 1329, 39-50, 0005-2736.

Mui, H., Ahkong, Q.F., Chow, L. & Hope, M.J. (2000). Membrane perturbation and the mechanism of lipid-mediated transfer of DNA into cells. *Biochim. Biophys. Acta*, 1467, 281-292, 0005-2736.

Newton, C., Pangborn, W, Nir, S. & Papahadjopoulos, D. (1978). Specificity of Ca2+ and Mg2+ binding to phsphatidylserine vesicles and resultant phase changes of bilayer membrane structure. *Biochim. Biphys. Acta*, 506, 281-287, 0005-2736.

Ostro, M.J., Giacomini, D., Lavelle, D., Paxton, W. & Dray, S. (1978). Evidence for traslation of rabbit globin mRNA after liposome mediated insertion into a human cell line. *Nature*, 274, 921-923, 0028-0836.

Papahadjopoulos, D., Vail, W.J., Pangborn, W. & Poste, G. (1976). Studies on membrane fusion. II. Induction in fusion in pure phospholipid membranes by calcium ions an other divalent metals. *Biochim. Biophys. Acta*, 448, 265-283, 0005-2736.

Papahadjopoulos, D.; Allen, T.M.; Gabizon, A.; Mayhew, E.; Matthay, K.; Huang, S.K.; Lee, K.D.; Woodle, M.C.; Lasic, D.D.; Redemann, C. & Martin, F.J. (1991). Sterically stabilized liposomes: improvements in pharmacokinetics and antitumor therapeutic efficacy. *Proc. Natl. Acad. Sci. USA*, 88, 11460-11464, 0027-8424.

Pardridge, W.M. (2007). shRNA and siRNA delivery to the brain. *Adv. Drug Del. Reviews*, 59, 141–152, 0169-409X..

Pisani, M., Bruni, P., Conti, C., Giorgini, E. & Francescangeli, O. (2005). Self.-asembled liposome-DNA metal complexes related to DNA delivery. *Mol. Cryst. Liq. Cryst.*, 434, 315-323, 1542-1406.

Pisani, M., Bruni, P., Caracciolo, G., Caminiti, R. & Francescangeli, O. (2006). Structure and phase behavior of self-assembled DOPC-DNA-Metal Cation complexes. *J. Phys. Chem. B*, 110, 13203-13211, 1520-6106.

Pisani, M., Fino, V., Bruni, P., Di Cola, E. & Francescangeli, O. (2008). Metal induced cubic phase in poly(ethyleneglycol)-functionalized dioleoyl phosphatidylethanolamine aqueous dispersions. *J. Phys. Chem. Letters B*, 112, 5276-5278, 1520-6106.

Pisani, M., Fino, V., Bruni, P. & Francescangeli, O. (2009). DNA condensation into inverted hexagonal phase in aqueous dispersion of poly(ethylene)-functionalized dioleoylphosphatidylethanolamine and metal cations. *Mol. Cryst. Liq. Cryst.*, 500, 132-143, 1542-1406.

Podgornik, R., Rau, D.C. & Parsegian, V.A. (1989). The action of interhelical forces on the organisation of DNA double helices: fluctuation-enhanced decay of electrostatic double-layer and hydration forces. *Macromolecules*, 22, 1780-1786, 0024-9297.

Portis, A., Newton, C., Pangborn, W. & Papahadjopoulos, D. (1979). Studies on the mechanism of membrane fusion: Evidence for an intermembrane Ca^{2+} phospholipid complex, synergism with Mg^{2+} and inhibition by spectrin. *Biochemistry*, 18, 780-790, 0006-2960.

Rädler, J.O., Koltover, I., Salditt, T. & Safinya, C.R. (1997). Structure of DNA-Cationic liposome complexes: DNA intercalation in multilateral membranes in distinct interhelical packing regimes. *Science*, 275, 810-814, 0036-8075.

Roush, W. (1997). Molecular biology: counterfeit chromosomes for humans. *Science*, 276, 38- 39, 0036-8075.

Safinya, C.R., (2001). Structures of lipid-DNA complexes: supramolecular assembly and gene delivery. *Curr. Opin. Struct. Biol.*, 11, 440-448, 0959-440X.

Safinya, C.R., Ewert, K., Ahmad, A., Evans, H.M., Raviv, U., Needleman, D.J., Lin, A.J., Slack, N.L., George, C. & Samuel, C.E. (2006). Cationic liposome-DNA complexes: from liquid crystal science to gene delivery applications. *Phil. Trans. R. Soc. A*, 364, 2573-2596, 1364-503X

Sessa, G. & Weissmann, G. (1968). Phospholipid spherules (liposomes) a model for biological membranes. *J. Lipid Res.*, 9, 310-318, 0022-2275.

Shi, G., Guo, W., Stephenson, S.M. & Lee, R.J. (2002). Efficient intracellular drug and gene delivery using folate receptor-targeted pH-sensitive liposomes composed of cationic/anionic lipid combinations. *J. Control Rel.*, 80, 309–319, 0168-3659.

Song, L.Y., Ahkong, Q.F.; Rong, Q., Wang, Z., Ansell, S., Hope, M.J. & Mui, B. (2002). Characterization of the inhibitory effect of PEG-lipid conjugates on the intracellular delivery of plasmid and antisense DNA mediated by cationic lipid liposomes. *Biochim. Biophys. Acta*, 1558, 1–13, 005-2736.

Szoka, F. & Papahadjopoulos, D. (1978). Procedure for preparation of liposomes with large internal aqueous space and high capture by reverse-phase evaporation. *Proc. Natl. Acad. Sci. USA*, 75, 4194-4198, 0027-8424.

Szoka, F.C.,Jr., Xu, Y. & Zelphati, O. (1996). How are nucleic acids released in the cell from cationic-nucleic acid complexes. *J. Liposome Res.*, 6, 567-587, 0898-2104.

Tarahowsky, Y.S., Khusainova, R.S., Gorelov, A.V., Nicolaeva, T.I., Deev, A.A., Dawson, A.K. & Ivanitsky, G.R. (1996).

DNA initiates polymorphic structural transitions in lecithin. *FEBS Lett.*, 390, 133-136, 0014- 5793.

Tardi, P.G., Boman, N.L. & Cullis, P.R. (1996). Liposomal doxorubicin. *J. Drug Target.*, 4, 129-140, 1061-186X.

Tresset, G., Cheong, W.C.D., Tan, Y.L.S., Boulaire, J. & Lam, Y.M. (2007). Phospholipid based artificial viruses assembled by multivalent cations. *Biophys. J.*, 93, 637-644, 0006-3495.

Tristram-Nagle, S.; Peytrache, H.I. & Nagle, J.F. (1998). Structure and interactions of fully hydrated dioleoylphosphatidylcholine bilayers. *Biophys. J.*, 75, 917-925, 0006-3495.

Turner, D.C. & Gruner, S.M. (1992). X-ray diffraction reconstruction of the inverted hexagonal (HII) phase in lipid-water systems. *Biochemistry*, 31, 1340-1355, 0006-2960.

Tyrrell, D.A., Heath, T.D., Colley, C.M. & Ryman, B.E. (1976). New aspects of liposomes. *Biochim. Biophys. Acta*, 457, 259-302, 0005-2736.

Uhrikova, D., Hanulova, M.; Funari, S.S., Khusainova, R.S., Sersen, F., Balgavy, P. (2005). The structure of DNA-DOPC aggregates formed in presence of calcium and magnesium ions: a small-angle synchrotron X-ray diffraction study. *Biochim: Biophys. Acta*, , 1713, 15-28, 005-2736.

Voinea, M. & Simionescu, M. (2002). Designing of intelligent liposomes for efficient delivery of drugs. *J. Cell. Mol. Med.*, 6, 465-474, 1582-1838.

Willard, H.F. (2000). Artificial Chromosomes Coming to Life. *Science*, 290, 1308-1309, 0036-8075.

Wilschut, J., Düzgünes, N., Fraley, R. & Papahadiopoulos, D. (1980). Studies on the mechanism of membrane fusion: kinetics of calcium ion induced fusion of phosphatidylserine vesicles followed by a new assay for mixing of aqueous vesicle contents. *Biochemistry*, 19, 6011-6021, 0006-2960.

Wilson, T., Papahadjopoulos, D. & Taber, R. (1979). The introduction of poliovirus DNA into cells via lipid vesicles (lipoosmes). *Cell*, 17, 77-84, 0092-8674.

Worthington, C.R. & Khare, R.S. (1978). Structure determination of lipid bilayers. *Biophys. J.*, 23, 407-425, 0006-3495.

Wrobel, I. & Collins, D. (1995). Fusion of cationic liposomes with mammalian cells occurs after endocytosis. *Biochim. Biophys. Acta*, 1235, 296-304, 0005-2736.

Wu, G.Y. & Wu; C.H. (1987). Receptor-mediated in vitro gene transformation by a soluble DNA carrier system.. *J. Biol. Chem.*, 262, 4429-4432, 0021-9258.

Yang, J.P. & Huang, L. (1997). Overcoming the inhibitory effect of serum on lipofection by increasing the charge ratio of cationic liposome to DNA. *Gen. Ther.*, 4, 950-960, 0969- 7128.

Zelphati, O. & Szoka F.C. (1996). Liposomes as a carrier for intracellular delivery of antisense olinucleotides: a real or magic bullet? *J. Control. Rel.*, 41, 99-119, 0168-3659.

Zhang, Y., Zhu, C. & Pardridge W.M. (2002). Antisense gene therapy of brain cancer with an artificial virus gene delivery system. *Mol. Ther.*, 6, 67-72, 1525-0016.

Zhou, Z., Sayer, B.G., Hughes, D.W., Stark, R.E. & Epand, R.M. (1999). Studies of phospholipid hydration by high-resolution magic-angle spinning nuclear magnetic resonance. *Biophys. J.*, 76, 387-399, 0006-3495.

Zuhorn, I.S., Oberle,V., Wisser, W.H., Engberts, J.B.F.N., Bakowski, U., Polushkin, E. & Hoekstra, D. (2002). Phase behaviour of cationic amphiphiles and their mixtures with helper lipids influences lipoplex shape, DNA traslocation and transfection efficiency. *Biophys. J.*, 83, 2096-2108, 0006-3495.

Zuhorn, I.S., Visser, W.H., Bakowsky, U., Engberts, J.B. & Hoekxtra, D. (2002). Interference of serum with lipoplex-cell intraction: modulation of intracellular processing. *Biochim. Biophys. Acta*, 1560, 25-36, 005-2736.

Zuhorn, I.S., Bakowski, U., Polushkin, E., Wisser, W.H., Stuart, M.C.A., Engberts, J.B.F.N. & Hoekstra, D. (2005). Nonbilayer Phase of Lipoplex-Membrane mIxture Determines Endosomal Escape of Genetic Cargo and Transfection Efficiency. *Mol. Ther.*, 11, 801- 810, 1525-0016.